PERSPECTIVES IN FLUID DYNAMICS

A Collective Introduction to Current Research

PERSPECTIVES IN FLUID DYNAMICS

A Collective Introduction to Current Research

Edited by

G. K. Batchelor

H. K. Moffatt

M. G. Worster

CAMBRIDGE UNIVERSITY PRESS

PUBLISHED BY THE PRESS SYNDICATE OF THE UNIVERSITY OF CAMBRIDGE
The Pitt Building, Trumpington Street, Cambridge, United Kingdom

CAMBRIDGE UNIVERSITY PRESS
The Edinburgh Building, Cambridge CB2 2RU, UK
40 West 20th Street, New York, NY 10011–4211, USA
10 Stamford Road, Oakleigh, VIC 3166, Australia
Ruiz de Alarcón 13, 28014 Madrid, Spain
Dock House, The Waterfront, Cape Town 8001, South Africa

http://www.cambridge.org

First published 2000

Printed in the United Kingdom at the University Press, Cambridge

Typeface Computer Modern 11/13pt   *System* LaTeX   [UPH]

A catalogue record for this book is available from the British Library

Library of Congress Cataloguing in Publication data

ISBN 0 521 78061 6 hardback

Contents

v

Preface

Research in fluid dynamics has been greatly stimulated over the past few decades, on the one hand by technical advances in analysis and experiment and a vast increase in available computational power, and on the other by the increasing recognition of an ever-expanding range of interdisciplinary fields in which fluid dynamics plays a pivotal role. Thus, for example, not only have there been significant developments in relation to our ability to predict and measure the complex nonlinear interactions within fluid flows that can lead to instability, chaos and turbulence, we have also witnessed the involvement of fluid dynamics and fluid-dynamical principles in a greatly expanded range of fields within both the physical and biological sciences.

At the undergraduate level, fluid dynamics is usually taught in courses in applied mathematics, physics or engineering. Yet at graduate and research level, the fluid dynamicist can become involved in astrophysics, biology, metallurgy, oceanography, meteorology, geophysics and much more, including the traditional branches of the engineering sciences. This has led to a tendency for fluid dynamics to become compartmentalized into specialities with different groups often unaware of advances being made in other areas of the subject. Beyond the undergraduate level, texts in fluid mechanics tend to be similarly specialized, and the graduate student is quickly channelled along a narrow path, which can diverge quite rapidly from paths being pursued by others.

In an attempt to counter this divisive tendency, we have in this book brought together eleven distinguished authors to provide under one cover an introduction to their speciality within fluid dynamics. Though the coverage of fluid-dynamical research is in no way exhaustive, the book does nevertheless include a wide range of topics of current interest. The material is intended to be introductory in that it does not assume any prior knowledge of the sub-fields, but the reader will benefit from having a general foundation in fluid dynamics as normally taught at undergraduate level. Authors were charged with being didactic rather than providing a comprehensive survey of the literature surrounding their subject. References have therefore been kept to a minimum, intended primarily to point the student either to seminal works or to review articles; and just occasionally to where more specialized results can be found. The book could be used as an accompaniment to a graduate-level course in fluid dynamics. More generally, we hope that it will be read and enjoyed by fluid-dynamicists of all levels as a means to learn about and appreciate the great breadth and wealth of our subject.

There is a sense of progression in the book: the first few chapters involve laminar flow, then instabilities and turbulence; these are followed by chapters that introduce other physical processes that act on fluids; and the final few chapters deal with geophysical phenomena, which involve many of the preceding ideas. However, the chapters are essentially self-contained and can be read in any order, while cross-references are included to help the reader to relate ideas met in different contexts.

Laminar flow is the mainstay of undergraduate texts and courses in fluid dynamics, so one might imagine that there is little new to say on the subject. However, when the boundaries of the fluid domain are deformable, either by being elastic or by forming an interface with another fluid, then many different and complex phenomena can occur. Such flows are described in the first three chapters.

The first chapter, by Stephen Davis, introduces the fundamental ideas of interfacial fluid dynamics on scales at which interfacial forces, such as surface tension, are dominant. This contrasts with traditional studies of surface waves, for example, in which the dominant restoring force (gravity) is internal. Many of the examples he uses to illustrate these ideas originate from modern technological processes, e.g. coating, for which the industrial importance of fluid mechanics is self-evident.

Yves Couder, in Chapter 2, introduces a fundamental instability of a moving fluid–fluid interface, the so-called Saffman–Taylor instability, and considers its nonlinear development. Even at small amplitude, it can have deleterious influence on the coating flows mentioned above and it is a major concern in secondary oil recovery, whereby attempts are made to force more oil from an aging reservoir by injecting a displacing fluid. At large amplitude, the instability causes the interface to adopt complex shapes and is a paradigm for many pattern-forming systems both within fluid mechanics and beyond.

Biology is but one example of a field of research in which fluid dynamics is having a huge impact outside of its traditional applications within the physical and engineering sciences. In Chapter 3, Tim Pedley focuses attention on the flows internal to arteries and veins. The fluid-mechanical novelty comes in part from the interactions between the blood flow and the elastic or collapsible properties of the blood vessels. Additionally the multiply-branched arterial and venous systems give rise to varieties of flow separation and shear enhancement that affect the physiology of the vessels and influence the progress of divers diseases.

Within pure hydrodynamics, modern developments have included a greater understanding of instabilities and further transitions to turbulence. Hydro-dynamic instabilities usually result from shear, and take place, therefore, within a background state that is flowing. In consequence, many unstable flows are only convectively unstable – disturbances grow downstream of where they are introduced but do not grow with time at any fixed location. In particular, they do not grow at the point at which they are introduced. Under certain conditions, however, temporal growth at a fixed location does occur. The flow is then said to be absolutely unstable. Absolute instability gives rise to turbulent bursts in pipe flow and to sudden increased mixing at a splitter plate, for example within some fuel-injection systems. In Chapter 4, Patrick Huerre gives an introduction to the concepts of convective and absolute instability and to the mathematical analysis governing such instabilities in the context of open shear flows.

Hydrodynamic instabilities almost invariably lead to turbulence. Turbulence is a vast area of research extending from investigations into its causes and fundamental nature to, at a very practical level, its influence on mechanical structures and on mixing. In Chapter 5, Javier Jiménez covers the essential ideas of how to characterize turbulence, how to measure it and how to simulate it numerically in order to make quantitative predictions in practical situations.

While the mixing of passive tracers by fluid flows is important in many contexts, much more complex interactions can occur when the quantity being transported itself drives or alters the flow. Such is the case in buoyancy-driven convection, described by Paul Linden in Chapter 6. The paradigm is convection driven by temperature gradients but, equally, convection can be driven by gradients in the concentration of dissolved or suspended impurities. A particular focus of the chapter is convection in confined regions and the development of stable stratification from isolated sources of buoyancy. In addition to many geophysical situations, this phenomenon has significant importance for the containment of reactive chemicals and for the natural ventilation of buildings, for example.

Magnetohydrodynamics (MHD), discussed in Chapter 7 by Keith Moffatt, involves flows that are driven by the Lorentz force resulting from a magnetic field, and the advection (transport) of the field by the flow. In this respect it is similar to buoyancy-driven convection, except that the buoyant agent is a vector rather than a scalar. Additional intrigue results from the fact that the magnetic field itself can be generated by the flow, so the possibility exists for self-sustaining dynamo action, although of course some energy source is required to maintain the flow against viscous dissipation and joule heating. Apart from its intrinsic interest and its application to industrial as well as astrophysical flows, MHD provides some insights into the understanding of turbulence, since the magnetic field is transported in the same way as vorticity.

During solidification, say of water into ice, the boundaries of the fluid domain are altered by phase changes resulting from the transport of heat. In Chapter 8, Grae Worster describes some of the fundamentals of solidification, which combines aspects of interfacial fluid dynamics and of convection, discussed in earlier chapters. In many circumstances the solid–liquid interface can become so convoluted as to form a porous medium in which continued heat and mass transfer, fluid flow and phase change can occur. These processes have significant influence on the fabrication of modern materials (alloys and semi-conductors for example) and on the evolution of many geophysical systems.

Geophysical fluid flows have long been a primary motivation for much fluid-mechanical research and involve many of the different phenomena described in earlier chapters. The most readily and extensively measurable of geophysical flows are those of the atmosphere, which affect our weather, our climate and, of more recent concern, the dispersal of anthropogenic pollutants. The oceans too, as great reservoirs and conveyors of heat and dissolved minerals and gases, have a major impact on climate. Modelling them provides huge challenges for fluid-mechanical prediction, particularly as measurements of their interior are much more difficult to obtain. But even the Earth beneath our feet is in motion, flowing over aeons in response to internal sources of heat and the periodic loading of the crust by successive ice ages.

Starting from the centre of the Earth, Herbert Huppert in Chapter 9 introduces us to many different flows within and of the Earth, from the rapid flows of molten iron in the outer core, which drive the geodynamo, to the eruptions of lava and volcanic ash, visible to us on the Earth's surface. A major focus of the chapter is a systematic discussion of the passage of magma as it percolates through the mantle, evolves within magma chambers, and erupts through volcanic vents either as lava flows or as vast ash-laden plumes into the atmosphere.

The dynamics of the ocean are introduced by Chris Garrett in Chapter 10. Although the global circulation is perhaps of primary concern, particularly in relation to climate prediction, it is effected and influenced by myriad small-scale processes that cause the mixing of different water masses. Garrett elucidates many of these and describes the challenges involved in making sensible interpretations of ocean measurements.

The mathematical prediction of atmospheric motion and properties is highly developed and relied upon daily in weather forecasts. Yet their global-scale evolution poses many predictive challenges, which are being tackled with increasing urgency as we try to understand and then to modify our own influence on climatic change. In Chapter 11, Michael McIntyre reveals the primary mechanisms giving rise to global circulations of the atmosphere that control some of the dominant characteristics of our climate.

The preparation of this volume would not have been possible without the expert help of Linda Drath who copy edited all the text, Mark Hallworth who prepared many of the figures, and Alison Harrison who gave essential secretarial support at all stages.

This book was the inspiration of George Batchelor, who set out a few years ago to prepare a sequel to his famous textbook *An Introduction to Fluid Dynamics*; a sequel that was in fact hinted at in his preface to that text. He realized that the task was beyond one person, the subject having broadened so much in the intervening thirty years, and invited us to help him edit a book that would serve this purpose. He played a full part in choosing the authors and defining the style of the book. It was also his intention to have written this preface. Sadly, George's health declined rapidly, and he died on 30 March 2000 without seeing the completion of the project. George inspired generations of fluid dynamicists through his founding of the *Journal of Fluid Mechanics* in 1956 and its Editorship for more than 40 years, and through his truly remarkable text. We speak for all the authors in wishing to dedicate this book to his memory.

HKM
MGW

George Batchelor
Prince Lectures, Arizona State University 1983

1

Interfacial Fluid Dynamics

STEPHEN H. DAVIS

1 Introduction

The Navier–Stokes equations for fluids that fill rigid containers govern an astonishing array of phenomena, in part due to the convective nonlinearity, $v \cdot \nabla v$, in the velocity vector v. The presence of this nonlinearity gives rise to non-uniqueness of the solutions, instability and bifurcation of solutions, the onset of turbulence, and separation. When the fluid is bounded by free boundaries, then the free-boundary nature of the problem introduces new nonlinearities that augment or compete with $v \cdot \nabla v$. The shapes, positions, and evolution of the boundaries couple with the velocity fields and pressure p, and all of these must be determined simultaneously. Problems in interfacial fluid dynamics are intrinsically free-boundary problems.

The solutions of free-boundary problems are intrinsically non-unique. For example, consider the partially filled bucket of water shown in figure 1. In the absence of gravity and in a steady state the interface would take the shape of the arc of a sphere in which the interface intersects the wall at a contact line. If the contact line is fixed, this state is locally stable, in the sense that if the bucket were slightly shaken and then stopped, the velocity would decay to zero, and the interface would return to the shape shown. However, if the liquid is strongly disturbed, enough that a droplet is expelled, then another locally stable state would be created. It would consist of the bulk water plus a drop suspended in the passive gas. Clearly, there is infinite non-uniqueness with the possibility of an arbitrary number of drops. All of these states are allowable if the total water volume is preserved. Thus, to make an interfacial-fluids problem well posed one must fix the volume and specify the number of 'pieces' of material one wishes to examine. Generally, interfacial states are not globally stable since a concentrated disturbance of large enough amplitude can lead to the creation of a disconnected 'piece'.

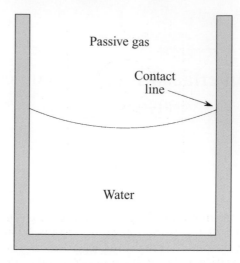

Figure 1. A sketch of a partially filled container in which the liquid–gas interface intersects the sidewall at a contact line.

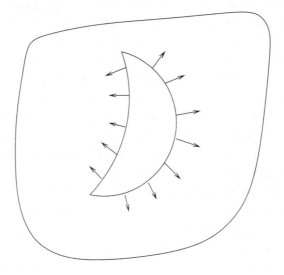

Figure 2. A sketch of an interface that is cut and opens due to the presence of surface tension.

The interface is generally taken to be a mathematical surface of zero thickness separating two phases. It is a free boundary but, furthermore, it is the site of localized forces, and is hence an *active boundary*. If one cuts an interface, as shown in figure 2, there is a force per unit length σ acting on the edge; σ is the surface or interfacial tension. The interface can, under certain circumstances, exhibit characteristics of surface viscosity and/or elasticity

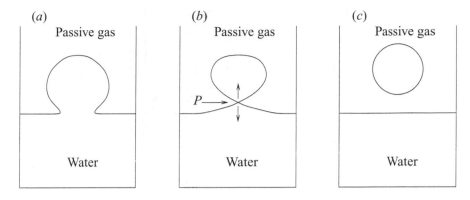

Figure 3. A sketch of the interface of figure 1 that is (*a*) given a strong disturbance at time $t = t_0$. (*b*) At $t = t_1$ the droplet contacts the bulk liquid at a single point P. (*c*) At $t = t_2$ long after the rupture, the disturbances have decayed, and the interfaces are now planar and spherical (or cylindrical in two dimensions) and locally stable.

independent of the properties of the bulk fluids. These interfacial forces can create interfacial motions, and, since the bulk liquid is viscous, these motions can be communicated to the bulk. Thus, bulk flow can drive the interface, and interfacial flow can drive the bulk; the two are tightly coupled.

The process of pinch-off of a droplet from the bulk is a subtle one that involves the rupture of a filament of liquid and the existence of a singularity at the point of splitting. Figure 3 shows the kinematics of such a scenario. As time t approaches a time t_1, a single point, P, forms the bridge between the two bodies. The point P has two identities: bulk liquid and droplet liquid. At $t = t_1^+$, point P splits and is the site of *trajectory splitting* as shown in figure 3(*b*). Whenever trajectory splitting occurs, the flow field is locally singular.

Finally, if the liquid in the bucket of figure 1 is disturbed, the interface, and usually the contact lines as well, will move. Whenever there is mutual displacement (here gas and liquid) at a solid boundary, the displaced fluid must be removed and the displacing fluid must replace it. This process involves a more subtle version of trajectory splitting, and the contact-line singularity is non-integrable (since there are infinite forces required to move the line) unless the local conditions of no slip are remodelled.

There are countless applications of interfacial fluid mechanics. The coating of a substrate by a liquid, and the rupture and removal of liquid films from substrates are ubiquitous sub-processes in engineering and nature. Heat transfer devices often depend on a liquid film separating a vapour and a hot solid to protect the substrate from a toxic vapour or to control the rate of transfer of heat. This may involve evaporative effects and flows driven by surface-tension gradients. If the substrate is cold, efficient condensation would depend on the liquid being in dropwise form, running off the site, to free the surface for further condensation. The human lung is lined with a mucous layer that expels foreign particles as it flows, and that is a barrier to pollutants or aerosol drugs entering the blood stream (for example see Grotberg 1994). To spray paint a surface, one would wish to have individual droplets spread and merge into a continuous film, while if the droplets were contaminants, one would wish the droplets to 'ball up' so that a clean-up process could remove them.

This chapter first discusses how the model of a thin interface is rationalized. Then thin films and spreading drops are discussed in terms of dynamics and instability. Finally, the singularities inherent in rupture and coalescence are discussed.

2 Interfacial regions

Consider a system that is in static equilibrium and consists of two bulk phases that are in contact. A neighbourhood of this contact region, called the *interfacial region*, in which the fluids mix to some extent, has anomalous physical properties (e.g. density, pressure, etc.) compared to those of the bulk phases. These anomalies take the form of rapid variations in property values normal to the interfacial region. In physical systems of common (fluid) phases at room temperature, the thickness of the interfacial region Δ_s can range from as little as a fraction of a molecular diameter (a few Å) to, perhaps, ten molecular diameters. However, when the temperature is raised towards the critical temperature (at which a liquid and its vapour lose their identities and become indistinguishable), Δ_s becomes large and finally, in principle, infinite. Let us denote as L the length scale that characterizes the gross system, e.g. the smallest geometrical dimension of the container in which the system lies. The limit $\Delta_s/L \to 0$ gives great simplification of the mathematical models of interfacial fluid mechanics. The discussion that follows was strongly influenced by the developments in Quinn & Scriven (1970).

The *dividing-surface* approach, developed by Gibbs in 1876 (see his collected works, Gibbs 1948), involves the replacement of the actual system by two bulk fluids separated by a 'dividing surface' called the interface. The bulk properties are assumed to continue their definitions smoothly right up to the interface where they may experience jump discontinuities from one bulk to the other. The interface must then be endowed with surface properties appropriate to the physical system and constitutive assumptions on the surface material must be made. This technique simplifies the system, and also applies to situations where Δ_s is so small that the interfacial region cannot be expected to display a three-dimensional continuum character.

2.1 Interfacial conditions in the continuum model

Consider systems having two bulk phases and an interfacial region. As discussed earlier, the bulk phases are regarded as continuum fluids that have well-defined bulk properties Q. Likewise, the interfacial region is regarded as an interface of zero thickness that has well-defined surface properties Q_s in the interfaces. Call the two fluids I and II, and assume that as the interface $\mathscr{S}$ is approached from either side, the respective bulk properties have well-defined limits. Denote by $[Q]$ the jump in the bulk quantities Q, i.e.

$$[Q] = Q_{II} - Q_I \qquad \text{on} \qquad \mathscr{S}. \qquad (2.1)$$

Likewise, require the unit normal vector v to $\mathscr{S}$ to be directed from II towards I. This convention on the normal vector is important to keep in mind since the mean curvature H of the interface, in particular its sign, is determined by v.

Suppose, for example, that the Navier–Stokes equations hold in fluids I and II and that at $\mathscr{S}$ jump conditions are given. One could derive the interfacial jumps in their most general forms (Aris 1962; Slattery 1990). This would involve the derivation of transport theorems and the investigation of differential geometry. Rather than delve into these, only the conditions at the interface necessary for the explanation of the phenomena discussed will be derived here. In what follows in this section it is assumed that phase transformation is absent.

The kinematic condition gives the continuity of normal velocity across $\mathscr{S}$. If the bulk velocity field is $v = (u, v, w)$ and V is the speed of the interface normal to itself, then $v \cdot v = V$, which for a single valued front $\mathscr{S}: z = h(x, y, t)$, takes the form

$$w = h_t + u h_x + v h_y \qquad \text{on} \qquad z = h(x, y, t). \qquad (2.2)$$

It is accepted on empirical grounds that, apart from cases of the flow of rarefied gases and certain instances of moving three-phase lines on solids (Dussan V. & Davis 1974), adjacent strata of viscous fluids have continuous tangential velocities. Therefore, on a fluid–fluid interface $\mathscr{S}$, the no-slip condition holds so that

$$[v \cdot t] = 0 \qquad \text{on} \qquad \mathscr{S}, \qquad (2.3)$$

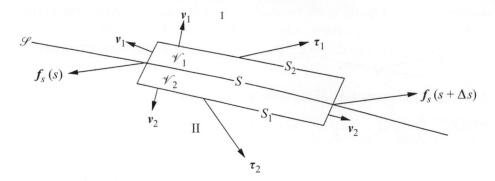

Figure 4. A sketch of a control volume $\mathscr{V}$ spanning an interface $\mathscr{S}$. The interfacial force f_s depends on arclength s, and v_i denote unit normals.

where t is the unit tangent vector to $\mathscr{S}$. The condition (2.3) holds on fluid–solid interfaces as well. However, when one wishes to use models involving inviscid fluids, condition (2.3) is abandoned.

   Consider the balance of linear momentum in which there is a force per unit length f_s acting on points of the interface. Figure 4 shows in a two-dimensional system a segment of the interface and a control volume spanning $\mathscr{S}$. (In general the interface would be endowed with a mass per unit area γ_s; in what follows for the sake of simplicity, this is taken to be zero.) Here τ_1 and τ_2 are stress vectors in the bulk fluids acting on areas S_1 and S_2 of the control volume, respectively. The interface $\mathscr{S}$ extends from s to $s + \Delta s$ where s is arclength, and the volume has width W_0 normal to the page. Finally, at the intersections of S_i with $\mathscr{S}$ there are localized interfacial forces f_s.

   Let us apply Newton's law to the material within $\mathscr{V} = \mathscr{V}_1 + \mathscr{V}_2$ and write for $i = 1, 2$, $\tau_i = T \cdot v_i$, and unit normals v_i,

$$\frac{\mathrm{d}}{\mathrm{d}t} \int_{\mathscr{V}} \rho v \mathrm{d}\mathscr{V} = \int_{S_1+S_2} T \cdot v \, \mathrm{d}S + [f_s(s + \Delta s) - f_s(s)] \, W + \int_{\mathscr{V}} \rho F \, \mathrm{d}\mathscr{V},$$

where F is the bulk body force per unit mass acting on fluids I and II, and T is the stress tensor. Since γ_s has been taken to be zero, there is no excess acceleration, or body force.

   Now let the volume $\mathscr{V}$ collapse onto S and thus $\mathscr{V} \to 0$, $S_1, S_2 \to S$ and a patch of surface remains, as shown in figure 5. In this case $v_1 \to n$ and $v_2 \to -n$ on the sides I and II, respectively. If ρ, v, t and F are smooth, then there is a local balance on $\mathscr{S}$,

$$-[T \cdot n] W_0 \Delta s + [f_s(s + \Delta s) + f_s(s)] \, W_0 = 0.$$

Figure 5. The control volume $\mathscr{V}$ of figure 4 becomes the surface S after the thickness approaches zero.

Now, divide by $W_0\Delta s$ and let $\Delta s \to 0$ to yield the point balance on $\mathscr{S}$,

$$-[\boldsymbol{T} \cdot \boldsymbol{n}] + \frac{\partial \boldsymbol{f}_s}{\partial s} = 0. \tag{2.4}$$

In order to proceed further one needs a constitutive equation for the interfacial force $\boldsymbol{f}_s$.

If the interface is passive, then $\boldsymbol{f}_s = \boldsymbol{0}$ and the bulk stresses are continuous across $\mathscr{S}$. If there is surface tension σ only on $\mathscr{S}$, then $\boldsymbol{f}_s = \sigma\boldsymbol{t}$ so that $\boldsymbol{f}_s$ is tangent to $\mathscr{S}$ with magnitude σ; $-\sigma$ acts like an interfacial pressure. As a result,

$$\frac{\partial \boldsymbol{f}_s}{\partial s} = \sigma\frac{\partial \boldsymbol{t}}{\partial s} + \frac{\partial \sigma}{\partial s}\boldsymbol{t},$$

which has a tangential component only if the surface tension varies from position to position on $\mathscr{S}$. There is a normal component (Aris 1962) given by $\partial\boldsymbol{t}/\partial s = \kappa\boldsymbol{n}$ (a Frenet formula) and in the three-dimensional case $\partial\boldsymbol{t}/\partial s = 2H\boldsymbol{n}$, where H is the mean curvature of the surface. Hence, the balance of linear momentum (2.4) becomes

$$-[\boldsymbol{T} \cdot \boldsymbol{n}] + 2H\sigma\boldsymbol{n} + \frac{\partial \sigma}{\partial s}\boldsymbol{t} = \boldsymbol{0}. \tag{2.5a}$$

The normal component of (2.5a) is

$$[\boldsymbol{n} \cdot \boldsymbol{T} \cdot \boldsymbol{n}] = 2H\sigma, \tag{2.5b}$$

which for inviscid fluids is the well-known Laplace relation,

$$p_2 - p_1 = -2H\sigma. \tag{2.5c}$$

where $2H = -\nabla \cdot \boldsymbol{n}$. There is a pressure excess on the concave side of the interface of magnitude equal to surface tension times twice the mean curvature.

The tangential component of (2.5a) is

$$[\boldsymbol{t} \cdot \boldsymbol{T} \cdot \boldsymbol{n}] - \frac{\partial \sigma}{\partial s} = 0, \tag{2.5d}$$

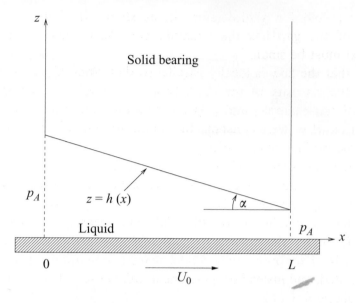

Figure 6. A sketch of a bearing supported by liquid dragged into the gap by the movement of the base plate.

which balances the jump in tangential stress with the surface-tension gradient: the Maragoni balance. For example if σ depends on temperature, then equation (2.5d) represents a balance between shear stress and surface-tension gradients. More complex rheological models for interfaces are discussed in Scriven (1960) and Slattery (1990).

In discussions to follow, if generalizations or augmentations of the above interfacial conditions are required, they will be determined *in situ*.

3 Thin films

Film dynamics and instability are central to a vast number of processes, as indicated in the Introduction. These will be analysed here by taking advantage of the geometrical disparity inherent in long-wave structures by using lubrication approximations.

3.1 Lubrication theory

The prototype problem in mechanics displaying geometrical disparity is the fluid-lubricated bearing as shown in figure 6. Here a heavy solid is supported by hydrodynamic forces when the fluid is forced beneath a slightly tilted surface and an underlying plate that drags the fluid into the gap; see

Schlichting (1968) for a discussion. If, as shown in figure 6, the upper boundary of the gap has the equation $z = h(x)$, then for a small tilt $\alpha = |dh/dx|$ must be small.

Assume that the flow is locally parallel so that, approximately, $\mu u_{zz} = p_x$. Here μ is the viscosity of the fluid, (u, w) is the velocity vector v in this two-dimensional example, and p is the pressure. Subscripts denote partial differentiation. If p_x were constant, this would represent parallel flow. When p_x varies, the profiles change with x.

The normal component of the Navier–Stokes equation is given approximately by $p_z = 0$ which makes the pressure including buoyancy independent of height. Finally, there is the continuity condition $u_x + w_z = 0$.

The boundary conditions on the plate and bearing are standard: for $0 < x < L$ (under the bearing) $u(x, 0) = U_0$, $w(x, 0) = 0$ and $u(x, h) = w(x, h) = 0$. In this approximate model realistic conditions at $x = 0$ and L are complicated, and instead of these one usually poses $p(0, z) = p(L, z) = p_A$, the atmospheric pressure.

Given that p depends on x only, one can integrate the first equation twice and use the conditions at the top and bottom to obtain the velocity profile

$$\mu u(x, z) = \tfrac{1}{2} p_x \left(z^2 - hz \right) + \mu U_0 \left(1 - \frac{z}{h} \right). \tag{3.1}$$

The flow is a linear combination of plane Couette flow (driven by the motion of the plate) and plane Poiseuille flow (driven by the pressure gradient p_x) induced by requiring that p have the same constant value at the exit and entrance.

Given that the flow is steady, the constant flow rate (in the x-direction) is given by $Q = \displaystyle\int_0^{h(x)} u(x, z)\,dz$, and using equation (3.1)

$$\mu Q = -\tfrac{1}{12} h^3 p_x + \tfrac{1}{2} \mu U_0 h. \tag{3.2}$$

Alternatively, equation (3.2) can be differentiated to give

$$\left(\tfrac{1}{12} h^3 p_x + \tfrac{1}{2} \mu U_0 h \right)_x = 0. \tag{3.3}$$

This is called the *Reynolds lubrication equation*. Given $h(x)$, it is an ordinary differential equation for p, subject to the given pressure conditions. Using either equation (3.2) or (3.3), one finds that

$$p(x) - p_A = 6 \mu U_0 \left[\int_0^x h^{-2} dx - \frac{\int_0^L h^{-2} dx}{\int_0^L h^{-3} dx} \int_0^x h^{-3} dx \right]. \tag{3.4}$$

The maximum pressure $p_m \sim \alpha^{-1}$ and so the upward force F_v exerted by

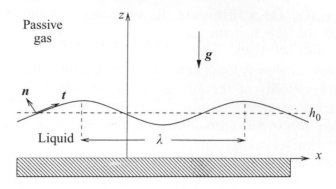

Figure 7. A sketch of a liquid film of mean thickness h_0 on a substrate. When the interface is disturbed (periodically in space) the wavelength $\lambda \gg h_0$. Gravity g acts vertically downward.

the fluid on the bearing scales as $F_v \sim \alpha^{-2}$, both as $\alpha \to 0$; a large weight can be supported by a fluid film. Such results can be generalized to three-dimensional and unsteady systems (see e.g. Oron, Davis & Bankoff 1997).

3.2 Liquid films on solid substrates

Consider a two-dimensional thin viscous film on a horizontal substrate as shown in figure 7. The mean thickness is h_0, the magnitude of the acceleration due to gravity is g, and the liquid has viscosity μ and density ρ, both constants. Further, the interface between the liquid and passive gas possesses a constant surface tension σ. The governing equations are Navier–Stokes

$$\rho(u_t + uu_x + wu_z) = -p_x + \mu(u_{xx} + u_{zz}), \tag{3.5a}$$

$$\rho(w_t + uw_x + ww_z) = -p_z + \mu(w_{xx} + w_{zz}) - \rho g, \tag{3.5b}$$

and continuity

$$u_x + w_z = 0. \tag{3.5c}$$

On the substrate, $z = 0$,

$$u = w = 0. \tag{3.5d}$$

On the interface, $z = h(x,t)$, there is the kinematic condition

$$w = h_t + uh_x, \tag{3.5e}$$

the condition of zero shear stress

$$(u_z + w_x)\left(1 - h_x^2\right) - 4h_x u_x = 0, \tag{3.5f}$$

and the Laplace condition of normal stress balance,

$$-p + 2\mu \left[w_z \left(1 - h_x^2\right) - h_x(u_z + w_x) \right] / n^2 = 2H\sigma, \qquad (3.5g)$$

where

$$2H = h_{xx}/n^3. \qquad (3.6)$$

Here the unit normal $\boldsymbol{n} = (-h_x, 1)/n$, and tangent $\boldsymbol{t} = (h_x, 1)/n$ have been used and $n = (1 + h_x^2)^{1/2}$.

In order to simplify the system and gain an understanding of the dynamics of the interface, let us assume that the scale of variation parallel to the substrate is $\lambda/2\pi$ and that in the normal direction is h_0. Then define new variables

$$X = \frac{2\pi x}{\lambda}, \qquad Z = \frac{z}{h_0}, \qquad (3.7a)$$

where one expects that $\partial/\partial X$, $\partial/\partial Z$ are unit-order quantities. From the continuity equation (3.5c), if the scale of u is U_0 then the scale of w must be $2\pi h_0 U_0/\lambda$. Then define

$$U = \frac{u}{U_0}, \qquad W = \frac{\lambda w}{2\pi h_0 U_0}. \qquad (3.7b)$$

Let time be measured by $\lambda/2\pi U_0$ so that

$$T = \frac{2\pi}{\lambda} U_0 t. \qquad (3.7c)$$

Now if $\lambda \gg h_0$ then one expects the flow in the film to be 'locally parallel', i.e. that $\mu u_{zz} \sim p_x$. In this case, the scale P_0 for p should be $\lambda \mu U_0/2\pi h_0^2$. Thus, write

$$P = p \left/ \left(\frac{\lambda \mu U_0}{2\pi h_0^2} \right) \right. . \qquad (3.8)$$

Define the small parameter ϵ,

$$\epsilon = \frac{2\pi h_0}{\lambda}, \qquad (3.9)$$

and rewrite system (3.6) in terms of the new variables:

$$\epsilon R(U_T + UU_X + WU_Z) = -P_X + U_{ZZ} + \epsilon^2 U_{XX}, \qquad (3.10a)$$

$$\epsilon^3 R(W_T + UW_X + WW_Z) = -P_Z + \epsilon^2(W_{ZZ} + \epsilon^2 W_{XX}) - G, \qquad (3.10b)$$

$$U_X + W_Z = 0, \qquad (3.10c)$$

$$U = W = 0 \quad \text{at} \quad Z = 0, \qquad (3.10d)$$

$$\left.\begin{aligned}
&W = H_T + U H_X \\
&\left(1 - \epsilon^2 H_X^2\right)\left(U_Z + \epsilon^2 W_X\right) - 4\epsilon^2 H_X U_X = 0 \\
&-P + 2\epsilon^2 \left[W_Z \left(1 - \epsilon^2 H_X^2\right) \right. \\
&\qquad \left. - H_X \left(U_Z + \epsilon^2 W_X\right)\right] / n^2 = C^{-1}\epsilon^3 H_{XX}/n^3
\end{aligned}\right\} \quad \text{at} \quad Z = H, \quad (3.10e)$$

$$H = h/h_0. \tag{3.10f}$$

The dimensionless parameters are the Reynolds, capillary, and gravity numbers:

$$R = \frac{U_0 h_0}{\nu}, \qquad C = \frac{U_0 \mu}{\sigma}, \qquad G = \frac{\rho g h_0^2}{\mu U_0}\epsilon. \tag{3.11}$$

Before taking limits, integrate the equation (3.10c) in Z from zero to $H(X, T)$, using integration by parts, the first of (3.10e), and the second of (3.10d) to obtain

$$H_T + \frac{\partial}{\partial X}\int_0^H U\,dz = 0. \tag{3.12}$$

This equation is in a more convenient form for the kinematic condition and ensures conservation of mass on a domain with a deflecting upper boundary.

Finally, let us pose approximate solutions using a regular perturbation in ϵ with R, C and G fixed, $(U, W, P) = (U_0, W_0, P_0) + \epsilon(U_1, W_1, P_1) + \cdots$. The leading-order system in this approximation is called lubrication theory, which applies to solutions with slow variations in x and t. Note from the third equation of (3.10e) as $\epsilon \to 0$, that if $C = O(1)$, then the surface tension is lost from the lubrication approximation. It turns out that ignoring surface tension makes various problems ill-posed; instability growth rates become unbounded as the wavenumber approaches infinity. Thus, one wishes to retain surface tension and so takes $C^{-1}\epsilon^3 = O(1)$ or equivalently writes

$$\bar{C} = \epsilon^{-3}C = O(1) \qquad \text{as} \qquad \epsilon \to 0. \tag{3.13}$$

This requires the surface tension σ to be large, $\sigma\epsilon^3 \sim 1$.

If the series is substituted into the governing dimensionless system, the $O(1)$ terms (dropping subscripts) are as follows:

$$U_{ZZ} = P_X, \qquad P_Z = -G, \qquad H_T + \frac{\partial}{\partial X}\int_0^H U\,dZ = 0, \qquad (3.14a\text{–}c)$$

$$U = 0 \qquad \text{at} \qquad Z = 0, \tag{3.14d}$$

$$U_Z = 0, \qquad -P = \bar{C}^{-1}H_{XX} \qquad \text{on} \qquad Z = H. \tag{3.14e, f}$$

For our purposes there is no need to find W although it can be obtained from continuity and the condition of zero flow normal to $Z = 0$.

Clearly, system (3.14) is vastly simpler than the original system. Let us seek solutions. Equation (3.14b) states that the pressure satisfies a hydrostatic balance: $P = -GZ$. It is convenient to define a reduced pressure $\bar{P}$, $\bar{P} = P + GZ$ so that the normal-stress balance, equation (3.14f) becomes $\bar{P} = GH - \bar{C}^{-1}H_{XX}$. One can solve equation (3.14a): the first integral gives $U_Z = \bar{P}_X(Z - H)$ which satisfies the shear-stress condition (3.14e). A further integral that satisfies the no-slip condition (3.14d) gives

$$U = \bar{P}_X \left(\tfrac{1}{2}Z^2 - HZ\right). \tag{3.15}$$

Conservation of mass, equation (3.14c), then gives $3H_T - \left(\bar{P}_X H^3\right)_X = 0$ and if $\bar{P}$ is eliminated gives

$$3H_T + \bar{C}^{-1}\left(H^3 H_{XXX}\right)_X - G\left(H^3 H_X\right)_X = 0. \tag{3.16}$$

Equation (3.16) is an evolution equation for the shape of the interface as a function of time: the Reynolds lubrication equation for the free-surface problem. It is a nonlinear partial differential equation and from its solution H one can determine U, W, and P by differentiation. What was a free-boundary problem has been converted in the limit $\epsilon \to 0$ to a single equation.

Equation (3.16) can be unscaled and written in primitive variables,

$$3\mu h_t + \sigma\left(h^3 h_{xxx}\right)_x - \rho g\left(h^3 h_x\right)_x = 0. \tag{3.17}$$

The middle term represents surface-tension effects while the last represents hydrostatic pressures.

The mass of the film is preserved as time progresses. To see this, integrate equation (3.17) over the appropriate x-interval. This would be one wavelength if the solution is periodic in x or $-\infty$ to ∞ if disturbances to the interface decay at large distances. In either case

$$\mu\frac{\mathrm{d}}{\mathrm{d}t}\int h\,\mathrm{d}x = 0.$$

Note also that equation (3.17) has the solution h is constant.

Equation (3.17) is a 'levelling' equation in that an initial bump near $x = 0$, say, will relax to the constant solution as $t \to \infty$. This can easily be seen for small disturbances by linearizing equation (3.17) about $h = h_0$ (constant). Let $h = h_0 + h'$ and discard products of the primed quantities to obtain

$$3\mu h'_t + \sigma h_0^3 h'_{xxxx} - \rho g h_0^3 h'_{xx} = 0. \tag{3.18}$$

If one uses normal modes, $h' = \bar{h}\exp(st + ik'x)$, then equation (3.18) becomes

$$3\mu s = - \left[\sigma(k')^4 + \rho g(k')^2\right] h_0^3 \qquad (3.19)$$

and $s < 0$ always; the planar solution returns as $t \to \infty$. If, rather than x-periodicity, Fourier transforms were applied to cases in which $|h'| \to 0$ as $|x| \to \infty$, the same conclusion would emerge.

The result is easily understood on a physical basis. If the planar interface is perturbed by a spatially periodic disturbance at $t = 0$, as shown in figure 7, surface tension tends to flatten the interface: a pressure field is induced by surface tension that drives fluid from beneath an elevation toward the valleys. Likewise, the hydrostatic pressure pumps fluid from beneath bumps toward valleys. These two pumpings are responsible for the stabilizations seen in equation (3.19).

The same evolution equation can be used to describe a liquid film beneath a plate. If z is still directed upwards, equation (3.17) applies if one replaces g by $-g$. In this case, the growth rates of equation (3.19) for small disturbances to the planar solution satisfy

$$3\mu s = - \left[\sigma(k')^4 - \rho g(k')^2\right] h_0^3 \qquad (3.20)$$

so that disturbances grow if $s > 0$, i.e. if

$$0 < (k')^2 < \frac{\rho g}{\sigma} \equiv (k'_c)^2 \qquad (3.21)$$

as shown in figure 8 and where k'_c is called the cutoff wavenumber. This is Rayleigh–Taylor instability (see e.g. Drazin & Reid 1981) in which heavy fluid over light fluid results in overturning. Thus, the range of wavenumbers for instability becomes very small when σ becomes large: surface tension cuts off (or stabilizes) small-scale variations since such shapes have large interfacial curvatures.

Equation (3.21) can be written in terms of the non-dimensional wave-number $k = k' h_0$ as follows:

$$0 \leqslant k^2 < \frac{\rho g h_0^2}{\sigma} \equiv B_0^{(h)}, \qquad (3.22)$$

where $B_0^{(h)}$ is the Bond number based on thickness.

The result (3.21), valid on the infinite domain in x, shows that the planar interface is unconditionally unstable since k' is arbitrary; all k' in the interval $(0, k'_c)$ grow with time.

Now consider the possibility of sidewalls being present at, say, $x = 0, L$. One would suppose that the presence of these would limit the wavenumbers allowed by the system. An estimate of this limitation is that at least one

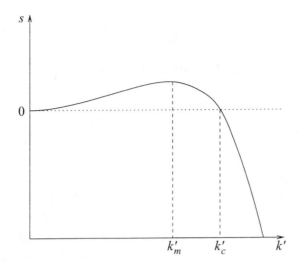

Figure 8. A typical curve of growth rate s versus x-wavenumber k' for interfacial instability. Unstable disturbances are confined to $0 < k' < k'_c$ and the maximum growth rate occurs at $k' = k'_m$.

wavelength λ should fit in the slot, so that if $\lambda > L$, then no eigenfunction would exist corresponding to $s > 0$. In this case from (3.21) the planar interface would be stable if $(2\pi/L)^2 \geqslant \rho g/\sigma$ or

$$4\pi^2 \geqslant \frac{\rho g L^2}{\sigma} \equiv B_0^{(L)}, \tag{3.23}$$

where $B_0^{(L)}$ is the Bond number based on gap width.

The gap estimate used above is generally only a rough approximation. The proper sidewall conditions should be derived by solving the Navier–Stokes equations in their neighbourhoods and matching these 'inner' solutions to the lubrication approximation of the 'outer' region.

Equation (3.17) governs the nonlinear development of the instability that leads to the dryout of the film (a zero-thickness region) as shown in figure 9. In a different situation in which the passive gas is replaced by another liquid of the same viscosity but small density, dryout does not occur. Instead, a thin film is generated between fingers, which themselves form droplets that can detach (Yantsios & Higgins 1991).

In all of the above the only physical effects acting are viscous, gravity, and pressure forces in the bulk and surface tension on the interface. When films become ultra thin, long-range molecular forces can be effective in augmenting the force balances discussed above.

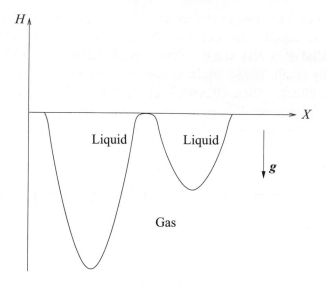

Figure 9. Film profile for Rayleigh–Taylor instability of a thin film under a horizontal plate for $G = 20$, $\bar{C}^{-1} = 4$ with initial shape $1 + 0.005 \sin k_m X$ according to Yantsios & Higgins (1991).

3.3 Rupture of ultra-thin films

The molecules in a macroscopic film interact with themselves or with the neighbouring gas or substrate. However, when a film has a thickness in the range 100–1000 Å, van der Waals attractions or repulsions can extend across the thickness of the film and contribute significant extra forces beyond those described by the Navier–Stokes equations. These act to pinch or expand the film at localities where there is a depression in the thickness, depending upon the sign of the forces present. If they pinch, a small depression will be deepened and the film will rupture in a finite time. If they expand, an initial depression will be thickened and the film will stabilize.

There are models, Ruckenstein & Jain (1984), for these forces in which they are represented as an extra body force $\nabla\phi'$ in the Navier–Stokes equation. Thus, the model has the balance of linear momentum given by

$$\rho(\boldsymbol{v}'_t + \boldsymbol{v}' \cdot \nabla \boldsymbol{v}') = -\nabla p' + \mu \nabla^2 \boldsymbol{v}' - \nabla \phi'. \tag{3.24}$$

For the special case of a film with parallel or quasi-parallel boundaries,

$$\phi' = \phi_r + \frac{A'}{6\pi h'^3}, \tag{3.25}$$

where ϕ_r is a reference constant and A' is the Hamaker constant. When $A' > 0$, the two interfaces attract each other and the film pinches. When

$A' < 0$, the interfaces repel each other. A rule of thumb is that pinching is associated with liquids that wet the substrate poorly and vice versa.

The potential ϕ' is very small unless h' is less than about 1000 Å since $|A'|$ is numerically small. If the liquid is deionized water (to eliminate electrical double-layer effects), then (Ruckenstein & Jain 1974) $A' \approx 10^{-13}$ erg, so that a 1000 Å film has $\phi' - \phi_r \approx 10^{-13}/6\pi(10^{-5})^3 \approx 5$ dyn cm^{-2}, which is comparable to the surface tension acting on a radius of curvature of 1 cm.

If one defines a reduced pressure $\bar{P}$, $\bar{P} = GH - \bar{C}^{-1}H_{xx} + A\Phi$, one obtains the evolution equation (Williams & Davis 1981)

$$3H_T + \bar{C}^{-1}\left(H^3 H_{XXX}\right)_X - G\left(H^3 H_X\right)_X + A(H^{-1}H_X)_X = 0, \qquad (3.26)$$

where

$$A = \frac{\rho A' \epsilon}{6\pi\mu^2 h_0} \qquad (3.27)$$

is the non-dimensional strength of the van der Waals attractions, and the constant ϕ_r has been incorporated into the pressure level.

One can see directly that if $A < 0$ then the van der Waals term represents forward nonlinear diffusion with diffusivity AH^{-1}; this represents a stable process. When $A > 0$, however, this represents backward diffusion which leads to rupture of the film. In dimensional form equation (3.26) is

$$3\mu h_t + \frac{A'}{2\pi}\left(h^{-1}h_x\right)_x - \rho g\left(h^3 h_x\right)_x + \sigma\left(h^3 h_{xxx}\right)_x = 0. \qquad (3.28)$$

Let h_0 be a constant solution and write $h = h_0 + h'$, linearize in h' and use normal modes, $h' = \bar{h}\exp(st + ik'x')$, to obtain the characteristic equation

$$3\mu s = \frac{(k')^2}{h_0^2}\left[\frac{A'}{2\pi h_0} - \sigma h_0(k')^2 - \rho g h_0^3\right].$$

Again, the growth rate has the form shown in figure 8 where now

$$(k_c')^2 = \left(\frac{A'}{2\pi h_0} - \rho g h_0^3\right)\Big/ \sigma h_0. \qquad (3.29)$$

There is instability as long as the van der Waals attractions for pinching ($A' > 0$) exceed the stabilizing effect of hydrostatic pressure. Since the film is thin, one would suppose that the hydrostatic pressure is negligible. This is the case only if $h_0^4 \ll A'/2\pi\rho g$ which for water would give

$$h_0 \ll 10^{-4} \text{ cm} = 1 \text{ μm}.$$

Result (3.29) suggests that every film is unstable for $g = 0$ due to van der

Stephen H. Davis

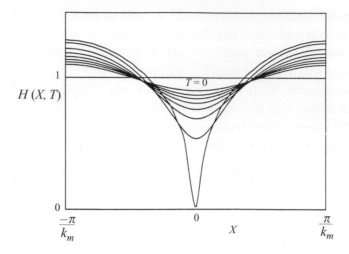

Figure 10. The film profile at various times T for an instabiity due to van der Waals attractions for $A = \bar{C} = 1$ and initial shape $1 + 0.1 \sin k_m X$. Copyright © 1988 Cambridge University Press. Reprinted with the permission of Cambridge University Press from Burelbach, Bankoff & Davis (1988).

Waals attractions. However, sidewalls sufficiently close to each other stabilize the longest waves.

Equation (3.26) governs the nonlinear behaviour of the film. Consider the case $g = 0$ in which k is fixed at the value k_m that maximizes the growth rate of linear theory (see figure 8). For the initial condition $H(X,0) = 1 + 0.1 \sin k_m X$, figure 10 shows the nonlinear evolution lead to rupture in a finite time. Moreover, the rate of film thinning, measured as the rate of decrease of the minimal thickness of the film, increases with time and becomes much larger than the disturbance growth rate given by the linear theory extrapolated to the time of breakup. This can be seen in equation (3.27). Notice that the 'effective' diffusion coefficient, proportional to h^{-1}, in the backward diffusion term increases indefinitely as $h \to 0$ and the film becomes thinner, while the local stabilization effect provided by surface tension weakens, proportionally to h^3. These models are based on the experimental observation by Sheludko (1967) that there is spontaneous breakup of thin static films due to the presence of van der Waals attractions.

3.4 Thermocapillarity

The surface tension σ acting on an interface $\mathscr{S}$ between two immiscible fluids usually depends on the scalar fields in the system (e.g. the electrical field and the temperature field), as well as on the concentration of foreign materials on the surface. Consider a single such field, the temperature θ, and pose an equation of state $\sigma = \sigma(\theta)$.

Surface tension enters the description of the dynamics of the system through the force balance on $\mathscr{S}$. On the one hand, the jump in normal stress at $\mathscr{S}$ balances surface tension times twice the mean curvature H of $\mathscr{S}$. Thus, thermocapillarity can alter the capillary pressure jump or make it vary from point to point, depending on θ. On the other hand, there is a jump in shear stress balanced by the surface-tension gradient. If one represents the equation of state by a linear law

$$\sigma = \sigma_0 - \gamma(\theta - \theta_0), \tag{3.30}$$

where θ_0 is a reference temperature, then the surface-tension gradients on $\mathscr{S}$ are proportional to γ: temperature gradients along $\mathscr{S}$ induce shear stresses on $\mathscr{S}$ that result in fluid motion. For common liquids $\gamma > 0$, so that there is surface flow from the hot end toward the cold end. Since the bulk fluids are viscous, they are dragged along; bulk-fluid motion results from interfacial temperature gradients. This is called the *thermocapillary effect*.

Define the non-dimensional temperature Θ by

$$\Theta = \frac{\theta - \theta_\infty}{\theta_0 - \theta_\infty}. \tag{3.31}$$

where θ_∞ is the temperature in the gas far from $\mathscr{S}$.

Consider the film of figure 7 where now the temperature at the base is fixed at $\theta = \theta_0$. The temperature at the interface is θ^I, which is unknown at the moment but is in part determined by a heat-transfer coefficient α_T, and the heat balance at $\mathscr{S}$, $k_T \nabla \theta \cdot \boldsymbol{n} + \alpha_T(\theta - \theta_\infty) = 0$. Here k_T is the thermal conductivity of the liquid. In terms of the lubrication scales defined earlier the energy equation and boundary conditions are

$$\epsilon R Pr[\Theta_T + U\Phi_X + W\Theta_Z] = \Theta_{ZZ} + \epsilon^2 \Theta_{XX}, \tag{3.32a}$$

$$\Theta = 1, \qquad Z = 0, \tag{3.32b}$$

$$\Theta_Z - \epsilon^2 \Theta_X H_X + B\Theta \left[1 + H_X^2\right]^{1/2} = 0, \qquad Z = H, \tag{3.32c}$$

where the Prandtl and Biot numbers are

$$Pr = \frac{\nu}{\kappa}, \qquad B = \frac{\alpha_T h_0}{k_T}, \tag{3.33}$$

and κ is the thermal diffusivity. The Prandtl number is a measure of the ratio of diffusivities while the Biot number is a measure of heat losses to the surrounding gas. As B becomes smaller, there is more heat available for thermocapillarity. For $Pr, B = O(1)$ as $\epsilon \to 0$, there is to first approximation only the vertical heat condition

$$\Theta_{ZZ} = 0 \qquad (3.34a)$$

with

$$\Theta = 1, \qquad Z = 0 \qquad (3.34b)$$

and

$$\Theta_Z + B\Theta = 0, \qquad Z = H. \qquad (3.34c)$$

The thermal field thus satisfies

$$\Theta = 1 - \frac{BZ}{1 + BH}. \qquad (3.35)$$

Notice, both for fixed flux ($B = 0$) and fixed temperature ($B \to \infty$) boundaries, that the temperature on the interface is constant so that there is no thermocapillary effect; only in the range $0 < B < \infty$ can surface-tension variations produce interfacial stresses.

One can find an evolution equation for H by solving the lubrication equations for the flow. Now, however, on $Z = H$ the shear stress is balanced by surface-tension gradients. In dimensional form $\mathbf{t} \cdot \mathbf{T} \cdot \mathbf{n} - \partial\sigma/\partial s = 0$, and $\sigma = \sigma(\theta)$ so that $\partial\sigma/\partial s = (d\sigma/d\theta)(\partial\theta/\partial s)$. Here s is arclength. Note here that

$$\frac{\partial}{\partial s} = \left(1 + h_x^2\right)^{-1/2} \left[\frac{\partial}{\partial x} + h_x \frac{\partial}{\partial z}\right]$$

so that in the lubrication approximation,

$$\frac{\partial}{\partial s} \sim \frac{\partial}{\partial x} + h_x \frac{\partial}{\partial z}.$$

The lubrication system to be solved (in non-dimensional terms) is thus

$$U_{ZZ} = P_X, \qquad (3.36a)$$

$$P_Z = -G, \qquad (3.36b)$$

$$U = 0 \qquad \text{at} \qquad Z = 0, \qquad (3.36c)$$

$$\left.\begin{array}{l} -P = \bar{C}^{-1} H_{XX} \\[2mm] U_Z - \dfrac{MH_X}{(1 + BH)^2} = 0 \end{array}\right\} \qquad \text{on} \qquad Z = H, \qquad (3.36d, e)$$

where the Marangoni number

$$M = \frac{(\Delta\sigma)\epsilon}{\mu U_0} \tag{3.37}$$

and $\Delta\sigma = \gamma(\theta_0 - \theta_\infty)$ is the surface-tension difference between $\Theta = \Theta_0$ and $\Theta = \Theta_\infty$, through equation of state (3.30).

The system (3.36) is solved as before and the evolution equation for H becomes

$$H_T + \left\{\left[-\tfrac{1}{3}GH^3 + \tfrac{1}{2}MB\frac{H^2}{(1+BH)^2}\right]H_X\right\}_X + \tfrac{1}{3}\bar{C}^{-1}\left[H^3 H_{XXX}\right]_X = 0, \tag{3.38a}$$

In dimensional form and for $B \ll 1$, which is typical for liquid/gas interfaces,

$$3\mu h_t + \frac{3\alpha_T \Delta\sigma}{2k_T}(h^3 h_x)_x - \rho g(h^3 h_x)_x + \sigma(h^3 h_{xxx})_x = 0. \tag{3.38b}$$

There is a uniform steady solution $h = h_0$ for any constant h_0. Disturb this solution as $h = h_0 + h'$, linearize in h' and use normal modes. The growth rate s is given by

$$3\mu s = \left[\frac{3\alpha_T \Delta\sigma}{2k_T} - \rho g h_0 - \frac{\sigma}{h_0}(k'h_0)^2\right](k'h_0)^2. \tag{3.39a}$$

Note again that the hydrostatic pressure and surface tension are stabilizing, but now the thermocapillarity is destabilizing. An imposed vertical temperature gradient has created surface-tension gradients along the interface when S is distorted.

There is an instability present when $\tfrac{3}{2}\alpha_T\Delta\sigma/k_T > \rho g h_0$ which is the case when, say for water, h_0 is less than 2–3 mm. In this case the neutral curve again resembles that in figure 8 where now

$$(k'_c h_0)^2 = \frac{h_0}{\sigma}\left[\frac{3}{2}\frac{\alpha_T(\Delta\sigma)}{k_T} - \rho g h_0\right]. \tag{3.39b}$$

The mechanism for this instability is clear since a corrugated interface in a fixed temperature gradient experiences hot regions at valleys and cool regions at peaks. The thermocapillarity effect then drives interfacial fluid from valleys to peaks and since the liquid is viscous, bulk fluid as well; this enhances the corrugations. Numerical solutions of equation (3.38) show that the thinning of the layer continues until a finite time t_d when the layer has zero thickness at one or more points. The film will then either be dried out or there will be an adsorbed layer on the substrate. Which of these occurs depends on whether the van der Waals forces (not included here) are always attractive or become repulsive at small scale.

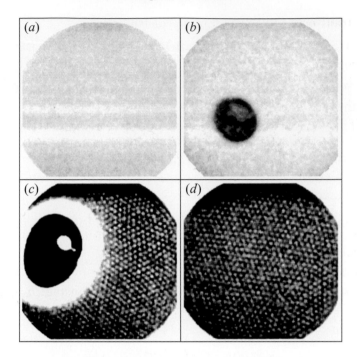

Figure 11. Infrared images from experiments with increasing fluid depth (increasing G). (a) 10% below onset of instability ($h_0 = 0.033$ cm). (b) For $h_0 = 0.011$ cm, the long-wavelength mode is the primary instability. The dark region is the dry spot. (c) At $h_0 = 0.22$ cm, the long-wavelength and hexagonal modes coexist. A droplet (light circle) is trapped within the dry spot (dark oval); the liquid layer is strongly deformed (light annulus) between the dry spot and the hexagonal pattern. (d) For $h_0 > 0.025$ cm. hexagons are the primary instability (here $d = 0.033$ cm). Copyright © 1995 American Institute of Physics. Reprinted with the permission of American Institute of Physics from van Hook *et al.* (1995).

Experiments by van Hook *et al.* (1995) on this interfacial Marangoni instability have shown precisely this behaviour. Figure 11 shows the 'dryout' of a 0.05–0.25 mm thick heated film of silicone oil. Notice that the dried-out region appears to be shiny suggesting the presence of an adsorbed film.

When numerical solutions of equations (3.38) in a bounded region are compared with the observations, the prediction of the critical thickness h_0 for instability is out by 20–30%. It is found that heat conduction in the gas above the liquid is of quantitative importance. Van Hook *et al.* (1995) solved the heat conduction problem for two layers and used this temperature in the thermocapillary-stress condition. They obtained an evolution equation like (3.38) but with the thermocapillary term slightly modified. This equation yields excellent quantitative agreement of the critical h_0 with experiment.

3.5 Phase transformation: evaporation/condensation

In the previous section, a liquid film was placed on a heated substrate and a thermal instability was studied. Since every liquid is to some degree volatile, it is worth examining the effects of this heating on the possibility of evaporation of the liquid.

Consider the two opposing cases of (*a*) an evaporating thin film of a pure, single-component liquid on a heated plane surface at constant temperature θ_0 that is higher than the saturation temperature at the given vapour pressure and (*b*) a condensing thin film of a pure, single-component liquid on a cooled plane surface at constant temperature θ_0 that is lower than the saturation temperature at a given vapour pressure. The speed of vapour particles is assumed to be sufficiently low that the vapour can be considered to be an incompressible fluid.

Let us first formulate boundary conditions appropriate for phase transformation at the film interface $z = h$. The mass balance at the interface is given by

$$j = \rho_g(v_g - v_i) \cdot n = \rho_f(v_f - v_i) \cdot n, \qquad (3.40)$$

where j is the mass flux normal to the interface, which is positive for evaporation and negative for condensation, ρ_g and ρ_f are, respectively, the densities of the vapour and the liquid, v_g and v_f are, respectively, the vapour and liquid velocities at $z = h$, and v_i is the velocity of the interface. In these cases the interface is no longer a boundary surface, but is in fact a source/sink of mass. The kinematic boundary condition can be generalized to

$$w = h_t + uh_x - j. \qquad (3.41)$$

Since $\rho_g/\rho_f \ll 1$, typically $\rho_g/\rho_f \approx 10^{-3}$, equation (3.40) shows that, relative to the interface, the magnitude of the normal velocity of the vapour at the interface is much greater than that of the corresponding liquid (though normally much less than Mach one). Hence, the phase transformation creates large accelerations of the vapour at the interface, where the back reaction, called the vapour thrust (or vapour recoil) represents a force on the interface. The dynamic pressure at the gas side of the interface is much larger than that at the liquid side,

$$\rho_g v_{g,e}^2 = \frac{j^2}{\rho_g} \gg \rho_f v_{f,e}^2 = \frac{j^2}{\rho_f}, \qquad (3.42)$$

where $v_{g,e}$ and $v_{f,e}$ are the normal components of vapour and liquid velocity

relative to the interface. Consider a corrugated interface during, say, evaporation. Points on the trough are closer to the hot plate than are points on
the crest, and so they have greater evaporation rates j and hence greater
momentum fluxes. As seen from equation (3.42), the sign of j is unimportant
in the destabilization (Burelbach, Bankoff & Davis 1988); evaporation and
condensation behave similarly.

The energy balance at $z = h$ is given by

$$j \left(L + \tfrac{1}{2}v_{g,e}^2 - \tfrac{1}{2}v_{f,e}^2\right) + k_T \nabla\theta \cdot \boldsymbol{n} - k_{T,g}\nabla\theta_g \cdot \boldsymbol{n} + 2\mu\epsilon_f \cdot \boldsymbol{n} \cdot \boldsymbol{v}_{f,e} - 2\mu_g\epsilon_g \cdot \boldsymbol{n} \cdot \boldsymbol{v}_{g,e} = 0,$$
$$(3.43a)$$

where L is the latent heat of vapourization per unit mass, $k_{T,g}$, μ_g, and θ_g
are, respectively, the thermal conductivity, viscosity, and temperature of the
vapour, and ϵ_f, ϵ_g are the rate-of-deformation tensors in the liquid and the
vapour (Burelbach *et al.* 1988).

The stress-balance boundary condition in the case of phase transformation
is given by

$$j(\boldsymbol{v}_{f,e} - \boldsymbol{v}_{g,e}) - (\boldsymbol{T} - \boldsymbol{T}_g) \cdot \boldsymbol{n} = \kappa\sigma(\theta)\boldsymbol{n} - \nabla_s\sigma, \qquad (3.43b)$$

where $\nabla_s\sigma$ is the surface gradient of interfacial tension and $\boldsymbol{T}_g$ is the stress
tensor in the vapour phase.

One needs to pose a constitutive equation relating the dependence of the
interfacial temperature θ_i and the interfacial mass flux (Palmer 1976; Plesset
& Prosperetti 1976; Sadhal & Plesset 1979). Its linearized form is

$$\tilde{K}j = \theta_i - \theta_s \equiv \Delta\theta_i, \qquad (3.44)$$

and

$$\tilde{K} = \frac{\theta_s^{3/2}}{\hat{\alpha}\rho_g L} \left(\frac{2\pi R_g}{M_w}\right)^{1/2};$$

θ_s is the absolute saturation temperature, $\hat{\alpha}$ is the accommodation coefficient, R_g is the universal gas constant, and M_w is the molecular weight of
the vapour. Note that the absolute saturation temperature θ_s serves now as
the reference temperature θ_∞ in the non-dimensionalization. When $\Delta\theta_i = 0$,
the phases are in thermal equilibrium with each other, i.e. their chemical
potentials are equal. In order for net mass transport to take place, a vapour
pressure driving force must exist, given for ideal gases by kinetic theory
(Schrage 1953), and represented in the linear approximation by the parameter $\tilde{K}$ (Burelbach *et al.* 1988). Departure of the parameter $\tilde{K}$ from ideal
behaviour is addressed by an accommodation coefficient depending on inter-
face/molecular orientation and steric effects, which represent the probability
of a molecule of vapour sticking after hitting the liquid–vapour interface.

The balances discussed above give rise to a 'one-sided' model for evaporation or condensation (Burelbach *et al.* 1988) in which the dynamics of the vapour are ignored, except that mass is conserved and one retains the effects of vapour thrust and the kinetic energy it produces. It is assumed that the density, viscosity, and thermal conductivity of the liquid are much greater than those in the vapour. Therefore the boundary conditions (3.43) are significantly simplified.

The energy balance relation (3.43a) becomes

$$k_T \theta_z = jL. \tag{3.45}$$

meaning that all the heat conducted to the interface in the liquid is converted to latent heat of vaporization.

Next, the normal and tangential stress conditions at the free surface, given by equation (3.44b), are reduced to

$$-\frac{j^2}{\rho_g} - \boldsymbol{n} \cdot \boldsymbol{T} \cdot \boldsymbol{n} = \kappa \sigma(\theta), \tag{3.46a}$$

$$\boldsymbol{t} \cdot \boldsymbol{T} \cdot \boldsymbol{n} = \boldsymbol{t} \cdot \nabla_s \sigma. \tag{3.46b}$$

Finally, the remaining boundary conditions are unchanged.

Consider now the non-dimensional formulation for the two-dimensional case. The dimensionless mass balance is modified by the presence of the non-dimensional evaporative mass flux J,

$$EJ = (H_T - UH_X + W)\left(1 + H_X^2\right)^{-1/2}, \tag{3.47a}$$

or at leading order of approximation

$$H_T + Q_X + EJ = 0, \tag{3.47b}$$

where

$$Q(X, T) = \int_0^H U dZ \tag{3.47c}$$

is the scaled volumetric flow rate per unit width parallel to the wall, $J = (h_0 L/k_T \Delta\theta) j$, $\Delta\theta \equiv \theta_0 - \theta_s$, and $\rho = \rho_f$. The parameter E is an evaporation number $E = k_T \Delta\theta / \rho v L$, which represents the ratio of the viscous time scale h_0^2/v to the evaporative time scale, $\rho h_0^2 L/k_T \Delta\theta$ (Burelbach *et al.* 1988). The latter is a measure of the time required for an initially stationary film to evaporate to dryness on a horizontal wall. For low evaporation rates E is small. The dimensionless versions of (3.45) and (3.46) are then

$$KJ = \Theta, \qquad \Theta_z = -J \qquad \text{at} \qquad Z = H, \tag{3.48a}$$

where

$$K = \tilde{K} \frac{k_T}{h_0 L}. \tag{3.48b}$$

Using the above to solve for θ, the full evolution equation can be written down (Oron *et al.* 1997).

Consider first the case of an evaporating/condensing thin liquid layer lying on a rigid plane held at a constant temperature. Mass loss or gain is retained, while other effects are neglected. One solves $\Theta_{zz} = 0$ along with boundary conditions $\Theta = 1$ at $z = 0$, the flux condition at $z = H$, and eliminates J yielding the dimensionless temperature field and the evaporative mass flux through the interface

$$\Theta = 1 - \frac{Z}{H + K}, \qquad J = \frac{1}{H + K}. \tag{3.49}$$

An initially flat interface will remain flat as evaporation proceeds and if surface tension, thermocapillary, and convective thermal effects are negligible, i.e. $M = \bar{C}^{-1} = \epsilon Re\, Pr = 0$, a scaled evolution equation of the form

$$H_T + \frac{\tilde{E}}{H + K} = 0 \tag{3.50}$$

will result, where $\tilde{E} = \epsilon^{-1} E$ and K, the scaled interfacial thermal resistance, is equivalent to the inverse Biot number B^{-1}. Physically, $K \neq 0$ represents a temperature jump from the liquid surface temperature to the uniform temperature of the saturated vapour, θ_s. The conductive resistance of the liquid film is proportional to H, and, assuming infinite thermal conductivity of the solid, the total thermal resistance is given by $(H + K)^{-1}$. For constant superheat temperature $\theta_0 - \theta_s$, equation (3.50) represents a volumetric balance, whose solution, subject to $H(0) = 1$, is

$$H = -K + \left[(K + 1)^2 - 2\tilde{E}T\right]^{1/2}. \tag{3.51a}$$

The film disappears in finite time

$$T_d = \frac{2K + 1}{2\tilde{E}} \tag{3.51b}$$

and the rate of disappearance of the film at $T = T_d$ remains finite:

$$\left.\frac{\mathrm{d}H}{\mathrm{d}T}\right|_{T=T_d} = -\frac{\tilde{E}}{K}. \tag{3.52}$$

The value of H_T remains finite because, as the film thins, the interface temperature θ_i, nominally at its saturation value θ_s, increases to the wall temperature. If $K = 0$, $\theta_i = \theta_s$ and the temperature gradient $(\theta_0 - \theta_s)/h \to \infty$, the thermal resistance vanishes and hence the mass flux will go to infinity as $h \to 0$. However, for large evaporation rates the interfacial temperature jump becomes significant, so that non-zero K is significant. Further, when the film becomes very thin, a thermal disturbance develops in the solid substrate, reflecting the fact that the thermal conductivity of the solid is finite. Hence the two thermal resistances, acting in series, prevent the evaporation rate from becoming infinite.

From Burelbach *et al.* (1988), the interfacial thermal resistance $K = 10$ for a 10 nm thick water film. Since K is inversely proportional to the initial film thickness, $K \sim 1$ for $h_0 = 100$ nm, so that $H/K \sim 1$ at this point. However, $H/K \approx 10^{-1}$ at $h_0 = 30$ nm, so the conduction resistance becomes small compared to the interfacial transport resistance shortly after van der Waals forces become appreciable.

The dimensionless vapour thrust gives an additional normal stress at the interface, $-\frac{3}{2}\tilde{E}^2 D^{-1} J^2$, where D is a unit-order scaled ratio between the vapour and liquid densities, $3\epsilon^{-3}\rho_g/2\rho$, and can be calculated using (3.49). The resulting scaled evolution equation for a film evaporating on an isothermal horizontal surface is obtained in the form

$$H_T + \tilde{E}(H+K)^{-1} + \left[\tilde{E}^2 D^{-1}\left(\frac{H}{H+K}\right)^3 H_X + \tfrac{1}{3}\bar{C}^{-1} H^3 H_{XXX}\right]_X = 0. \quad (3.53)$$

It is often the case that $\tilde{E}$ can itself be a small number and it can be used as an expansion parameter for the description of slow evaporation compared to the non-evaporating base state (Burelbach *et al.* 1988) appropriate to very thin evaporating films.

Taking into account van der Waals forces, Burelbach *et al.* (1988) gave the complete evolution equation for a heated or cooled thin film on a horizontal plane surface in the form

$$\begin{aligned}
H_T + \tilde{E}(H+K)^{-1} + \tfrac{1}{3}\bar{C}^{-1}(H^3 H_{XXX}) + \{[AH^{-1} + \tilde{E}^2 D^{-1} H^3 (H+K)^{-3} \\
+ KMPr^{-1}(H+K)^{-2}]_X H_X\} = 0. \quad (3.54)
\end{aligned}$$

Here the first term represents the rate of volumetric accumulation, the second the mass loss, the third the stabilizing capillary term, and the fourth, fifth, and sixth the van der Waals, vapour thrust, and thermocapillary terms, respectively, all destabilizing. This is the first full statement of the possible competition among various stabilizing and destabilizing effects on a horizontal plate, with scaling to make them present at the same order.

3.6 Falling films

In the above examples, the liquid layer is motionless until a disturbance is applied. In more general applications such as solar-power receivers or rain running down a windscreen, the film possesses mean shear, which itself can lead to surface-wave instabilities.

Consider a thin liquid layer flowing down a plane inclined to the horizontal by angle α. The equations are consistent with a uniform film of depth h_0 in parallel flow with the profile

$$\bar{u}(z) = \frac{\rho g \sin \alpha}{\mu} \left(h_0 z - \tfrac{1}{2} z^2 \right) \tag{3.55a}$$

and the hydrostatic pressure distribution

$$\bar{p}(z) = p_A + \rho g (h_0 - z) \cos \alpha. \tag{3.55b}$$

This layer is susceptible to long-surface-wave instabilities, as discovered by Yih (1955, 1963) and Benjamin (1957) using linear stability theory. Benney (1966) extended the theory into the nonlinear regime by deriving a nonlinear evolution equation for the interface shape $z = h(x, t)$.

The application of long-wave theory gives rise to the evolution equation

$$H_T + Re\, H^2 H_X = 0, \tag{3.56}$$

where

$$Re = G \sin \alpha, \qquad G = \frac{\rho^2 h_0^3 g}{\mu^2}. \tag{3.57}$$

Equation (3.56) is a first-order wave equation that can be solved by the method of characteristics. The solution evolves from an initial distribution and describes nonlinear waves that travel and steepen. This can be easily seen by linearizing the equation about $H = 1$. Let $H = 1 + H'$, and obtain $H_T' + ReH_X' = 0$ whose solution is $H'(X, T) = H_0 \exp(X - ReT)$. There are no instabilities present in equation (3.56). Figure 12 shows, for $\alpha = 6.4°$, surface waves that result from an interaction of the shear and the surface deflection. In order to describe the instabilities, one must consider the next term in the asymptotic expansion in ϵ. If one does so, then the augmented equation takes the form

$$H_T + Re\, H^2 H_X + \epsilon \left[\tfrac{2}{15} Re^2 H^6 H_X - \tfrac{1}{3} H^3 H_X (G \cos \alpha) + S H^3 H_{XXX} \right]_X = 0, \tag{3.58}$$

where $S = (\rho \sigma h_0 / 3\mu^2)\epsilon^2 = O(1)$.

If equation (3.58) is linearized about $H = 1$, the waves that occur travel

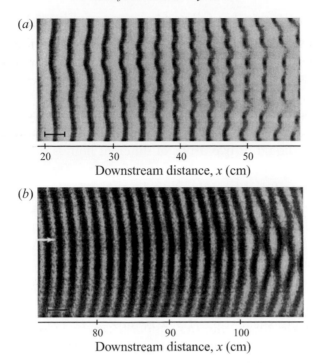

Figure 12. Photographs of three-dimensional patterns arising in falling films. The film flows on an inclined plane and is perturbed at the upstream end ($x = 0$). Visualization is by fluorescence imaging; the film thickness is proportional to the brightness, i.e. the thick region is bright and the thin region is dark. The direction of the flow is from the left to the right. Lines seen on the photograph are of equal heights: (*a*) Synchronous three-dimensional instability of two-dimensional periodic waves. A snapshot taken at the inclination angle of 6.4°, Reynolds number of 72, and imposed perturbation frequency of 10.0 Hz. (*b*) A herringbone (or checkerboard) pattern due to three-dimensional subharmonic instability. A shapshot taken at the inclination angle of 4.0°, Reynolds number of 50.5, and imposed perturbation frequency of 14 Hz. Copyright © 1995 American Institute of Physics. Reprinted with the permission of American Institute of Physics from Liu, Schneider & Gollub (1995).

at speed Re, as above, and the growth rate s satisfies

$$s = \epsilon k^2 \left[\tfrac{2}{15} Re^2 - \tfrac{1}{3} G \cos \alpha - S k^2 \right]. \tag{3.59}$$

Thus, there is an instability of the plane interface when $\tfrac{2}{15} Re^2 > \tfrac{1}{3} G \cos \alpha$ or when $Re^2 > \tfrac{5}{2} \cot \alpha$; hydrostatic pressures delay the onset of instability.

Equation (3.58) can be integrated numerically to describe the nonlinear evolution of initial shapes in two dimensions. See Oron *et al.* (1997) for extensions of equation (3.58) to three dimensions. A simulation of this would show wave trains that falling films support. Figure 12 shows them at higher Re that is governed by equation (3.58).

3.7 Free films

Free films of liquid are bounded by two interfaces between liquid and gas or liquid and liquid. Examples of such a configuration may be provided by two bubbles in a liquid or two drops of different liquids suspended in a third liquid. In order to study the behaviour of such a free film, one needs to formulate the interfacial boundary conditions at both the interfaces, given by $z = h^{(1)}(x, t)$ and $z = h^{(2)}(x, t)$. These are standard but complicated.

Consider the case when surface tension and viscous forces are present and van der Waals attractions occur (Erneux & Davis 1993). At leading order in ϵ, the governing system would be

$$U_{ZZ} = -P_X \tag{3.60}$$

subject to, say, zero shear stress, $U_Z = 0$, on $Z = H^{(1)} = -H^{(2)}$, the varicose case.

One integral of equation (3.60) is $U_Z = -P_X Z + C$. In order to satisfy both shear conditions one must take $C = 0$ and $P_X = 0$, i.e. the incorporation of the pressure gradient must be postponed to the next order. If this is done, then $U_Z \equiv 0$ so that $U = U(X, T)$ only and U is unknown at the moment. The condition of mass conservation gives

$$H_T + (UH)_X = 0. \tag{3.61a}$$

This is a single equation in two unknowns, H and U. If one then goes to next order in ϵ, one can determine another relation, namely

$$\overline{Re}\, H(U_T + UU_X) = -H[\Phi_X - \bar{C}^{-1} H_{XXX}] + 4(HU_X)_X. \tag{3.61b}$$

Here $\overline{Re} = \epsilon Re = O(1)$, $\bar{C} = C\epsilon^{-1} = O(1)$. Equation (3.61b) resembles the one-dimensional Navier–Stokes equation with 'density' H, 'viscosity' $4H$ and kinematic viscosity 4. The 'pressure gradient' contains contributions from van der Waals and surface-tension forces.

If one linearizes relations (3.61) about $H = 1$ and $U = 0$, one obtains the characteristic equation

$$\overline{Re}\, s^2 + 4k^2 s + k^2[\bar{C}^{-1}k^2 - 3A] = 0.$$

The planar film is unstable only if $3A > \bar{C}^{-1}k^2$, i.e. when van der Waals attractions, $A > 0$, outweigh the stabilization produced by surface tension.

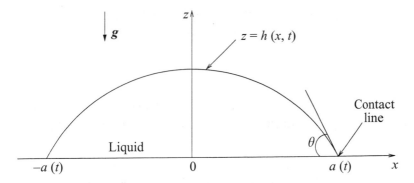

Figure 13. A sketch of a two-dimensional liquid drop with interface at $z = h(x,t)$, contact lines at $x = \pm a(t)$, and contact angle θ.

In the infinite-layer case, $A > 0$ implies unconditional instability though the presence of sidewalls can lead to stable films.

The coupled system (3.61) can be integrated numerically for periodic boundary conditions. The rupture takes place in a finite time, $T = T_R$, and one can find local similarity solutions at the point of rupture. See §5 for a discussion of singularities. This film can be a means of creating droplets that result from the fragmentation of the layer.

4 Contact lines

In §§2 and 3, interfaces between immiscible fluids were discussed with emphasis on the effects of interfacial forces. In the present section a more complex system will be considered, in which interfaces intersect. Figure 13 shows a fluid–fluid interface meeting a fluid–solid interface. Since each interfacial region is considered to be a mathematical interface, the intersection is a curve called the *contact* or *common line*, or a *tri-junction*.

Contact lines occur in many important applications. A *static* contact line is one that is motionless relative to the solid. Figure 14(*a*) shows the side view of a liquid droplet attached to an inclined plane. Owing to the presence of gravity, the apparent contact angle, the angle measured within the liquid, at the front and back appear to be unequal in order to balance the component along the plate of the weight of the drop.

Figure 14(*b*) shows the phenomenon of the die swell in which the liquid is forced from an orifice, and expands after exiting the channel. Here the bulk fluid is in motion though the contact lines are stationary.

Figure 14(*c*) shows a gas bubble rising in a liquid in a frame of reference fixed to the bubble. If there are small amounts of contaminants in the

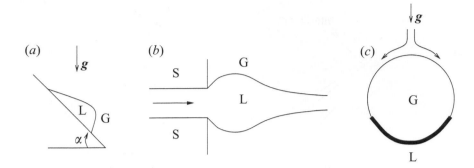

Figure 14. Sketches of several contact-line problems: (*a*) a liquid drop on an inclined plane, (*b*) liquid forced through a die, and (*c*) a gas bubble with contaminant, rising in a liquid. L denotes liquid, S solid and G gas.

liquid, they can be attracted to the interface, become surface active, and reduce the surface tension proportionally to their surface concentration. In the configuration shown, the translation of the bubble induces surface flows from the top to the bottom of the bubble, convecting the contaminant and altering its distribution. If this surface material is concentrated enough, it behaves like a rigid cap which strongly affects the drag of the bubble. The edge of the cap is a static contact line. Static contact lines are the sites of flow singularities that affect the ability to solve numerically for the flow fields.

The droplet in figure 14(*a*) will begin to move down the plate after the tilt angle of the plate has been increased enough. Now the contact lines at the front and back may be moving. The die swell system of figure 14(*b*) can have the fluid–fluid interface dislodged from the sharp corner and spread outward. The contact line may come to rest and again be static. Alternatively, it can oscillate up and down and remain non-stationary.

Moving contact lines are also the sites of flow singularities. However, now, if the usual modelling is used, the singularities are non-integrable in that they correspond to the need to apply infinite forces to move the contact lines. Clearly, such models are physically unacceptable. In what follows, moving contact lines will be analysed kinematically and dynamically and new models will be introduced.

4.1 Kinematics

In order to model moving contact lines one has to understand qualitatively what occurs locally. To this end Dussan V. & Davis (1974) placed a drop of honey on a smooth horizontal plate of glass and marked a point on

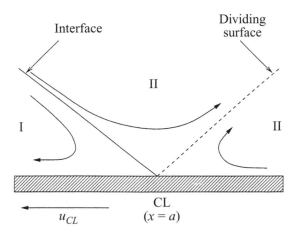

Figure 15. The kinematics near a moving contact line CL in a reference frame moving with u_{CL} according to Dussan V. & Davis (1974). The interface rolls over the solid and in fluid II there is a dividing streamline emerging from CL.

the interface with a non-surface-active dye. The plate was then tilted from the horizontal enough to cause the drop to move downward. The drop was viewed at successive times and the marker was seen to move toward the contact line, reach it, and then be overrun by the drop. Looking through the bottom of the plate, one could see the marker apparently in contact with the plate though it had been distorted by the shear. This experiment demonstrates that interfacial points arrive at the contact line in finite times (unlike the behaviour near the nose of a bluff body where an infinite time is needed in order for far-field points to arrive at the stagnation point). It also shows that the contact line is not a material line: it is composed of different material points at different times. Finally, it shows that the drop rolls down the plate.

The local flow field near a moving contact line is quite complicated. Figure 15 shows a sketch of a small local region. Material near the fluid–fluid interface enters the volume as shown, and material on the right attached to the solid–fluid boundary also enters as shown. If there is no leakage of fluid beneath the drop, then there must be at least one dividing streamline in fluid II as shown. This has been verified experimentally.

The experiment shows another property of the moving contact line, namely that it is the site of trajectory splitting. To see this consider a point on the fluid–fluid interface. It has two identities: its fluid I identity rolls under the advancing fluid, while its fluid II identity is ejected along the dividing streamline. It is worth noting that a point on the contact line has triple identity: fluids I and II, and solid.

As an example of the implications of trajectory splitting, consider a wedge of liquid, fluid II, say, local to the contact line. The velocities along the boundaries are distinct near the origin, and in fact have different senses. Let us seek a solution for creeping flow in the wedge in polar coordinates (r, θ) with corresponding velocity (u, v). Introduce a stream function ψ such that $u = -(1/r)\psi_\theta$ and $v = \psi_r$. Then ψ satisfies a biharmonic equation subject to the no-slip conditions $\psi = \psi_1$ on $\theta = 0$ and $\psi = \psi_2$ on θ_α and the zero-penetration conditions $\psi_\theta = 0$ on $\theta = 0, \alpha$. There is a simple solution to this local flow with bounded ψ at $r = 0$, namely $\psi = (\psi_2 - \psi_1)\theta/\alpha + \psi_1$, independent of r. Note that if $\psi_1 \neq \psi_2$, i.e. $u_1 \neq u_2$, then $v \equiv 0$ and the radial velocity $u = (1/r)(\psi_2 - \psi_1)/\alpha$, which is singular at $r = 0$. This demonstrates the well-known contact-line singularity. If $u \sim 1/r$ near the contact line, then there are infinities in the stress components and these are non-integrable. The force necessary to move the contact line is logarithmically infinite when Stokes flow is assumed and the no-slip condition is enforced.

It is straightforward to estimate the terms in the Navier–Stokes equations that were neglected by the Stokes flow assumption and show that these are negligible compared to the viscous terms as $r \to 0$. Thus, the force necessary to move the contact line of a Newtonian fluid is infinite when the no-slip condition is applied.

4.2 Dynamics

The discussion in the previous section showed that at a contact line a Newtonian fluid must exhibit local slip in order to move. This has led to the remodelling of the flow near the contact line to allow slip locally on the fluid–solid interface. This should be regarded as an effective slip, not molecular slip, that enables one to define mathematical models that give reasonable predictions for macroscopic flow quantities. It is a device that makes the macroscopic dynamics well defined though it is also an admission that the microscopic physics near the contact line is not well understood nor well described.

In order to have a well-posed problem at a moving contact line two ingredients are needed. One must pose a slip law, which relieves the force singularity. One must also pose a condition on the contact angle θ; one knows empirically that measured contact angles depend on the speed and sense of motion, u_{CL}, of the contact line.

The slip law is an *ad hoc* relationship at contact line $x = a(t)$ often taken in the form

$$w = 0, \qquad u = \beta u_z \qquad \text{at} \qquad z = 0. \qquad (4.1a, b)$$

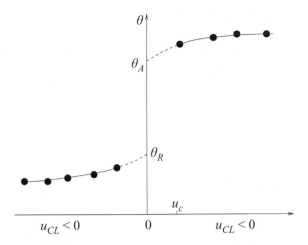

Figure 16. Typical experimental results for the dependence of the dynamic contact angle, θ, on the speed of the contact line. When $u_{CL} > 0$ ($u_{CL} < 0$) the contact line is advancing (receding). u_c denotes the lowest speed at which an experimental measurement has been made.

Equation (4.1) states that the slip velocity is proportional to the shear stress, essentially u_z, through a numerically small coefficient β (called the slip length). Near the contact line u_z can become large (but finite) so the slip becomes appreciable there.

The second ingredient is the function $\theta(u_{CL})$, which is obtained empirically. Figure 16 (see Dussan V. 1979) shows typical measured values of θ for the cases of liquid displacing gas, $u_{CL} > 0$, and gas displacing liquid, $u_{CL} < 0$, and there is a discontinuity at $u_{CL} = 0$. This indicates that a liquid displacing gas at $u_{CL} > 0$ must stop as $\theta \to \theta_A^+$; the line must remain stationary as θ decreases from θ_A to θ_R, at which time the gas can displace liquid. The jump in the θ vs. u_{CL} characteristic induces what is called contact-angle hysteresis.

It is now possible to calculate the macroscopic spreading of a liquid drop on a plate; the system is shown in figure 13.[1] Aside from the Navier–Stokes equation and continuity, one poses conditions (4.1) on the solid–liquid interface, and zero shear stress and normal stress equal to surface tension times curvature and the kinematic condition on the fluid–fluid interface, $z = h(x, t)$. In addition there is the condition of contact

$$h(\pm a(t), t) = 0 \tag{4.2}$$

---

[1] It is the choice of the slip law, here equation (4.1b), and the function θ, which relates the angle to the contact-line speed, say, that models the small-scale physics near the contact line and such things as the roughness and the chemical inhomogeneity of the substrate. Equations (4.1b) and (4.2) then relate the small scales to the macroscopic behaviour of the system.

and the contact-angle condition

$$h_x(\pm a(t), t) = \mp \tan \theta. \tag{4.3}$$

Finally, the drop volume is known:

$$\int_{-a}^{a} h \, \mathrm{d}x = V_0. \tag{4.4}$$

When contact angles are measured through a light microscope, it is found that the data are well described by the relations (Dussan V. 1979)

$$u_{CL} = K_A(\theta - \theta_A)^m, \qquad \theta > \theta_A, \tag{4.5a}$$

$$u_{CL} = K_R(\theta_R - \theta)^m, \qquad \theta < \theta_R, \tag{4.5b}$$

where K_A, K_R, and m are positive and θ_A and θ_R are called the advancing and receding contact angles, respectively, each measured in the liquid.

A major contribution of Dussan V. and her coworkers was to argue and show, both theoretically and experimentally, that measured angles, given by (4.5) are not material constants but are affected by the outer flows and hence the geometries of the systems. The departures of measured angles from the microscopic angles are due to viscous bending of the interface near the contact line. For small capillary numbers $C = \mu u_{CL}/\sigma$, these departures are small. In order to obtain a geometry-invariant θ, one has to go to smaller scale and use asymptotic matching to find a new angle, the intermediate angle θ_I. Ngan & Dussan V. (1989) and Dussan V., Ramé & Garoff (1991) excise a small neighbourhood of the contact line and on the arc of the sector prescribe the flow at $r = R$ and $\theta = \theta_I$, which results from an asymptotic theory for $\beta \to 0$. In all the descriptions, the local region near the contact line supplies the outer field with a one-parameter family of solutions. They can be labelled by the value of slip length β and by the angle θ_I at $r = R$.

De Gennes (1985) and Troian *et al.* (1989) considered perfect spreading, $\theta = 0$, and envisioned the existence of a precursor film on the solid ahead of the front. Then in effect one has macroscopic spreading over a prewetted surface. The precursor film may have finite or infinite extent but in either case its existence removes the problem of a contact line. Thus, rather than a slip coefficient β, one has a precursor film thickness δ that labels the local solution at the front. Comparable outer fields result from the two approaches if $\beta \approx \delta$ (Spaid & Homsy 1996).

In what follows it will be supposed that the apparent and microscopic contact angles have similar functional forms of contact-line speed and hence equations (4.5) will be used for the discussion. Of course, the coefficients K_A, K_R, and m are not directly known but have to be inferred. Hocking

(1983) assumed that the microscopic contact angle contains hysteresis but is independent of speed u_{CL} and deduced forms like equations (4.5) for the apparent angle. His model can be obtained as a limiting case, $K_A, K_R \rightarrow \infty$, in (4.5).

Consider the model outlined above for a two-dimensional liquid drop spreading on a horizontal plate; the gas is considered passive. One can simplify the problem greatly by taking the contact angles to be small and the drop to have small slopes. One can then use the scaling and asymptotics of lubrication theory of §3 to obtain an evolution equation for the drop (Greenspan 1978).

When surface tension is constant and body forces are absent, one obtains

$$\mu h_t + \sigma \left[\left(\tfrac{1}{3} h^3 + \beta h^2 \right) h_{xxx} \right]_x = 0 \qquad (4.6)$$

subject to conditions (4.2)–(4.5).

This system, which can be solved numerically subject to initial conditions, has two distinct time scales. One, from equation (4.6), is associated with viscosity and surface tension, $t_1 = \ell \mu / \sigma$ where ℓ is a length scale. Another scale comes from the edge condition (4.5) and is associated with the pulling at the contact line. Let K_{CL} be either K_A or K_R and θ_s be either θ_A or θ_R, respectively. The second scale is $t_2 = \ell \theta_s^{-m} / K_{CL}$. In the case of small θ_s, the appropriate ratio t_1 / t_2 is $\bar{C}_m = \mu K_{CL} / (\sigma \theta_s^{3-m})$. If $\bar{C}_m \rightarrow 0$, then the evolution equation (4.6) becomes quasi-steady, i.e. the time derivative is negligible for times greater than $O(\bar{C}_m)$, and has the solution $h_{xx} = -\kappa(t)$ where κ is the as-yet-unknown curvature dependent only on time t. Thus, the interface is the arc of a parabola, the lubrication limit of the constant-curvature circle.

The solution subject to contact conditions (4.2) is $h = \tfrac{1}{2} \kappa (a^2 - x^2)$. The curvature κ is related to volume V_0 by equation (4.4), so that $\kappa = \tfrac{3}{2} V_0 / a^3$ and hence

$$h = \frac{3}{4} \frac{V_0}{a^3} \left(a^2 - x^2 \right). \qquad (4.7)$$

Finally, one substitutes (4.7) into the contact-angle characteristic (4.5) to obtain

$$\left. \begin{aligned}
\dot{a} &= K_A \left[\frac{3V_0}{2a^2} - \theta_A \right]^m, \qquad \theta > \theta_A, \\[2mm]
\dot{a} &= K_R \left[\theta_R - \frac{3V_0}{2a^2} \right]^m, \qquad \theta < \theta_R.
\end{aligned} \right\} \qquad (4.8)$$

Equations (4.8) can be used to monitor drops that expand, contract, or a combination of these. For example, if there is only spreading, $\theta \geqslant \theta_A > 0$, then one can obtain the approach to the steady state with $\theta = \theta_A$, $a_\infty = \sqrt{3V_0/2\theta_A}$. If $\theta_A = 0$, the droplet will spread forever and one can obtain simple power laws valid for long times $a \sim t^q$. For example, for $\theta_A = 0$ and $m = 3$, one obtains $q = 1/7$. A list of such theoretical and experimental power laws is given in table I of Erhard & Davis (1991). The theory and experiment agree well in terms of the exponents, though the multiplicative constant is not well tested.

An alternative approach is that of Hocking (1983) in which θ is taken to be either θ_A or θ_R but in either case is independent of u_{CL}. This corresponds to $\bar{C}_m \to \infty$ in system (4.2)–(4.6). The result is (4.6) with the time derivative retained and the contact-angle condition

$$\theta = \theta_A \qquad \text{for} \qquad a > 0, \qquad \text{and} \qquad \theta = \theta_R \qquad \text{for} \qquad a < 0. \qquad (4.9)$$

For constant $\beta \to 0$, i.e. $\beta \ll \bar{C}_m$, Hocking (1995a) found that the apparent angle θ satisfies

$$\theta^3 = \theta_0^3 + u_{CL} \left[K_A^{-1} + 9\bar{C}_m \ln(h_m/\beta) \right] . \qquad (4.10)$$

This is obtained at second order in β through matched asymptotic expansions with a double boundary layer near the contact line. Thus, if $K_A^{-1} \gg 9\bar{C}_m \ln(h_m/\beta)$, then the results given above hold since then the effects of slip appear only as a first correction and spreading is controlled by the angle versus speed characteristic. On the other hand, if $K_A^{-1} \ll 9\bar{C}_m \ln(h_m/\beta)$, the Hocking theory applies and the spreading is limited by the slip.

If the Hocking model is used, it is found that the power laws for large t correspond to the same exponents that correspond to $m = 3$; only the constant multipliers differ between the predictions of the two theories. Hocking (1995a) argued that the multiplier for spreading oils is numerically closer to that of the Hocking theory.

If one takes the view that the local physics near the contact line is 'unknown' as seen by the macroscopic viewer, one can 'excise' a neighbourhood of the contact line and instead match an outer solution, which solves the evolution equation with a local (singular) wedge flow that has the additional information $\theta_I = \theta_I(R)$. Here the angle θ_I is an intermediate angle inferred from the asymptotic of Hocking for $\beta \to 0$ as a function of distance R from the contact line. Ngan & Dussan V. (1989) obtained this relation in the form

$$\theta_I \sim \theta_A + C \left[\frac{2 \sin \theta_A}{\theta_A - \sin \theta_A \cos \theta_A} \left(\ln \frac{R}{\beta} + 1 \right) + l(\theta_s) \right] + \cdots, \qquad (4.11)$$

where $R/\beta \to \infty$, $l(\theta_A)$ depends directly on the form of the slip boundary condition, and C is capillary number

$$C = \frac{\mu u_{CL}}{\sigma}. \tag{4.12}$$

Note that (4.11) is valid in the limit of $\beta \ll C \ll 1$. Dussan V. *et al.* (1991) show good agreement with predictions from this theory by measurements in mutual-displacement systems.

Another approach to spreading for $\theta_A = 0$ is that of de Gennes (1985), who examined the small-scale physics of contact lines. He reasoned that with perfect spreading there should be a nearly uniform precursor foot ahead of the droplet in which attractive van der Waals forces are operative. In such a 'foot' he takes ϕ given by equation (3.25) as the potential for these forces. He analyses the thick drop that smoothly blends into the foot, which extends far forward along the substrate. On the one hand there is no longer a contact line nearby to consider, and, on the other hand, the actual contact line at the edge of the foot is not considered. This model is also able to predict an apparent contact angle θ satisfying equation (4.11) with $\theta_A = 0$ and $m = 3$.

When $\theta_A > 0$, the model (3.25) no longer holds and one must use a van der Waals model appropriate to a wedge-shaped region. Hocking (1995a) showed that in this case

$$\phi = \phi_R + \frac{A'}{6\pi h^3} \left[\theta_A^4 - (h_x')^4 \right], \qquad A' > 0, \tag{4.13}$$

where A' is a modified Hamaker constant. In Hocking's theory θ_A is constant, and when he solves the spreading-drop problem with such a force potential, he finds that the contact-angle condition emerges as a natural boundary condition; the power laws of table I in Ehrhard & Davis (1991) remain unchanged, though the multipliers do change slightly. His conclusion is that for $\theta_A > 0$ the presence of the van der Waals potential does not have a significant effect on the spreading process.

As discussed above, the macroscopic dynamics of spreading drops, and hence spreading in general, is connected to the microscale physics and chemistry of the system by the conditions at the contact line, namely the slip condition (4.1) and the contact-angle condition (4.3). Great progress has been made in recent years, beginning with the work of de Gennes (1985), on the underlying mechanisms present in a small region near the contact line. This work was recently reviewed by Leger & Joanny (1992), who discussed the origin of contact-angle hysteresis by roughness (defects) and chemical inhomogeneity of the surface, including the possibility of irregular jump motions of the contact line.

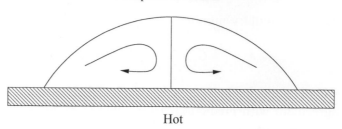

Hot

Figure 17. A sketch of the thermocapillary flow in a heated droplet.

4.3 Hydrostatic pressure

There is a quite interesting effect of gravity on the spreading of drops. Assume that gravity *g* acts vertically downward. In this case the evolution equation takes the form

$$\mu h_t + \left\{ \left(\tfrac{1}{3}h^3 + \beta h^2 \right) (\sigma h_{xx} - \rho g)_x \right\}_x = 0. \tag{4.14}$$

When the drop is initially 'thick', hydrostatic pressure will of course tend to flatten it and enhance its spreading rate. When the drop thins as it spreads, hydrostatic effects become negligible. However, at yet later times, hydrostatic effects re-emerge as being important, since at very long times the curvature $h_{xx} \to 0$ more quickly than $h \to 0$. This result of Ehrhard & Davis (1991) is in agreement with the observations of Cazabat & Cohen-Stuart (1986) and Ehrhard (1993). Ehrhard shows that for silicone-oil droplets the change in behaviour occurs at about 10^3 s.

4.4 Thermocapillarity

It turns out that the heating or cooling of the plate can strongly affect the rate of spreading of a liquid drop. Consider such spreading on a uniformly heated horizontal plate where the surface tension $\sigma = \sigma(\theta)$. Thermocapillarity creates a flow in the drop that retards the spreading. One can solve first for the temperature as a functional of *h* and hence determine the thermocapillary shear stress $d\sigma/dx \equiv \sigma_x$ on the interface. One can then obtain the evolution equation (Ehrhard & Davis 1991)

$$\mu h_t + \sigma \left[\left(\tfrac{1}{3}h^3 + \beta h^2 \right) (h_{xx} - \rho g h)_x \right]_x + \frac{\alpha_T \Delta\sigma}{2k_T} \left[\left(\tfrac{1}{2}h^2 + \beta h \right) h_x \right]_x = 0 \tag{4.15}$$

valid for Biot number $B = \alpha_T h_0 / k_T \ll 1$.

In the quasi-static case, $\bar{C}_m \to 0$, heating retards spreading as shown by the sense of the streamlines in figure 17. Even when $\theta_A = 0$, any amount of heating will cause the drop to cease spreading at a finite width. Of course

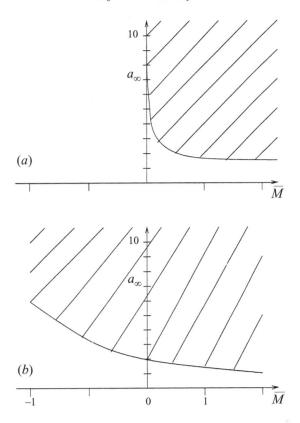

Figure 18. Final spreading: final drop widths a_∞ as functions of Marangoni number $\bar{M} \propto \Delta\sigma$ for $G = 0$ and for two different advancing contact angles (a) $\theta_A = 0$ and (b) $\theta_A = 0.5$.

the final state is a dynamical one even though the contact line is stationary. Figure 18 shows how heating, $\Delta\sigma > 0$, retards spreading while cooling, $\Delta\sigma < 0$, promotes it. The experiments by Ehrhard (1993) show quantitative agreement with these predictions. In neither case does the spreading follow a power law.

4.5 Evaporation or condensation

If a drop is composed of a volatile liquid, the spreading and mass loss/gain can compete to determine the dynamics of the drop.

Consider now only viscous and surface-tension forces and the effect of mass loss. One can then derive an evolution equation analogous to (4.14) of the form

$$\mu h_t + \frac{k_T \Delta \theta}{\rho L} (h + \tilde{K})^{-1} + \sigma \left[\left(\tfrac{1}{3} h^3 + \beta h^2 \right) h_{xxx} \right]_x = 0, \qquad (4.16)$$

again with edge conditions $h = 0$, $h_x = \mp \tan \theta$ at $x = \pm a(t)$.

Anderson & Davis (1995) further allowed the apparent contact angle to depend explicitly on the mass transport due to phase transformation and hypothesized the edge condition

$$\frac{da}{dt} = -\frac{1}{\hat{K} \theta(t)} + \bar{\eta} f(\theta), \qquad (4.17)$$

where $\bar{\eta}$ is constant and $f(\theta) = (\theta - \theta_A)^m$ for $\theta > \theta_A$, $f(\theta) = 0$ for $\theta_R < \theta < \theta_A$, and $f(\theta) = (\theta_R - \theta)^m$ for $\theta < \theta_R$, in which the contact line moves by the joint effects of spreading and mass loss. This results in an increase in the apparent contact angle as a function of the rate of heating. Thus, heating directly affects both the evolution equation and the edge condition.

One result of following the evolution in time of a spreading, evaporating drop is that, for nearly the full lifetime of the drop, there is a balance at the contact line between the mass loss and the spreading: the right-hand side of equation (4.17) is zero. The result is that $\theta \neq \theta_A$ given by thermodynamics, but θ is equal to a dynamically determined value much greater than θ_A.

Hocking (1995b) took the microscopic angle to be constant (unaffected by mass transport) and analysed a steady version of the above system with mass loss present in the evolution equation but absent in the edge condition. He found the same qualitative effects of evaporation as Anderson & Davis (1995), although the magnitudes of the steepening are smaller.

5 Singularities, corners and cusps

In the derivation of the continuum equations of viscous fluid mechanics one begins with integral balances of mass, momentum and energy and deduces point forms at interior points of the flow domain. In doing so, one assumes that all the fields, such as pressure, velocity and temperature, are sufficiently smooth. This smoothness precludes the presence of singularities in the fields, though such singularities may exist on the boundary of the domain. Singularities in the interior of a domain are sometimes allowed when the fluid is inviscid. There are advantages in considering point or line sources, vortices, etc. as idealizations of some more complex flows.

It is not surprising that there are singularities in water waves modelled by inviscid fluids and surface-tension-free interfaces: the Stokes wave of maximum height has a corner of angle $2\pi/3$. What is less intuitive is that free-boundary problems with surface tension and viscous fluid flow also can display singularities. These singularities are present in part because the standard model of viscous-fluid dynamics is used, and the interfaces are taken to be mathematical surfaces of zero thickness. Clearly, one must do some remodelling if one wishes to relax such singular behaviour.

In the present section two types of behaviour will be discussed. First, there are singularities due to the splitting of trajectories, as illustrated by moving contact lines and by rupture and coalescence mechanics. Secondly, there are cusps and corners on interfaces produced either by external drive or phase transformation.

5.1 Trajectory splitting

As discussed in §4.1 on contact-line kinematics, in the standard model of viscous, incompressible fluid subject to the no-slip condition, the local flow field can be regarded as a wedge flow with impenetrable boundaries whose tangential (radial) velocities at the contact line are multivalued. For the present purposes, the most important observation is that the contact line is the site of trajectory splitting. A point on the fluid I–fluid II interface has a dual identity. As it moves toward the contact line and then arrives it splits, and part of it 'rolls under' fluid I, and part of it is expelled along the dividing streamline. The splitting gives rise to a strong singularity in the velocity field; the velocity gradients behave like r^{-1}, where r is the distance to the contact line. Whenever trajectories split in this model, the velocity vector is multivalued and the gradients are infinite.

The trajectories of fluid points are governed by the system

$$\frac{\mathrm{d}x}{\mathrm{d}t} = v(x, t), \qquad x(0) = x_0, \tag{5.1}$$

where v is the Eulerian velocity, t is time and $x(t)$ is the position of a point. When v is Lipschitz continuous, the trajectory through each point is unique. When v is less smooth, uniqueness is not guaranteed. The simple example in one dimension,

$$\frac{\mathrm{d}x}{\mathrm{d}t} = x^{1/2}, \qquad x(0) = 0,$$

shows that the continuous but non-differentiable velocity $x^{1/2}$ has two trajectories through the origin: $x(t) \equiv 0$, and $x(t) = \frac{1}{4}t^2$.

Trajectory splitting is likewise present when one mass of liquid divides into two as shown in figure 3. This example of rupture involves a single point,

P, dividing into two. Likewise, the reverse process, coalescence, involves the merger of two points into one, again involving the same singular behaviour.

There has been a great deal of recent activity in examining rupture processes for fluids, in two dimensions or in axisymmetric geometries; see Eggers (1997) for a detailed review. Usually, one uses slender-body theory (two orders of the lubrication approximation) to generate quasi-one-dimensional equations governing a thread or sheet of liquid, and then seeks similarity solutions in a neighbourhood of a rupture point. This procedure characterizes the local singular behaviour of the splitting.

As an example, consider the capillary instability of a cylindrical mass of viscous liquid. If the cylinder has radius a, viscosity μ, density ρ, and the interface has surface tension σ, then linear stability theory (Rayleigh 1879) tells us that only axisymmetric perturbations grow and then when the wavenumber k' lies in the interval $0 < k'a < 1$. As the instability proceeds, the neck at $z = z^*$ decreases in diameter and approaches zero at $t = t^*$ where rupture might occur.

Observations show that the final stage of instability before breakup behaves in a manner independent of the length of the jet, or other macroscopic quantities, but is characterized by only material parameters. From these one can construct a length scale $\ell_s = \mu^2/\rho\sigma$ and a time scale $t_s = \mu^3/\rho\sigma^2$. One can then hope to describe the final stages of instability in terms of

$$t = (t^* - t')/t_s \tag{5.2}$$

and

$$z = (z' - z^*)/\ell_s. \tag{5.3}$$

For slender bodies, $|\partial/\partial z'| \ll |\partial/\partial r'|$, one can then show that the radius $h'(z', t')$ and the axial velocity $w'(z', t')$ satisfy the continuity and linear momentum balances as follows:

$$(h'^2)_{t'} + (h'^2 w')_{z'} = 0, \tag{5.4a}$$

$$\rho(w'_{t'} + w' w'_{z'}) = -p'_{z'} + \frac{3\mu}{h'^2}(h'^2 w'_{z'})_{z'}, \tag{5.4b}$$

where

$$p' = \sigma/h'. \tag{5.5}$$

Now, define similarity solutions of the form

$$h' = \ell_s H(\eta)t, \tag{5.6a}$$

$$w' = \frac{\sigma}{\mu} W(\eta)t^{-1/2}, \tag{5.6b}$$

where

$$\eta = \frac{z}{\ell_s \sqrt{t}}. \tag{5.6c}$$

If forms (5.6) are substituted into the system (5.4), then one obtains a set of ordinary differential equations for H and W,

$$\frac{H'}{H} = \frac{2 - V'}{2V + \eta}, \tag{5.7a}$$

$$H^2 [\eta V + V^2]' = 2H' + 6(H^3 V')'. \tag{5.7b}$$

System (5.7) is of third order and so three boundary conditions are necessary. Far from the pinch-off point, the solution should match to the relatively slowly varying capillary jet; it can be shown that this corresponds to

$$H(\eta) \sim \eta^2 \qquad \text{as} \qquad \eta \to \pm\infty. \tag{5.8a}$$

Finally, a regular solution far from the pinch-off point is required and from equation (5.7a) if there exists a point η_0 for which

$$2V(\eta_0) + \eta_0 = 0, \tag{5.8b}$$

then one must require that

$$V'(\eta_0) = 2. \tag{5.8c}$$

Eggers (1997) has shown that there exists only one such η_0.

This numerical solution of system (5.7)–(5.8) is a subtle one; it turns out that $\eta_0 = -1.5699...$ which corresponds to $H(0) = 0.0304...$. From equation (5.6a) $h'_{\min} = \ell_s H(0)t$. From equation (5.6b), one sees the singularity in the velocity field as $t \to 0$.

Analyses such as these are principally mathematical necessities that allow one to obtain accurate numerical solutions for systems with pinch-off.

5.2 Cusps and corners

When interfaces with surface tension are forced appropriately, the curvature of the interface can get very large at certain points. Joseph *et al.* (1991) observed that two co-rotating cylinders beneath a liquid–liquid interface can distort the interface into an apparent cusp as shown in figure 19(*a*). Jeong & Moffatt (1992) proposed a simplified model in which a vortex dipole of strength α is submerged a distance d below the interface; the fluid satisfies

(a)

(b)

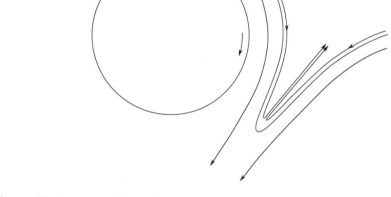

Figure 19. A cusped interface due to roller rotation. (a) 12 500 cS silicone oil according to Joseph *et al.* (1991), Copyright © 1991 Cambridge University Press. Reprinted with the permission of Cambridge University Press from Joseph *et al.* (1991). (b) The sketches of the likely streamlines in the air according to Joseph (1992), Copyright © 1992 Elsevier Science. Reprinted with the permission of Elsevier Science from Joseph (1992).

the Stokes equations and the dynamics of the second fluid is neglected. In two dimensions they find that the radius of curvature $\mathscr{R}$ at the 'cusp' satisfies

$$\mathscr{R}/d \sim \tfrac{256}{3}\exp(-32\pi\mu\alpha/d^2\sigma), \tag{5.9}$$

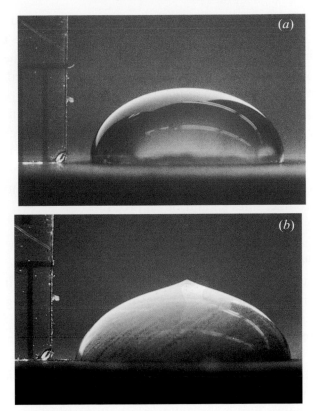

Figure 20. (*a*) A frozen water droplet on a plate. (*b*) The drop undercooled from below and frozen according to Anderson *et al.* (1996). Copyright © 1995 Elsevier Science. Reprinted with the permission of Elsevier Science from Anderson *et al.* (1996).

an exponentially small effect of surface tension as $\sigma \to 0$. The 'cusp' is not an actual one for non-zero σ. However, when an estimate of $\mathscr{R}$ is made for typical situations with typical fluids, $\mathscr{R}$ is in the Angstrom range well beyond the continuum limit. Thus, whenever continuum theory applies, the interface is cusped.

Figure 19(*b*), reproduced from Joseph (1992), sketches the presumed streamlines in the two-fluid problem. Kinematically, this is identical to that of the moving contact line with the contact angle π. The reflux in the second fluid suggests the presence of the trajectory-splitting singularity. Hence, one suspects that the inclusion of the dynamics of the second fluid would influence the result of analysis.

Another example of interfacial cusps arises in phase transformations. Anderson, Worster & Davis (1996) analysed a droplet of water on a solid plate; the fluid is static. When the substrate is undercooled and the drop is frozen, the ice has a cusp at its peak, figure 20. Despite the fluid–fluid interface having surface tension, the solid drop, which is the locus of the contact line (water, ice, gas) as the drop is frozen, is cusped. Anderson *et al.* (1996) showed that it is the contact-line dynamics in the partially frozen drop that gives rise to the cusp. When the contact angle θ, measured in the liquid, is zero, as it should be for pure water, then cusps never occur. When θ is allowed to depend on the speed of the contact line, which means on the rate of freezing, then the cusp appears and the bi-concave shape of the frozen drop is reproduced. This result is evidence that such non-equilibrium conditions at contact lines should be enforced.

5.3 Conclusions

Trajectory-splitting singularities and cusps or corners can occur on interfaces with surface tension in liquids with viscosity. An understanding of when such structures are present is not merely of intrinsic interest, but is essential to the accuracy of numerical codes that give convergent, reliable, approximate solutions.

When singular behaviour does arise, the model of the physical problem is often very useful for understanding the 'outer' flow outside a neighbourhood of the singular point. The 'inner' solution usually requires one to remodel the physics in order to make the point less singular or even regular. Recall the case of the moving contact line. When the no-slip condition is enforced on all boundaries, the contact line is the site of a non-integrable singularity. If one is interested in macroscopic dynamics only, then one can insert effective slip and solve boundary-value problems, whose predictions compare well with experimental data. The contact line is still a singular point but the singularity is now integrable and so integral quantities such as force, torque, heat transfer, etc. are well defined.

If one wanted to explore further, then one should realize that, as discussed in §2, interfaces may have non-zero thicknesses Δ_s on a continuum level and the contact line would then be a contact region. If one wished to retain terms of $O(\Delta_s)$, then one would be making fuzzy the point force of surface tension over a finite interval and hence one could obtain finite values of all field quantities. Shikhmurzaev (1997) has posed such a model in which surface tension is allowed to vary spatially in an $O(\Delta_s)$ neighbourhood of the contact line. He finds that there is a slip law enforced locally, though of a more complicated type than the one discussed here.

When one encounters cusps or corners, again there is the violation of the assumption that the interfacial curvature is much smaller than Δ_s. One can again alter the model as above and obtain some insight into the 'rounding' of the interface (e.g. Shikhamuzaev 1997).

6 Discussion

Interfacial dynamics is an intrinsically nonlinear part of fluid dynamics. Large distortions of interfaces lead to strongly nonlinear systems that generally are examined by approximate methods.

In § 3 thin viscous films were subjected to surface tension, thermocapillary forces, van der Waals attractions and phase transformation. Dynamical and instability phenomenon were described in the long-scale limit which precludes the creation of droplets that pinch-off from the layer.

In § 4 long-scale methods were again used, this time to examine the spreading of viscous drops on substrates. This involves the kinematics and dynamics of moving contact lines.

In § 5 singular behaviour associated with phenomena such as pinching-off and coalescence were discussed, specifically in the context of viscous films with surface tension.

Clearly, this discussion gives the reader only a glimpse of the areas comprising interfacial dynamics. Missing from this discussion are instability phenomena that have scales of unit order such as Marangoni convection, and 'wind-generated' instabilities of liquid films, as well as the coupling of such phenomena (as in the stabilization of a capillary jet by wind stress). Also missing are detailed discussions of phenomena involving singularities: fragmentation, coalescence, impact, etc. Many such applications arise in the process analysis of micromechanical devices, microgravity environments, and complex-fluid systems such as colloids and foams. In order to analyse such more involved systems, one must utilize more sophisticated analytical and numerical methods, as well as remodelling of the governing systems. The discussion in § 2 on the underlying nature of interfacial models should help in this regard.

Acknowledgements

The author is pleased to thank his secretary, Ms Judy Piehl, without whose help this chapter would never have been completed. This work was partially supported by a grant from the United States Department of Energy Engineering Research Program of the Office of Basic Energy Sciences.

References

ANDERSON, D. M. & DAVIS, S. H. 1995 The spreading of volatile liquid droplets on heated surfaces. *Phys. Fluids* A **7**, 248–265.

ANDERSON, D. M., WORSTER, M. G. & DAVIS, S. H. 1996 The case for a dynamic contact angle in containerless solidification. *J. Cryst. Growth* **163**, 329–338.

ARIS, R. 1962 *Vectors, Tensors and the Basic Equations of Fluid Mechanics*. Prentice-Hall.

BENJAMIN, T. B. 1957 Wave formation in laminar flow down an inclined plane. *J. Fluid Mech.* **2**, 554–574.

BENNEY, D. J. 1966 Long waves on liquid films. *J. Math. Phys.* **45**, 150–155.

BURELBACH, J. P., BANKOFF, S. G. & DAVIS, S. H. 1988 Nonlinear stability of evaporating/condensing films. *J. Fluid Mech.* **195**, 463–494.

CAZABAT, A. M. & COHEN-STUART, M. A. 1986 Dynamics of wetting: effect of surface roughness. *J. Chem. Phys.* **90**, 5845–5849.

DRAZIN, P. G. & REID, W. H. 1981 *Hydrodynamic Stability*. Cambridge University Press.

DUSSAN V., E. B. 1979 On the spreading of liquids on solid surfaces: static and dynamic contact lines. *Ann. Rev. Fluid Mech.* **11**, 371–400.

DUSSAN V., E. B., RAMÉ, E. & GAROFF, S. 1991 On identifying the appropriate boundary condition at a moving contact line: an experimental investigation. *J. Fluid Mech.* **230**, 97–116.

DUSSAN V., E. B. & DAVIS, S. H. 1974 On the motion of a fluid–fluid interface along a solid surface. *J. Fluid Mech.* **65**, 97–95.

EGGERS, J. 1997 Theory of drop formation. *Phys. Fluids* A **7**, 941–953.

EHRHARD, P. & DAVIS, S. H. 1991 Non-isothermal spreading of liquid drops on horizontal plates. *J. Fluid Mech.* **229**, 365–388.

EHRHARD, P. 1993 Experiments on isothermal and non-isothermal spreading. *J. Fluid Mech.* **257**, 463–483.

ERNEUX, T. & DAVIS, S. H. 1993 Nonlinear rupture of free films. *Phys. Fluids* A **5**, 1117–1122.

GENNES, P. G. DE 1985 Wetting: statics and dynamics. *Rev. Mod. Phys.* **57**, 827–863.

GIBBS, J. W. 1948 On the equilibrium of heterogeneous surfaces. In *The Collected Works of J. Willard Gibbs*, Vol. 1, Ch. 3, pp. 55–354. Yale University Press.

GREENSPAN, H. P. 1978 On the motion of a small viscous droplet that wets a surface. *J. Fluid Mech.* **84**, 125–143.

GROTBERG, J. B. 1994 Pulmonary flow and transport phenomena. *Ann Rev. Fluid Mech.* **26**, 529–571.

HOCKING, L. M. 1983 The spreading of a thin drop by gravity and capillarity. *J. Mech. Appl. Maths* **36**, 55–69.

HOCKING, L. M. 1995a The wetting of a plane surface by a fluid. *Phys. Fluids* **7**, 1214–1220.

HOCKING, L. M. 1995b On contact angles in evaporating liquids. *Phys. Fluids* **7**, 2950–2955.

HOOK, S. J. VAN, SCHATZ, M. F., McCORMICK, W. A., SWIFT, J. B. & SWINNEY, H. L. 1995 Long-wave length instability in surface-tension driven Bénard convection. *Phys. Rev. Lett.* **75**, 4397–4400.

JEONG, J. T. & MOFFATT, H. K. 1992 Free-surface cusps associated with flow at low Reynolds number. *J. Fluid Mech.* **241**, 1–22.

JOSEPH, D. D. 1992 Understanding cusped interfaces. *J. Non-Newtonian Fluid Mech.* **44**, 127–148.

JOSEPH, D. D., NELSON, J., RENARDY, M. & RENARDY, Y. 1991 Two-dimensional cusped interfaces. *J. Fluid Mech.* **223**, 383–409.

LEGER, L. & JOANNY, J. F. 1992 Liquid spreading. *Rep. Prog. Phys.* **55**, 431–486.

LIU, J., SCHNEIDER, J. B. & GOLLUB, J. P. 1995 Three-dimensional instabilities of film flows. *Phys. Fluids* **7**, 55–67.

NGAN C. G. & DUSSAN V., E. B. 1989 On the dynamics of liquid spreading on solid surfaces. *J. Fluid Mech.* **209**, 191–226.

ORON, A., DAVIS, S. H. & BANKOFF, S. G. 1997 Long-scale evolution of thin liquid films. *Rev. Mod. Phys.* **69**, 931–980.

PALMER, H. J. 1976 The hydrodynamic stability of rapidly evaporating liquids at reduced pressure. *J. Fluid Mech.* **75**, 487–511.

PLESSET, M. S. & PROSPERETTI, A. 1976 Flow of vapour in a liquid enclosure. *J. Fluid Mech.* **78**, 283–290.

QUINN, J. A. & SCRIVEN, L. E. 1970 *Interfacial Phenomena*. 14th Advanced Seminar, American Institute of Chemical Engineers, New York.

RAYLEIGH, LORD 1879 On the instability of jets. *Proc. Lond. Math. Soc.* **x**, 4–13. (Also *Scientific Papers* vol. 1, pp. 361–371.)

RUCKENSTEIN, E. & JAIN, R. K. 1974 Spontaneous rupture of thin liquid films. *Chem. Soc. Faraday Trans.* **270**, 132–137.

SADHAL, S. S. & PLESSET, M. S. 1979 Effect of solid properties and contact angle in dropwise condensation and evaporation. *J. Heat Transfer* **101**, 48–54.

SCHLICHTING, H. 1968 *Boundary Layer Theory*. McGraw Hill.

SCHRAGE, R. N. 1953 *A Theoretical Study of Interphase Mass Transfer*. Columbia University Press.

SCRIVEN, L. E. 1960 The dynamics of a fluid interface. *Chem. Engng Sci.* **12**, 98–108.

SHIKHMURZAEV, Y. D. 1997 Moving contact lines in liquid/liquid/solid systems. *J. Fluid Mech.* **334**, 211–250.

SHELUDKO, A. 1967 Thin liquid films. *Adv. Colloid Interface Sci.* **1**, 391–463.

SLATTERY, J. C. 1990 *Interfacial Transport Phenomena*. Springer.

SPAID, M. A. & HOMSY, G. M. 1996 Stability of Newtonian and viscoelastic dynamic contact lines. *Phys. Fluids* **8**, 460–478.

TROIAN, S. M., HERBOLZHEIMER, E., SAFRAN, S. A. & JOANNY, J. F. 1989 Fingering instabilities of driven spreading films. *Europhys. Lett.* **10**, 25–30.

WILLIAMS, M. B. & DAVIS, S. H. 1981 Nonlinear theory of film rupture. *J. Colloid Interface Sci.* **90**, 220–228.

YANTSIOS, S. G. & HIGGINS, B. G. 1991 Rupture of thin films: nonlinear stability analysis. *J. Colloid Interface Sci.* **147**, 341–350.

YIH, C. S. 1955 Stability of parallel laminar flow with a free surface. *Proc. 2nd US Congress Appl. Mech.*, pp. 623–628, ASME.

YIH, C. S. 1963 Stability of parallel flow down an inclined plane. *Phys. Fluids* **6**, 321–333.

Department of Engineering Sciences and Applied Mathematics,
Northwestern University, Evanston, IL 60208, USA

2

Viscous Fingering as an Archetype for Growth Patterns

YVES COUDER

1 Introduction

Perhaps because man has had to survive in the wild for so long, our eye (or rather our brain) is extremely well trained in detecting and recognizing patterns. We also have a spontaneous tendency to relate them to each other. When describing sand dunes we compare them to sea waves. When speaking of ice crystals we use a vocabulary, e.g. dendrites, branches etc., borrowed from botany. The book *On Growth and Form* by D'Arcy Thompson (1917) was the first written attempt to list such analogies and seek their possible underlying physical bases. For this reason it has become an essential reference in the field of morphogenesis – the appearance and evolution of shape.

In a wide variety of situations the formation of patterns results from a growth process. Growth gives form to living entities but also generates patterns by a variety of physical processes. Such processes include viscous fingering, dendritic growth of crystals, electrodeposition, growth of bacterial colonies, propagation of flame fronts, fractures in brittle solids, dielectric breakdown, diffusion-limited aggregation, corrosion of solids, etc. All these problems belong to a class often called free-boundary problems.

These phenomena have been investigated for fundamental reasons but also because of the needs of industrial research. From this point of view the morphogenetic instabilities are often considered a nuisance and studied to find ways to avoid their occurrence. For instance the petroleum industry has been consistently trying to find ways of inhibiting viscous fingering because it limits oil recovery in porous media (see §2.4). Viscous instabilities are also a limitation in coating processes where a fluid has to be spread evenly onto a solid surface (see §6). Similarly, engineers seeking to obtain a good metal plating try to avoid any fractal or dendritic growth in electrodeposition. These instabilities can also be beneficial: the intricacies of the dendrites in

53

the solidification of steel are essential for its mechanical properties. Fractal growth, because it builds up structures with a large interface, can be used to improve catalytic effects and is vital for the formation of our lungs for instance.

During the last twenty years the investigation of a hydrodynamic instability discovered by Saffman & Taylor in 1958 has served as a reference in the field of pattern formation. It has the advantage of being experimentally simple. The theory, complex as it may seem, still provides the best understanding of a pattern-forming system. Saffman–Taylor fingering is thus studied for reasons which go beyond its pure hydrodynamic interest, and will be the guiding theme of this chapter.

Another pattern-forming phenomenon is solidification. Although the two fields of research – viscous fingering and solidification – developed independently (at least initially), there is a remarkable parallel between the successive steps of research in the two areas. The linear stability analysis of a plane front was published for viscous fingering by Chuoke, Van Meurs & Van der Pol in 1959 and for solidification by Mullins & Sekerka in 1964. The existence (in the absence of surface tension) of families of solutions having the shape of curved fronts was demonstrated by Saffman & Taylor (1958) for viscous fingers. It had been preceded by a similar demonstration, by Ivantsov (1947), for parabolic needle crystals. The selection of the observed solutions was understood much later, at a time when everyone had become aware of the similarities between the two problems. Both were solved progressively in the 1980s by the converging efforts of many researchers.

Early reviews on the development of this area are in Saffman (1986), Bensimon *et al.* (1986) and Homsy (1987). More recent ones, presented from a variety of viewpoints, are in Pelcé (1988), Kessler, Koplik & Levine (1988), Couder (1991), Tanveer (1991), Pomeau & Ben Amar (1992) and McCloud & Maher (1995).

1.1 Growth in a Laplacian field

The best way to find out what viscous fingers look like is to do an experiment. Take two circular glass plates, one of them having a central injection hole. Set the plates horizontally, clamped together but separated by thin spacers (of typical thickness 0.3 mm). First fill the cell with a viscous fluid, e.g. silicon oil, then blow air into the central hole so as to push the oil out radially. An example of what you will see is given in figure 1. The air does not penetrate the oil regularly to form a circular region but forms a pattern which becomes increasingly complex. This experiment is the axisymmetric variant

Figure 1. A superposition of successive states of a radial Saffman–Taylor pattern.

of viscous fingering introduced by Bataille (1968) and by Paterson (1981). In this instability, the two fluids and their interface move due to the applied pressure. As shown below (see §2.1) the fluids have a Poiseuille flow driven by the applied pressure. The experimental situation can thus be described by a two-dimensional model where the spatial distribution of pressure forms a Laplacian field and the fluids move with a velocity proportional to the gradients of pressure. Thus schematized, this experiment is reduced to a general problem sometimes called the inverse Stefan problem. It turns out that other morphogenetic systems, such as for instance diffusion-limited aggregation (see §3.1), can be reduced to the same mathematical problem. The mathematical ingredients for this type of growth are the following.

An interface (figure 2) separates two regions of a plane respectively labelled (1) and (2). There exists a scalar field $P(x, y, t)$ in this plane. In the case of Saffman–Taylor fingering this field P is the pressure p. In the simplest cases P is constant in region 1 and in region 2 satisfies the Laplacian:

$$\nabla^2 P = 0. \tag{1.1}$$

At a given time, the determination of the field P depends on the boundary conditions defined at infinity and at the interface between regions 1 and 2. We are interested in situations in which this interface moves in the local gradient of P with a normal velocity

$$V_n \propto \boldsymbol{n} \cdot \nabla P, \tag{1.2}$$

where $\boldsymbol{n}$ is the unit vector normal to the interface. The displacement of the interface, because it changes the location of a boundary, modifies the field P and thus, in turn, the interface velocity. This process leads to an instability which generates a pattern. The whole process is strongly non-local, due to the long-range interactions introduced by Laplace's law.

Two remarks can be made

(i) These dynamical equations are also obtained for other processes. As we will see in §3.1, the growth in diffusion-limited aggregation (DLA) is also described by equations (1.1) and (1.2) where P is a different type of quantity (a probability density). In both Saffman–Taylor fingering and DLA the dynamics defined by these equations results in the spontaneous formation of a ramified pattern where the growth occurs essentially at the protruding extremities.

(ii) It can be noted that equations (1.1) and (1.2) are mathematically the fundamental equations of electrostatics. In this analogy, *at a given time* the velocity corresponds to the electric field and P to the electric potential. Though in the present case these equations are dynamical we shall use this analogy to obtain an intuitive understanding of the origin of the instability.

Returning to the Saffman–Taylor problem we consider (figure 2) a situation where the two fluids are initially separated by a flat interface and we assume this interface to be disturbed by a small protrusion of typical size L. The curves of constant values of P will be parallel lines, only distorted in front of the bump. It is a property of Laplacian fields that this distortion will only affect these curves to a "depth" of the order of L. As a result the gradient of P is locally larger in front of the protrusion (see figure 2) and so is the velocity. The amplitude of the protrusion will thus grow: the interface is unstable. This amplification of the velocity is equivalent to the amplification of the electric field by the point effect in electrostatics. Note that if the fluids were moving in the opposite direction, i.e. if the more-viscous fluid was forcing the less-viscous one to recede, the change in velocity would serve to reduce the protrusion and the front would be stable.

Ideally, in the absence of any other factor, the more pointed the protrusion

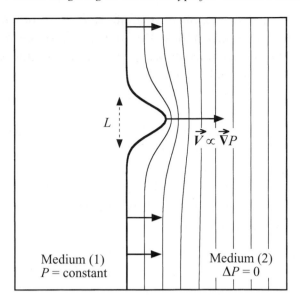

Figure 2. Sketch of the lines of constant values of the scalar field P in front of a local disturbance of a plane front (in the absence of surface tension).

the larger the gradient and the velocity. For this reason several works have suggested that in this case the instability will lead to the formation of cusp-like singularities in finite time (Shraiman & Bensimon 1984; Howison 1986; Bensimon & Pelcé 1986). However, in real experiments, additional effects due to the interfacial capillarity effect are present which stabilize the interface at small scales.

In order to account quantitatively for the various effects, it is necessary to perform a linear stability analysis (as done for Saffman–Taylor fingering in §2). The validity of the linear stability analysis is, in principle, limited to small-amplitude perturbations. In reality it provides a length scale which retains its validity for large-amplitude fingers. As the protrusions continue growing they interact non-locally. Two situations can eventually be reached. steady curved fronts can be obtained when the system is confined (§4) or fractal structures build up when it is not (§5). For both types of growth three factors are determinant: the length scales of the problem; the geometry of the cell; the isotropy or the non-isotropy of the system.

1.2 Growth in a diffusive field

In a related family of problems the field P is diffusive and equation (1.1) is replaced by

$$\frac{\partial P}{\partial t} = D\nabla^2 P, \tag{1.3}$$

while equation (1.2) remains the same. In solidification for instance, the growth is linked to either the diffusion of heat or that of impurities (see chapter 8). Similarly, the growth in bacterial colonies is limited by the diffusion of a nutrient in a gel. The front is unstable and the previous intuitive argument (figure 2) can be used again here. However, at large amplitudes, the interaction of different structures is different from the Laplacian case. With a diffusive field the thickness of the disturbed region in the vicinity of an interface moving at velocity V is bounded by the diffusion length $l_D = D/V$. Regions of the interface separated from each other by more than l_D will behave independently. For this reason l_D will be a lower limit to the scales on which the pattern can have a fractal structure. Only at very small velocities will l_D be very large and the behaviour of the system close to what it would be with a Laplacian field.

1.3 Directional growth

Finally, in a third type of situation, the experimental set-up is designed so as to impose a fixed gradient on the scalar P. The mean direction of the growth is then fixed by the orientation of this gradient. The archetype of such growth is the directional growth of crystals where the solidification front is confined by an imposed mean temperature gradient. There is also a directional type of viscous fingering in which a pressure gradient is imposed. In general, the system, being stabilised at small scales by the capillary effects and at large scales by the gradient, can have a stable plane front at low velocities. Beyond a finite threshold it destabilizes into a cellular pattern having a periodicity predicted by the linear stability analysis. The further evolution of the pattern is completely different from that of free growth. The gradient generates stabilizing nonlinear terms so that the amplitude of the cells saturates. These systems are one-dimensional and are investigated for two reasons. From a practical point of view they are of interest to the coating industry, because this instability imposes a limit on the velocity at which it is possible to spread a fluid onto a solid. From a fundamental point of view they serve as archetypes of the dynamics of one-dimensional spatially extended dynamical systems. They exhibit spatio–temporal chaos, propagating waves, solitary waves and other nonlinear processes. A few of their characteristics will be presented in §6.

2 The basis of Saffman–Taylor viscous fingering instability

2.1 Potential flow in a Hele-Shaw cell

We first consider a single fluid flowing in a Hele-Shaw cell formed by two parallel plates separated by a narrow gap of thickness b. The frame of

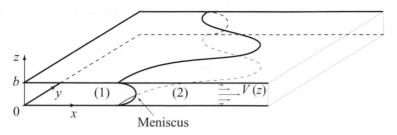

Figure 3. Sketch of the Hele-Shaw cell.

reference is given in figure 3. The motion is driven by pressure and limited by viscous friction so that the Navier–Stokes equation reduces to

$$\nabla p = \mu \nabla^2 \boldsymbol{u}. \tag{2.1}$$

Since the thickness is small, only the velocity components parallel to the plates are considered. Their dominant variation is in the z-direction so that (2.1) can be written

$$\nabla p = \mu \frac{\partial^2 \boldsymbol{u}}{\partial z^2}, \tag{2.2}$$

which can be integrated to give the parabolic velocity field of plane Poiseuille flow:

$$\boldsymbol{u} = \frac{1}{2\mu} z(z - b)\nabla p. \tag{2.3}$$

Integrating the flow through the cell thickness, we define an average velocity $\langle \boldsymbol{u} \rangle$ by

$$Q = \int_0^b \boldsymbol{u}\,\mathrm{d}z = \langle \boldsymbol{u} \rangle b \tag{2.4}$$

and find that $\langle \boldsymbol{u} \rangle$ is governed by the potential law

$$\langle \boldsymbol{u} \rangle = -\frac{b^2}{12\mu}\nabla p. \tag{2.5}$$

This relation between mean velocity and pressure is a particular case of Darcy's law, which holds more generally for fluids moving through porous media such as e.g. rocks or piles of glass beads. It is usually written

$$\langle \boldsymbol{u} \rangle = -K\nabla p, \tag{2.6}$$

where K is the permeability of the medium.

Finally, in both a Hele-Shaw cell and in a porous medium, since the fluid is incompressible, $\nabla \cdot \langle u \rangle = 0$ and the pressure is a Laplacian field,

$$\nabla^2 p = 0. \tag{2.7}$$

In the following we will adopt the two-dimensional approximation where the velocities of the fluid are the average ones satisfying relation (2.6) and we will omit the average symbol.

2.2 Linear stability analysis of the front between two fluids of different viscosity

The Saffman–Taylor instability occurs in a Hele-Shaw cell at the interface between two fluids of different viscosities when the fluid of low viscosity displaces the fluid of higher viscosity. In order to neglect the possible effect of the difference of density, we will suppose the cell to be horizontal. The flow of each fluid satisfies relation (2.5) so that

$$u_i = K_i \nabla p_i \tag{2.8}$$

where $i = 1$ or 2 respectively for the two fluids and where

$$K_i = -\frac{b^2}{12\mu_i}. \tag{2.9}$$

The pressure distribution on the two sides of the interface satisfies

$$\nabla^2 p_i = 0. \tag{2.10}$$

The linear stability analysis is used to predict whether a flow is stable. In the case of a spatially extended system this analysis has to be performed with regard to perturbations of all spatial periodicities. For this purpose we apply to a straight interface a periodic disturbance of arbitrary wave-vector k. We then compute, in a linear approximation, whether this disturbance will grow. For viscous fingering this linear analysis was first performed by Chuoke *et al.* (1959).

2.2.1 The basic flow

The Hele-Shaw cell is represented by an infinite plane in which the two different fluids occupy two half-planes separated by a linear interface (dashed line on figure 4). In the basic state the two fluids move with the same average velocity V along the x-axis:

$$u_i^0 = V. \tag{2.11}$$

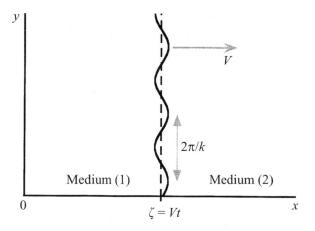

Figure 4. Definition of the axes for the stability analysis. For Chuoke's analysis of viscous fingering the front separates fluid (1) on the left from fluid (2). For solidification (§ 3.2) (1) is the solid, (2) is the liquid.

The pressure distribution is found by integrating (2.8) and choosing the pressure to be zero on the interface located at $x = \zeta = Vt$,

$$p_i^0 = \frac{V}{K_i}(x - Vt). \qquad (2.12)$$

2.2.2 The disturbance of the interface

A sinusoidal disturbance of wave-vector $\boldsymbol{k}$ is then imposed on the interface (figure 4). The aim of our calculation will be to find whether the amplitude of the disturbance will grow in time (unstable situation) or decrease in time (stable situation). The amplitude of the disturbance is thus chosen as $\epsilon\exp(\sigma t)$ where ϵ is small. Should σ be negative for all k the front will be stable. If it is positive for a range of values of k the flow will be unstable for perturbations having these wave-vectors. The interface at time t is now given by

$$\zeta = Vt + \epsilon e^{\sigma t} \sin ky. \qquad (2.13)$$

The velocity field and the pressure field are disturbed and can be written as

$$p_i^T = p_i^0 + p_i, \qquad u_i^T = u_i^0 + u_i. \qquad (2.14)$$

Darcy's law (equation (2.8)) and the incompressibility constraint (2.10) are valid for both the undisturbed fields (u_i^0 and p_i^0) and the total disturbed field (u_i^T and p_i^T). They thus also hold for the disturbance (u_i and p_i). Using them we can compute the correction to the pressure distribution. We seek solutions of the form

$$p_i = p_i'(x,t)\sin ky \tag{2.15}$$

having the same spatial periodicity. The amplitude p_i' must grow or decrease in time with the interface disturbance itself. Its dependence on x must be such that on moving away from the interface the disturbance of the pressure field decreases to zero. As the pressure is a Laplacian field (2.10) it must decrease exponentially with a length scale fixed by the wavelength of the disturbance. So we must have

$$p_1 = A_1 e^{\sigma t + kx}\sin ky, \qquad p_2 = A_2 e^{\sigma t - kx}\sin ky. \tag{2.16}$$

2.2.3 Continuity of the normal velocities at the interfaces

The continuity of the normal components of the velocity at the interface,

$$u_{n_1} = u_{n_2}, \tag{2.17}$$

can be approximated, in the limit of small deformations, by the continuity of the velocities along Ox

$$u_{x_1} = u_{x_2} = \frac{\partial \zeta}{\partial t} - V, \tag{2.18}$$

so that

$$K_i \left(\frac{\partial p_i}{\partial x} \right)_{x=\zeta} = \epsilon \sigma e^{\sigma t}\sin ky. \tag{2.19}$$

Using (2.16) and keeping only the first-order terms by taking $\zeta = Vt$ we thus determine A_1 and A_2. The pressure on each side is now given by

$$p_1 = \frac{\epsilon \sigma}{K_1 k} e^{\sigma t + k(x-Vt)}\sin ky, \qquad p_2 = -\frac{\epsilon \sigma}{K_2 k} e^{\sigma t - k(x-Vt)}\sin ky. \tag{2.20}$$

2.2.4 The pressure jump at the interface due to surface tension

The problem is completed by the interfacial effects due to capillarity. They create a pressure jump Δp given by Laplace's law between the two sides of a curved interface,

$$\Delta p = \gamma(\kappa_1 + \kappa_2), \tag{2.21}$$

where γ is the surface tension and κ_1 and κ_2 are the two main curvatures of the interface. In the most common Saffman–Taylor experimental situation, a gas forces oil into motion; the oil wets the solid walls so that the two fluids are separated by a meniscus as sketched in figure 3. The strongest curvature of the interface is perpendicular to the plane of the cell. But the analysis of the Saffman–Taylor instability is two-dimensional. The curvature

κ_2 is thus considered to be of the order of $\kappa_2 \approx 2/b$ and constant along the interface, so that its dynamical effect can be neglected. Only κ_1 is taken into account. This is a bold assumption as it neglects the fact that when the viscous fluid is pushed out of the cell it leaves on the solid surfaces two wetting films having a thickness which is a function of the normal velocity of the interface (Landau & Levich 1942; Bretherton 1961). This assumption is not always valid as has been shown experimentally (Tabeling, Zocchi & Libchaber 1987) and theoretically (Reinelt 1987). Tanveer (1990) has treated the complete three-dimensional system and shown within which limits the two-dimensional assumption is valid.

Returning to the two-dimensional stability analysis and neglecting the higher order terms in the expression for the curvature, we can write,

$$p_2^T(\zeta) - p_1^T(\zeta) = \gamma \frac{\partial^2 \zeta}{\partial y^2}, \tag{2.22}$$

where

$$p_2^T(\zeta) - p_1^T(\zeta) = p_2^0(\zeta) - p_1^0(\zeta) + p_2(\zeta) - p_1(\zeta). \tag{2.23}$$

We use equations (2.12) and (2.20) for the different pressures. In order to retain only the first-order terms we take ζ equal to Vt in the last two terms $(p_2(\zeta) - p_1(\zeta))$. After a simple calculation we obtain an expression for the amplification rate:

$$\sigma = \left(\frac{K_1 - K_2}{K_1 + K_2}\right) V k + \gamma \left(\frac{K_1 K_2}{K_1 + K_2}\right) k^3 \tag{2.24}$$

or, returning to the physical variables,

$$\sigma = \left(\frac{\mu_2 - \mu_1}{\mu_1 + \mu_2}\right) V k - \gamma \frac{b^2}{12} \left(\frac{1}{\mu_1 + \mu_2}\right) k^3. \tag{2.25}$$

In the case of a gas displacing a high-viscosity fluid such as oil, $\mu_2 \gg \mu_1$, so that taking $\mu_1 = 0$ and $\mu_2 = \mu$, equation (2.25) simplifies to:

$$\sigma = V k - \gamma \frac{b^2}{12\mu} k^3. \tag{2.26}$$

Figure 5 shows a graph of $\sigma(k)$ for two values of V in the case where $\mu_2 > \mu_1$. For a given velocity V the disturbance of maximum growth rate σ has a wave-vector

$$k_c = \frac{2}{b} \sqrt{\frac{\mu V}{\gamma}}, \tag{2.27}$$

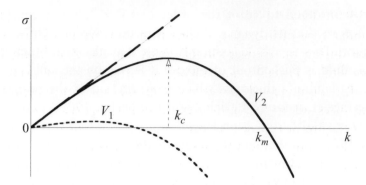

Figure 5. The wave-vector dependence of the amplification rate predicted by the linear analysis of viscous fingering due to Chuoke *et al.* (1959) (equation (2.26)). A similar result is found in the analysis by Mullins & Sekerka (1964) of a solidification front (equation (3.7)).

corresponding to the most unstable wavelength

$$l_c = \pi b \sqrt{\frac{\gamma}{\mu V}}, \tag{2.28}$$

usually called the capillary length of the instability. The interface is unstable to all perturbations of wave-vectors smaller than $k_{\max} = \sqrt{3}k_c$. For all positive velocities there is a range of k for which $\sigma > 0$: when the fluid of smaller viscosity forces the fluid of larger viscosity to recede, the interface is always unstable. In the reverse situation (i.e. when $V < 0$), σ is negative for all k and the interface is stable.

2.3 Miscible fluids

As demonstrated by the linear stability analysis the existence of the limiting small scale l_c results from the capillary forces at the interface between two immiscible fluids. What happens in the absence of surface tension? To answer this question experiments have been performed using two miscible fluids (e.g. water and glycerol). In this case there is still formation of fingers but they have a fixed scale of the order of $3b$, where (b is the cell's thickness). This means that the system spontaneously locks onto a new small scale but the flow has become fully three-dimensional. None of the hypotheses on which the above stability calculations are based can be used in this case (Paterson 1985).

2.4 Fingering in porous media

A very similar type of instability occurs in porous media, which is of great importance during oil recovery in oil fields (see Po Zen Wong 1999 and references therein). Usually, immediately after the drilling of a new well, oil flows out of it spontaneously. However when this spontaneous process stops most of the oil is still locked in the deposit as it is absorbed in porous rocks. If several wells are drilled near to each other it is possible then to force the oil to flow out of one of them by injecting high pressure water in the others. Unfortunately the flow of the two fluids in the porous medium satisfies equations (1.1) and (1.2) and water is less viscous than oil. As a result the front is unstable and fingers of water will rapidly reach the extraction well and short circuit the motion of the oil. Various tricks with additives in the water are used to try to avoid this situation. Even though Saffman–Taylor fingering can be used here as a model, the natural situation is more complex because of the variable permeability of the rocks and of the role of wetting.

3 The basis of the instability of other physical systems

3.1 Diffusion-limited aggregation

The aggregation of particles limited by diffusion is a physical process (observed in soot for example) which inspired a frequently investigated numerical model system introduced by Witten & Sander (1981, 1983). In a plane system, particles are created, one at a time, far away from a central seed. They perform a random walk on the underlying square lattice. If a particle visits any of the sites neighbouring the central seed, it remains there. The process is repeated a large number of times, each particle sticking whenever it visits a site neighbouring a site already occupied. The growth is therefore radial and results in a fractal aggregate, which has been frequently investigated. In a variant of this model particles come from far away to stick to a linear baseline. Figure 6 shows a fractal DLA resulting from such a simulation. The similarity of DLA with viscous fingering was pointed out by Paterson (1984). The scalar field is here the probability P_v of visit by a randomly walking particle to a given lattice site. With a steady flux of particles from a source far away it obeys a Laplacian law

$$\nabla^2 P_v = 0. \tag{3.1}$$

The average speed at which a region of the aggregate grows is proportional to the local gradient of P_v

$$V_n = \nabla P_v. \tag{3.2}$$

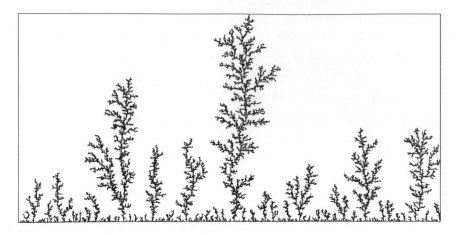

Figure 6. A DLA grown on linear base (courtesy A. Arnéodo). At this stage of the growth only the largest structures are still growing, the others are screened off.

In the standard model there is no surface tension and the growth is unstable at all scales. However, the numerical technique itself, because it always uses a lattice or particles of finite size, introduces a length scale l_u: the lattice mesh or the particle size. The role of this discretization is evident at the first stages of the growth shown on figure 6: the first layer deposited on the baseline is lacunary at a scale of the order of l_u. Several works (Viczek 1984; Sarkar 1985; Sander, Ramanlal & Ben Jacob 1985; Kadanoff 1985; Liang 1986) have been devoted to variants of DLA in which the rules are changed so as to introduce an effective surface tension. The small length scale then becomes larger than l_u and the patterns are more similar to Saffman–Taylor fingering.

3.2 Solidification instability

The solidification problem is treated in Chapter 8. We present here only its main results in order to underline the similarities and differences of this instability with viscous fingering. Crystals can be grown either from a melt or from a solution (Langer 1980 and Caroli, Caroli & Roulet 1992). Experimentally the growth of a crystal in either of these situations occurs when the system loses equilibrium through undercooling. We limit ourselves here to considering the growth in a melt.

The solidification of a pure melt occurs when the temperature T_∞ imposed on the fluid is below the temperature T_f of equilibrium between the liquid and the solid. The undercooling is $\Delta = T_f - T_\infty$. As the solid grows into the liquid, it releases latent heat: a diffusive temperature field is formed in front of the crystallization front. It is controlled by two Fourier equations of diffusion, one on each side of the interface:

$$\frac{\partial T}{\partial t} = D_T^i \nabla^2 T. \tag{3.3}$$

The scalar field here is the temperature and D_T^i is the heat diffusion coefficient, for the liquid ($i = 1$) or the solid ($i = 2$) respectively.

The heat produced by the solidification is diffused on both sides of the interface; this provides a relation between the growth velocity and the temperature gradients:

$$LV_n = \left[D_T^1 C_p^1 (\nabla T)_1 - D_T^2 C_p^2 (\nabla T)_2 \right] \cdot \boldsymbol{n}, \tag{3.4}$$

where L is the latent heat, V_n the normal velocity of the front, and C_p^1 and C_p^2 the heat capacity of the solid and the liquid respectively. The two diffusion constants are usually taken to be equal (symmetrical model).

Finally, the melting temperature of the solid is modified when the interface is curved by the pressure induced by capillarity. This is the Gibbs–Thomson effect. The equilibrium temperature is changed:

$$T = T_f \left(1 - \frac{\gamma \kappa}{L} \right), \tag{3.5}$$

where γ is the surface tension and κ the curvature of the interface ($\kappa > 0$ if the centre of curvature is in the solid). It is found that only at undercooling $\varDelta = L/C_p^1$ is the plane front a solution for a forced growth at constant V_n.

The solidification process is *a priori* associated with two length scales: the diffusion length scale $l_D^T = D_T/V$ and the capillary length scale d_0. This latter scale is the size of the smallest crystal in equilibrium in the melt at the given temperature: it is related to L, the latent heat of solidification, γ, the surface tension, and C_p, the heat capacity through

$$d_0 = \frac{\gamma C_p T_f}{L^2}. \tag{3.6}$$

The linear analysis of the stability of the plane solidification front in a pure melt is due to Mullins & Sekerka (1964). It shows that it is unstable at all velocities but that small scales are stabilized by capillary effects through the Gibbs–Thomson effect. As demonstrated in Chapter 8, in the situation where equation (3.3) can be approximated by Laplace's equation, the amplification rate of a perturbation of wave-vector k is

$$\sigma = V|k| \left(1 - \frac{D_T}{V} d_0 k^2 \right), \tag{3.7}$$

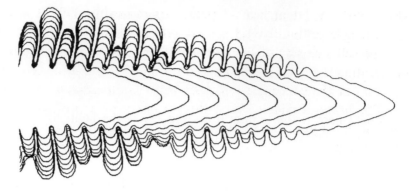

Figure 7. The growth of a parabolic dendritic monocrystal (from Dougherty & Gollub 1988).

a relation comparable to that found by Chuoke *et al.* (1959) for viscous fingering (see equation (2.26) and figure 5). The maximum instability of a front growing at velocity V occurs at

$$l_c^s = 2\pi\sqrt{\frac{3D_T d_0}{V}} = 2\pi\sqrt{3l_D^T d_0}, \tag{3.8}$$

which is the length scale of the instability; it is proportional to the geometrical mean of the two physical scales l_D^T and d_0.

In most real solidification processes the impurities rejected by the solid form a diffusive field in front of the crystal. As the molecular diffusion constant D_c is much smaller than D_T most crystallization processes are controlled by the impurity diffusion field. An important difference with the above case is that while the diffusion of heat is almost equal in the solid and the liquid, the diffusion of impurities is large in the solution and small in the solid. A linear stability analysis was performed by Langer (1980) in this case. Its results are similar to those obtained by Mullins & Sekerka. The maximum instability occurs at a length scale l_c^{sc} proportional to the geometrical mean between the impurity diffusion length $l_D^c = D_C/V$ and a chemical capillary length d_0'. Like viscous fingering, this linear analysis only gives a length scale for the problem. Experimentally, the plane growth only exists transitorily but the length scale l_c^{sc} retains its importance in the case of the needle crystals called dendrites. One of them is shown on figure 7. They will be described in §4.2.

3.3 Other systems

As already stated, many phenomena belong to this class of processes. We shall here quote two more. The growth of bacterial colonies (Fujikawa &

Matsushita 1991; Ben Jacob *et al.* 1992) takes place in a Petri dish filled with an agar–agar gel containing a weak concentration of nutrient. In the middle of the cell a small number of bacteria are deposited. As the bacteria start proliferating they deplete the concentration of nutrient in their vicinity. This depletion of the concentration tends to be compensated by molecular diffusion. The bacteria which are at the periphery of a protrusion will be located in a stronger gradient, will receive more food and will thus proliferate. Once again the scheme of figure 2 is relevant: the interface is unstable and the tips grow faster.

Other systems generate other types of morphologies as they belong to a different class. For instance the breaking of a window pane generates a pattern of propagating cracks. The dynamics of fractures, widely investigated recently (see Hull 1999 for a review), occurs in a stress field which is modified by its progression. It is thus also a free boundary problem. However, the morphologies observed are very different. For instance the breaking of a glass plate is characterized by the formation of both radial and orthoradial cracks meeting at right angles. The origin of the difference in the morphologies lies in the fact that the stress field in which the cracks propagate is tensorial instead of scalar.

4 The existence of stable curved fronts

The instability of the plane fronts in all these systems does not preclude the possibility of existence of stable steady solutions of different shapes. Such fronts are observed experimentally and the search for stable curved fronts actually preceded the linear stability analysis of the plane ones. Two cases have been the focus of interest: the Saffman–Taylor finger and the parabolic needle crystal.

4.1 Isotropic case

Most of the knowledge we have on non-fractal isotropic growth comes from viscous fingering and was obtained in two geometries. In the first, introduced by Saffman & Taylor (1958), the two fluids move in a Hele-Shaw cell in the shape of a long linear channel closed on both lateral sides. In the second, used by Thomé *et al.* (1989), the fluids move between two walls forming a wedge.

4.1.1 *Linear channel*

This configuration has translational invariance along the cell. The nature of the pattern observed is controlled by a parameter *B* proportional to the

Yves Couder

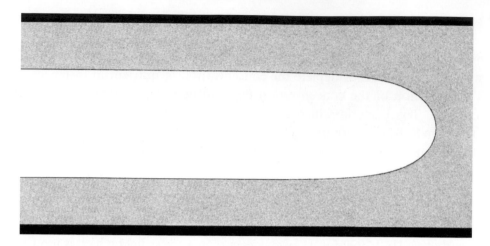

Figure 8. Photograph of Saffman–Taylor finger of width $\lambda = 0.5$ obtained in a linear channel at $B \approx 2 \times 10^{-4}$.

square of the ratio of the small scale (l_c, the length scale of the instability (2.28)) to the large scale (the width W of the channel). It is traditionally chosen as:

$$B = \frac{\gamma}{12\mu V} \left(\frac{b}{W} \right)^2 = 8.45 \times 10^{-3} \left(\frac{l_c}{W} \right)^2. \qquad (4.1)$$

For a large range of values of B one single finger is observed to move steadily along the cell with a well-defined width and shape as shown on figure 8. As B decreases, the ratio λ of the finger width to the channel width tends towards $\lambda = 0.5$. This situation persists (Tabeling *et al.* 1987) down to values of $B \approx 1.4 \times 10^{-4}$ i.e. for $W \approx 8l_c$. For $B < 1.4 \times 10^{-4}$ the finger becomes unstable as described in §5.1.2.

Saffman & Taylor (1958) showed that it was possible, if surface tension was neglected, to find by conformal transform techniques a family of analytical solutions for finger-shaped interfaces. Their theory can be summarized as follows. The finger and the channel are drawn in figure 9(a). The origin of the axes is at the finger tip, the walls are respectively at $y = \pm 1$ and the finger has two asymptotes at $y = \pm \lambda$ (the channel and the finger width are here 2 and 2λ respectively). The velocity of the finger is U and the velocity of the fluid far from the finger is V. If the finger width is λ, then $V = \lambda U$. The velocities given by equation (2.5) derive from potentials

$$\phi_i = -\frac{b^2}{12\mu_i} p_i. \qquad (4.2)$$

It is possible to define stream functions ψ_i such that

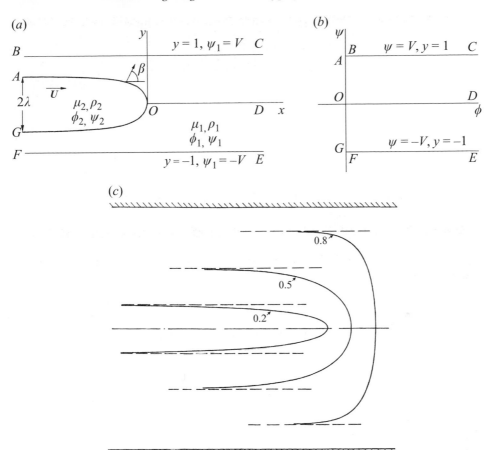

Figure 9. (a, b) Diagrams (from Saffman & Taylor 1958) showing the boundaries respectively in physical space and in the (ϕ, ψ)-plane. (c) Three finger profiles of the continuous family of solutions given by equation (4.8) (from Saffman & Taylor 1958).

$$u_i = \frac{\partial \phi_i}{\partial x} = \frac{\partial \psi_i}{\partial y}, \qquad v_i = \frac{\partial \phi_i}{\partial y} = -\frac{\partial \psi_i}{\partial x}. \tag{4.3}$$

Since the flow corresponds to a Laplacian field, the complex potential $\omega = \phi + i\psi$ is an analytic function of $z = x + iy$. The continuity of the normal components of the velocity at the interface can be written:

$$\frac{\partial \phi_1}{\partial n} = \frac{\partial \phi_2}{\partial n} = U \cos\beta, \tag{4.4}$$

where β is the angle between the x-axis and the normal n to the interface. On the surface of the finger, $\partial \phi / \partial n = \partial \psi / \partial s$ where s is the curvilinear position along the interface. Since $\cos\beta = \partial y/\partial s$, we have on the surface of the finger,

$$\psi_1 = \psi_2 = Uy. \tag{4.5}$$

In the absence of surface tension the continuity of pressure through the interface gives

$$\phi_1 = \phi_2 = 0. \tag{4.6}$$

On the walls $(y = \pm 1)$, $\psi = \pm V$ and at infinity in front of the finger $\phi = Vx$. The problem is then transferred into the (ϕ, ψ)-plane (figure 9b) and the trick is to consider $x + iy$ to be an analytic function of $\phi + i\psi$. On this plane the surface of the finger is mapped onto the segment AG so that there is no longer a free boundary, and the walls of the channel map onto $\psi = \pm V$ with $\phi > 0$. The function satisfying the boundary conditions is

$$z = \frac{\omega}{V} + \frac{2}{\pi}(1 - \lambda)\ln\frac{1}{2}\left[1 + \exp\left(-\frac{\pi\omega}{V}\right)\right], \tag{4.7}$$

which determines the potential as a function of z. The shape of the interface is found by setting $\phi = 0$ in equation (4.7). Written in dimensional variables, W being the width of the channel,

$$x = \frac{W(1 - \lambda)}{2\pi}\ln\frac{1}{2}\left[1 + \cos\left(\frac{2\pi y}{\lambda W}\right)\right]. \tag{4.8}$$

These fingers are parameterized by their relative width λ which can take any value from 0 to 1 (figure 9c). The shape of the solution of width 0.5 fits the experimentally observed finger (figure 8) extremely well.

The remaining problem was that no reason was found at the time for the selection of this particular solution. The answer to this question came much later and was obtained by investigating the role of surface tension. This analysis is too long to be reproduced here; we will only summarize its results. The selection was first obtained in numerical investigations by McLean & Saffman (1981), Schwartz & Degregoria (1987) and Vanden-Broeck (1983) and fully understood in analytic works by Combescot *et al.* (1986, 1988), Hong & Langer (1986), and Shraiman (1986). Surface tension cannot be treated as a perturbation because it introduces higher-order terms in the integro–differential equation defining the interface. When the transcendentally small terms due to surface tension are taken into account, the requirement that the finger should be smooth at the tip introduces a solvability condition. For a given value of B there is only one discrete set of solutions of width λ_n for which this condition is satisfied; they all tend to have a width 0.5 for vanishing B. In the limit of very low B, their dependence

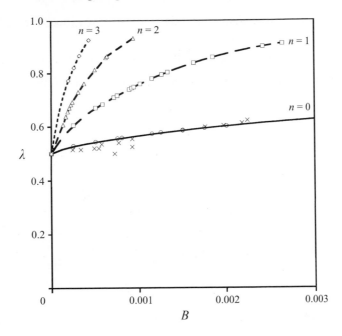

Figure 10. The B-dependence of the widths of the discrete set of solutions selected by surface tension. The open symbols result from numerical simulation (courtesy M. Ben Amar), the crosses are the finger widths measured by Saffman & Taylor (1958).

on B is given analytically by

$$\left(\lambda_n - \tfrac{1}{2}\right) = \left(\tfrac{1}{8}\right) a_n \left(16\pi^2 B\right)^{2/3}, \tag{4.9}$$

where the coefficients a_n are related to the successive integers n by

$$a_n = 2(n + 4/7)^2. \tag{4.10}$$

This dependence can also be found numerically. Figure 10 shows the evolution of these solutions as a function of B. The stability of the various branches was investigated by Bensimon, Pelcé & Shraiman (1987), Tanveer (1987b) and Kessler et al. (1988): only the lower branch is stable, the others are unstable through tip splitting.

The singular characteristic of this selection is demonstrated by the fact it can be removed experimentally by various types of very localized and small disturbances of the finger tip (Couder et al. 1986b and Rabaud, Couder & Gerard 1988). When this is done the normal Saffman–Taylor finger is replaced by a narrower and faster finger. This is a direct experimental confirmation of the singular role ascribed by the theory to the finger tip. The new structure, often called an anomalous finger, will be discussed in § 4.2.2.

Finally we can note that it is possible to introduce for DLA growth an equivalent to surface tension (by using specific rules for the sticking of particles on the aggregate). Kadanoff (1985) and Liang (1986) have shown that it was thus possible to recover the stable isotropic Saffman–Taylor finger in a linear channel.

4.1.2 Sector-shaped cells

In a sector-shaped cell the viscous finger moves between two lateral walls forming an angle θ_0. By convention the angle of the wedge θ_0 is considered positive when the finger moves in the divergent direction, the situation mainly considered here. The parameter controlling the growth is chosen, by analogy with the linear channel, as $B = 8.45 \times 10^{-3} (l_c/\theta_0 r)^2$ where $r\theta_0$ is the curvilinear local width $W(r)$. For moderate values of B a single finger is observed which occupies a finite fraction $\lambda_{\theta_0} > 0.5$ of the angular width of the wedge (figure 11*a*). This fraction λ_{θ_0} is approximately a linear function of the wedge angle (figure 11*b*). The finger grows as a self-similar structure if and only if B is kept constant, i.e. if the velocity V decreases as r^{-2}. In most practical cases V is constant, so the finger is observed to become spontaneously unstable (figure 11*a*).

The theoretical treatment of this situation followed the same path as that of the linear case. Neglecting surface tension, a set of self-similar solutions was first found for $\theta = 90°$ (V. Hakim in Thomé *et al.* 1989). Then for any value of θ_0, families of solutions given by hypergeometric functions and again parameterized by their width λ_{θ_0} were found by Ben Amar (1991*a*). The selection due to isotropic surface tension was investigated both numerically (Ben Amar *et al.* 1991, Ben Amar 1991) and analytically by Combescot & Ben Amar (1991). For a given value of B there is again selection of a discrete set of solutions. The evolution of this set as a function of B is given for the wedge with $\theta_0 = 20°$ in figure 12 where it can be compared with that of the parallel channel. The comparison shows that the selected widths λ_{θ_0} are shifted to larger values. Furthermore the levels $n = 0$ and $n = 1$ coalesce at a value B_1 so that they form a loop and do not exist for $B < B_l$. The experimental observations show that when moving at constant velocity the finger adapts constantly to the evolution of the local parameter B as defined in the tip region. The finger is first stable, with a slowly varying width given by that of the level $n = 0$. When the local B reaches the limit value B_1, the finger becomes unstable.

Taken together, the results obtained in the linear and sector-shaped cells show that: (i) for an isotropic interface the selection process defines the finger width, and the selection occurs at the large scale of the system; (ii)

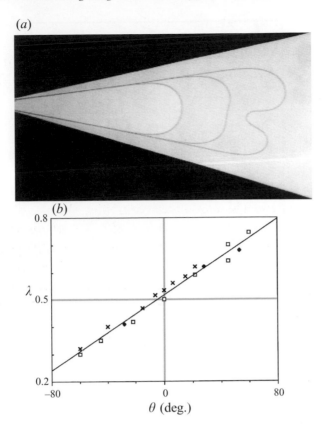

Figure 11. (*a*) Three superposed photographs of the evolution of a Saffman–Taylor finger of width $\lambda = 0.66$ obtained in a sector-shape-cell of angle $\theta_0 = 30°$. (*b*) The crosses are the width of the stable fingers as a function of the angle θ_0 at $B \sim 10^{-3}$. The open squares are the width of ensemble–averaged unstable Saffman–Taylor fingers. The black diamonds are the width of ensemble-averaged DLA grown in wedges.

the global curvature imposed by the geometry affects the selection and the stability of the finger

4.1.3 Circular geometry

In the axisymmetric geometry (see figure 1), at the very beginning of the air injection, the interface is circular. When the perimeter of this circle becomes large enough the front destabilizes spontaneously at the scale l_c and a small number (usually 5 to 7) of radially growing fingers are formed. Each of these fingers resembles the fingers obtained in wedges. This analogy can be made quantitative (Thomé *et al.* 1989). Each finger grows as if it were enclosed in a cell, having virtual walls formed by the bisectors of the region separating

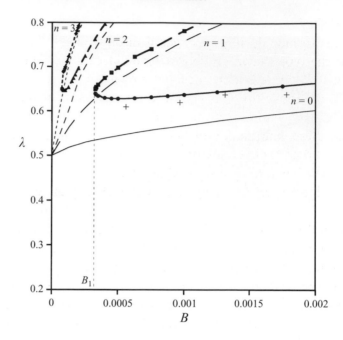

Figure 12. The *B*-dependence of the widths of the discrete set of solutions in a wedge of angle $\theta_0 = 20°$ (thick lines). The thin lines show for comparison the discrete set in the linear cell. The crosses are experimental results.

this finger from its two neighbours. With this rule the shape, the width and the stability of each finger can be predicted by the results obtained for sector-shaped cells. Since in practical situations the growth occurs at constant velocity, the local *B* decreases and there will be a constant unsteady evolution of the pattern. As they continue to grow the fingers destabilize and tend to form increasingly complex 'trees'. However, even when the growth has formed a fractal object, the initial breaking of the symmetry still appears to persist and to be at the origin of the formation of independent trees.

4.2 Non-isotropic situation

Non-isotropic growth can be obtained in the three types of experiments described here. We will first briefly describe its archetype which is the dendritic growth of crystals (for reviews see Huang & Glicksman 1981 and Pomeau & Ben Amar 1992). Then we will show that anomalous Saffman–Taylor fingering in the presence of localized disturbances is a similar type of growth.

4.2.1 *The dendrite*

The structures formed during the crystallization of a solution or a melt are dendrites of the type shown on figure 7 (provided this crystallization is not too slow). Each dendrite is a monocrystal growing in a well-defined direction and having a parabolic tip. This growth direction is one of the main axes of the crystal lattice. At some distance from the tip the curved front becomes unstable and there is formation of lateral branches which grow along other main crystallographic directions of the crystal. The theoretical analysis of dendritic growth has a history somewhat similar to that of Saffman–Taylor fingering. We limit ourselves here to a brief recalling of results in the case closest to the Laplacian problem, namely a two-dimensional situation where there is only diffusion on one side of the interface, a condition well satisfied in the case of impurity diffusion. This also limits considerations to cases where the diffusion length is very large compared to the crystal size and where the boundaries are far away. It was shown by Ivantsov (1947) that, in the absence of surface tension, parabolas were possible stationary solutions of the shape of the interface separating the solid and the liquid. As in the Saffman–Taylor problem the argument gave no hint as to the selection of the experimentally observed solution. In the case of dendrites there is still no selection mechanism known when isotropic surface tension is taken into account (Kessler, Koplik & Levine 1986a and Caroli, Caroli & Roulet 1987). However, crystals are anisotropic and the surface tension γ of the interface is a function of the angle Θ of the normal to the interface with the main axis of the crystal. For a cubic crystal

$$\gamma = \gamma_0(1 + \epsilon\cos\Theta). \tag{4.11}$$

If this anisotropy of the surface tension is taken into account, a discrete set of parabolas is selected for a given undercooling (Ben Amar 1990). The observed solution is the fastest one. It is defined by its radius of curvature r_d which is proportional to the instability length scale l_c^{sc} (§ 3.2). The coefficient of proportionality is a function of the anisotropy ϵ. In the limit of very weak anisotropy the theoretical result can be written

$$r_d = 0.5\epsilon^{-7/8}l_c^{sc}. \tag{4.12}$$

This type of selection has been confirmed by Dougherty & Gollub (1988) and Maurer, Perrin & Tabeling (1991) but the experimental measurement of ϵ is very difficult so that the dependence on ϵ has not yet been confirmed experimentally.

It is worth mentioning that Honjo, Ohta & Matsushita (1986) have shown that the presence of very strong external noise could destroy the effect of anisotropy. No dendritic growth of the crystal is then observed and the growth appears isotropic.

4.2.2 *The anomalous Saffman–Taylor finger*

A comparison of the two types of growth, viscous fingers or dendrites, described above leads one to ask whether their difference is linked with the difference in their physical nature or simply to a difference in their symmetries. Is it possible to obtain viscous fingers which look like dendrites? A positive answer to this question is given when preferential directions of growth are imposed on viscous fingering. A global disturbance can be imposed by using glass plates deeply engraved with a periodic lattice of grooves (Ben Jacob *et al.* 1985; Chen & Wilkinson 1985) or replacing the viscous fluid by an anisotropic liquid crystal (Buka, Kertesz & Vicsek 1986). In fact, since the selection is a singular effect occurring at the tip, a single disturbance localized in this region is sufficient to upset the growth (Couder *et al.* 1986*a*, *b*). In the radial configuration all these disturbances result in fingers which grow faster than normal, have stable tips and exhibit dendritic-like side branches. In linear channels the difference is more striking and easier to study quantitatively. Figure 13(*a*) shows a finger obtained in a linear channel in which the local disturbance is created by a thin groove engraved along the axis of each of the two glass plates (Couder *et al.* 1986). The observed fingers are much narrower than usual (for instance $\lambda = 0.22$ on figure 13*a*). Their shapes are however still very well fitted by the Saffman–Taylor analytical solutions of the same width λ (4.8) so that a quantitative study is possible.

The selection of these anomalous fingers (Rabaud *et al.* 1988) is clearly different from that in the isotropic case. Fingers grown at the same velocity in channels of different width have very different λ but a similar tip. As the finger shapes are given by equation (4.8) the radius of curvature at the tip r_{ST} can be deduced from the measured λ:

$$r_{ST} = \frac{\lambda^2 W}{\pi(1 - \lambda)}. \tag{4.13}$$

Anomalous fingers grown in various cells at the same velocity with the same disturbance have the same r_{ST}. On varying the velocity, r_{ST} is found to be proportional to the capillary length scale (figure 13*b*):

$$r_{ST} = \alpha l_c. \tag{4.14}$$

The value of the coefficient α is observed to be dependent on the strength of the applied disturbance. Relation (4.14) is similar to (4.12), meaning that

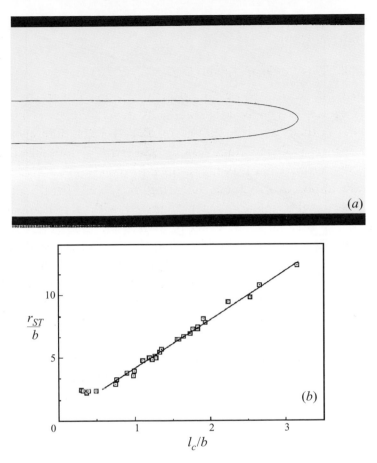

Figure 13. (*a*) A photograph of an anomalous finger of width $\lambda = 0.22$ obtained in a linear channel when one of the glass plates has a single groove etched along its axis. (*b*) A graph of the radius of curvature of the tip r_{ST}/b versus l_c/b (Rabaud *et al.* 1988). The saturation occurs when $r_{ST} \approx 2.5b$ where the two-dimensionality breaks down.

anomalous fingers are not only visually similar to dendrites, they also obey the same selection rules. This situation has now been widely investigated experimentally (Kopf-Sill & Homsy 1987; Zocchi *et al.* 1987; Rabaud *et al.* 1988), numerically (Dorsey & Martin 1987; Sarkar & Jasnow 1989) as well as theoretically (Hong & Langer 1986, 1987; Combescot & Dombre 1989).

The numerical simulations show that in the presence of anisotropy parallel to the channel, the lower level of the discrete set does not tend towards $\lambda = 0.5$ but towards $\lambda = 0$ as shown on figure 14. This regime is observed in a limited domain of values of the parameter B. There are two crossovers, respectively at very low and very high velocities. A very slowly growing

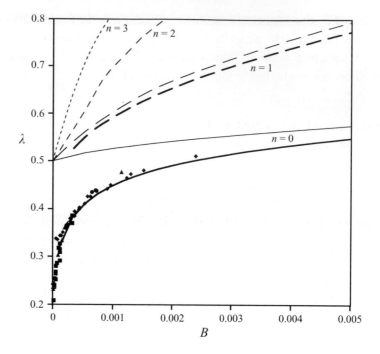

Figure 14. The *B*-dependence of the width of the discrete set of solutions in the presence of anisotropy (thick lines). Black symbols: measured width of fingers obtained with a given disturbance of the tip. The thin lines show the discrete set in the linear cell.

finger is only weakly affected by anisotropy and its width remains close to that of the isotropic one. (This crossover is similar to that investigated for crystals grown in a narrow channel (Kessler *et al.* 1986*b*).) For very fast growing fingers when the tip radius of curvature becomes of the order of the thickness of the cell (at $r_{ST} \approx 2.5b$) saturation is observed (figure 13*b*). This corresponds to the breakdown of the two-dimensional behaviour.

The selection process of normal Saffman–Taylor fingers has been removed mainly by disturbances that affect the isotropy or the homogeneity of the interface; this is the most common situation and the most relevant to an analogy with crystal growth. However, since this selection is due to a delicate relation between an isotropic surface tension and a hydrodynamic flow, it can also be removed by a disturbance of the flow only, the interface remaining isotropic. This was suggested theoretically by Combescot & Dombre (1989) and experimentally confirmed by Thomé, Combescot & Couder (1990) in experiments in which the flow, ahead of the finger, is disturbed by a moving disk.

Taken together the results for both the anomalous Saffman–Taylor fingers and the dendritic growth show that either a local disturbance or anisotropy results in a selective process which defines the radius of curvature at the finger tip. The selection thus occurs at the small scale of the system.

4.3 The stability of the curved fronts

Both the Saffman–Taylor and the solidification instabilities generate structures, fingers or dendrites, which are curved fronts that can have steady solutions but can also become unstable. It is thus necessary to examine specifically the problem of the stability of curved fronts. This was first undertaken by Zel'dovich *et al.* (1980) in the context of flame fronts. A review of this problem can be found in Pelcé (1988). The results presented above have shown that the selection mechanisms at work in the isotropic or anisotropic cases lead respectively to the tip being of small or strong curvature respectively compared to the typical length scale. For this reason the type of instability observed in the two cases is very different.

4.3.1 Isotropic case

Isotropic curved fronts are selected to be on the scale of the channel width. Naively, decreasing B means that the finger tip becomes wider and wider in relation to l_c. The finger front will thus be increasingly similar to a straight front and tend to be unstable at the capillary length l_c corresponding to the local normal component of the velocity. This is observed in wedges where the experiments (Thomé *et al.* 1989) show that the finger is destabilized more and more easily through tip splitting for increasing angles. For each angle, the finger is destabilized (figure 11a) when the local value of B at the finger tip reaches the value B_1 for which the lower level $n = 0$ ceases to exist (figure 12). By contrast, in the linear channel the finger is stabilized by its curvature and Bensimon *et al.* (1987) and Tanveer (1987a) found the lower branch $n = 0$ of figure 10 to be linearly stable while the other branches were unstable. Experimentally, however, for B values smaller than 1.4×10^{-4}, the observed finger is unstable to tip splitting (figure 15a). This can be ascribed to the fact that as B tends to zero the branches become more closely spaced and the natural noise is enough to induce a subcritical transition to one of the neighbouring unstable states.

4.3.2 Anisotropic case

For non-isotropic fronts, dendrites or anomalous fingers, the finger tip is defined by the small scale. On decreasing B, the tip radius of curvature r_{ST}

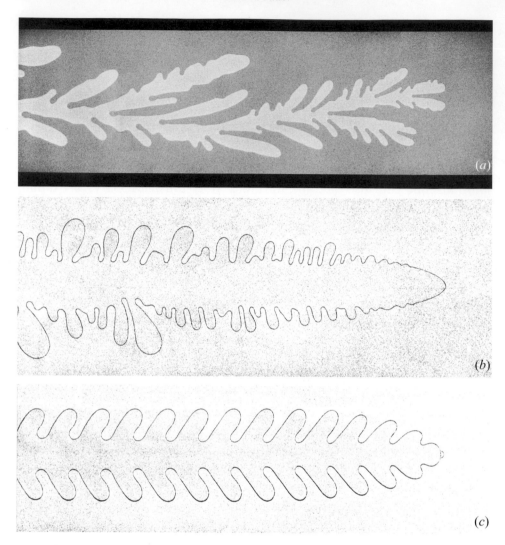

Figure 15. (*a*) A very unstable isotropic finger in a linear cell. (*b*) The natural instability of an anomalous Saffman–Taylor finger grown in a wide channel (Rabaud *et al.* 1988) to be compared with the dendrite on figure 7. (*c*) An anomalous Saffman–Taylor finger subjected to a periodic forcing.

always remains in the same ratio to the capillary length scale; when the anisotropy is large the tip is thus observed to be stable. The same is not true of the lateral sides. They are always unstable to dendrites (figure 7). For anomalous viscous fingers, as the scale of r_{ST} moves away from the scale of W, the relative width of the finger λ decreases and the finger profile becomes parabolic in a larger and larger region behind the tip. As a result a lateral instability grows (figure 15*b*). It is identical in nature to the lateral instability of crystalline dendrites (figure 7).

The specificity of the growth of an instability advected along a curved front was first pointed out by Zel'dovich *et al.* (1980) in the context of flame fronts. The growing wave, being advected along a curved front, is stretched by a kinematic effect due to the growing tangential velocity. This effect was investigated by Pelcé & Clavin (1987) and Kessler & Levine (1987). A second effect was pointed out by Caroli *et al.* (1987): each region of the front is unstable through a process described by the linear stability analysis of equation (2.26) or (3.7). Along the curved fronts the normal velocity decreases continuously away from the tip. As a result there is a continuous shift of the maximum amplification rate towards larger wavelengths. Both effects tend to increase the wavelength away from the tip.

The most important point that underlies these arguments is that the side branching is an instability of a convective type. This means that it has a spatial growth rather than a temporal one (for a discussion of absolute and convective instabilities see Chapter 4). This was first demonstrated in anomalous Saffman–Taylor fingers because they can be manipulated easily. Rabaud *et al.* (1988) gave a sudden local disturbance to the tip of an anomalous finger and showed that it resulted in the growth of a wave packet which, in the frame of reference of the finger, was advected away from the unstable region. No waves remained in the front region so that any later disturbance was uncorrelated with an earlier one. This is precisely the characteristic of a convective instability. Furthermore this experiment permitted the direct observation of the wavelength change along the profile due to both the Zel'dovich stretching and the shift in wavelength selection. The same demonstration was done in the case of dendritic growth by Quian & Cummins (1990) using a laser pulse to disturb the tip.

The main characteristic of convectively unstable media is that they behave as a selective amplifier of the noise. This is precisely the characteristic that Dougherty & Gollub (1988) demonstrated experimentally for the dendritic side branching. But in convective instabilities noise can be replaced by a periodic forcing of larger amplitude. The instability is thus made strictly periodic at the imposed frequency. A strictly periodic side branching is observed when the input pressure is modulated and acts as a forcing frequency. It is worth noting that similar periodic side branches are obtained spontaneously with a small bubble at the finger tip because it forms a small local oscillator (figure 15*c*). A similar temporal forcing was applied to a growing dendrite by Bouissou *et al.* (1990) by imposing a modulated flow around the dendrite tip.

5 Fractal structures

We can now return briefly to open geometries in which the large scale is not
fixed but is of the order of the size of the pattern itself, becoming constantly
larger as it grows. Figures 16(a) and 16(b) show two large patterns grown in a
DLA simulation and in a Saffman–Taylor fingering experiment respectively.
Both exhibit a very complex structure with a hierarchy of branches of
different sizes. A classical question about complex structures is whether or
not they have a fractal structure.

The fractal objects first investigated were mathematically defined self-
similar structures built with an iterative scaling construction rule (for reviews
see for instance Mandebrodt 1982 and Vicsek 1989). One of the characteristics of such mathematical fractals embedded in a space of dimension d
is their fractal dimension d_f. This dimension is usually obtained by the
box-counting method. The number $N(\epsilon)$ of d-dimensional boxes necessary
to cover the whole pattern is measured as a function of the size ϵ of these
boxes. We then plot $\log(N(\epsilon))$ as a function of $\log(\epsilon)$. For a mathematical
fractal this plot gives a linear dependence showing that the ε-dependence of
N is of the form

$$N(\epsilon) \propto \epsilon^{-d_f}, \qquad (5.1)$$

where d_f is the structure's fractal dimension.

It has become habitual now, whenever a complex physical structure is
observed, to do some image processing and to seek a scaling of the type of
equation (5.1). In the case shown on figure 16(a, b) the plots of $\log(N(\epsilon))$
as a function of $\log(\epsilon)$ exhibit a well-defined linear behaviour in a range
roughly falling between L and l_c (or l_u). In this range the slope of the plot
is approximately $d_f = 1.67$ so that these patterns are said to be fractals.
Muthukumar (1983) and Tokuyama & Kawasaki (1984), using different
theoretical arguments, proposed that DLA grown in a plane should have
$d_f = 5/3$, a value close to what is found numerically. It should be underlined
that in all physical fractals the domain of fractal behaviour is bounded. In
the present case, for length scales larger than L the slope of the $\log - \log$
plot tends to zero because on very large scales the pattern is similar to a
point. For very small scales there is another crossover to a slope -2: the
pattern on these scales is a two-dimensional object. The fractal behaviour
in growth patterns is thus best observed when the two typical length scales
are furthest from each other. For this reason DLA serves as a good model
for fractal growth because it provides the largest fractal range. It is mostly
investigated in the radial configuration and can be either isotropic as shown
in figure 16 or anisotropic as in figure 18.

(*a*)

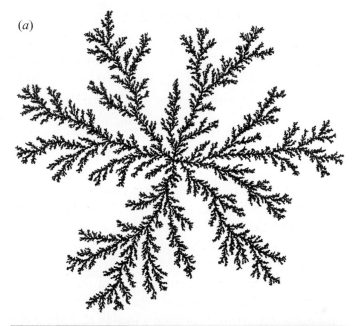

(*b*)

Figure 16. (*a*) An isotropic DLA grown off-lattice and having 10^6 particles (courtesy A. Arnéodo). (*b*) A radial fractal Saffman–Taylor fingering pattern.

In the following we will describe successively complex patterns obtained in isotropic and anisotropic situations. We will concentrate mainly on two of their characteristics.

(i) The scaling properties which give them a fractal character. Though understanding why patterns grown in a Laplacian field have a fractal structure is certainly a major problem, no complete theory exists about it yet.

(ii) The existence of a well-defined ensemble average for these fractal structures which, in a given geometry, is related to the corresponding stable solutions.

5.1 The isotropic case

5.1.1 Open growth: the fractal properties of DLA and viscous fingers

For some time the main issue in the field of complex patterns has been the measurement of their fractal dimension. The characterization of the fractal structures resulting from Laplacian growth was initially performed on radially grown DLA clusters (figure 16a) and several techniques were used. For DLA good scaling laws covering several decades were obtained. It was only later, after Paterson (1984) had pointed out the similarities of the equations of the two systems, that the fractal properties of viscous fingering were sought. This fractal structure was first observed in porous cells (Nittman, Daccord & Stanley 1985 and Chen *et al.* 1985) or with non-Newtonian fluids (Daccord, Nittmann & Stanley 1986; Lenormand, Touboul & Zarcone 1988 and Van Damme *et al.* 1988). It is more difficult to observe in the case of the pure Saffman–Taylor fingers (figure 16b). Several secondary effects can lead to spurious effects. Figure 16(b) is about the best that can be achieved; it was obtained in a small cell having a very thin spacing b, and oil with a very large viscosity. The applied pressures are high so that the glass plates have to be very thick (4 cm in figure 16b). If these conditions are met the growth is fractal and similar to DLA (Rauseo, Barnes & Maher 1987; Couder 1988; and May & Maher 1989). In spite of a limited range of scales between the large and the small scale the measured fractal dimension of the pattern shown in figure 16(b) is in good agreement with that of DLA.

The overall dynamics of fractal growth has received no general theoretical treatment as yet. At present it is only possible to describe the various processes at work in its built up. We can describe it in the geometry in which the linear stability analyses were performed, where an initial straight front separates two half-planes. As described above, the initial destabilization of the front occurs at the characteristic small scale of the system. However, when the amplitude of the disturbances increases, the different protrusions of the front interact non-locally (as shown for DLA on figure 6), the fastest growing ones tending to screen off the others so that their growth is slowed down and ultimately stopped. Simultaneously these fastest protrusions have more space to grow so that they become unstable either by tip splitting or by side branching. The two opposing trends build up a fractal structure. The screening-off is responsible for the formation of larger and larger, increasingly widely spaced, trees (figure 6), a process often called coarsening. Simultaneously, tip splitting and side branching keep constantly generating the smaller scales of the fractal distribution.

5.1.2 Confined growth: the ensemble-average properties

We can now return to confined experimental configurations that limit the freedom of the system. The question we address concerns the structure of the very unstable patterns obtained when the two characteristic length scales of the problem are very different from each other but either fixed or controlled. This study concerns very unstable patterns obtained in the geometries in which stable smooth fronts have been found.

(i) The linear channel When fingers are grown in linear channels at large velocities (i.e. very small B, in the range from 10^{-4} to 10^{-6}) the fingers are strongly unstable (see figure 15a) with a large number of branches which have on average a width which scales on the capillary length at this velocity (Kopf-Sill & Homsy 1988 and Meiburg & Homsy 1988).

It can however be noted that on average a steady regime of growth is reached which needs to be characterized statistically (Arnéodo *et al.* 1989). For this purpose a large number N of identical runs of the same experiment can be performed at the same very small value of B so that N independent realizations of patterns of the type shown on figure 15(a) are obtained. If a transverse section of the cell is chosen and discretized it is possible to build up a histogram representing, for each point of the section, how many times during the N runs it has been occupied by air. A division by N gives the mean occupancy $r(y)$ in the transverse section. This function is zero on the two walls and maximum in the centre of the cell. It is a remarkable feature that the width λ' at mid-height of these histograms is 0.5, i.e. the same as that of the stable fingers.

This investigation can be pushed further by studying the analogous problem of the growth of DLA in a channel geometry (Arnéodo *et al.* 1989). A linear strip is used having a width W_s in units of the grid mesh size. The

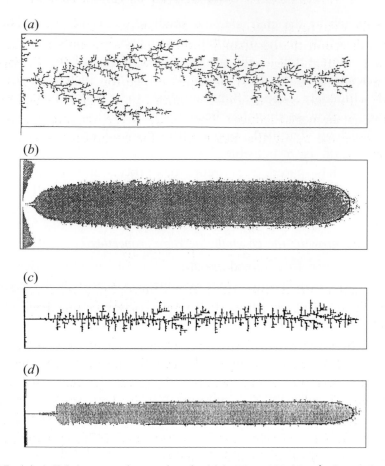

Figure 17. (*a*) A DLA grown in a strip of width $W = 128$ (Arnéodo *et al.* 1989). (*b*) The region of the channel with an occupancy rate above average in the strip in (*a*). (*c*) An anisotropic DLA grown in a strip of width $W = 64$. (*d*) The region of the channel with an occupancy rate above average in an anisotropic case.

system is confined by the lateral walls which reflect the randomly walking particles. The strip is open at one end and closed at the other. The particles stick onto the closed end of the strip. A large number of aggregates ($N = 512$) having the same total number M of particles are grown. One of them is shown on figure 17(*a*). For each point of the grid the number of realizations for which it has been occupied by a particle of an aggregate is counted. This number, divided by N, gives $\rho(x, y)$, the mean occupancy of this point.

The histogram of the occupancy along the axis of the linear strip shows that, except in the initial region and in the tip region, ρ is independent of x. This means that in the regions where the growth has ceased, the cell's translational invariance imposes itself on the occupancy profile. In the transverse direction the occupancy profiles have a maximum value ρ_{max} at the centre ($y = 0$) and decrease to zero at the walls ($y = \pm W_s/2$). For viscous fingers, the profiles (a sharp step-shaped profile for stable fingers) become smoother with decreasing B, and the width of the regions which are never visited (along the walls) reduces. The profile of the histogram obtained for DLA far from the tip (in the region where the evolution is finished) is well fitted by

$$\rho(x, y) = \rho_{max}\cos^2(\pi y/W_s). \tag{5.2}$$

We obtain an average finger shape by seeking for each x the value y_m of y such that:

$$y_m = \frac{1}{\rho_{max}} \int_0^\infty \rho(x, y)\mathrm{d}y. \tag{5.3}$$

The important result is that the profile defined by $y_m(x)$ has (with some statistical fluctuation) the shape of Saffman–Taylor finger of width $\lambda = 0.5$ (figure 17*b*).

What relation is there between the fractal structure and the steady averages obtained here? When confined to the cell the mean growth has become translationaly invariant along the channel. Across the channel the patterns have retained their fractal character on all scales between W and l_u (or l_c). By varying either W (in the DLA experiments) or l_c (in the Saffman–Taylor experiments) it is found (Arnéodo *et al.* 1989 and 1996) that the maximum mean occupancy of the cell width varies as

$$\rho_{max} \propto (W_s/l_u)^{-0.33}. \tag{5.4}$$

It is a property of a fractal structure of dimension d_f that its section along a line has a dimension $d_f - 1$. This means that the resulting mean density along this line scales as $d_f - 2$. The experimental exponent -0.33 corresponds to $d_f = 1.67 \pm 0.01$, in agreement with the usual fractal dimension of DLA.

(ii) Sector-shaped cells A similar result is obtained in sector-shaped cells. Using the same procedures, unstable Saffman–Taylor fingers and DLA aggregates can be grown in wedges with angles θ_0. The relative angular width $\lambda_m(\theta_0)$ and the profile given by $y_m(x)$ can then be measured. Figure 11(b) shows that for all convergent cells and for weakly divergent ones, $\lambda_m(\theta_0)$ coincides remarkably well with the limiting width of the stable fingers obtained in these cells at small B. However, the mean profiles show a tendency to split and the envelope shape differs from that of the stable fingers (Levine & Tu 1992 and Arnéodo *et al.* 1996).

5.2 Non-isotropic case

Even though anisotropic patterns can be obtained with viscous fingering this is not the best system for the investigation of anisotropic fractal growth. This type of study is mostly done on anisotropic DLA or on dendrites.

5.2.1 Open growth: the fractal properties of anisotropic DLA and of dendrites

Preferential directions of growth can be given to diffusion-limited aggregates by several means. One is noise reduction, which reveals the anisotropy of the underlying lattice (Julien, Kolb & Botet 1984; Nittman & Stanley 1986; Meakin, Kertèsz & Vicsek 1986; Meakin 1987). The on-lattice procedure has a weak anisotropy, which is enhanced if the rules of the simulation are changed by requiring that a site should be visited m times before a particle sticks there. The larger the imposed m, the larger the effective anisotropy. Other rules permit the introduction of anisotropy in simulations performed without an underlying lattice (Eckmann *et al.* 1989). In both cases the resulting patterns (figure 18*a*) have the general aspect of dendrites and are very different from the usual DLA because tip splitting is inhibited. Only two processes are present here: screening which leads to coarsening; and side branching which keeps generating small-scale structures. In the simulations the fractal dimension decreases with increasing anisotropy. It is generally believed that, asymptotically, the very anisotropic clusters are self-affine fractals of dimension $d_f = 3/2$.

In order to obtain a comparable situation in crystal growth it is necessary to grow a single dendrite in a very thin cell (so as to have quasi-two-dimensionality). It is also necessary to be as close as possible to a Laplacian situation, i.e. have a very large diffusion length. In this limit the interaction between side branches is long range and similar to what it would be for a Laplacian field. A series of experiments on slowly growing dendrites was performed in a solution of ammonium bromide by Couder *et al.* (1990). Figure 18(*b*) is a photograph of such a dendrite. Though the crystal is compact along its axis, the competition between its side branches gives it the structure of a self-affine fractal. Far away from the tip the fractal dimension of this crystal is $d_f = 1.58 \pm 0.03$. Note that for faster growing crystals the structure is only fractal-like up to approximately the diffusion length scale l_D^c. At larger scales the morphology is still complex but the box-counting method shows that at those scales the fractal dimension becomes equal to 2. There is thus a transition to a dense branching morphology (Uwaha & Saito 1989).

(*a*)

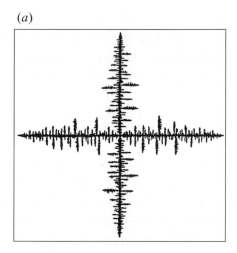

(*b*)

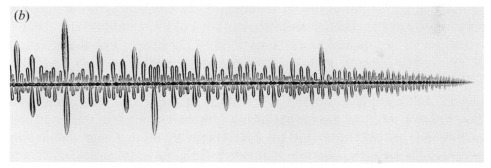

Figure 18. (*a*) An anisotropic DLA obtained with the noise-reduced technique with $m = 10$ (courtesy A. Arnéodo) (*b*) Photograph of a length of 1.25 mm of an ammonium bromide dendrite with a tip radius of curvature $r_d = 2.5\,\mu m$ (Couder *et al.* 1990).

5.2.2 *The ensemble-average properties*

(a) Anomalous viscous fingers and anisotropic DLA grown in a strip A generalization of the previous result about ensemble averages can be sought. (Couder *et al.* 1990; Arnéodo *et al.* 1991) in cells having plates engraved with a periodic square structure. It is also possible to grow anisotropic DLA patterns in a strip using the noise-reduced on-lattice simulation. Both in viscous fingering and in DLA when the easiest growth path is along the axis of the channel the unstable pattern is markedly narrower than for the isotropic case (figure 17*c*).

When the ensemble-averaging procedure is used again it produces transverse profiles which are narrower and sharper than in the isotropic case. For large m the value of $\rho_{\max}$ becomes independent of W_s. The regions of high occupancy are again in the shape of Saffman–Taylor fingers (figure 17*d*), and analogous to the stable anomalous fingers. The profiles given by $y_m(x)$ are well fitted by Saffman–Taylor solutions, so the tip radius of curvature r^f_{ST} can be deduced from the observed λ. For a given m the mean profiles obtained in cells of different width W have different λ but the same r^f_{ST}. The radius of curvature is scaled on the lattice scale l_u, the coefficient of proportionality depending on the anisotropy (fixed by m):

$$r^f_{ST} = \alpha l_u. \tag{5.5}$$

This is precisely the characteristic of the selection of the stable anomalous Saffman–Taylor finger and of dendrites ((4.12) and (4.14)). Inspection of the patterns shows qualitatively that increasing m increases the anisotropy. There is an empirical relation (Arnéodo *et al.* 1991) between α and m: measurements of r^f_{ST} for m from 2 to 15 give $\alpha(m) \propto m^{-1.5}$. This relation is reminiscent of the power law of equation (4.12). Unfortunately the relation between m and the effective anisotropy is not known quantitatively.

The mean occupancy profile thus reflects the shape and the selection of the stable solution in both the isotropic and anisotropic cases. This correspondence can be pushed even one step further: Arnéodo *et al.* (1991) showed that there is a continuous transition in the fractal dimension from $d_f = 5/3$ to $d_f = 3/2$ when the anisotropy is increased and that this transition corresponds to the continous crossover between normal and anomalous fingers observed on figure 14.

(b) The dendrite For dendrites an average profile can also be defined (Couder *et al.* 1990) and this can be done on a single dendrite of the type shown on figure 18(*b*). The area $S(x)$ occupied by the two-dimensional pattern from the tip to a distance x along the axis is measured. A logarithmic plot of this area as a function of x shows a power law $S(x) \propto x^{1.5}$. This means that the area occupied by the dendrite is the same as that of a dense parabola. This virtual parabola has the same radius of curvature as the tip. In other words, the dendrite shown on figure 18 occupies, on average, the same area as if it had remained stable without any side branching. This property is apparent on one single realization because the stability of the tip creates a rate of occupancy $\rho_{\max} = 1$ along the axis of the dendrite (a situation also obtained in DLA for large m and in Saffman–Taylor fingering when the imposed anisotropy is large).

5.3 Mean-field theory

The first attempt at a mean-field theory for the mean occupation by fractal aggregates was due to Witten & Sander (1983). They proposed in the continuum limit the equations

$$\frac{\partial \rho}{\partial t} = \nabla^2 w, \qquad \frac{\partial \rho}{\partial t} = w(\rho + l_u^2 \nabla^2 \rho), \tag{5.6}$$

in which ρ is the mean density of the aggregate, w the mean density of the randomly walking particles and l_u the lattice mesh size. The first equation is conservation of mass, the second represents the law that the DLA only grows when a site next to the existing aggregate is visited by a particle. V. Hakim showed that this set of equations could not be correct, as it leads, for the growth in a channel, to a density ρ decreasing as x^{-2}, and to the front going to infinity in a finite time. This effect is due to the absence, in the equation, of any lower threshold for the densities. In reality, because of the existence of the small scale, there is a finite threshold.

In order to take into account the existence of this threshold, Brenner, Levine & Tu (1991) suggested a heuristic modification in which the first term of the second equation is modified:

$$\frac{\partial \rho}{\partial t} = \nabla^2 w, \qquad \frac{\partial \rho}{\partial t} = w(\rho^\eta + l_u^2 \nabla^2 \rho), \tag{5.7}$$

the exponent being $\eta > 1$. This modification artificially introduces a cutoff in the growth rate. By numerical simulation Brenner *et al.* recovered the behaviour of the averages that were obtained numerically and experimentally and in particular the two selection mechanisms characteristic of the isotropic and anisotropic situations respectively. For the mean aggregates the Laplacian term can be written locally as $\partial^2/\partial n^2 + \kappa \partial/\partial n$ where $\boldsymbol{n}$ is the direction of the gradient and κ the local front curvature. As suggested by Brenner *et al.* this term, since it is curvature-dependent, can play a role similar to that of surface tension in the stable case. Further developments can be found in Levine & Tu (1992, 1993). However, in spite of the success of this set of equations, several problems remain, as pointed out by Arnéodo *et al.* (1996). There is no clear way to choose the cutoff parameter η introduced in (5.7). More fundamentally, the mean-field theory, in contradiction with reality, assumes the aggregate to be transparent to the diffusing particles.

6 Directional growth

Directional growth is not central to the present chapter, which is devoted to the morphogenetic aspects of viscous fingering, but we will discuss it briefly because of the interest in comparing it with free growth and also because it corresponds to several important industrial processes.

To each of the systems that have been examined here, it is possible to associate a variant in which the field P has a fixed imposed gradient. For viscous fingering, for instance, it corresponds to situations where a meniscus separating the fluid from air is located between two solid surfaces moving away from each other. Most such experiments model industrial processes in which viscous fluids are spread on solid surfaces. Experiments have been done between a plane and a cylinder (Pearson 1960; Savage 1977), between two co-rotating solid rollers (Pitts & Greiler 1961), and in the narrow passages of journal bearings (Taylor 1963). The same phenomenon occurs in the peeling of an adhesive tape (McEwan & Taylor 1966). In all these situations the space between the two solid surfaces forms a kind of Hele-Shaw cell, but of non-constant thickness. In the frame of reference of the boundaries, a low-viscosity fluid (i.e. air) is penetrating a high-viscosity fluid. The front is unstable to viscous fingering but the widening gap induces a pressure gradient that adds a stabilizing factor which has two effects. The fingering instability only appears above a finite instability threshold. Above this threshold the Saffman–Taylor instability is saturated by a nonlinear term so that a front is formed with fingers of finite and constant amplitude. These fingers leave behind them an uneven fluid layer so that in coating processes it is this unstability which limits the velocity at which fluids can be deposited on a solid surface.

From a theoretical point of view these coating instabilities are similar (Hakim *et al.* 1990) to directional solidification (in the limit of small Péclet number). In this latter type of experiment a liquid is pulled at constant velocity across a linear temperature gradient between two ovens at fixed temperatures. The temperatures are chosen so that the material solidifies in this gradient. In this situation the Mullins–Sekerka (1964) instability has a different dynamics (see Jackson & Hunt 1966; Langer 1980; Caroli *et al.* 1992 and references therein). There is now a return force towards a linear front fixed in the laboratory frame of reference. The crystallization front is stable at low velocities and only destabilizes above a critical pulling speed. It then forms a widely studied cellular pattern in which the cells have an amplitude limited by nonlinear terms. This instability is also of practical importance; it limits the pulling speed in the preparation of monocrystals.

Both of these instabilities generate linear fronts with a large number of identical cells. For this reason they have become model systems for the study of the nonlinear behaviour of extended one-dimensional systems. Some results in this area are summarized below.

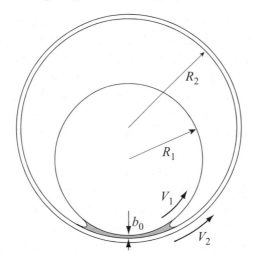

Figure 19. The printer's instability experimental set-up.

6.1 Isotropic case

From the point of view of nonlinear dynamics the directional viscous fingering experiment, having the widest variety of dynamical behaviours, is now known as 'the printer's instability' (Rabaud, Michalland & Couder 1990). It is composed of two horizontal cylinders of different radii (R_1 and R_2) placed one inside the other (figure 19). The inner cylinder has its axis parallel to the outer one, but off-centre so that at the bottom they are separated from each other by a small gap b_0 along one generatrix. A small amount of oil, sufficient to fill the gap, is introduced into the cell. A situation comparable to Saffman–Taylor fingering is obtained when the two cylinders are co-rotating so that the two walls have the same velocity in the gap region. The interesting meniscus here is located where the surfaces move away from each other. It remains stable up to a threshold value of the mean rotation. The instability appears as a supercritical process in which the front becomes a sinusoid of wavelength λ_c. As the velocities increase, the amplitude of the deformation grows and the front becomes a series of parallel fingers separated by thin oil walls (figure 20a). If the thickness gradient is weak the shape of these fingers is close to that of normal Saffman–Taylor fingers. The fundamental difference with the free growth is that the gradient limits the amplitude of these fingers and suppresses the long-range interaction responsible for the screening-off. A given finger has an amplitude limited by nonlinear terms and does not inhibit the growth of its neighbours anymore. In the case of a strict corotation of the two cylinders the pattern is not steady, but chaotic (figure 20b). Continuously some cells split and others are pinched off.

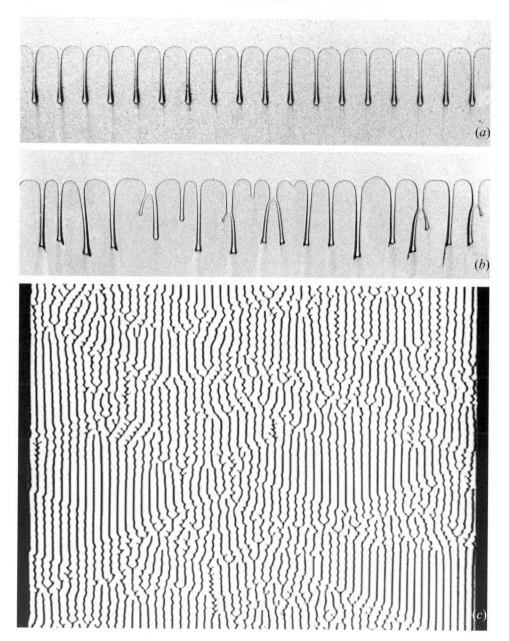

Figure 20. (*a*) Photograph of a detail of the cellular front of the printer's instability in a stable case ($\Omega_2 = 0$). (*b*) Photograph of a detail of a chaotic region of the cellular front. (*c*) Spatio–temporal evolution of the array of cells of the printer's instability in a regime of spatio–temporal intermittency. The time increases from the top to the bottom of the graph (Michalland *et al.* 1993).

The two cylinders can also be rotated with different tangential velocities V_1 and V_2. In the Saffman–Taylor fingering equivalent, this would imply that the two plates slide relative to each other in the direction of the finger motion, a situation never investigated. The nonlinear behaviour of the array of fingers, is that V_1 and V_2 form two independent control parameters. When only one of the cylinders is rotating the fingers are stable. When the second cylinder is set into co-rotation, temporal and spatial defects appear in the linear array of cells. As the co-rotation is increased, there is coexistence of chaotic domains (having constant formation and destruction of cells) with quiescent domains of steady, spatially periodic, cells. It is possible, by image processing, to reconstruct an image, each horizontal line of which gives the position, at a given time, of the walls separating the cells and where the evolution of these positions with time is displayed along the vertical axis (figure 20*c*). The two types of domain are clearly visible. The domains themselves are not stable: they appear, extend, then shrink and vanish and in their interplay they do not have symmetrical roles. The chaotic domains are active and emit waves which travel in the quiescent regions. When the waves travelling along the front reach a slightly disturbed region, a chaotic burst appears. So chaotic domains are related through space and time by the waves. For a given value of the control parameter the statistics of the sizes of the chaotic domains is steady. On increasing the second velocity the chaotic regions increase in size and ultimately the front becomes chaotic everywhere (Michalland, Rabaud & Couder 1993). The system thus becomes chaotic through spatio–temporal intermittency, a process characterized (Chaté & Maneville 1987) in the one-dimensional model phase equations introduced by Kuramoto & Tsuzuki (1976) and Sivashinsky (1985). In the experiment, as in the model, the transition to spatio-temporal chaos is similar to a second-order phase transition and exhibits critical exponents.

Finally, in the situation where the two surfaces move in opposite directions the front is still unstable, but through a subcritical bifurcation in which solitary waves are observed. Far from the threshold the front is formed entirely of left or right travelling cells separated by sources or sinks (Cummins, Fourtune & Rabaud 1993).

6.2 Anisotropic case

In the standard experiments on directional growth of crystals there is formation of a cellular front but it has a very different type of dynamics due to the preferential directions of growth imposed by the crystallographic anisotropy (Trivedi 1984). In most cases, as tip splitting is inhibited, the formation of

BEN AMAR, M., HAKIM, V., MASHAAL, M. & COUDER, Y. 1991 Self-dilating fingers in wedge shaped Hele-Shaw cells. *Phys. Fluids* A **3**, 2039–2042.

BEN JACOB, E., GODBEY, R., GOLDENFELD, N. D., KOPLIK, J., LEVINE, H., MUELLER, T. & SANDER, L. M. 1985 Experimental demonstration of the role of anisotropy in interfacial pattern formation. *Phys. Rev. Lett.* **55**, 1315–1318.

BEN JACOB, E., SCHMUELI, H., SHOCHET, O. & TENENBAUM, A. 1992 Adaptative self-organisation during growth of bacterial colonies. *Physica* A **187**, 378–424.

BENSIMON, D., KADANOFF, L. P., LIANG, S., SHRAIMAN, B. I. & TANG, C. 1986a Viscous flow in two dimensions. *Rev. Mod. Phys.* **58**, 977–999.

BENSIMON, D. & PELCÉ, P. 1986b Tip splitting solutions to a Stefan problem. *Phys. Rev.* A **33**, 4477–4478.

BENSIMON, D., PELCÉ, P. & SHRAIMAN, B. I. 1987 Dynamics of curved fronts and pattern selection. *J. Phys. Paris* **48**, 2081–2087.

BOUISSOU, P., CHIFFAUDEL, A., PERRIN, B. & TABELING, P. 1990 Dendritic side branching forced by an external flow. *Europhys. Lett.* **13**, 89–94.

BRENNER, E., LEVINE, H. & TU, Y. 1991 Mean field theory for the diffusion-limited aggregation in low dimensions. *Phys. Rev. Lett.* **66**, 1978–1981.

BRETHERTON, F. P. 1961 The motion of long bubbles in tubes. *J. Fluid Mech.* **10**, 166–188.

BUKA, A., KERTESZ, A. J. & VICSEK, T. 1986 Transitions of viscous fingering patterns in nematic liquid crystals *Nature* **323**, 424–425.

CAROLI, B., CAROLI, C., MISBAH, C. & ROULET, B. 1987 On velocity selection for needle crystals in fully non-local model of solidification *J. Phys. Paris* **48**, 547–552.

CAROLI, B., CAROLI, C. & ROULET, B. 1987 On the linear stability of needle crystals: evolution of a Zel'dovich localised front deformation. *J. Phys.* **48**, 1423–1437.

CAROLI, B., CAROLI, C. & ROULET, B. 1992 Instabilities of planar solidification fronts. In *Solids Far From Equilibrium* (ed. C. Godrèche). Cambridge University Press.

CHATÉ, H. & MANEVILLE, P. 1987 Transition to turbulence via spatiotemporal intermittency. *Phys. Rev. Lett.* **58**, 112–115.

CHEN, JING-DEN & WILKINSON, D. 1985 Pore scale viscous fingering in porous media. *Phys. Rev. Lett.* **55**, 1892–1818.

CHUOKE, R. L., VAN MEURS, P. & VAN DER POL, C. 1959 The instability of slow immiscible viscous liquid-liquid displacements in permeable media. *Petrol. Trans. AIME* **216**, 188–194.

COMBESCOT, R. & BEN AMAR, M. 1991 Selection of Saffman–Taylor fingers in the sector geometry. *Phys. Rev. Lett.* **67**, 453–456.

COMBESCOT, R. & DOMBRE, T. 1989 Selection in the anomalous Saffman–Taylor fingers induced by a bubble. *Phys. Rev.* A **39**, 3525–3535.

COMBESCOT, R., DOMBRE, T., HAKIM, V., POMEAU, Y. & PUMIR, A. 1986 Shape selection of Saffman–Taylor fingers. *Phys. Rev. Lett.* **56**, 2036–2039.

COMBESCOT, R., DOMBRE, T., HAKIM, V., POMEAU, Y. & PUMIR, A. 1988 Analytic theory of the Saffman–Taylor fingers. *Phys. Rev.* A **37**, 1270–1283.

COUDER, Y., CARDOSO, O., DUPUY, D., TAVERNIER, P. & THOMÉ, H. 1986a Dendritic growth in the Saffman–Taylor experiment. *Europhys. Lett.* **2**, 437–443.

COUDER, Y. GERARD, N. & RABAUD, M. 1986b Narrow fingers in the Saffman–Taylor instability. *Phys. Rev.* A **34**, 5175–5178.

COUDER, Y. 1988 Viscous fingering in a circular geometry. In *Random Fluctuations and Pattern Growth* (ed. H. E. Stanley & N. Ostrowsky). Kluwer.

COUDER, Y., ARGOUL, F., ARNÉODO, A., MAURER, J. & RABAUD, M. 1990 Statistical

Properties of fractal dendrites and anisotropic diffusion-limited aggregates. *Phys. Rev.* A **42**, 3499–3503.

COUDER, Y. 1991 Growth Patterns: from the stable curved fronts to fractal structures. In *Chaos, Order and Patterns* (ed. R. Artuso, P. Cvitanovic & G. Casati). Plenum.

CUMMINS, H. Z., FOURTUNE L. & RABAUD, M. 1993 Successive bifurcations in directional viscous fingering. *Phys. Rev.* E **47**, 1727–1738.

D'ARCY THOMPSON, W. 1917 *On Growth and Form*, vols. 1 and 2. Cambridge University Press (reissued in 1952).

DACCORD, G., NITTMANN J. & STANLEY, H. E. 1986 Radial viscous fingers and diffusion-limited aggregation fractal dimension and growth sites. *Phys. Rev. Lett.* **56**, 336–339.

DORSEY, A. T. & MARTIN, O. 1987 Saffman–Taylor fingers with anisotropic surface tension. *Phys. Rev.* A **35**, 3989–3992.

DOUGHERTY, A. & GOLLUB, J. P. 1988 Steady state dendritic growth in NH_4Br from solution. *Phys. Rev.* A **38**, 3043–3053.

ECKMANN, J. P., MEAKIN, P., PROCACCIA, I. & ZEITAK, R. 1989 Growth and form of noise reduced diffusion-limited aggregation. *Phys. Rev.* A **29**, 3185–3195.

FLESSELES, J. M., SIMON, A. J. & LIBCHABER, A. J. 1991 Dynamics of one-dimensional interfaces: an experimentalist's view. *Adv. Phys.* **40**, 1–51.

FUKIKAWA, H. & MATSUSHITA, M. 1991 Bacterial fractal growth in the concentration field of nutrient. *J. Phys. Soc. Japan* **60**, 88–94.

HAKIM, V., RABAUD, M., THOMÉ, H. & COUDER, Y. 1990 Directional growth in viscous fingering. In *New Trends in Nonlinear Dynamics and Pattern Forming Phenomena* (ed. by P. Coullet & P. Huerre). Plenum.

HOMSY, G. M. 1987 Viscous fingering in porous media. *Ann. Rev. Fluid Mech.* **19**, 271.

HONG, D. C. & LANGER, J. S. 1986 Analytical theory of the selection mechanism in the Saffman–Taylor problem. *Phys. Rev. Lett.* **56**, 2032–2035.

HONG, D. C. & LANGER, J. S. 1987 Pattern selection and tip perturbation in the Saffman–Taylor problem. *Phys. Rev.* A **36**, 2325–2332.

HONJO, H., OHTA, S. & MATSUSHITA, M. 1986 Irregular fractal like crystal growth of ammonium chloride. *J. Phys. Soc. Japan* **55**, 2487–2490.

HOWISON, S. D. 1986 Fingering in Hele-Shaw cells, *J. Fluid Mech.* **167**, 439–453.

HUANG, S. C. & GLICKSMAN, M. E. 1982 Fundamentals of dendritic solidification I. Steady-state tip growth. *Acta Metall.* **29**, 701–715; II. Development of sidebranch structure. *Acta Metall.* **29**, 717–734.

HULL, D. 1999 *Fractology*, Cambridge University Press.

IHLE, T. & MÜLLER KRUMBAAR, H. 1994 Fractal and compact growth morphologies in phase transitions with diffusion transport. *Phys. Rev.* E **49**, 2972–2991.

IVANTSOV, G. P. 1947 Temperature field around spherical, cylindrical and acicular crystal growing in a supercooled melt. Translated from *Dokl. Acad. Nauk.* **58**(4), 567–569.

JACKSON, K. A. & HUNT, J. D. 1966 Transparent compounds that freeze like metals. *Trans. Metal. Soc. AIME* **236**, 1129–1215.

JULIEN, R., KOLB, M. & BOTET, M. 1984 Diffusion limited aggregation with directed and anisotropic diffusion. *J. Phys. Paris* **45**, 395–399.

KADANOFF, L. P. 1985 Simulating hydrodynamics: a pedestrian model. *J. Statist. Phys.* **39**, 267–283.

KESSLER, D., KOPLIK, J. & LEVINE, H. 1986a Steady-state dendritic crystal growth. *Phys. Rev.* A **33**, 3352–3357.

SHRAIMAN B. & BENSIMON, D. 1984 Singularities in non-local interface dynamics. *Phys. Rev.* A **30**, 2840–2842.

SIVACHINSKY, G. I. 1985 Weak turbulence in periodic flows. *Physica* D **17**, 243–255.

TABELING, P., ZOCCHI, G. & LIBCHABER, A. 1987 An experimental study of the Saffman–Taylor instability. *J. Fluid Mech.* **177**, 67–82.

TANVEER, S. 1987a Analytic theory for the selection of a symmetric Saffman–Taylor finger in a Hele-Shaw cell. *Phys. Fluids* **30**, 1589–1605.

TANVEER, S. 1987b Analytic theory of the linear stability of the Saffman–Taylor finger. *Phys. Fluids* **30**, 2318–2329.

TANVEER, S. 1990 Analytic theory for the selection of Saffman–Taylor fingers in the presence of thin film effects. *Proc. R. Soc. Lond.* A **428**, 511–545.

TANVEER, S. 1991 Viscous displacement in a Hele-Shaw cell. In *Asymptotics Beyond all Orders* (ed. H. Segur, S. Tanveer & H. Levine), p. 131. Plenum.

TAYLOR, G. I. 1963 Cavitation of a viscous fluid in narrow passages. *J. Fluid Mech.* **16** 595–619.

THOMÉ, H., RABAUD, M., HAKIM V. & COUDER, Y. 1989 The Saffman–Taylor instability: from the linear to the circular geometry. *Phys. Fluids* A **1**, 224–240.

THOMÉ, H., COMBESCOT, R. & COUDER, Y. 1990 Controlling singularities in the complex plane: experiments in real space. *Phys. Rev.* A **41**, 5739–5742.

TOKUYAMA, M. & KAWASAKI, K. 1984 Fractal dimension for diffusion-limited aggregates. *Phys. Lett.* **100A**, 337–340.

TRIVEDI, R. 1984 Interdendritic spacing, Parts I and II *Metall. Trans.* A **15**, 977–982.

UWAHA, M. & SAITO, Y. 1989 Aggregation growth in a gas of finite density: velocity selection via fractal dimension of diffusion-limited aggregation. *Phys. Rev.* A **40**, 4716–4723.

VAN DAMME, H., ALSAC, E., LAROCHE, C. & GATINEAU, L. 1988 On the relative roles of low surface tension and non-newtonian rheological properties in fractal fingering. *Europhys. Lett.* **5**, 25–30.

VANDEN BROECK, J. M. 1983 Fingers in a Hele-Shaw cell with surface tension. *Phys. Fluids* **26**, 2033–2034.

VICSEK, T. 1984 Pattern formation in diffusion-limited aggregation. *Phys. Rev. Lett.* **53**, 2281–2284.

VICSEK, T. 1989 *Fractal Growth Phenomena*. World Scientific.

WITTEN, T. & SANDER, L. M. 1981 Diffusion-limited aggregation, a kinetic critical phenomenon. *Phys. Rev. Lett.* **47**, 1400–1403.

WITTEN, T. & SANDER, L. M. 1983 Diffusion-limited aggregation. *Phys. Rev.* B **27**, 5686–5697.

ZEL'DOVICH, YA. B., ISTRATOV, A. G., KIDIN, N. I. & LIBROVICH, V. B. 1980 Flame propagation in tubes: hydrodynamics and stability. *Combust. Science Technol.* **24**, 1–13.

ZOCCHI, G., SHAW, B. E., LIBCHABER, A. & KADANOFF, L. P. 1987 Finger narrowing under local perturbations in the Saffman–Taylor problem. *Phys. Rev.* A **36**, 1894–1900.

Laboratoire de Physique Statistique*, Ecole Normale Supérieure,
24 rue Lhomond, 75231 Paris Cedex 05, France
* associé au C.N.R.S. et aux Universités Paris 6 et Paris 7

3

Blood Flow in Arteries and Veins

T. J. PEDLEY

1 Introduction

Material transport from one part of the body to another, in vertebrates or other large animals, involves fluid (liquid or gas) flowing along and across the walls of systems of tubes. The most widely studied tube systems are the mammalian cardiovascular and ventilatory systems. The purposes of studying the mechanics of physiological flows can be summed up under four headings:

(1) pure physiology, or understanding how animals work;
(2) pathophysiology, or understanding how they go wrong, i.e. the origins and development of diseases;
(3) diagnosis, or working out how to infer what has gone wrong from (relatively) simple and non-traumatic measurements, for example of blood pressure;
(4) cure, which in the mechanical context usually means bioengineering, and which here includes vascular and other surgery as well as the design of prosthetic devices.

Different scientists and practitioners will have different motivations, but the basic mechanical principles are clearly the same for all. In the medical context (2, 3 and 4 above) it is worth remarking that disease of the arteries, notably atherosclerosis, causes around 50% of the deaths in modern western society, through heart attacks and strokes, and drastically lowers the quality of life of those elderly patients who suffer reduced blood flow to the legs through constriction or blockage (stenosis) of the femoral (thigh) arteries. Fluid mechanical factors are strongly implicated in the development of atherosclerosis (atherogenesis) as well as its effects, as explained further in §4 below.

For steady flow in a long straight rigid circular tube, every element of a Newtonian viscous fluid flows in a straight line with a constant speed, and the velocity profile is parabolic. The relationship between the driving pressure gradient and the flow rate through the tube is linear, the ratio of the two (the viscous resistance) being inversely proportional to the fourth power of the tube diameter (Poiseuille's Law). Such a smooth, laminar flow will break down into turbulence if the Reynolds number,

$$Re = \rho \bar{u} d / \mu, \tag{1.1}$$

exceeds a critical value of about 2000, where ρ and μ are the fluid density and viscosity, d is the tube diameter and $\bar{u}$ is the average fluid velocity within it ($\bar{u} = 4Q/\pi d^2$, where Q is the volume flow rate).

Unfortunately, Poiseuille flow is not to be found anywhere in the human body. Blood vessels (and airways) are typically short, curved, branched and elastic (figure 1), the flow is not steady and blood is not a Newtonian fluid, though this last is not of major importance in the large blood vessels of concern here. In short tubes, the flow does not become fully developed, and high velocity gradients occur in thin boundary layers near their entrance, which enhances the rate at which energy is dissipated and therefore increases the pressure drop for a given flow rate (Prandtl 1952). In curved tubes, secondary motions are generated, which distort the primary velocity profile (Dean 1927, 1928) and also enhance the energy dissipation (White 1929). Flow in a bifurcation combines the main features of entry flow and curved tube flow (Schroter & Sudlow 1969, and figure 16 below).

The details of unsteady viscous flow are significantly different from those of steady flow, whatever the geometry, when the dimensionless frequency parameter α is not small, where

$$\alpha^2 = \rho \omega d^2 / 4\mu \tag{1.2}$$

and ω is the dominant radian frequency of the unsteadiness (Womersley 1955). Moreover, a time-dependent pressure, such as that generated by a pumping heart, causes the dimensions of an elastic tube to vary with time, and this is responsible for the propagation of pressure waves along the tube (Young 1809). Physiological fluid flows, in fact, are very complicated and it is a major challenge to measure or calculate the full velocity field.

In the context of large arteries, the principal topics that have been subjected to fluid dynamic study are as follows: (a) pulse-wave propagation and reflection, leading to a determination of the load against which the heart must act, the diagnosis of severe constrictions (stenoses), and the design of cardiac assist devices (McDonald 1974; Elzinga & Westerhof 1991);

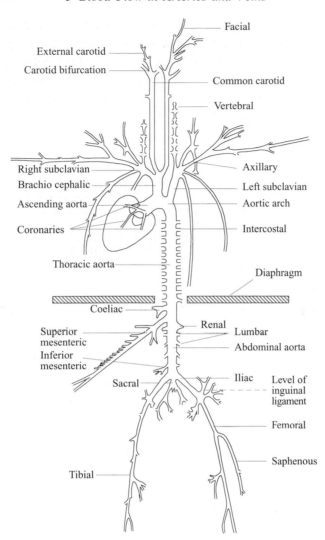

Figure 1. A diagrammatic representation of the major arteries in the mammalian arterial tree (after McDonald 1974).

(*b*) the distribution in space and time of the viscous shear stress exerted by the flowing blood on the artery wall, because of its importance in atherogenesis (Fry 1987; Giddens, Zarrins & Glagov 1993; Friedman 1993); (*c*) mass transport across the arterial endothelium and within the vessel wall, the detailed mechanisms for which are also of fundamental importance in atherogenesis (Yuan, Chien & Weinbaum 1991; Weinbaum & Chien 1993); (*d*) the functioning of natural and artificial heart valves (Lee & Talbot 1979); and (*e*) fluid–structure interaction phenomena associated with flow in a col-

T. J. Pedley

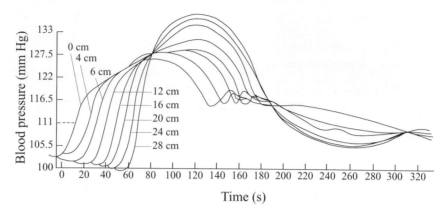

Figure 2. Instantaneous blood pressure records made at a series of sites along the aorta in a dog; 0 cm is at the start of the descending aorta (after Olson, R. M. *J. Appl. Physiol.* vol. 24, 1968, pp. 563–569).

lapsible tube under external compression. Vessel collapse is not normally of relevance in arteries, whose internal pressure considerably exceeds atmospheric and hence external pressure (except possibly some coronary arteries, squeezed by contracting heart muscle), but it does become important if the artery is actively compressed (e.g. by a blood-pressure-measuring cuff) and it can be dominant in veins above the level of the heart because of the gravitational fall of pressure with height.

It is topics (*a*), (*e*) and (*b*) that I shall concentrate on here, in that order: pulse wave propagation (§2), flow in collapsible tubes (especially the jugular vein of the giraffe!) (§3), and arterial wall shear stress (§4). More mathematical detail on topics (*a*) and (*b*) can be found in Pedley (1980), as well as the many papers referred to here; however, the chapter of that book dealing with collapsible tubes has been superseded by more recent work and reference should only be made to the original papers.

2 Pulse propagation

2.1 Observations

When the aortic valve opens at the beginning of the cardiac ejection phase (systole), the pressure rises at the entrance to the aorta and a volume of blood (about 80 ml in man) is ejected. Because the vessel is elastic, the pressure rise locally distends it, rather than setting the blood into motion in all vessels as it would if the system were rigid. The distended segment then contracts (with overshoot because of inertia), accelerating the blood a bit

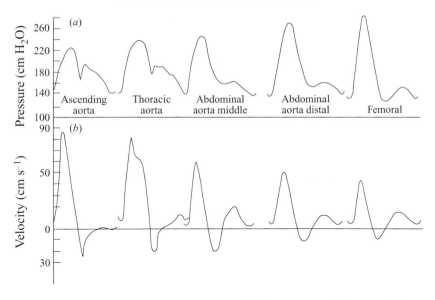

Figure 3. Matched records of (*a*) pressure and (*b*) average velocity at different sites in the arteries of a dog (from McDonald 1974).

further downstream, and so on. In fact, a wave is generated and propagates downstream. The restoring force is provided by the elasticity of the walls, and the inertia by the mass of the blood.

Figure 2 shows the pressure wave form measured at a succession of equally spaced sites down a canine aorta. Propagation of the wave can be clearly seen, the front propagating at a speed of about 5 m s^{-1}, as can two other phenomena: *peaking*, i.e. an increase in amplitude as the wave propagates; and *steepening* of the wave front. The wave form of flow velocity has a different shape from that of the pressure, and its amplitude decreases with distance from the heart (figure 3). The objective of theoretical modelling is to explain the above pressure-wave phenomena – propagation (and its speed), peaking, steepening – and the corresponding velocity-wave phenomena.

2.2 Basic theory

As with all mathematical modelling, it is sensible to start with the simplest possible model, and add complications only when it becomes clear in what respects the simple model is inadequate. We therefore begin by considering an infinitely long straight horizontal elastic tube, with uniform undisturbed cross-sectional area A_0 and uniform external pressure p_e, containing an inviscid incompressible fluid of constant density ρ which is initially at rest.

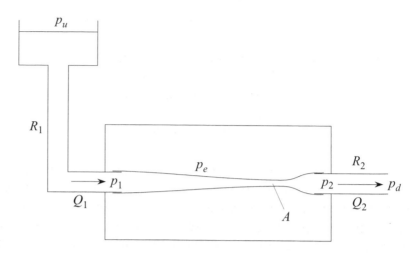

Figure 7. Sketch of a standard laboratory experiment. p_1, Q_1 are pressure and flow rate upstream of the collapsible segment; p_2, Q_2 are pressure and flow rate downstream; p_u, p_d are total pressure far upstream and downstream; p_e is pressure in the chamber surrounding the collapsible segment. R_1 and R_2 represent the rigid pipes up- and downstream, whose resistance can be prescribed.

the lung during a forced expiration, cough or sneeze, because an increase in alveolar air pressure, intended to increase the expiratory flow rate, is also exerted on the outside of the airways. In this case, increasing alveolar pressure above a certain level does not increase the expiratory flow rate, a process known as *flow limitation.* (iv) Urine flow in the urethra during micturition, where flow limitation is again commonplace. These and other examples are discussed in greater detail by Shapiro (1977a, b). Note that in all the cases mentioned the Reynolds number of the flow (Re) is in the hundreds or higher. Moreover the flow is essentially steady, except in (ii): the beating of the heart is not transmitted back along the veins in the limbs, for example, because of the presence of valves in those veins (Caro *et al.* 1978); there are also valves in the jugular veins of giraffes.

3.1.1 Experiments

Many workers have performed laboratory experiments on nominally steady flow through collapsible tubes. In the standard experiment a segment of collapsible (e.g. rubber) tube is mounted at its ends on rigid tubes and contained in a chamber whose pressure, p_e, can be independently controlled; the behaviour of the system depends on two independent pressure differences, e.g. $p_u - p_d$ and $p_e - p_d$, where p_u, p_d are the fluid pressures far up- and downstream (see figure 7). Some early experimental studies sought to characterize the collapsible tube by plotting the pressure difference along it ($\Delta p = p_1 - p_2$) against the flow rate, Q. There was some confusion in the literature because it was not always clear which controlled pressure difference was being varied, as flow rate was varied, and which was held constant. Three different examples, in each of which the shape of the $\Delta p, Q$ curve is quite different, are shown in figure 8(a, b, c), taken from Brecher (1952), Bertram (1986) and Conrad (1969), respectively. The following explanations of the three different curves are taken from Kamm & Pedley (1989).

For the case depicted in figure 8(a), $p_1 - p_2$ is increased while $p_1 - p_e$ is held constant. This can be accomplished either by reducing p_2 with p_1 and p_e fixed, or by simultaneously increasing p_1 and p_e while p_2 is held constant. With either manoeuvre, Q at first increases but above a critical value it levels off and exhibits flow limitation: however much the driving pressure is increased the flow rate remains constant or may even fall (so-called 'negative effort dependence' (Mead *et al.* 1967)), as a result of increasingly severe tube collapse. This version of the experiment is directly relevant to forced expiration from the lung (Elad & Kamm 1989; Lambert 1989) to venous return (Guyton 1962) and to micturition (Griffiths 1971).

Different results are obtained if $p_1 - p_2$ or Q is increased while $p_2 - p_e$ is held constant at some negative value (Fry 1958; Brower & Noordergraaf 1973; Bonis & Ribreau 1978). In this case the tube is collapsed at low flow rates, but starts to open up from the upstream end as Q increases above a critical value, so that the resistance falls and $p_1 - p_2$ ceases to rise; so-called 'pressure-drop limitation' (figure 8b). This experiment is not directly applicable to any particular physiological condition, but it turns out that $p_2 - p_e$ is a natural control parameter for at least one of the types of theoretical model that have been proposed (Shapiro 1977a). Figure 8(b) represents some experimental results of pressure–flow relations for several values of $p_e - p_2$.

In a third type of experiment, $p_1 - p_2$ is held constant while $p_2 - p_e$ is decreased from a large positive value. The tube first behaves as though it were rigid and the flow rate is nearly constant. Then, as $p_2 - p_e$ becomes sufficiently negative to produce partial collapse, the resistance rises and Q begins to fall. This is analogous to what happens in the pulmonary capillaries toward the apex of the lung (Permutt, Bromberger-Barnea & Bane 1963). One notable variation on these experiments is that pioneered by Conrad (1969) who held p_e and the pressure downstream of a flow resistance constant; upstream of the flow resistance the pressure, p_2, varies with Q as does the degree of tube collapse. Thus at high flow rates the tube is distended and its resistance is low, but as the flow rate is reduced below a critical value the tube starts to

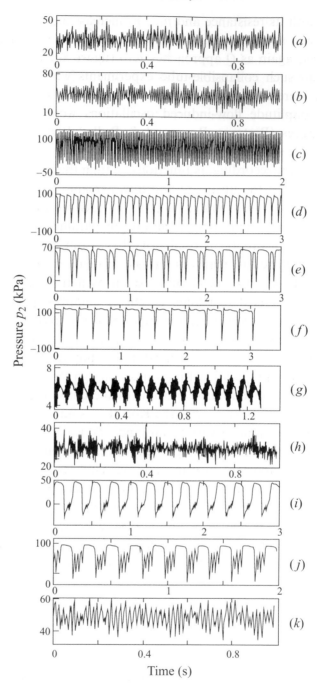

Figure 9. Pressure p_2 at the downstream end of the collapsible segment, plotted against time t during self-excited oscillations for various values of the governing parameters (from Bertram *et al.* 1991).

pressures quoted here are taken relative to atmospheric. One might ask why the pressure generated by the heart needs to be so high, since a higher pressure p_a at the root of the aorta has at least two disadvantages. There is both a greater tendency to oedema in the feet ('swollen ankles'), which has to be countered by anatomical and physiological adaptations such as tight skin and fascia in the legs (an 'anti-gravity suit'; Hargens *et al.* 1987), and greater energy demands on the left ventricle of the heart: if the mean volume flow rate (cardiac output) is Q, the work done by the ventricle per unit time is $W = p_a Q$. Hence a greater muscle mass is required: heart mass is 2.3% of body mass in adult giraffes, compared to about 0.5% in other mammals (Mitchell & Skinner 1993; see also Goetz *et al.* 1960), with a correspondingly greater oxygen requirement.

It might be supposed (and, indeed, has often been proposed, e.g. Badeer & Rietz 1979), that a siphon mechanism could operate in the head and neck of an upright giraffe. Then p_a would not have to be much larger than in man, but the effect of gravity would mean that the internal pressure would be subatmospheric in the upper regions, and that is physiologically undesirable. Moreover, blood vessels are elastic, not rigid, so subatmospheric internal pressure, with approximately atmospheric pressure externally, would result in collapse of the vessels.

Indeed, the giraffe jugular vein is normally partially collapsed, as indicated both by direct observation (Goetz *et al.* 1960) and, especially, by inference from measurements of intravascular pressure, which is positive and increasing with height above the heart (Hargens *et al.* 1987), not negative and decreasing as the gravitational gradient and the siphon concept would lead one to expect (Pedley 1987). The measurements of Hargens *et al.* (1987) show the internal pressure to be about 7 mm Hg at 0.3 m above the heart, rising to 16 mm Hg at 1.2 m above the heart. If the vein were an uncollapsed cylinder, of diameter 2.5 cm, the measured pressure distribution would require an absurdly high flow rate of 27 l s^{-1}, by Poiseuille's Law. The necessary increase in resistance can be achieved only if the vessel is quite severely collapsed. Laboratory studies using highly collapsible tubes (Hicks & Badeer 1989) show internal pressure to be uniformly zero, i.e. atmospheric. That this is approximately true for the human jugular vein has also been well-known for many years (Guyton 1962).

Pedley, Brook & Seymour (1996) gave a thorough discussion of why the siphon concept could not be valid for a laboratory or living system modelled as an inverted U-tube with a collapsible downflow arm. Their steady-flow model was very simplistic, in that blood inertia was totally neglected, but we outline it briefly here, with reference to figure 10. Suppose that the inverted

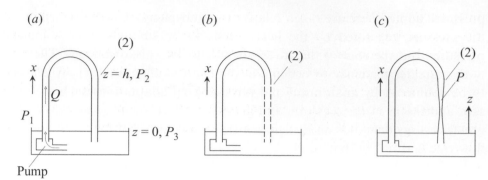

Figure 10. Sketch of possible experiments with an inverted U-tube, supplied with flow rate Q by a pump at $z = 0$ generating pressure P_1. The point (2) is at height $z = h$ and the pressure there is P_2; the pumped fluid is collected in a reservoir at level $z = 0$, where pressure is atmospheric, P_3. (a) Complete rigid tube; P_2 is subatmospheric. (b) Tube is cut off at (2); P_2 is atmospheric. (c) Tube replaced by collapsible drain tubing below (2); P_2 is atmospheric. (See text for explanation.) From Pedley et al. (1996).

U-tube has uniform viscous resistance R per unit length, so that the variation of pressure p with distance x along it is given by

$$\frac{dp}{dx} = -RQ - \rho g \frac{dz}{dx}, \qquad (3.1)$$

where Q is the flow rate, g is the gravitational acceleration, and z is the height above the inlet pump, where $x = z = 0$ and pressure $p = P_1$. If the U-tube is rigid everywhere (figure 10a) and the outlet, where pressure is atmospheric (zero), is at the same level as the inlet, so that $z = 0$ while $x = L + h$, then (3.1) gives

$$P_1 - 0 = (L + h)RQ.$$

This means that some flow can be generated whenever $P_1 > 0$. However, at point (2), where $z = h$ and $x = L$, (3.1) also gives

$$P_1 - P_2 = LRQ + \rho g h, \qquad (3.2)$$

so P_2 is subatmospheric if $RQ < \rho g$. If the downflow arm is replaced by a highly collapsible segment (figure 10c) or is absent altogether (figure 10b) then P_2 must be zero, and (3.2) shows that a positive flow rate is achievable only if $P_1 > \rho g h$. In other words, the heart has to 'pump the blood uphill' if the jugular vein can collapse.

Pedley et al. (1996) also discussed the possible reasons for the difference between the simple laboratory experiment (atmospheric pressure everywhere in the collapsed downflow arm) and the giraffe jugular vein (non-uniform pressure, slightly above atmospheric, rising with height). The main difference is probably non-uniform elastic properties in the vein. However, they also considered the effect of inertia on steady flow, following the one-dimensional model of Shapiro (1977a), and predicted a possible form of flow limitation that had apparently not been considered before. We now turn to the one-dimensional model, retaining fluid inertia as well as permitting non-uniform tube properties.

3.2 One-dimensional models

The traditional basic equations for one-dimensional flow in smoothly-varying elastic tubes are those already introduced, equations (2.1)–(2.3), except that (a) the external pressure $p_e(x)$ and the tube-law function $\tilde{P}(A, x)$ may independently vary with x, and (b) the momentum equation (2.2) is modified by the explicit inclusion of gravity and the addition of a term representing viscous resistance (Shapiro 1977a):

$$\frac{\partial u}{\partial t} + u\frac{\partial u}{\partial x} = -\frac{1}{\rho}\frac{\partial p}{\partial x} - g\frac{dz}{dx} - \frac{R(A, u)uA}{\rho}. \tag{3.3}$$

The convective inertia term does not include the contribution from integrating the non-flat velocity profile across the tube, but this can in general be incorporated into the resistance term RuA; R is assumed to be positive, and increases rapidly as A decreases. The equations (2.1), (2.3) and (3.3) are exactly analogous to those for water flow in a shallow channel with a free surface.

3.2.1 Steady flow

Let us first consider steady flow in a horizontal tube with uniform elastic properties and external pressure. Equation (2.1) gives

$$uA = Q, \tag{3.4}$$

where Q is the constant volume flow rate. If we now combine the remaining equations into a single equation for $\partial A/\partial x$, we obtain

$$\frac{1}{A}\frac{\partial A}{\partial x} = \frac{-RQ}{\rho\left(c^2 - u^2\right)}, \tag{3.5}$$

where c^2 is again the speed of propagation of small-amplitude long waves, as given by (2.5).

Now consider how A will vary with x, starting from an upstream location at which the tube is distended (i.e. the area A_1 is such that $\tilde{P}(A_1)$ is greater than the value, close to zero, at which the cross-section buckles – see figure 4 – so that $d\tilde{P}/dA = \tilde{P}'(A_1)$ is relatively large) and the fluid velocity is less than the wave speed; the flow is said to be *subcritical*. Since RQ is positive, it follows that $\partial A/\partial x$ is negative. The decrease of A with x will then have two consequences: u will increase because uA is constant (equation (3.4)); and c will decrease, because A and $\tilde{P}'(A)$ will both decrease (equation (2.5) and figure 4). Hence $c^2 - u^2$ will decrease and $(1/A)\partial A/\partial x$ will become more negative.

Eventually, if the tube is long enough, one of two things will happen. If inertia is weak, so that $u_1(= Q/A_1)$ is very much less than $c(A_1)$, the area may become small enough for c to increase again while u is still less than c. Equation (3.5) can then be integrated forward in x and the collapse process would be smooth; this process is termed *viscous collapse*. On the other hand, if u_1 is large enough compared with $c(A_1)$, a point will be reached at which $c^2 = u^2$ and $\partial A/\partial x$ is predicted to be infinitely negative. Clearly this is impossible. What it means is that steady flow at the flow rate Q with upstream area A_1 cannot be achieved: either the flow will remain unsteady or the upstream conditions will change so that u does not exceed c anywhere. In the latter case, the flow is said to be *choked*, by analogy with compressible gas flow in a nozzle.

Are there circumstances in which supercritical flow ($u > c$) can occur in a tube which has subcritical flow upstream? Clearly not according to equation (3.5), but it is possible in tubes which are not horizontal and not intrinsically uniform. To allow the potential phenomena to be investigated in general, it is helpful to follow Shapiro (1977a) and recast the tube law (5) to reflect the facts that the undisturbed cross-sectional area may vary with x ($A_0(x)$, say) and that the elastic properties may also vary, i.e.

$$p - p_e(x) = K_p(x)\tilde{P}(\alpha), \tag{3.6}$$

where

$$\alpha = A/A_0(x) \tag{3.7}$$

is the scaled area, and K_p is the 'stiffness' of the tube. This is not quite the most general form of tube law, because the shape of $\tilde{P}(\alpha)$ could also depend on x, but it is sufficiently general to describe all relevant phenomena. We also introduce the *speed index*,

$$S = u/c, \tag{3.8}$$

where (cf. (2.5))

$$c^2 = c_0^2(x)\alpha\tilde{P}'(\alpha) \tag{3.9}$$

and

$$c_0^2(x) = K_p(x)/\rho. \tag{3.10}$$

The speed index is analogous to the Froude number in free-surface channel flow. Combining these equations with (3.3) and (3.4), we obtain, in place of (3.5),

$$(1 - S^2)\frac{\alpha_x}{\alpha} = \frac{S^2}{A_0}A_{0x} - \frac{1}{\rho c^2}\left[p_{ex} + g\rho z_x + K_{px}\tilde{P}(\alpha) + RQ\right], \tag{3.11}$$

where the suffix x represents differentiation. We can also derive the following equation for the variation of the speed index:

$$(1 - S^2)\frac{(S^2)_x}{S^2} = \frac{A_{0x}}{A_0}\left[-2 + (2 - M)S^2\right]$$

$$+ \frac{K_{px}}{\rho c^2}\left[M\tilde{P}(\alpha) - (1 - S^2)\,\alpha\tilde{P}'(\alpha)\right] + \frac{M}{\rho c^2}\left[p_{ex} + g\rho z_x + RQ\right], \tag{3.12}$$

where

$$M(\alpha) = 3 + \frac{\alpha\tilde{P}''(\alpha)}{\tilde{P}'(\alpha)} = \frac{1}{\alpha^2\tilde{P}'}\left(\alpha^3\tilde{P}'\right)'. \tag{3.13}$$

It can be seen that, although the right-hand sides of (3.11) and (3.12) are different in general, that of (3.12) is precisely $-M$ times that of (3.11) when $S = 1$. It follows that it is possible for α_x and $(S^2)_x$ to be non-zero when $S = 1$, as long as the right-hand sides are zero at $S = 1$. Since RQ is always positive, a necessary condition is that at least one of $(p_e + g\rho z)_x$, K_{px} or $-A_{0x}$ should be negative, i.e. the external pressure, the height or the stiffness should decrease with x, or the undisturbed cross-sectional area should increase.

In the case of a given flow rate Q in a vertical collapsible tube, or the jugular vein of an upright giraffe (figure 10c), it is clear that dz/dx is negative ($= -1$), whether or not the other quantities vary. Certainly, in the laboratory experiment with a uniform collapsible tube subjected to a uniform (atmospheric) external pressure, the flow in the highly collapsed tube is supercritical ($S > 1$). In this case equation (3.11) gives

$$(1 - S^2)\frac{\alpha_x}{\alpha} = \frac{g\rho - RQ}{\rho c^2}, \tag{3.14}$$

and a steady state is possible in which the cross-sectional area is given by the gravity–viscous balance $R(\alpha)Q = g\rho$. Note too that the steady state is stable

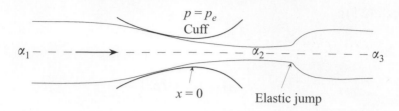

Figure 11. Smooth transition from sub- to supercritical flow in a collapsible tube when a cuff is inflated round it, followed by a transition back to subcritical flow at an elastic jump.

when $S > 1$, in the sense that an increase of α above its equilibrium value leads to negative α_x because R increases as α decreases; the corresponding subcritical ($S < 1$) steady state would be unstable, implying that α_x would not become zero in subcritical flow.

3.2.2 An elastic constriction

It is instructive to investigate the transition from sub- to supercritical flow in the particular example of a cuff being inflated over an initially distended elastic tube (figure 11). We suppose that the tube has uniform intrinsic properties and is horizontal, and that the cuff is sufficiently short for the resistance term to be negligible. The cuff itself is represented by an external pressure,

$$p_e = P_e(1 - x^2/l^2), \qquad |x| < l$$
$$= 0, \qquad\qquad |x| < l \,; \tag{3.15}$$

the only non-zero term on the right-hand side of (3.11) or (3.12) is that involving p_{ex}. The tube law (3.6) is taken (for illustration not realism) to be one that becomes stiffer as it is distended, but does not incorporate the increased stiffness at very small cross-sectional area that is shown in figure 4:

$$\tilde{P}(\alpha) = \tfrac{1}{2}\alpha^2 + \tilde{P}_0, \tag{3.16}$$

so from (3.9)

$$c = \alpha c_0,$$

where c_0 is now a constant. We let the (steady) flow rate be $Q = A_0 c_0 q$, and suppose that the cross-sectional area of the tube is $A_1 = A_0 \alpha_1$. The upstream flow is subcritical if $u_1 (= Q/A_1)$ is less than $c_1 = \alpha_1 c_0$, i.e. if $q < \alpha_1^2$; indeed, in general

$$S = q/\alpha^2. \tag{3.17}$$

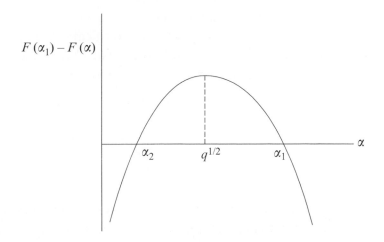

Figure 12. Sketch of the function $F(\alpha)$ defined by equation (3.18*b*).

Substituting (3.17) into (3.11) we obtain the following equation for α:

$$\left(1 - \frac{q^2}{\alpha^4}\right) \alpha \alpha_x = -\frac{p_{ex}}{\rho c_0^2}.$$

This can be integrated, subject to the upstream boundary condition that $\alpha = \alpha_1$ when $p_e = 0$, to give

$$p_e(x) = \rho c_0^2 \left[F(\alpha_1) - F(\alpha)\right], \qquad (3.18a)$$

where

$$F(\alpha) = \frac{1}{2}\left(\alpha^2 + \frac{q^2}{\alpha^2}\right). \qquad (3.18b)$$

The right-hand side of equation (3.18*a*) is sketched in figure 12. Note that the maximum of the curve occurs at $\alpha = q^{1/2}$ (i.e. $S = 1$) so α_1 is the larger of the two zeros. If the maximum value of p_e (P_e from (3.15)) is less than the maximum value on the curve, i.e. $\rho c_0^2[F(\alpha_1) - q]$ which we will call P_{max}, then the point on the curve representing the state at a given value of x will move up from the point α_1 (so α will fall), will fail to reach the maximum so α_x will be zero where $p_{ex} = 0$, and will return to α_1 downstream. However, if P_e is precisely equal to P_{max}, then the point will just reach the maximum and thereafter can move down the other side.

To put it more analytically, $S = 1$ at the same point as $p_{ex} = 0$, so equation (3.11) leaves α_x undetermined. In fact, if the limit is taken carefully, it can be seen that α_x is non-zero at the maximum of the cuff, and the dimensionless area must continue to decrease, towards the other zero,

$\alpha_2 = q/\alpha_1$. Downstream of the cuff the flow is supercritical. As Griffiths (1971) pointed out, a transition from sub- to supercritical flow requires there to be an elastic constriction. It should be noted that changes in A_0 or K_p with x can have a similar effect to changes in p_e, and the above arguments can readily be generalized to such cases.

Two further questions immediately arise.

(1) What happens if P_e is greater than P_{max}?

(2) How does the supercritical downstream flow return to a subcritical state if downstream conditions demand it?

(1) $P_e > P_{max}$ leads to the condition referred to above as choking, in which S^2 is predicted to equal 1 while the right-hand side of (3.11) or (3.12) is still non-zero. Steady flow is impossible. If the cuff were slowly inflated so that P_e increased up to P_{max} and then beyond it, a wave would propagate upstream (this is possible, because $u < c$ in subcritical flow) and change the upstream conditions so that the steady state immediately upstream of the cuff would be consistent with the flow being exactly critical at the point of maximum p_e. If the upstream pressure and hence area were prescribed, so α_1 would be given, then the dimensionless flow rate q would be determined by solving

$$P_e = \rho c_0^2 \left[F(\alpha_1) - q \right],$$

where $F(\alpha)$ is given by (3.18b). Writing $\overline{P}_e = P_e/\rho c_0^2$, we solve the quadratic equation for q to obtain

$$q = \alpha_1 \left(\alpha_1 - \overline{P}_e^{1/2} \right), \tag{3.19}$$

where the negative square root is taken because the upstream flow is subcritical. For a given maximum external pressure P_e and a given upstream area α_1, equation (3.19) provides the maximum (steady) value of the flow rate through the cuff. In other words, the system exhibits *flow limitation*. The above analysis of choking is now generally agreed to give a correct description of flow limitation during forced expiration from the lungs (Dawson & Elliott 1977; Elad, Kamm & Shapiro 1987, 1988) and during male micturition (Griffiths 1971).

There may, of course, be circumstances in which q is prescribed as well as α_1 far upstream of the cuff. In that case the steady state would not be set up, and the upstream propagating wave would be reflected back towards the cuff, and so on: the flow would be permanently unsteady.

(2) Upstream wave propagation is prohibited in supercritical flow, and a smooth transition to subcritical flow downstream of the cuff is impossible.

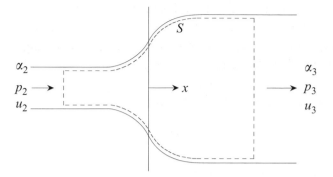

Figure 13. Control surface, S, used to derive the jump conditions for an elastic jump; axes are fixed in the jump.

The only possibility is for there to be an abrupt transition, i.e. an elastic jump (cf. §2.4). We proceed to analyse such a jump.

3.2.3 Elastic jumps

Conditions on either side of an elastic jump are related by conservation of mass and momentum (energy is lost, however, through flow separation and turbulence), and can be derived in the same way as those for a hydraulic jump in free-surface channel flow. In the context of an elastic tube, the jump conditions were first derived by Griffiths (1971). For a jump that is fixed, with supercritical flow upstream (area A_2, pressure p_2 and velocity u_2) and (presumably) subcritical downstream (A_3, p_3, u_3) the (dimensional) jump conditions are (see figure 13):

mass
$$u_2 A_2 = u_3 A_3 \quad (= Q),$$
(3.20)

momentum
$$A_2 \left(p_2 + \rho u_2^2\right) + \int_2^3 p \, dA = A_3 \left(p_3 + \rho u_3^2\right).$$
(3.21)

We now suppose that the jump is short enough for the x-dependence in the tube law (3.6) to be neglected, and we suppose that p is given by the tube law (3.16) even in the jump. From the dimensionless forms of (3.20) and (3.21) we can then deduce that a jump is possible from dimensionless area α_2 to a new area α_3, where

$$\alpha_2 \alpha_3 \left(\alpha_2^2 + \alpha_2 \alpha_3 + \alpha_3^2\right) = 3q^2.$$
(3.22)

It can readily be seen that if the upstream flow is supercritical, as it will be downstream of the cuff in the example of the previous subsection, so that $\alpha_2 < q^{1/2}$ (from (3.17)), then the downstream flow must be subcritical, $\alpha_3 > q^{1/2}$.

Thus conditions downstream of the cuff plus jump will again be subcritical, though with a different area ($\alpha_3 < \alpha_1$) because of the energy loss at the jump. In a general system with a transition to subcritical flow, the far-downstream dimensionless area will not be equal to α_3. This is not normally a problem, because in general there are non-uniformities and in particular viscous losses (RQ in equation (3.11)). The location of the elastic jump is not fixed *a priori* and in such systems it is located in precisely the right place to give the correct conditions at the downstream end (see Elad *et al* 1987, 1988). Such arguments are familiar from shallow-water theory and shock-wave analysis.

In their analysis of the giraffe jugular vein, Pedley *et al.* (1996) came across circumstances in which the prescribed subcritical state at the downstream end could not be achieved for any elastic jump location, leading to a new mechanism of steady flow limitation. However, computations of time-dependent flow from realistic initial conditions, by Brook (1997), showed that the difficulty could be removed by permitting rather small-amplitude oscillations at the location of the jump.

3.3 Modelling laboratory experiments

We return briefly to the experiment depicted in figure 7, in which the collapsible tube has finite length, L. Even if choking is not predicted, the model leading to equation (3.5) must break down near the downstream end, because dA/dx would have to become positive again near $x = L$. Following Cancelli & Pedley (1985), we add two new features to the model, both of which should be important in the region downstream of the narrowest point. One is longitudinal tension T in the tube wall, the simplest model for which causes equation (3.6) to be replaced by

$$p - p_e = K_p \tilde{P}(\alpha) - T \frac{d^2 A}{dx^2} . \tag{3.23}$$

In the highly collapsed region, the tube wall resembles two flattish membranes under tension, with longitudinal curvature roughly proportional to d^2A/dx^2. It was felt that the addition of extra x-derivatives would enable more boundary conditions to be applied, such as $A(L) = A(0) = A_0$. The other new feature is the recognition that flow through a constriction will separate, a process leading to enhanced energy loss and therefore substantially incomplete pressure recovery in the region downstream of the narrowest point. The energy loss near the narrowest point had already been identified as important in lumped-parameter models (Pedley 1980). Crude momentum arguments can be used to suggest that a reasonable, yet still simple, model of the energy loss in steady flow can be achieved by replacing equation (3.3), downstream of the narrowest point, by

$$\chi u \frac{du}{dx} = -\frac{1}{\rho}\frac{dp}{dx},$$
(3.24)

where χ is a non-negative quantity, less than 1; in their (unsteady) calculations Cancelli & Pedley took $\chi = 0.2$. It proves not to be crucial what model for energy loss is assumed, as long as there is some; equation (3.24) is more amenable to analysis than (3.3), say.

The steady flow model described by equations (3.4), (3.23) and (3.24) with

$$\tilde{P}(\alpha) = (1 - \alpha^{-3/2}) \qquad \text{for} \qquad \alpha < 1$$
$$= k(\alpha - 1) \qquad \text{for} \qquad \alpha > 1$$
(3.25)

can be analysed using phase-plane methods. Jensen & Pedley (1989), for example, took $\chi = 1$ for $0 < x < x_s$ (the unknown point of flow separation, taken to be identical with the narrowest point at sufficiently high Reynolds number) and $\chi = \text{constant} \leqslant 1$ for $x_s < x < L$. The principal results can be summarized as follows.

(i) When $\chi = 1$ everywhere, i.e. there is no energy loss in the collapsible tube downstream of the narrowest point, then there exists a critical value of the flow rate Q, dependent on the longitudinal tension T, above which the steady problem has no solution. In other words, the presence of longitudinal tension alone does not abolish choking. This should not have been a surprise; in the same way, surface tension does not abolish critical behaviour in shallow-water channel flow.

(ii) However, whenever there is any downstream energy loss, i.e. $\chi < 1$ for $x_0 < x < L$, then a steady solution exists for all positive values of flow rate Q and tension T. Since some such energy loss is inevitable, it follows that the breakdown of steady flow is not caused by choking, i.e. the non-existence of a steady flow at the chosen parameter values, but must arise through instability of the steady solution.

Jensen (1990) gave a detailed linear, and weakly nonlinear, analysis of the instability of the steady flow. He used the same one-dimensional model but with the time derivatives, $\partial A/\partial t$ and $\partial u/\partial t$, restored. The elasticity equations (3.23) and (3.25) remained unchanged. When appropriately non-dimensionalized, Jensen's model has two principal governing dimensionless parameters, in addition to χ (which was fixed at a value of 0.2 in all numerical computations): $\hat{Q}$, which is proportional to the flow rate Q; and

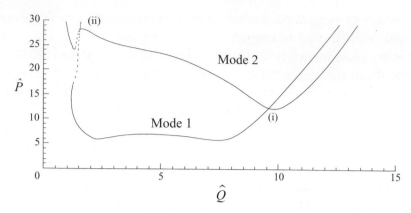

Figure 14. Stability boundaries for the first two modes of instability, plotted on the dimensionless $(\widehat{P}, \widehat{Q})$ plane $(\widehat{P} \propto p_e - p_2; \widehat{Q} \propto q)$, as predicted by the one-dimensional model of Jensen (1990). The Hopf bifurcations are subcritical where the curves are dashed, supercritical elsewhere; (i) and (ii) are mode-crossing points.

$\widehat{P}$, proportional to the transmural pressure, $p_e - p_2$, at the downstream end of the collapsible segment when the flow is steady. Other parameters describe the resistance and inertance of the upstream and downstream rigid segments; these were kept fixed throughout. Figure 14 shows the computed stability boundaries in the $(\widehat{P}, \widehat{Q})$-plane, for the first two instability modes found. The general shape, showing stable steady flow for sufficiently small $\widehat{P}$ at all $\widehat{Q}$ (the tube remaining effectively open), and for sufficiently small $\widehat{Q}$ at all $\widehat{P}$ (the tube being collapsed when $\widehat{P}$ is large enough), is in qualitative agreement with the control diagrams plotted by Bertram *et al.* (1990). So too are the presence of mode-crossing points (i) and (ii) and the existence of several regions in parameter space in which different behaviour of the system is to be expected. Jensen's weakly nonlinear analysis showed that both modes become unstable through supercritical Hopf bifurcations everywhere except for the small segments of the stability boundaries marked as dashed in figure 14, where they are subcritical Hopf bifurcations.

In a subsequent paper Jensen (1992) showed some results of a numerical integration of the fully nonlinear one-dimensional equations, at a few selected points in parameter space, near the upper-left mode-crossing point in figure 14. Some of the computed time series, of $p_2(t) = p(L, t)$ for example, look quite similar to the measurements of Bertram *et al.* (1990) shown in figure 9. It is clear that this one-dimensional model contains much that is relevant to the self-excited oscillations of real collapsible tubes in the laboratory. It would be possible to extend Jensen's (1992) fully nonlinear computations to cover the whole of parameter space, and map out the behaviour in as much (or more) detail as has been done experimentally.

However, this has not been done, and should not, because of the severe oversimplifications in the one-dimensional model, both in the crude solid mechanics of equations (3.23) and (3.25) and in the crude *ad hoc* model of flow separation and the associated energy loss (equation (3.24)), which is especially weak in unsteady flow. What is required is a solution of the unsteady three-dimensional Navier–Stokes equations, coupled to the equations for the unsteady three-dimensional large-deformation theory of highly compliant shells, but numerical codes for the solution of such problems are not yet available in any branch of computational mechanics, and would require resources in excess of any available to us. Instead, the reader is referred to the three-dimensional shell theory with low Reynolds number flow (Heil 1997), or the two-dimensional Navier–Stokes/membrane computations of Luo & Pedley (1996, 1998). The former reveals a mode of instability that cannot be included in a one-dimensional model – snap-through buckling – while the latter, on the other hand, predicts a supercritical Hopf bifurcation that is remarkably similar to the one arising from the one-dimensional model. However, neither has yet been taken far enough to cover much of the parameter space.

4 Arterial wall shear stress and flow separation

4.1 Introduction

The distribution of wall shear stress in arteries is a topic in which the fluid mechanical details are important but are not yet fully understood. The evidence linking wall shear to atherogenesis has been reviewed many times, and I shall not rehearse it in detail here; the articles by Giddens *et al.* (1993), Friedman (1993) and Fry (1987) are to be especially recommended.

The original 'low wall shear stress' hypothesis, by Caro, Fitz-Gerald & Schroter (1971), was based on a visual correlation between sites in large arteries susceptible to atherogenesis and places where the wall shear stress was expected to be low. The sites in question, mostly in the aorta and its major branches, were on the inside of bends, on the outer walls of bifurcations, on the wall opposite a side branch, and in regions of cross-sectional area expansion such as the 'carotid sinus' in the internal carotid artery just downstream of its bifurcation from the common carotid artery.

Simple steady flow experiments in a cast of the aorta confirmed that the wall shear is low and that flow separation may occur at the susceptible sites (Caro *et al* 1971). Intuition about flow separation is often based on

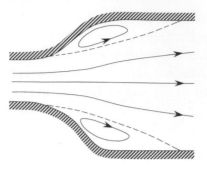

Figure 15. Steady flow separation at an abrupt expansion in a two-dimensional channel, illustrated by a symmetrical expansion, at which a symmetrical flow may be possible.

steady two-dimensional flow. For example, at a realistic arterial value of the Reynolds number ($300 < Re < 2000$), the flow through a simple expansion in a two-dimensional channel (figure 15), experiences a deceleration which is associated with a pressure rise (Bernoulli's theorem). This causes the flow to separate, leaving a closed recirculating eddy beneath; the velocity is low and hence so is the wall shear stress. Moreover, fluid elements remain in the eddy for a long time, so particles can escape only by diffusion and this may help to enhance the uptake of undesirable chemicals into the wall.

The geometry of arteries is three-dimensional, and the flow resembles its two-dimensional counterpart only on planes of symmetry, if they exist. Off the plane of symmetry, however, the flow is quite different in three dimensions. Secondary motions arise in a uniform curved tube, since the faster-moving fluid near the centre is swept towards the outside of the bend, to be replaced at the inside by the slower-moving fluid near the walls. The consequence is a relatively high wall shear near the outside wall, but a much reduced wall shear on the inside part of the wall. Steady flow in a symmetrical bifurcation resembles that in two curved tubes stuck together: the wall shear stress at the outside of the bends (the inner wall of the bifurcation) is enhanced further because new boundary layers with steep velocity gradients develop downstream from the flow divider as it splits the oncoming velocity profile (figure 16). Separation may be seen if the area ratio (of combined daughter tubes to parent tube) exceeds 1, so that there is an overall expansion, or if the curvature of the outer wall is relatively sharp, or if the distribution of flow rate in the daughter tubes is markedly unequal (Zeller, Talukder & Lorenz 1970; Walburn & Stein 1980). When the geometry of the bifurcation is not symmetrical, flow separation is more likely. The extreme case is that of a perpendicular side branch of a parent artery (figure 17). Regions of

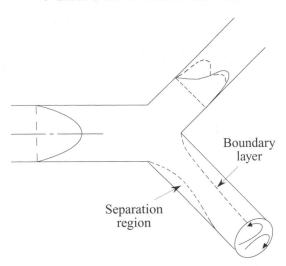

Figure 16. Sketch of flow downstream of a symmetrical bifurcation with Poiseuille flow in the parent tube. Secondary motion, a new boundary layer and a possible separation region are indicated in the lower branch; distorted velocity profiles in the plane of the junction (solid curve) and in the perpendicular plane (broken curve) are sketched in the upper branch. (From Pedley, Schroter & Sudlow 1977.)

separated flow are seen just inside the daughter tube and on the wall of the parent tube opposite the branch. The trajectories of individual fluid elements can be very complex (Karino, Kwong & Goldsmith 1979), as can the shear stress distribution (Lutz *et al* 1977). Another asymmetric geometry that has been widely studied is the carotid bifurcation, where the carotid sinus is particularly prone to separation and to atheroma (Ku 1988).

In all the above cases, the location of flow separation and the wall shear stress distribution may be sensitively dependent on small geometrical details, but within separated eddies the fluid velocity and wall shear stress remain low. In other circumstances that fact cannot be relied on. One example is under the longitudinal 'horseshoe' (or 'hairpin') vortices that stretch away downstream of an abrupt protuberance on an otherwise smooth boundary (Fukushima & Azuma 1982). Other examples arise in non-uniform conduits, particularly when the flow is time-dependent and even (or especially) when it is two-dimensional. The time-dependence arises either from wall motion (Pedley & Stephanoff 1985; Ralph & Pedley 1988, 1990) or from an unsteady driving pressure gradient (Sobey 1985; Tutty & Pedley 1993). In these cases waves (so called 'vorticity waves') are generated in the channel downstream of an indentation and separated eddies occur under the wave crests and above the troughs, at locations quite distant from the indentation itself. Moreover,

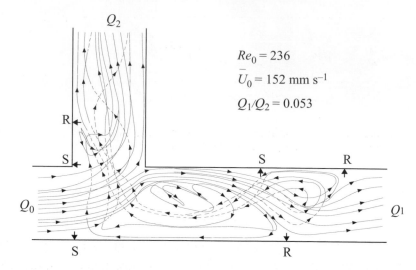

Figure 17. Flow patterns in an asymmetric three-dimensional branch. Projections of three-dimensional particle paths in a right-angled branch when most of the flow goes down the side branch. The subscript '0' denotes parameter values in the parent tube. S and R denote respectively separation and reattachment (From Karino *et al.* 1979.)

the wall shear stress is largest, not smallest, under the recirculating flow in the eddies. Neither of these flow features is consistent with steady-flow intuition, providing a further demonstration of how hard it will be to assess wall shear stress and predict sites of atherogenesis in individual subjects.

As an approach to the analysis of such phenomena, the rest of this chapter introduces the methods of high Reynolds number asymptotics as applied to two-dimensional, steady or unsteady, internal flow. The methods were pioneered by K. Stewartson and F. T. Smith in the 1960s and 1970s in a series of important papers; the following analysis is based in particular on Smith (1976 *a,b*).

4.2 Governing equations and scaling

Consider high Reynolds number flow in a two-dimensional channel with a long slender indentation in one wall (figure 18). The channel width is a, the upstream flow is steady Poiseuille flow with average velocity $\overline{U}$, the incompressible fluid has density ρ and kinematic viscosity v, and the indentation has length λa ($\lambda \gg 1$), height $O(\epsilon a)$ ($\epsilon \ll 1$) and may be time-dependent with time scale $\overline{T}$. We introduce Cartesian coordinates $a(\lambda x, y)$, corresponding velocity components $\overline{U}(u, v/\lambda)$, pressure $\rho \overline{U}^2 p$, and time $\overline{T}t$.

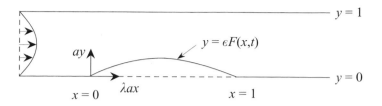

Figure 18. Sketch of a two-dimensional channel with time-dependent indentation $y = \epsilon F(x,t)$ in one wall. Steady Poiseuille flow enters far upstream.

The shape of the wall with the indentation is given by

$$y = \epsilon F(x,t), \tag{4.1}$$

where $F(x,t) = 0$ for $x \leqslant 0$ and $x \geqslant 1$, while the upper wall is at $y = 1$ for all x. The governing continuity and Navier–Stokes equations, in dimensionless form, are

$$u_x + v_y = 0, \tag{4.2}$$

$$St\, u_t + \lambda^{-1} \left(u u_x + v u_y \right) = -\lambda^{-1} p_x + Re^{-1} \left(\lambda^{-2} u_{xx} + u_{yy} \right), \tag{4.3}$$

$$St\, v_t + \lambda^{-1} \left(u v_x + v v_y \right) = -\lambda p_y + Re^{-1} \left(\lambda^{-2} v_{xx} + v_{yy} \right), \tag{4.4}$$

where suffixes represent partial differentiation and the Reynolds and Strouhal numbers are respectively defined by

$$Re = \overline{U} a / v \gg 1, \tag{4.5}$$

$$St = a / \overline{U}\, T \ll 1. \tag{4.6}$$

The boundary conditions are those of no slip and no penetration,

$$u = v = 0 \quad \text{on} \quad y = 1 \tag{4.7}$$

$$u = 0, \quad v = \epsilon \left(u F_x + St F_t \right) \quad \text{on} \quad y = \epsilon F, \tag{4.8}$$

together with the condition of Poiseuille flow upstream,

$$u \sim u_0(y) \equiv 6y(1 - y) \tag{4.9}$$

as $x \to -\infty$. The $u F_x$ term is retained in (4.8) because we shall need the boundary conditions for an inviscid fluid.

The objective of this section is to find the orders of magnitude of ϵ, λ and St, in terms of Re, for which self-consistent and non-trivial solutions can be found. We take the Reynolds number to be sufficiently large that $Re \gg \lambda \gg 1$, so that the viscous terms in (4.3) and (4.4) are negligible

to be solved subject to the boundary conditions

$$U = V = 0 \qquad \text{at} \qquad z = 0, \tag{4.21a}$$

and

$$U \sim 6(z - A) \qquad \text{as} \qquad z \to \infty, \tag{4.21b}$$

where (4.21b) comes from matching to the core flow.

The boundary layer on the indented wall $y = \epsilon F$ can be similarly analysed, if a Prandtl transformation is first made. The new variables here are given by

$$y - \epsilon F = \epsilon \tilde{z}, \qquad u = \epsilon \tilde{U}, \qquad v = \epsilon^2 \tilde{V} + \epsilon^2 \tilde{U} F_x + \lambda St \epsilon F_t, \qquad p = \epsilon^2 \tilde{P}(x, t),$$

where $\tilde{P}$ is already known in terms of P and A (4.13). The resulting boundary-layer problem is the same as that on the other wall, (4.20a, b) and (4.21a, b), except that the variables are replaced by those with a tilde over them, and the outer boundary condition (4.21b) is replaced by

$$\tilde{U} \sim 6(\tilde{z} + F + A) \qquad \text{as} \qquad \tilde{z} \to \infty. \tag{4.21c}$$

Two comments can be made here. First, if the boundary-layer problems were 'classical' ones, with P and $\tilde{P}$ prescribed in advance, then the relevant outer boundary condition in each case would be

$$U_z \to 6 \qquad \text{as} \qquad z \to \infty, \qquad \text{or} \qquad \tilde{U}_{\tilde{z}} \to 6 \qquad \text{as} \qquad \tilde{z} \to \infty. \tag{4.22}$$

Solution of the problem would then determine the displacement functions $(-A)$ and $(F + A)$. Since A appears in both of these, however, and in (4.13), the problems are coupled and 'interactive', and $P, \tilde{P}$ and A will normally be determined only after the boundary-layer problems are solved.

Secondly, A can be determined explicitly without solving the boundary-layer problem in some circumstances, namely when the cross-stream pressure gradient is negligible, $\sigma \ll 1$. In that case $P = \tilde{P}$ from (4.13) so the two boundary-layer problems, with (4.22) in place of (4.21b) or (4.21c), are really identical. Thus, if the boundary-layer problem has a unique solution, it follows that (4.21b) and (4.21c) are also identical, i.e.

$$A = -\tfrac{1}{2} F. \tag{4.23}$$

The displacement of the core-flow streamlines is the average of the displacements of the two walls. Smith (1976a) derived this result for steady flow, extending it to unsteady flow in Smith (1976b). He assumed that the boundary-layer problem would have a unique solution, which was confirmed in his linearized analysis for a very small indentation (see §4.3 below). However, Borgas & Pedley (1990) demonstrated, in a similarity solution for steady geometries in which $F \propto x^{1/3}$, that the solution of the boundary-layer problem is not unique if F is negative and decreases sufficiently rapidly, i.e. the channel expands with a sufficiently adverse pressure gradient. This was confirmed by a fully numerical solution in a similar geometry by Bujurke, Pedley & Tutty (1996) and is related to the well-known non-uniqueness of separated flow in a symmetrically expanding channel (or 'diffuser' – see Reneau, Johnston & Kline 1967).

4.3 Linearized solution for small time-independent indentations

We now proceed to an explicit solution of the time-independent boundary-layer problems for very small indentations with $F \ll 1$. The longitudinal velocity is taken to be a small perturbation to the oncoming shear flow, so that

$$ U = 6z + U', \qquad \tilde{U} = 6\tilde{z} + \tilde{U}', $$

where $U' \ll 6z$, $\tilde{U}' \ll 6\tilde{z}$ everywhere in the boundary layers. All unknowns (U', V, P, A etc.) are taken to be of the same, small, order of magnitude as F. Then the problem near $y = 1$ becomes

$$ U'_x + V_z = 0, \tag{4.24a} $$

$$ 6z U'_x + 6V = -P_x + U'_{zz}, \tag{4.24b} $$

$$ U' = V = 0 \quad \text{on} \quad z = 0, \quad U' \sim -6A \quad \text{as} \quad z \to \infty. \tag{4.24c} $$

The problem can be solved using Fourier transforms in x, for which we define

$$ \widehat{U}(k,z) = \int_{-\infty}^{\infty} e^{-ikx} U'(x,z) \mathrm{d}x, \tag{4.25} $$

and similar transforms $\widehat{V}$, $\widehat{P}$, $\widehat{A}$, $\widehat{F}$ for V, P, A, F. The transforms of (4.24a,b) then give

$$ ik\widehat{U} + \widehat{V}_z = 0, \tag{4.26a} $$

$$ 6ikz\widehat{U} + 6\widehat{V} = -ik\widehat{P} + \widehat{U}_{zz}. \tag{4.26b} $$

Eliminating $\widehat{V}$ and $\widehat{P}$ results in a single equation for $\widehat{U}_z$ which is essentially Airy's equation:

$$ \widehat{U}_{zzz} - 6ikz\widehat{U}_z = 0 ; $$

the solution that satisfies $\widehat{U}_z \to 0$ as $z \to \infty$ is

$$ \widehat{U}_z = \widehat{B}(k)\mathrm{Ai}(\zeta), \tag{4.27} $$

where $\zeta = (6ik)^{1/3}z$ and $\widehat{B}$ is an unknown function. (Note that to ensure that the Airy function $\mathrm{Ai}(\zeta) \to 0$ as $z \to \infty$ for all k on the inversion contour between $-\infty$ and $+\infty$ it is necessary to choose the correct branch of $(6ik)^{1/3}$. This is achieved by means of a branch cut in the k-plane along the positive imaginary axis, so that $-3\pi/2 < \arg k < \pi/2$.)

Applying (4.26b) at $z = 0$ gives

$$ik\widehat{P} = \widehat{U}_{zz}|_{z=0} = (6ik)^{1/3}\widehat{B}(k)\mathrm{Ai'}(0).$$

Moreover

$$\widehat{U}(\infty) = \frac{\widehat{B}(k)}{(6ik)^{1/3}} \int_0^\infty \mathrm{Ai}(\zeta)\mathrm{d}\zeta = \frac{\widehat{B}(k)}{3(6ik)^{1/3}}$$

and (4.24c) tells us that this is equal to $-6\widehat{A}$. Eliminating $\widehat{B}$, we obtain one relationship between $\widehat{P}$ and $\widehat{A}$:

$$\widehat{P} = \frac{-108\mathrm{Ai'}(0)}{(6ik)^{1/3}}\widehat{A}. \tag{4.28}$$

A similar analysis for the boundary layer on the other wall gives

$$\widehat{\widetilde{P}} = \frac{+108\mathrm{Ai'}(0)}{(6ik)^{1/3}}\left(\widehat{F} + \widehat{A}\right), \tag{4.29}$$

while (4.13) gives

$$\widehat{P} - \widehat{\widetilde{P}} = -\sigma k^2 \widehat{A}. \tag{4.30}$$

Thus all unknowns can be determined in terms of $\widehat{F}(k)$. In particular

$$\widehat{A} = -\frac{\widehat{F}}{2 + (\sigma k^2/\gamma)(ik)^{1/3}}, \tag{4.31}$$

where $\gamma = -18 \times 6^{2/3}\mathrm{Ai'}(0) \approx 15.383$. Clearly $\widehat{A} = -\frac{1}{2}\widehat{F}$ when $\sigma = 0$, as expected from (4.23). One further quantity of interest is the wall shear rate; on the wall opposite the indentation, for example, this is equal to

$$-u_y|_{y=1} = 6 + U_z'|_{z=0}.$$

Writing $\tau = U_z'|_{z=0}$, we see from (4.27) that

$$\widehat{\tau} = \widehat{B}(k)\mathrm{Ai}(0) = \frac{\gamma'(ik)^{1/3}\widehat{F}}{2 + (\sigma k^2/\gamma)(ik)^{1/3}},$$

where $\gamma' = 18 \times 6^{1/3}\mathrm{Ai}(0) \approx 11.612$.

The inverses of the Fourier transforms, with and without cross-stream pressure gradient, can be expressed as convolutions and evaluated by standard methods (analytical or numerical) for particular functions $F(x)$. The

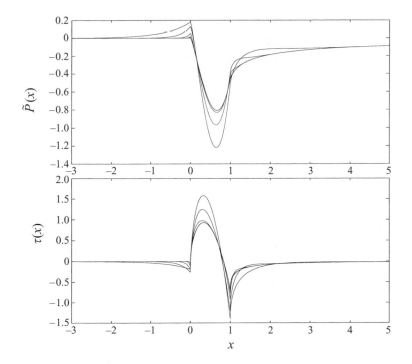

Figure 19. Graphs of perturbations to (*a*) pressure and (*b*) shear rate at the indented wall for various values of σ, according to linear theory (curves computed by J. C. Guneratne). The ordinate scale is arbitrarily set to correspond to $h = 0.5$.

results for $\tilde{P}(x)$ and $\tilde{\tau}(x)$ (pressure and shear rate perturbation on the wall with the indentation) are plotted for several values of σ in figure 19, for the case in which

$$F(x) = hx(1 - x),$$

where $h \ll 1$. Unsurprisingly, the pressure rises and the shear rate falls ahead of the maximum indentation, and vice versa downstream. The reduction in shear rate downstream would tend to cause flow reversal if h were not small. Note also the increasing extent of upstream influence as σ increases. All these features are confirmed by numerical solution of the nonlinear boundary-layer problem. Similar results were presented by Smith (1976*a*).

Linearized solutions can be obtained in a similar way for time-dependent indentations. Suppose that the indentation height, and hence all perturbations to the oncoming flow, are proportional to e^{it} and define the Fourier transform in x by

$$e^{it}\widehat{U}(k, z) = \int_{-\infty}^{\infty} e^{-ikx} U'(x, z, t)\mathrm{d}x$$

etc., instead of (4.25). Then the solution for $\widehat{U}_z$ is again given by (4.27) except that ζ is redefined as

$$\zeta = (6\mathrm{i}k)^{1/3}z + \zeta_0,$$

where

$$\zeta_0 = \mathrm{i}\beta/(6\mathrm{i}k)^{2/3}. \tag{4.32}$$

The final result for $\widehat{A}$ can again be written in the form of (4.31), but now with

$$\gamma = \frac{-6^{5/3}\mathrm{Ai}'(\zeta_0)}{\displaystyle\int_{\zeta_0}^{\infty}\mathrm{Ai}(\zeta)\mathrm{d}\zeta}. \tag{4.33}$$

This is much more difficult to invert than before, because of the infinite number of poles associated with the zeros in the k-plane of the denominator of (4.31) (see Bogdanova & Ryzhov 1983). We discuss the solution no further here, except to say that, when $\beta > 21.7\sigma^{-2/7}$ (in the present notation), one of the poles has a negative imaginary part, which means that non-decaying oscillations will be found downstream of the moving indentation. The downstream oscillations take the form of propagating waves, the physics of which are more easily understood in the context of indentations that oscillate with relatively large amplitude, for then the waves appear in the inviscid core flow.

4.4 Vorticity waves

In this section we suppose that the indentation height scale ϵ is much greater than the boundary-layer thickness δ while remaining small compared with 1. We then perform a weakly nonlinear analysis of the inviscid core flow, keeping $\sigma = O(1)$, so the cross-stream pressure gradient is important in (4.12), and taking $St = \beta\epsilon/\lambda$ (equation (4.17)), where $\beta = O(1)$, so that the time-dependence is important at $O(\epsilon^2)$ in the x-momentum equation (4.3). That equation then gives

$$u_0u_{2x}+u_0'v_2+\beta A_tu_0'+AA_x\left(u_0'^2-u_0u_0''\right)=-\tilde{P}_x-\frac{5\sigma}{6}A_{xxx}\int_0^y u_0^2(y')\mathrm{d}y', \tag{4.34}$$

after use is made of (4.11) and (4.12). The kinematic boundary conditions on v, from (4.7) and (4.8), give

$$v_2(1) = 0$$

and

$$v_2(0) = u_0'(0)\left(FA_x + AF_x + FF_x\right) + \beta F_t.$$

We can therefore evaluate (4.34) at $y = 1$ and $y = 0$. Putting $y = 1$ gives

$$-\tilde{P}_x - \sigma A_{xxx} = -6\beta A_t + 36AA_x, \tag{4.35}$$

where we recall that $u_0'(1) = -u_0'(0) = -6$. Evaluating (4.34) at $y = 0$ gives another equation involving $\tilde{P}$ and A, and if $\tilde{P}$ is eliminated between that equation and (4.35), a single partial differential equation is obtained for $A(x, t)$:

$$A_{xxx} - 2\beta' A_t = \beta' F_t + \frac{36}{\sigma}(FA_x + AF_x + FF_x), \tag{4.36}$$

where $\beta' = 6\beta/\sigma$.

It will be noticed that equation (4.36) is a linear equation for A despite the fact that the derivation is weakly nonlinear. The AA_x term in (4.35) cancels the corresponding term at $y = 0$, because its coefficient, $[u_0'(1)]^2$, has the same value at $y = 0$. If the magnitudes of the unperturbed shear rates at the two walls were different, as would be the case in an annular channel for example, the nonlinear term would not disappear. In the regions where $F = 0$ ($x < 0$ or $x > 1$) the equation would then be the Korteweg–de Vries (KdV) equation, according to which disturbances could propagate upstream as well as downstream (Borgas 1986).

As it is, however, equation (4.36) becomes the linearized KdV equation when $F = 0$, with zero on the right-hand side. This supports only downstream propagating waves: if A is taken proportional to $\exp[\mathrm{i}(\omega t - kx)]$ then

$$\omega = k^3/2\beta'. \tag{4.37}$$

This result also tells us that the group velocity, $\mathrm{d}\omega/\mathrm{d}k$, is three times the phase velocity, ω/k. Thus, if the indentation were started moving from rest, the speed of the wave front would be three times the phase speed of any disturbances. The prediction of such waves in inviscid core flow was first made by Secomb (1979).

Equation (4.36) was solved numerically by Pedley & Stephanoff (1985) for a particular function $F\left[= \frac{1}{2}(1 - \tanh(\alpha(x - x_1)))\right]$, and with a particular value of α, chosen because it was a close approximation to the downstream end of the moving indentation in laboratory experiments performed by those authors, as described below.

Results of the computations for $\sigma = (2.17)(2\beta')^{-2/3}$ (also chosen to match one of the experimental runs), and with x rescaled to $x' = x(2\beta')^{-1/3}$, are shown in figure 20. The development of a wave train can be clearly seen. If a displacement becomes observable when $|A| = 0.02$ (say) then the first wave crest will be noticed at $t \approx 0.2$ ($t = 0$ is the start of the cycle, when the

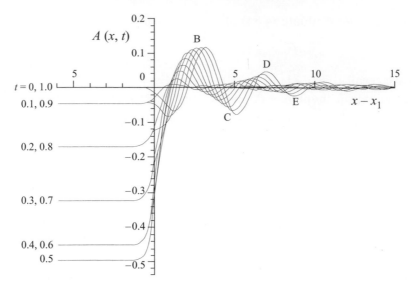

Figure 20. Graphs of A against $(x - x_1)(2\beta')^{-1/3}$ at different times t during a cycle, computed from (4.36), with $F(x, t)$ given by a tanh profile and with $\sigma = 2.17(2\beta')^{-2/3}$. From Pedley & Stephanoff (1985). B, C, and D are referred to on figures 22 and 23.

wall is flat, and $t = 1$ at the end). As time goes on, this crest will grow and propagate downstream, and a new trough will become observable further downstream. Later still, another crest will be seen, and so on: the wave front clearly propagates faster than the crests/troughs.

Equation (4.36) can be derived from the vorticity equation rather than the momentum equation, and that derivation makes it clear that the essential feature of the oncoming flow $u_0(y)$ is a non-zero vorticity gradient u_0''. The waves can therefore be called vorticity waves. The mechanism of their generation is essentially the same as that of Rossby waves in the atmosphere (see Chapter 11, figure 10 and accompanying text), for which the gradient of background vorticity is provided by the variation with latitude of the vertical component of the Earth's angular velocity.

In the experiments of Pedley & Stephanoff (1985), the flow conduit (figure 21) was a long rectangular channel (height $a = 1$ cm; width $= 10$ cm). Far from the entrance a section of one wall was replaced by a square piston, of side nearly 10 cm, wrapped in a stiff elastic membrane which permitted it to be pushed in and out by a motor in a prescribed sinusoidal oscillation. Steady (Poiseuille) flow was introduced upstream and it was confirmed that the flow across the middle of the channel was close to two-dimensional. The flow was visualized using small pearl-essence flakes (Mearlmaid AA) and the centreplane at and downstream of the piston was illuminated and

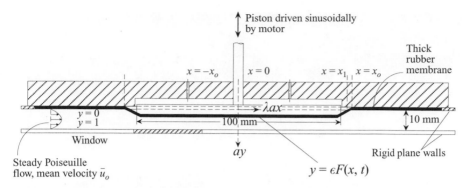

Figure 21. Experimental apparatus for observing flow in an approximately two-dimensional channel with a section of one wall which can be moved in and out; there is steady Poiseuille flow upstream. Dimensionless variables are also marked (see text). From Pedley & Stephanoff (1985).

photographed; one set of photographs, taken at different times during the cycle, is shown in figure 22. A wave train is indeed generated in the core, and the wave front clearly propagates more quickly than the wave crests, in a manner very similar to that shown in figure 20. Confirmation that the theory, albeit small-amplitude and non-viscous, is relevant to the experiments is provided by figure 23, where the positions and times of the first few crests/troughs are plotted, the curves arising from the solution of (4.36) and the points from the experiments, for two particular sets of parameters. Agreement is quite good for the first crest and first trough, deteriorating further downstream where viscous effects are presumably more significant. Agreement became less good for lower frequencies (smaller St).

A feature of the experiments that is not predicted by the simple theory is the recirculating flow in the separated flow zones beneath the wave crests and above the troughs. These do not appear to be regions of very sluggish flow, as in steady two-dimensional separation, but instead consist of a tightly rotating vortex near the downstream end of the separated flow region, sometimes accompanied by another, co-rotating, vortex just upstream. These, and all other, features of the observed flow were faithfully reproduced in solutions of the two-dimensional Navier–Stokes equations by Ralph & Pedley (1988). Unlike the experiments, however, the computations were able to provide values for the wall shear stress distributions, and revealed that the walls next to the tightly rotating vortices under the waves experienced the largest values of shear stress in the whole flow, greater even than the maximum value above the moving indentation. This finding is totally inconsistent with intuition based on steady two-dimensional separated flow and reinforces the

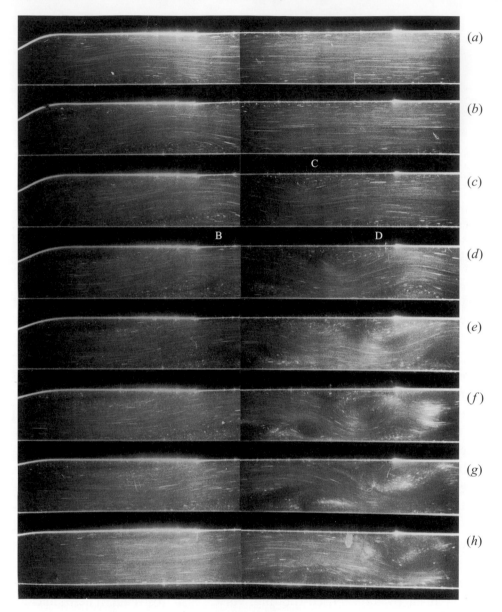

Figure 22. Photographs of the midplane, taken from above, showing the development of the flow field downstream of the oscillating indentation. The mean flow is from left to right, and the governing parameters are $Re = 610$, $St = 0.038$, $\epsilon = 0.38$. The dimensionless times at which the photographs were taken are (a) 0.41; (b) 0.48; (c) 0.55; (d) 0.62; (e) 0.69;(f) 0.76; (g) 0.84; (h) 0.91. The second eddy B is first visible in (a), the third C in (b) and the fourth D in (c). Eddy doubling has occurred in (d) (eddy B) and (e) (eddy C). From Pedley & Stephanoff (1985.)

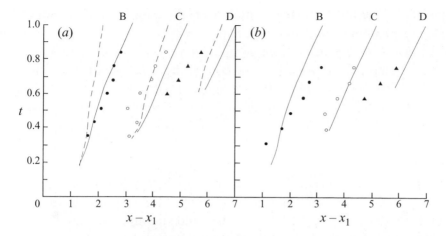

Figure 23. Predicted and measured positions of the wave crests and troughs, plotted against time, for two runs, both with $\epsilon = 0.38$ and (a) $St = 0.057$; (b) $St = 0.037$. The solid curves are theoretical, beginning at the time when the magnitude of the relevant maximum or minimum of A equals 0.02. The points are experimental: $\bullet$, wave B; $\circ$, C; $\triangle$, D. The broken curves in (a) represent the predicted positions when $F(x,t) \propto t^2 f(x)$. From Pedley & Stephanoff (1985).

warning given in §4.1, that intuition cannot be relied on for assessing the distribution of wall shear stress in flow-conveying conduits such as arteries.

5 Conclusion

It is obvious that an analysis of unsteady flow in slightly indented two-dimensional channels cannot be the last word on arterial fluid dynamics. The purpose of the previous section was to show how mathematical analysis of an idealized model can be used to explain certain features of the flow and wall shear stress distribution in complex conduits – here the generation of vorticity waves and the consequent peaks in wall shear stress at locations far from the disturbance that generates them. Non-uniqueness of the steady flow (Borgas & Pedley 1990) also implies sensitivity to initial conditions and hence to the time course of the incoming flow. Whether similar effects are to be found in three dimensions remains to be seen: neither the theory nor the computation has yet been performed. The most popular model problem in three dimensions has been that of flow in a uniform weakly curved tube, where there is not yet a complete theory even for fully-developed steady flow (Berger, Talbot & Yao 1983; Pedley 1980, chap. 4). Computations show the steady flow to be non-unique at high enough Reynolds number (Daskopoulos & Lenhoff 1989). There have in addition been many computational and experimental studies of steady and unsteady flow in model bifurcations, both idealized ones and casts of real arteries (see Pedley 1995). However, as the geometry becomes more complex, one has to ask what fundamental understanding can be gained from computing the flow at a small number of parameter values? If there is indeed great sensitivity to geometrical variation, it follows that the detailed flow pattern will vary considerably between individual subjects. If you want to predict the pattern for a particular subject (following possible surgery, say) then you had better use that subject's geometry in your CFD code. Fortunately, magnetic resonance imaging is currently becoming good enough, and computers big enough, for that soon to be possible, and several research groups are working towards it (e.g. Taylor, Hughes & Zarins 1996). For understanding the fluid mechanics, however, model problems remain the most informative.

Understanding venous blood flow and venous diseases is the main cardio-vascular motivation for studying flow in collapsible tubes, though as stated in §3 there are numerous other physiological applications. Considerable current research is being devoted to analysing the self-excited oscillations that arise in the laboratory (figure 9). A recent development has been the realiza-tion that the internal boundary-layer methods of §4 can be coupled to the equations of membrane mechanics to provide an analytical complement to the computations of Luo & Pedley (1996). Guneratne (1999) has made some exciting discoveries of multiple non-uniqueness of steady flow, and there is much more to be done.

Of course, the cardiovascular system is only one (albeit the most generously funded) of the physiological transport systems on which fluid dynamical research can shed useful light, and large vessels are only a part of it. The microvascular, respiratory, lymphatic, excretory, alimentary, aural and ocular systems have all been and continue to be sources of fascinating fluid dynamical problems. It is hoped that future volumes in this new series can devote chapters to at least some of them.

Acknowledgements

The author is deeply indebted to all his present and former students, post-docs and colleagues whose work forms the basis of most of this chapter. He is also grateful to the EPSRC for the award of a Senior Fellowship, during which the chapter has been written.

References

BADEER, H. S. & RIETZ, R. R. 1979 Vascular hemodynamics: deep-rooted misconceptions and misnomers. *Cardiology* **64**, 197–207.

BERGER, S. A., TALBOT, L. & YAO, L. S. 1983 Flow in curved pipes. *Ann. Rev. Fluid Mech.* **15**, 461–512.

BERTRAM, C. D. 1986 Unstable equilibrium behaviour in collapsible tubes. *J. Biomech.* **19**, 61–69.

BERTRAM, C. D., RAYMOND, C. J. & PEDLEY, T. J. 1990 Mapping of instabilities during flow through collapsed tubes of differing length. *J. Fluids Struct.* **4**, 125–154.

BERTRAM, C. D., RAYMOND, C. J. & PEDLEY, T. J. 1991 Application of nonlinear dynamics concepts to the analysis of self-excited oscillations of a collapsible tube conveying a flow. *J. Fluids Struct.* **5**, 391–426.

BOGDANOVA, E. V. & RYZHOV, O. S. 1983 Free and induced oscillations in Poiseuille flow. *Q. J. Mech. Appl. Maths* **36**, 271–287.

BONIS, M. & RIBREAU, C. 1978 Etude de quelques propriétés de l'écoulement dans une conduite collabable. *La Houille Blanche* **3/4**, 165–173.

BORGAS, M. S. 1986 Waves, singularities and non-uniqueness in channel and pipe flows. PhD Thesis, Cambridge University.

BORGAS, M. S. & PEDLEY, T. J. 1990 Non-uniqueness and bifurcation in annular and planar channel flow. *J. Fluid Mech.* **214**, 229–250.

BRECHER, G. A. 1952 Mechanism of venous flow under different degrees of aspiration. *Am. J. Physiol.* **169**, 423–433.

BROOK, B. S. 1997 The effect of gravity on the haemodynamics of the giraffe jugular vein. PhD Thesis, University of Leeds.

BROWER, R. W. & NOORDERGRAAF, A. 1973 Pressure flow characteristics of collapsible tubes: a reconciliation of seemingly contradictory results. *Ann. Biomed. Engng* **1**, 333–335.

BUJURKE, N. M., PEDLEY, T. J. & TUTTY, O. R. 1996 Comparison of series expansion and finite-difference computations of internal flow separation. *Phil. Trans R. Soc. Lond.* A **354**, 1751–1773.

CANCELLI, C. & PEDLEY, T. J. 1985 A separated flow model for collapsible tube oscillations. *J. Fluid Mech.* **157**, 375–404.

CARO, C. G., FITZ-GERALD, J. M. & SCHROTER, R. C. 1971 Atheroma and arterial wall shear: observation, correlation and proposal of a shear dependent mass transfer mechanism for atherogenesis. *Proc. R. Soc. Lond.* B **177**, 109–159.

CARO, C. G., PEDLEY, T. J., SCHROTER, R. C. & SEED, W. A. 1978 *The Mechanics of the Circulation*. Oxford University Press.

CONRAD, W. A. 1969 Pressure-flow relationships in collapsible tubes. *IEEE Trans. Bio-Med. Engng* **BME-16**, 284–295.

DANAKY, D. T. & RONAN, J. A. 1974 Cervical venous hums in patients on chronic hemodialysis. *New Engl. J. Med.* **291**, 237–239.

DASKOPOULOS, P. & LENHOFF, A. M. 1989 Flow in curved ducts: bifurcation structure for stationary ducts. *J. Fluid Mech.* **203**, 125–148.

DAWSON, S. V. & ELLIOTT, E. A. 1977 Wave-speed limitation on expiratory flow – a unifying concept. *J. Appl. Physiol.* **43**, 498–515.

DEAN, W. R. 1927 Note on the motion of fluid in a curved pipe. *Phil. Mag.* (7) **4**, 208–223.

DEAN, W. R. 1928 The streamline motion of fluid in a curved pipe. *Phil. Mag.* (7) **5**, 673–695.

ELAD, D. & KAMM, R. D. 1989 Parametric evaluation of forced expiration using a numerical model. *Trans. ASME: J. Biomech. Engng* **111**, 192–199.

ELAD, D., KAMM, R. D. & SHAPIRO, A. H. 1987 Choking phenomena in a lung-like model. *Trans. ASME: J. Biomech. Engng* **109**, 1–9.

ELAD, D., KAMM, R. D. & SHAPIRO, A. H. 1988 Mathematical simulation of forced expiration. *J. Appl. Physiol.* **65**, 14–25.

ELZINGA, G. & WESTERHOF, N. 1991 Matching between ventricle and arterial load: an evolutionary process. *Circulation Res.* **68**, 1495–1500.

FRIEDMAN, M. H. 1993 Atherosclerosis research using vascular flow models: from 2-D branches to compliant replicas. *Trans. ASME: J. Biomech. Engng* **115**, 595–601.

FRY, D. L. 1958 Theoretical considerations of the bronchial pressure-flow-volume relationships with particular reference to the maximum expiratory flow volume curve. *Phys. Med. Biol.* **3**, 174–194.

FRY, D. L. 1987 Mass transport, atherogenesis and risk. *Arteriosclerosis* **7**, 88–100.

FUKUSHIMA, T. & AZUMA, T. 1982 The horseshoe vortex: a secondary flow generated in arteries with stenosis, bifurcations and branchings. *Biorheology* **19**, 143–154.

GIDDENS, D. P., ZARINS, C. K. & GLAGOV, S. 1993 The role of fluid dynamics in the localisation and detection of atherosclerosis. *Trans. ASME: J. Biomech. Engng* **115**, 588–594.

GOETZ, R. H. & KEEN, E. N. 1957 Some aspects of the cardiovascular system in the giraffe. *Angiology* **8**, 542–564.

GOETZ, R. H., WARREN, J. V., GAVER, O. H., PATTERSON, J. L., DOYLE, J. T., KEEN, E. N. & MCGREGOR, M. 1960 Circulation of the giraffe. *Circulation Res.* **8**, 1049–1058.

GRIFFITHS, D. J. 1971 Hydrodynamics of male micturition I. Theory of steady flow through elastic-walled tubes. *Med. Biol. Engng* **9**, 581–588.

GUNERATNE, J. C. 1999 High Reynolds number flow in collapsible channels. PhD Thesis, Cambridge University.

GUYTON, A. C. 1962 Venous return. In *Handbook of Physiology, Section 2, Circulation*, Vol. II (ed. W. F. Hamilton & P. Dow). American Physiological Society.

HARGENS, A. R., MILLARD, R. W., PETTERSSON, K. & JOHANSEN, K. 1987 Gravitational haemodynamics and oedema prevention in the giraffe. *Nature* **329**, 59–60.

HEIL, M. 1997 Stokes flow in collapsible tubes: computation and experiment. *J. Fluid Mech.* **353**, 285–312.

HICKS, J. W. & BADEER, H. S. 1989 Siphon mechanism in collapsible tubes: application to circulation of the giraffe head. *Am. J. Physiol.* **256**, R567–R571.

JAN, D. L., KAMM, R. D. & SHAPIRO, A. H. 1983 Filling of partially collapsed compliant tubes. *Trans. ASME: J. Biomech. Engng* **105**, 12–19.

JENSEN, O. E. 1990 Instabilities of flow in a collapsed tube. *J. Fluid Mech.* **220**, 623–659.

JENSEN, O. E. 1992 Chaotic oscillations in a simple collapsible-tube model. *Trans. ASME: J. Biomech. Engng* **114**, 55–59.

JENSEN, O. E. & PEDLEY, T. J. 1989 The existence of steady flow in a collapsed tube. *J. Fluid Mech.* **206**, 339–374.

KAMM, R. D. & PEDLEY, T. J. 1989 Flow in collapsible tubes: a brief review. *Trans. ASME: J. Biomech. Engng* **111**, 177–179.

KAMM, R. D. & SHAPIRO, A. H. 1979 Unsteady flow in a collapsible tube subjected to external pressure or body forces. *J. Fluid Mech.* **95**, 1–78.

KARINO, T., KWONG, H. H. M. & GOLDSMITH, H. L. 1979 Particle flow behaviour in models of branching vessels. I. Vortices in 90° T-junctions. *Biorheology* **16**, 231–248.

KORTEWEG, D. J. 1878 Über die Fortpflanzungesgeschwindigkeit des Schalles in elastischen Rohren. *Ann. Phys. Chem.* (3) **5**, 525–542.

KU, D. N. 1988 A review of carotid duplex scanning. *Echocardiography* **5**, 53–69.

LAMBERT, R. K. 1989 A new computational model for expiratory flow from non-homogeneous human lungs. *Trans. ASME: J. Biomech. Engng* **111**, 200–205.

LEE, C. S. F. & TALBOT, L. 1979 A fluid-mechanical study of the closure of heart valves. *J. Fluid Mech.* **91**, 41–63.

LIGHTHILL, J. 1975 *Mathematical Biofluiddynamics.* SIAM, Philadelphia.

LIGHTHILL, J. 1978 *Waves in Fluids.* Cambridge University Press.

LUO, X.-Y. & PEDLEY, T. J. 1996 A numerical simulation of unsteady flow in a 2-D collapsible channel. *J. Fluid Mech.* **314**, 191–225 (and corrigendum **324**, 408–409.)

LUO, X.-Y. & PEDLEY, T. J. 1998 The effects of wall inertia on flow in a 2-D collapsible channel. *J. Fluid Mech.* **363**, 253–280.

LUTZ, R. J., CANNON, J. N., BISCHOFF, K. B. & DEDRICK, R. L. 1977 Wall shear stress distribution in a model canine artery during steady flow. *Circulation Res.* **41**, 391–399.

MCDONALD, D. A. 1974 *Blood Flow in Arteries*, 2nd Edn. Edward Arnold.

MEAD, J., TURNER, J. M., MACKLEM, P. T. & LITTLE, J. 1967 Significance of the relationship between lung recoil and maximum expiratory flow. *J. Appl. Physiol.* **22**, 95–108.

MITCHELL, G. & SKINNER, J. D. 1993 How giraffe adapt to their extraordinary shape. *Trans. R. Soc. S. Africa* **48**, 207–218.

MOENS, A. I. 1878 *Die Pulskurve.* Brill, Leiden.

PEDLEY, T. J. 1980 *The Fluid Mechanics of Large Blood Vessels.* Cambridge University Press.

PEDLEY, T. J. 1987 How giraffes prevent oedema. *Nature* **329**, 13–14.

PEDLEY, T. J. 1995 High Reynolds number flow in tubes of complex geometry with application to wall shear stress in arteries. In: *Biological Fluid Dynamics* (ed. C. P. Ellington & T. J. Pedley). Soc. Exp. Biol. Symposium 49, pp. 219–241.

PEDLEY, T. J., BROOK, B. S. & SEYMOUR, R. S. 1996 Blood pressure and flow rate in the giraffe jugular vein. *Phil. Trans. R. Soc. Lond.* B **351**, 855–866.

PEDLEY, T. J. & LUO, X.-Y. 1998 Models of flow and oscillations in collapsible tubes. *Theor. Comput. Fluid Dyn.* **10**, 277–294

PEDLEY, T. J., SCHROTER, R. C. & SUDLOW, M. F. 1977 Gas flow and mixing in the airways. In *Bioengineering Aspects of the Lung* (ed. J. B. West), chap 3, pp. 163–265. Marcel Dekker.

PEDLEY, T. J. & STEPHANOFF, K. D. 1985 Flow along a channel with a time-dependent indentation in one wall: the generation of vorticity waves. *J. Fluid Mech.* **160**, 337–367.

PERMUTT, S., BROMBERGER-BARNEA, B. & BANE, H. N. 1963 Hemodynamics of collapsible vessels with tone. The vascular waterfall. *J. Appl. Physiol.* **18**, 924–932.

PRANDTL, L. 1952 *The Essentials of Fluid Mechanics*, (English translation of 3rd Edn). Blackie & Son.

RALPH, M. E. & PEDLEY, T. J. 1988 Flow in a channel with a moving indentation. *J. Fluid Mech.* **190**, 87–112.

RALPH, M. E. & PEDLEY, T. J. 1990 Flow in a channel with a time-dependent indentation in one wall. *Trans. ASME: J. Fluids Engng* **112**, 468–475.

RENEAU, L. R., JOHNSTON, J. P. & KLINE, S. J. 1967 Performance and design of straight, two-dimensional diffusers. *Trans. ASME: J. Basic Engng* **89**, 141–150.

SCHROTER, R. C. & SUDLOW, M. F. 1969 Flow patterns in models of the human bronchial airways. *Respir. Physiol.* **7**, 341–355.

SECOMB, T. W. 1979 Flows in tubes and channels with indented and moving walls. PhD Thesis, Cambridge University.

SHAPIRO, A. H. 1977a Steady flow in collapsible tubes. *Trans. ASME: J. Biomech. Engng* **99**, 126–147.

SHAPIRO, A. H. 1977b Physiologic and medical aspects of flow in collapsible tubes. *Proc. 6th Can. Congr. Appl. Mech.*, pp. 883–906.

SMITH, F. T. 1976a Flow through constricted or dilated pipes and channels: Part I. *Q. J. Mech. Appl. Maths* **29**, 343–364.

SMITH, F. T. 1976b Flow through constricted or dilated pipes and channels: Part II. *Q. J. Mech. Appl. Maths* **29**, 365–376.

SOBEY, I. J. 1985 Observation of waves during oscillatory channel flow. *J. Fluid Mech.* **151**, 395–426.

TAYLOR, C. A., HUGHES, T. J. R. & ZARINS, C. K. 1996 Computational investigations in vascular disease. *Computers in Physics* **10**, 224–232.

TUTTY, O. R. & PEDLEY, T. J. 1993 Oscillatory flow in a stepped channel. *J. Fluid Mech.* **247**, 179–204.

WALBURN, F. J. & STEIN, P. D. 1980 Flow in a symmetrically branched tube simulating the aortic bifurcation: the effects of unevenly distributed flow. *Ann. Biomed. Engng* **8**, 159–173.

WEINBAUM, S. & CHIEN, S. 1993 Lipid transport aspects of atherogenesis. *Trans. ASME: J. Biomech. Engng* **115**, 602–610.

WHITE, C. M. 1929 Streamline flow through curved pipes. *Proc. R. Soc. Lond.* A **123**, 645–663.

WOMERSLEY, J. R. 1955 Method for calculation of velocity, rate of flow and viscous drag in arteries when the pressure gradient is known. *J. Physiol. Lond.* **127**, 553–563.

WOMERSLEY, J. R. 1957 The mathematical analysis of the arterial circulation in a state of oscillatory motion. *Wright Air Development Center, Tech. Rep.* WADC-TR 56-614.

YOUNG, T. 1809 On the functions of the heart and arteries. *Phil. Trans. R. Soc. Lond.* **99**, 1–31.

YUAN, F., CHIEN, S. & WEINBAUM, S. 1991 A new view of convective-diffusive transport processes in the arterial intima. *Trans. ASME: J. Biomech. Engng* **113**, 314–329.

ZELLER, H., TALUKDER, N. & LORENZ, J. 1970 Model studies of pulsating flow in arterial branches and wave propagation in blood vessels. In *Fluid Dynamics of Blood Circulation and Respiratory Flow* AGARD Conf. Proc. 65.

Department of Applied Mathematics and Theoretical Physics,
University of Cambridge, Silver Street, Cambridge CB3 9EW, UK

4

Open Shear Flow Instabilities

PATRICK HUERRE

1 Introduction

Shear flows encompass a wide variety of configurations commonly encountered in the engineering and natural sciences: mixing layers generated by bringing into contact two streams of different velocity, wakes behind bluff bodies, jets, boundary layers near solid surfaces, etc. Such flows typically exhibit strong shear in the cross-stream direction and are spatially developing along the stream from a well-defined spatial origin, say the trailing edge of a splitter plate, an obstacle, a nozzle exit plane, the leading edge of a plate, etc. The scope of this presentation is limited to *open* shear flows, i.e. situations where fluid particles enter and leave the domain of interest in finite time without being recycled, in contrast to closed flows in a finite box.

Open shear flows are known to be prone to instabilities: temporally or spatially growing waves are generated which travel along the stream and lead to the formation of unsteady vortical structures. According to experimental observations, some spatially developing shear flows, e.g. co-flowing mixing layers, homogeneous jets, wakes at Reynolds numbers below the onset of the Kármán vortex street, boundary layers on a flat plate, behave as *noise amplifiers* which are sensitive to external noise. Others, e.g. mixing layers with a sufficiently large counterflow, hot or swirling jets, wakes at Reynolds numbers above the onset of the Kármán vortex street, behave as *flow oscillators* with a well-defined frequency that is insensitive to low noise levels. The goal of the chapter is to present a general framework based on instability theory which *a priori* establishes a clear distinction between the extrinsic dynamics of noise amplifiers and the intrinsic dynamics of flow oscillators. In other words, we wish to predict whether a given shear flow displays one or the other type of dynamics solely from the properties of the instability waves it can support. Such an issue is not only of fundamental

159

interest in order to understand the various processes leading from laminar flow to turbulence, it is also of crucial importance if one is to devise efficient control strategies well suited to the dynamical nature of the flow under consideration.

The theory to be presented below relies on the fundamental distinction between convective and absolute instability introduced in the context of plasma physics by Briggs (1964) and Bers (1983). A parallel shear flow of given velocity profile is said to be *convectively unstable* if the growing wavepacket produced in response to an impulsive source localized in space and time is advected away. It is *absolutely unstable* if the growing wavepacket expands around the source to contaminate the entire medium. In the case of parallel flows that are invariant under Galilean transformations, this distinction appears at first sight to be preposterous: a simple change of reference frame transforms a flow from convectively unstable to absolutely unstable and vice versa, and the 'laboratory frame' is not properly defined. However, when Galilean invariance is broken, e.g. in spatially developing flows, in flows with a definite origin, or in flows forced at a specific streamwise station, the laboratory frame is singled out and it is precisely in these instances that the distinction between convective and absolute instability becomes of interest. It should be emphasized that, in order for these concepts to be relevant, one must enforce a *scale-separation* assumption: the flow under consideration must be slowly evolving along the stream over a typical instability wavelength. This strong hypothesis is made throughout the ensuing theoretical developments in order to recover the locally parallel flow instability properties as a leading-order approximation at each streamwise station.

Under such conditions, whenever a flow is convectively unstable everywhere, it is shown to behave as a noise amplifier and to be preferentially described in terms of spatially growing or decaying waves that evolve according to the known *local dispersion relation* pertaining to the instability properties at each streamwise station. When a flow is absolutely unstable within some streamwise interval it may support time-periodic intrinsic oscillations. However, absolute instability is only a necessary condition for the appearance of synchronized oscillations.

The next step in the formulation involves the theoretical description of the resulting self-sustained time-periodic state in terms of a so-called *global mode*, i.e. an extended wavepacket that lives in the underlying spatially developing flow and beats at the same global frequency at all streamwise locations. The global frequency typically satisfies an eigenvalue problem in the streamwise direction. The eigenfunction distribution along the stream is composed of spatially amplified and decaying waves that have to be suitably chosen to comply with specific upstream and downstream boundary conditions. Here again, the scale separation assumption is invoked to construct a global mode in terms of known local instability properties. Global frequency selection criteria are then derived, which, in many instances, depend on instability properties at a single streamwise location. This station effectively acts as a wavemaker which imposes its frequency on the entire flow. Provided that the scale-separation assumption is retained, the same methodology may in principle be applied in the linear and nonlinear régimes. In the latter instance, the basic ingredient becomes a *local nonlinear dispersion relation* describing the finite-amplitude states at each streamwise station, from which nonlinear global modes may be constructed.

This chapter cannot do justice to the many interesting applications of absolute/convective instability concepts to a variety of shear flows, as reviewed by Huerre & Monkewitz (1990) and Monkewitz (1990). A recent extended account of linear and nonlinear shear flow instabilities is presented in Huerre & Rossi (1998). We have deliberately kept to a minimum the presentation of classical linear instability results in parallel shear flows. For further background information on instability theory viewed from other perspectives, the reader is invited to consult the books by Lin (1955), Drazin & Reid (1981), Swinney & Gollub (1981) and Godrèche & Manneville (1998), the review articles by Maslowe (1981, 1986), Orszag & Patera (1981), Ho & Huerre (1984), Bayly, Orszag & Herbert (1988) and Herbert (1988). As emphasized by Jimenez in Chapter 5, instability processes play an important part not only in laminar–turbulent transition, but also in the dynamics of fully turbulent flows. Finally, it should be noted that many of the theoretical notions to be presented in the following sections have been introduced concurrently in the radically distinct context of astrophysical dynamos and sun spot cycles, as reviewed by Soward (2001).

2 Open shear flows: amplifiers versus oscillators

The objective of this section is to present supporting evidence for the assertion that open shear flows may be classified as either *amplifiers* with *extrinsic dynamics* or as *oscillators* with *intrinsic dynamics*. Two specific spatially developing open shear flows are selected for consideration: the mixing layer and the wake behind a circular cylinder. Throughout, x, y and z respectively denote the streamwise, cross-stream and spanwise coordinates with unit vectors e_x, e_y and e_z, and t is the time variable.

2.1 The mixing layer as a prototype noise amplifier

A mixing layer is formed by bringing into contact two uniform co-flowing streams of the same fluid but at different velocities U_1 and $U_2 < U_1$ at the trailing edge of a splitter plate, as sketched in figure 1(a). Viscous diffusion produces a shear or mixing region between the two flows characterized by a streamwise velocity $U(y;x)$ which exhibits much steeper gradients in the y-direction than in the x-direction. At a given streamwise station x, the velocity profile $U(y;x)$ is inflectional. According to classical instability theory (Drazin & Reid 1981), this feature is responsible for the onset of a powerful inviscid instability resulting in the formation of quasi-two-dimensional spanwise Kelvin–Helmholtz vortices (figure 1b) which are convected at the average velocity $\overline{U} = (U_1 + U_2)/2$ and grow, in the co-moving frame, proportionally to the net shear $\Delta U = U_1 - U_2$ (Brown & Roshko 1974; Browand & Weidman 1976, Ho & Huang 1982). As vortices roll up and are advected downstream, they become susceptible to a secondary pairing instability which induces their merging to form larger vortical structures. This recurrent pairing process is the dominant mechanism by which the local shear layer thickness increases with downstream distance. As a result of these nonlinear interactions the local wavelength between spanwise vortices approximately scales with the local shear layer thickness. The quasi-two-dimensional structures are also subjected to secondary three-dimensional instabilities which lead to the formation of streamwise vortices (Bernal & Roshko 1986; Lasheras & Choi 1988). The reader is referred to Ho & Huerre (1984) for a survey of mixing layer dynamics from both an experimental and theoretical point of view.

It is convenient to characterize the growth of the vortical layer by the local vorticity thickness $\delta_\omega(x)$ geometrically defined in figure 1(a) and such that

$$\delta_\omega(x) = \frac{\Delta U}{(\mathrm{d}U/\mathrm{d}y)_{\max}}. \tag{2.1}$$

Typical variations of $\delta_\omega(x)$ are depicted in figure 3(b, d). Two non-dimensional control parameters may be constructed: the Reynolds number Re and velocity ratio R, respectively defined as

$$Re = \frac{\Delta U\, \delta_\omega(0)}{v}, \qquad R = \frac{\Delta U}{2\overline{U}}. \tag{2.2a, b}$$

The velocity ratio R quantifies the relative magnitude of the destabilizing shear ΔU with respect to the average advection velocity $\overline{U}$. When $R = 0$, as in figure 2(a), there is no net shear and the flow becomes uniform far downstream. A coflow mixing layer is generated in the range $0 < R < 1$

(a)

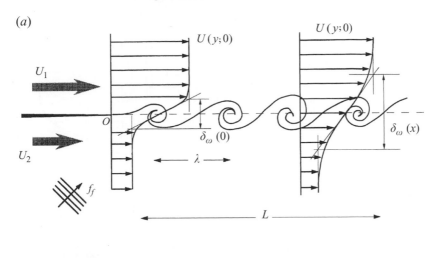

(b)

Figure 1. (a) Spatially developing coflow mixing layer ($R < 1$); (b) Kelvin–Helmholtz spanwise vortices in mixing layer experiment by Lasheras & Choi (1988).

(figure 2b). When $R = 1$, the lower stream is at rest (figure 2c). Finally a counterflow mixing layer is produced if $R > 1$ (figure 2d). In coflow mixing layers, the spreading rate $d\delta_\omega/dx$ is found to increase proportionally with R. The velocity ratio acts as a dilatation parameter: the closer R is to unity, the faster is the streamwise development. By contrast, the Reynolds number Re has a minimal impact on the dynamics of the large vortical structures, provided that it is large enough, say $Re > 100$.

The above considerations fail to capture the most important hidden control parameter, namely *external noise*. Spatially developing coflow mixing layers ($0 < R < 1$) are observed to be extremely sensitive to outside perturbations, whether naturally present as noise, or intentionally applied as external forcing.

In *unforced* naturally evolving mixing layers, power spectra measured in the roll-up region immediately downstream of the splitter plate are generally broadband with a maximum at the natural shedding frequency f_n such that the Strouhal number $St_n = f_n \delta_\omega(0)/\overline{U}$ remains approximately equal to 0.03. The natural wavelength between spanwise vortices is then $\lambda_n = \overline{U}/f_n$. Further

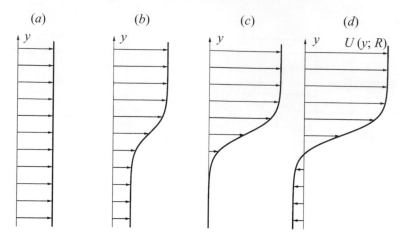

Figure 2. Typical mixing-layer velocity profiles as a function of velocity ratio R when $U_1 \geqslant |U_2|$: (a) $R = 0$; (b) $0 < R < 1$; (c) $R = 1$; (d) $R > 1$.

downstream, successive pairings or triplings occur somewhat randomly along the stream and lead to the appearance of broad peaks at $f_n/2$, $f_n/3$, etc.

In *forced coflow mixing layers*, the content of external perturbations is specified by imposing well-defined excitations. The flow response to different forcing frequencies f_f can be conveniently documented by the streamwise evolution of the perturbation energy $E_f(x)$ for various spectral components (figure 3a, c) and the streamwise growth $\delta_\omega(x)$ of the mean vorticity thickness (figure 3b, d). If $f_f \approx f_n$ (figure 3a, b), the roll-up process becomes perfectly synchronized: the fundamental component $E_{f_f}(x)$ reaches a well-defined maximum at the roll-up station, at which point the subharmonic component $E_{f_f/2}(x)$ starts to grow, to become maximum at the pairing station further downstream. Forcing in the vicinity of f_n inhibits pairing when compared with the natural case. If $f_f \approx f_n/2$ (figure 3c, d), a different streamwise evolution is observed: the subharmonic component $E_{f_f}(x)$ increases right from the trailing edge and its maximum is reached further upstream. Forcing in the vicinity of $f_n/2$ promotes pairing when compared with the natural case. Variations in forcing frequency also significantly affect the shear layer spreading rate as evidenced by comparing figure 3(b) and figure 3(d). The spatially developing coflow mixing layer acts as an *amplifier* of external perturbations.

As the velocity ratio R increases above unity, the dynamics of mixing layers undergoes qualitative changes that have been exemplified in ingenious experiments by Strykowski & Niccum (1991). When suction is applied to the annular outer region of a conventional circular jet, a counterflow mixing

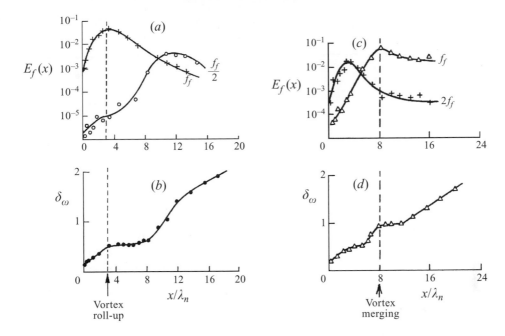

Figure 3. Streamwise evolution of perturbation energy $E_f(x)$ contained in various frequencies f (a, c) and of vorticity thickness $\delta_\omega(x)$ (b, d). In (a, b) forcing frequency is $f_f \approx f_n$. In (c, d) forcing frequency is $f_f \approx f_n/2$. After Ho & Huerre (1984) based on experiments of Ho & Huang (1982).

layer is produced of appropriate velocity ratio $R > 1$. When R increases above unity, the initially broadband power spectrum gives way to a discrete spectrum at a well-defined frequency (figure 4a). The jet shear layer dynamics effectively become synchronized into a self-sustained periodic régime. More specifically, let any physical perturbation quantity $\psi(\boldsymbol{x}, t)$ be written in the form

$$\psi(\boldsymbol{x}, t) = \mathrm{Re}\left\{\phi(\boldsymbol{x})\,a(t)\,e^{-\mathrm{i}\omega_g t}\right\}, \qquad (2.3)$$

where ω_g is a real frequency, $\phi(\boldsymbol{x})$ is the shape function characterizing the spatial distribution of the flow oscillator, Re denotes the real part, and $a(t)$ is a complex amplitude function. In the vicinity of a supercritical Hopf bifurcation point at $R = R_{G_c}$ (Guckenheimer & Holmes 1983; Manneville 1990), the amplitude $a(t)$ is known to be governed by the *Landau equation*

$$\frac{\mathrm{d}a}{\mathrm{d}t} = \sigma\,a - b|a|^2 a, \qquad (2.4)$$

where σ and b are complex constants. The real part σ_r is proportional to the departure from criticality, $R - R_{G_c}$, and the real part b_r is positive. According

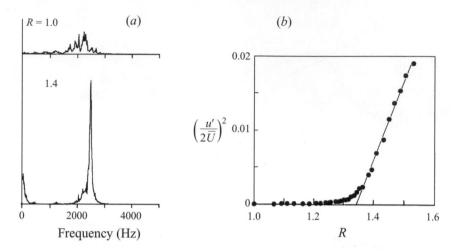

Figure 4. (*a*) Velocity power spectra (linear scale) measured in the jet shear layer of Strykowski & Niccum (1991) drawn for different values of the velocity ratio *R*. (*b*) Square of saturation amplitude at fixed spatial position $x/d = 0.25$ versus velocity ratio *R*, where *d* is the jet diameter at the exit plane.

to (2.4), the amplitude $|a(t)|$ reaches a saturation level $|a(\infty)| = (\sigma_r/b_r)^{1/2}$. Since $\sigma_r \sim (R - R_{Gc})$ one therefore must have $|a(\infty)| \sim (R - R_{Gc})^{1/2}$. Such a scaling law is the hallmark of a supercritical Hopf bifurcation. According to figure 4(*b*), the amplitude of the associated limit cycle effectively increases proportionally to $(R - R_{Gc})^{1/2}$ from the critical value $R_{Gc} = 1.34$. This bifurcation, which affects the entire flow within the first few diameters, signals the appearance of a *self-sustained global mode*, a notion that is precisely defined in §4. The spatio-temporal behaviour of mixing layers therefore undergoes a drastic qualitative change from noise amplifier to oscillator, as the velocity ratio *R* increases above the bifurcation point $R_{Gc} = 1.34$. This transition from extrinsic to intrinsic dynamics is confirmed by monitoring the response of the flow to external time-periodic perturbations applied at the frequency of the limit cycle: below R_{Gc} in the amplifier régime, the response increases with forcing amplitude, whereas above R_{Gc} in the oscillator régime, it becomes relatively insensitive to forcing amplitude.

2.2 The bluff body wake as a prototype flow oscillator

When a circular cylinder of diameter *d* and length *L* is placed in a uniform upstream flow U_∞, a wake is generated which is controlled by two non-dimensional parameters: the Reynolds number $Re = (U_\infty d)/\nu$ and the aspect ratio L/d. In this short presentation, we focus solely on the purely two-dimensional cylinder wake whereby $L/d = \infty$, and the only control parameter left is the Reynolds number Re. The following sequence of régimes is observed as Re increases. In the very low Reynolds number range, $Re \ll 1$, steady creeping flow prevails as described by the Stokes solution (Batchelor 1967). When Re exceeds 5 or 6, the flow remains steady but a recirculation bubble composed of two counter-rotating vortices appears behind the cylinder. When Re crosses the critical value $Re_{Gc} = 48.5$, the flow experiences a transition to a time-periodic state which takes the form of the celebrated Kármán vortex street: two travelling rows of counter-rotating vortices are shed from the cylinder (figure 5). As in counterflow mixing layers, frequency spectra within the wake exhibit a sharp peak at a fundamental frequency f. The associated Strouhal number $St = fd/U_\infty$ increases with Re (Williamson 1989) from the value $St = 0.2$ at the transition point Re_{Gc}. As discussed in §3.5, the wake has in effect undergone a transition from noise amplifier to flow oscillator when Re increases through Re_{Gc}. The underlying spatio-temporal structure responsible for this qualitative change of behaviour is again a self-sustained time-periodic global mode. According to the experimental observations of Provansal, Mathis & Boyer (1987), the onset of the Kármán vortex street is associated with a Hopf bifurcation at $Re_{Gc} = 48.5$, which is analogous to that occurring in counterflow mixing layers. As in the former case, the amplitude of the limit-cycle oscillations is observed to scale as $(Re - Re_{Gc})^{1/2}$. The reader is referred to Williamson (1996) for a recent review of wake dynamics.

3 Absolute/convective instabilities in parallel flows

The objective of this section is to introduce instability concepts pertaining to steady parallel unidirectional shear flows. In keeping with the general philosophy outlined in §1, the streamwise development of the basic flow is entirely neglected. In other words, the velocity profile is extracted at a given streamwise station x and replicated indefinitely upstream and downstream to generate a strictly parallel shear flow. The same idealization may be performed at all streamwise stations.

3.1 The dispersion relation: temporal/spatial instability waves

It is convenient to introduce from the start non-dimensional dependent and independent variables based on length and velocity scales ℓ and V that are characteristic of the basic flow under study. The study of two-dimensional incompressible flows of a viscous fluid of constant density ρ and shear viscosity μ reduces to the determination of a stream function Ψ governed by the vorticity equation

$$\left(\frac{\partial}{\partial t} + \frac{\partial \Psi}{\partial y}\frac{\partial}{\partial x} - \frac{\partial \Psi}{\partial x}\frac{\partial}{\partial y}\right)\nabla^2\Psi = \frac{1}{Re}\nabla^4\Psi, \tag{3.1}$$

where $Re = \rho V\ell/\mu$ is the Reynolds number. The stability of a parallel basic flow $U(y)$ is examined by decomposing the total stream function Ψ into basic and perturbation contributions according to

$$\Psi(x, t) = \int^y U(y')\,dy' + \psi(x, y, t). \tag{3.2}$$

After substitution of (3.2) into (3.1) and linearization around the basic state $U(y)$, one obtains the equation governing the perturbation stream function:

$$\left(\frac{\partial}{\partial t} + U(y)\frac{\partial}{\partial x}\right)\nabla^2\psi - U''(y)\frac{\partial\psi}{\partial x} = \frac{1}{Re}\nabla^4\psi. \tag{3.3}$$

Since this partial differential equation is invariant under arbitrary translations $x \mapsto x+\text{const.}$, $t \mapsto t+\text{const.}$, it is natural to seek so-called *normal-mode* solutions of complex frequency ω and complex wavenumber k in the form

$$\psi(x, y, t) = \text{Re}\left\{\phi(y)\,e^{i(kx-\omega t)}\right\}, \tag{3.4}$$

where the unknown complex eigenfunction $\phi(y)$ characterizes the distribution of fluctuations in the y-direction. A perturbation velocity field is associated with (3.4), with x- and y-components

$$u(x, y, t) = \frac{\partial\psi}{\partial y} = \text{Re}\left\{\phi'(y)\,e^{i(kx-\omega t)}\right\},$$

$$v(x, y, t) = -\frac{\partial\psi}{\partial x} = -\text{Re}\left\{ik\,\phi(y)\,e^{i(kx-\omega t)}\right\}. \tag{3.5a, b}$$

Upon substituting the normal-mode decomposition (3.4) into (3.3), one is led to the classical *Orr–Sommerfeld equation*

$$[U(y) - c][\phi'' - k^2\phi] - U''(y)\,\phi = \frac{1}{ik\,Re}\left(\frac{d^2}{dy^2} - k^2\right)^2\phi, \tag{3.6}$$

where $c = \omega/k$ denotes the complex phase velocity. For a basic flow $U(y)$ bounded by solid walls at $y = y_1, y_2$, the perturbation velocity field (3.5a, b) vanishes at $y = y_1, y_2$ and $\phi(y)$ must satisfy the boundary conditions

$$\phi(y_1) = \phi(y_2) = 0, \qquad \phi'(y_1) = \phi'(y_2) = 0. \tag{3.7a, b}$$

If the basic flow is unbounded, as in the case of free shear layers, the above conditions have to be replaced by exponential decay boundary conditions

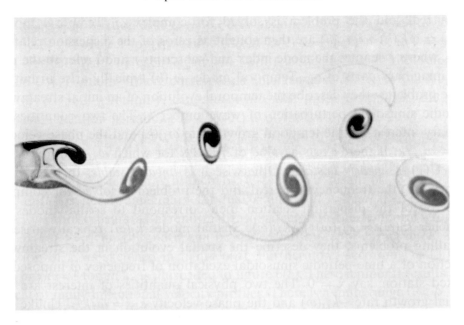

Figure 5. A Kármán vortex street in the wake of a circular cylinder, $Re = 80$. After Perry, Chong & Lim (1982).

at $y = \pm\infty$. The Orr–Sommerfeld equation (3.6) and associated boundary conditions (3.7a, b) admit non-trivial solutions for $\phi(y)$ if and only if the complex wavenumber k and frequency ω satisfy a dispersion relation of the form

$$D(k, \omega; Re) = 0. \tag{3.8}$$

Thus, to each complex pair k, ω there corresponds a complex eigenfunction $\phi(y)$ solution of (3.8).

In the limit of vanishing viscosity $Re = \infty$, the Orr–Sommerfeld equation (3.6) reduces to the *Rayleigh equation*

$$[U(y) - c][\phi'' - k^2\phi] - U''(y)\phi = 0. \tag{3.9}$$

Since (3.9) is second order, the no-slip conditions (3.7b) must be dropped and $\phi(y)$ can only be required to satisfy the impermeability conditions (3.7a) or to decay exponentially in the case of an infinite y-domain. Here again, the two-point boundary-value problem (3.9), (3.7a) has no solutions unless k and ω satisfy a dispersion relation $D(k, \omega) = 0$.

Two distinct approaches have traditionally been chosen to treat the viscous and inviscid linear instability problems specified by systems (3.6, 3.7a, b) and (3.9, 3.7a) respectively. In the *temporal* framework, the wavenumber k is

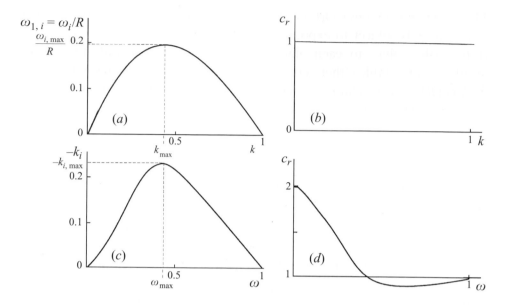

Figure 6. Hyperbolic-tangent mixing layer instability characteristics: (a) normalized temporal growth rate ω_i/R versus real wavenumber k; (b) real phase velocity $c_r \equiv \omega_r/k$ versus k; (c) spatial growth rate $-k_i$ of amplified branch $k_1^+(\omega)$ versus real frequency ω at $R = 1$; (d) $c_r \equiv \omega/k_r$ versus ω at $R = 1$. Curves such as (c) and (d) only apply when the flow is convectively unstable, $0 < R < 1.315$. After Michalke (1964, 1965).

This property is used in §3.5 to scale the spatial instability characteristics pertaining to different velocity ratios conveniently.

3.2 The linear impulse response: absolute/convective instability

According to the discussion above, the properties of normal modes sustained by a given parallel shear flow $U(y)$ are encapsulated in a dispersion relation of the form

$$D(k, \omega; R) = 0, \tag{3.13}$$

where R collectively denotes the set of control parameters. The cross-stream distribution of fluctuations is entirely contained in the eigenfunction $\phi(y)$ that is *slaved* to each eigenvalue pair k, ω. It is tempting to consider the dispersion relation (3.13) as being associated with a partial differential equation involving solely the streamwise variable x and time t:

$$D\left(-\mathrm{i}\frac{\partial}{\partial x}, \mathrm{i}\frac{\partial}{\partial t}; R\right)\psi(x, t) = S(x, t), \tag{3.14}$$

for a complex scalar field $\psi(x,t)$.[1] The source term $S(x,t)$ specifies the forcing imposed on the system in some localized interval of x and t. In the absence of forcing $S(x,t) = 0$, the normal-mode decomposition $\psi(x,t) = A\,e^{i(kx-\omega t)}$ leads to non-trivial solutions $A \neq 0$ if and only if k and ω satisfy $D(k,\omega;R) = 0$. By embedding the dispersion relation into the one-dimensional x,t system (3.14), the y-structure has been entirely projected out and all pertinent notions may be introduced succinctly with a minimum of calculations. This short cut will be resorted to repeatedly in the course of this presentation. In this context, simple analytical models may be explored to illustrate the main ideas. Consider for instance the linear complex Ginzburg–Landau equation

$$\left(\frac{\partial}{\partial t} + U\frac{\partial}{\partial x}\right)\psi - \mu\psi - (1 + ic_d)\frac{\partial^2\psi}{\partial x^2} = 0, \tag{3.15}$$

where U, μ and c_d are real control parameters. Applying the transformations $\partial/\partial t \mapsto -i\omega, \partial/\partial x \mapsto ik$ leads to the dispersion relation

$$\omega = Uk + c_d k^2 + i(\mu - k^2). \tag{3.16}$$

The Ginzburg–Landau equation is seen to exhibit the minimum structure necessary to include the effects of advection (U), dispersion (c_d) and instability (μ). Its temporal instability properties (k real, ω complex) are summarized in figure 7(a,b) in the unstable domain $\mu > 0$ of parameter space. There is a finite band $|k| < \sqrt{\mu}$ of unstable wavenumbers and the phase velocity $c_r = \omega_r/k$ varies linearly with k. Spatial instability properties (ω real, k complex) are presented in figure 7(c,d). Note that for ω real, equation (3.16) admits two complex roots, respectively denoted $k^+(\omega)$ and $k^-(\omega)$. In order to properly interpret the spatial behaviour, it is essential to know how to assign the k^+ and k^- branches on either side of the time-periodic source that generated them. The resolution of this issue is postponed to §3.4.

Following the lead of Bers (1983), we are now in a position to introduce general definitions in the context of a flow described by the partial differential equation (3.14). Let $G(x,t)$ be the Green function or impulse response defined by

$$D\left[-i\frac{\partial}{\partial x}, i\frac{\partial}{\partial t}; R\right]G(x,t) = \delta(x)\,\delta(t), \tag{3.17}$$

where δ denotes the Dirac delta function. The fundamental solution G contains all the information regarding the spatio-temporal dynamics of the perturbation field. It is a general property of linear systems that the response $\psi(x,t)$ to a given forcing $S(x,t)$ as stated in (3.14) is simply obtained by

---

[1] The field $\psi(x,t)$ is unrelated to the perturbation stream function $\psi(x,y,t)$ introduced in (3.2).

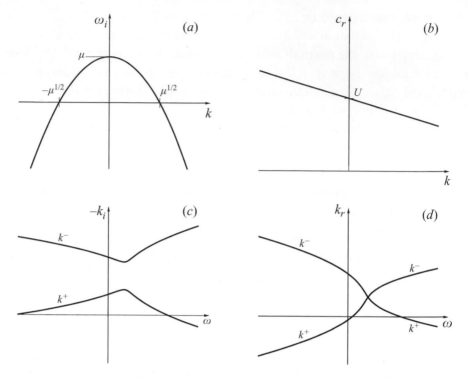

Figure 7. Linear instability characteristics of complex Ginzburg–Landau equation (3.15) for $\mu \geqslant 0$: (a) temporal growth rate ω_i versus real wavenumber k; (b) real phase velocity $c_r \equiv \omega_r/k$ versus k; (c) spatial growth rate $-k_i$ of k^+ and k^- branches versus real frequency ω; (d) $k_r^{\pm}$ versus ω. Curves such as (c) and (d) only apply in the convectively unstable range $0 \leqslant \mu \leqslant U^2/[4(1+c_d^2)]$.

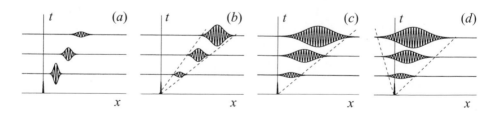

Figure 8. Linear impulse response $G(x,t)$. (a) Linearly stable flow; (b) linearly convectively unstable flow; (c) marginally convectively/absolutely unstable flow; (d) absolutely unstable flow.

convolution of $S(x,t)$ and $G(x,t)$. Several types of impulse response behaviour are then possible depending on the value of the control parameter R, as sketched in figure 8(a–d). One may first distinguish between stable and unstable flows:

A flow is *linearly stable* if

$$\lim_{t\to\infty} G(x,t) = 0 \quad \text{along all rays} \quad \frac{x}{t} = \text{const.} \tag{3.18}$$

The impulse response then consists of a decaying wavepacket (figure 8a).

Otherwise, the flow is *linearly unstable*, i.e.

$$\lim_{t\to\infty} G(x,t) = \infty \quad \text{along at least one ray} \quad \frac{x}{t} = \text{const.} \tag{3.19}$$

The impulse response then typically consists of an unstable wavepacket confined within a wedge or several wedges in the (x,t)-plane (figure 8b–d). Among unstable flows, one must further differentiate between absolute instability and convective instability.

An unstable flow is said to be linearly *convectively unstable* if

$$\lim_{t\to\infty} G(x,t) = 0 \quad \text{along the ray} \quad \frac{x}{t} = 0. \tag{3.20}$$

The unstable wavepacket is then advected away to leave the source ultimately undisturbed (figure 8b).

An unstable flow is said to be linearly *absolutely unstable* if

$$\lim_{t\to\infty} G(x,t) = \infty \quad \text{along the ray} \quad \frac{x}{t} = 0. \tag{3.21}$$

The unstable wavepacket surrounds the source and gradually contaminates the entire medium (figure 8d). For an unstable flow that is marginally convectively/absolutely unstable, the trailing-edge of the wavepacket remains at the source location as depicted in figure 8(c).

3.3 Analysis in the complex wavenumber and frequency plane: absolute/convective instability criterion

How do these various definitions translate to Fourier space? In particular, are there criteria directly applicable to the dispersion relation $D(k,\omega;R) = 0$, which allow one to discriminate between stable, unstable, convectively unstable, absolutely unstable flows? In order to address these issues, it is enlightening to derive the solution of equation (3.14) for various source functions $S(x,t)$. The following analysis relies in large measure on the *causality principle*: the response $\psi(x,t)$ to a source $S(x,t)$ switched on at $t = 0$ remains zero for $t < 0$. 'Effect cannot precede cause'. In other words

$$S(x,t) = 0, \quad \psi(x,t) = 0, \quad \text{for} \quad t < 0. \tag{3.22}$$

The partial differential operator in (3.14) does not involve x- or t-dependent coefficients and the solution is conveniently obtained by going

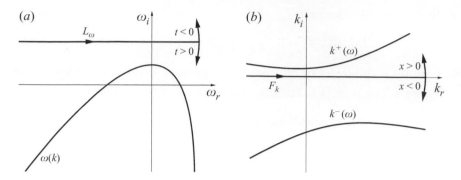

Figure 9. Integration contours (*a*) in the complex ω-plane (*b*) in the complex k-plane. Curve labelled $\omega(k)$ in (*a*) represents the locus of the temporal mode as k travels along F_k; curves labelled $k^+(\omega)$ and $k^-(\omega)$ in (*b*) represent the locus of the spatial branches as ω travels along L_ω. Sketches refer to the specific case of the Ginzburg–Landau equation with a single temporal mode (3.16) and two spatial branches.

into Fourier space. The perturbation field is expressed as the double Fourier integral

$$\psi(x,t) = \frac{1}{(2\pi)^2} \int_{L_\omega} \int_{F_k} \psi(k,\omega)\,e^{i(kx-\omega t)}dk\,d\omega \qquad (3.23)$$

along straight line contours L_ω and F_k in the complex ω- and k-planes, as sketched in figure 9(*a, b*). The contour F_k is initially taken to lie on the horizontal k_r-axis while L_ω is selected so as to comply with causality: it is a horizontal line lying above all the singularities of $\psi(k,\omega)$ in the complex ω-plane as k travels along the k_r-axis. To understand the latter assertion, it is convenient to anticipate the procedure involved in the calculation of the L_ω-integral in (3.23), i.e. the inverse Fourier transform

$$\psi(k,t) = \frac{1}{2\pi} \int_{L_\omega} \psi(k,\omega)\,e^{-i\omega t}d\omega. \qquad (3.24)$$

The calculation of $\psi(k,t)$ relies on first closing the contour L_ω by a semicircle at infinity and then appealing to the residue theorem (Ablowitz & Fokas 1997). Whether the contour is closed above or below L_ω is dictated by the requirement that the integral along the semicircle make a zero contribution. When $t < 0$ (resp. $t > 0$), the integrand in (3.24) decays exponentially fast along a semicircle at infinity lying above (resp. below) L_ω. The closing of the contour L_ω is therefore as indicated in figure 9(*a*). The residue theorem may then be invoked to calculate $\psi(k,t)$ for $t < 0$ and $t > 0$. The causality condition (3.22) requires that $\psi(k,t) = 0$ for $t < 0$. As a result, there cannot be any singularities of $\psi(k,\omega)$ arising in the residue calculation of $\psi(k,t)$ for

$t < 0$. All singularities of $\psi(k, \omega)$ must therefore be confined to the lower half-ω-plane, below L_ω.

For later use, the closing of the F_k-contour in the complex k-plane is determined by anticipating the equivalent procedure in the calculation of the F_k-integral in (3.23), i.e. the inverse Fourier transform

$$\psi(x, \omega) = \frac{1}{2\pi} \int_{F_k} \psi(k, \omega) \, e^{ikx} dk. \tag{3.25}$$

When $x > 0$ (resp. $x < 0$), the integrand in (3.25) decays exponentially along a semi-circle at infinity lying above (resp. below) F_k and the closing of the F_k-contour must be as sketched in figure 9(b).

Once the integration contours have been specified, it is a straightforward matter to apply the Fourier transform

$$\psi(k, \omega) = \int_{-\infty}^{+\infty} \int_{-\infty}^{+\infty} \psi(x, t) \, e^{-i(kx - \omega t)} dx \, dt \tag{3.26}$$

to (3.14) by making the substitutions $\partial/\partial t \mapsto -i\omega$, $\partial/\partial x \mapsto ik$. The governing equation in spectral space is simply

$$D(k, \omega)\psi(k, \omega) = S(k, \omega), \tag{3.27}$$

where the control parameter R is hereafter omitted for brevity.

In the absence of a source, $S(k, \omega) = 0$ and normal modes with $\psi(k, \omega) \neq 0$ are recovered provided that $D(k, \omega) = 0$. In the presence of a source, $S(k, \omega) \neq 0$ and

$$\psi(k, \omega) = \frac{S(k, \omega)}{D(k, \omega)}. \tag{3.28}$$

For simplicity, assume that both $S(k, \omega)$ and $D(k, \omega)$ are analytic in k and ω. Standard complex-variable techniques (Ablowitz & Fokas 1997) may now be used to evaluate the inverse Fourier transform (3.24), i.e.

$$\psi(k, t) = \frac{1}{2\pi} \int_{L_\omega} \frac{S(k, \omega)}{D(k, \omega)} \, e^{-i\omega t} d\omega. \tag{3.29}$$

Here k plays the role of a parameter confined to the F_k-contour on the real k_r-axis (figure 9b). The only singularities of the integrand in (3.29) are poles associated with zeros of $D(k, \omega)$, i.e. temporal modes $\omega_j(k)$. Assume that there exists a single temporal mode $\omega(k)$. In the case of the Ginzburg–Landau equation, $\omega(k)$ is given explicitly by (3.16). Otherwise, it may have to be computed numerically as for mixing layers. As k travels along F_k, the pole $\omega(k)$ describes a locus in the complex ω-plane which is by construction below L_ω to comply with causality. When $t < 0$, the L_ω-contour is closed in the upper half-ω-plane and, as expected, $\psi(k,t) = 0$ by Cauchy's theorem. When $t > 0$, the L_ω-contour is closed in the lower half-ω-plane, and application of the residue theorem yields

$$\psi(k,t) = -\mathrm{i}\frac{S[k,\omega(k)]}{(\partial D/\partial \omega)[k,\omega(k)]} \mathrm{e}^{-\mathrm{i}\omega(k)t}. \tag{3.30}$$

The temporal mode $\omega(k)$ is seen to arise naturally in the calculation of the response $\psi(x,t)$ to a source $S(x,t)$: each $\omega(k)$ appears as a pole singularity in the complex ω-plane. The image of F_k by $\omega(k)$ in the complex ω-plane is a curve which necessarily lies below the L_ω-contour (figure 9a). For future reference, note that, as ω travels along L_ω, the integrand in (3.23) with $\psi(k,\omega)$ specified in (3.28), displays pole singularities in the complex k-plane at zeros of $D(k,\omega)$ associated with spatial branches $k^+(\omega)$ and $k^-(\omega)$. A typical configuration is illustrated in figure 9(b). These loci, which constitute the image of L_ω by $k(\omega)$ in the complex k-plane, can be divided into two disjoint sets: the k^+ (resp. k^-) branches are, by definition, solely confined to the upper (resp. lower) half-k-planes, as ω travels along L_ω. Any crossing of the contour F_k, i.e. k_r-axis, by spatial branches is by construction prohibited. If such an intersection were to occur, the locus of $\omega(k)$ in the complex ω-plane would cross L_ω, which is excluded by causality. *The causality principle therefore requires that $\omega(k)$ lie below L_ω in the complex ω-plane and that $k^+(\omega)$ and $k^-(\omega)$ branches be confined well within their respective half-k-planes.* For well-posed problems, such a requirement can always be satisfied by positioning L_ω sufficiently high in the complex ω-plane.

It remains to perform the inverse Fourier transform with respect to k, namely

$$\psi(x,t) = -\frac{\mathrm{i}}{2\pi} \int_{F_k} \frac{S[k,\omega(k)]}{(\partial D/\partial \omega)[k,\omega(k)]} \mathrm{e}^{\mathrm{i}[kx-\omega(k)t]} \, \mathrm{d}k. \tag{3.31}$$

This integral, which represents the response wavepacket to a given source function $S(x,t)$, is not in general amenable to an explicit calculation. However, according to the definitions of stability, instability, convective instability, absolute instability introduced in (3.18)–(3.21), it is sufficient to determine the long-time behaviour of the Green function $G(x,t)$ along different spatio-temporal rays $x/t = \mathrm{const}$. From here on, we therefore focus on the dynamics of $G(x,t)$ defined in (3.17). Under such conditions, the source function reduces to $S(x,t) = \delta(x)\delta(t)$, i.e. $S(k,\omega) = 1$ and the wavepacket integral (3.31) becomes

$$G(x,t) = -\frac{\mathrm{i}}{2\pi} \int_{F_k} \frac{1}{(\partial D/\partial \omega)[k,\omega(k)]} \mathrm{e}^{\mathrm{i}[kx-\omega(k)t]} \mathrm{d}k. \tag{3.32}$$

This expression falls into the general class of integrals of the form

$$G(x,t) = -\frac{i}{2\pi} \int_{F_k} f(k)\, e^{\rho(k;x/t)t}\, dk, \qquad (3.33)$$

where, in the present instance,

$$f(k) = \frac{1}{(\partial D/\partial \omega)[k, \omega(k)]}, \qquad \rho\left(k;\frac{x}{t}\right) = i\left[k\frac{x}{t} - \omega(k)\right]. \qquad (3.34a,b)$$

Note that in (3.33), time t is a large parameter, and that x/t is the particular spatio-temporal ray under consideration. The characteristic feature of integrals of this type is the presence in the integrand of a fast exponential associated with the large parameter t. The method of steepest descent (Bender & Orszag 1978; Bleistein & Handelsman 1986) is then well suited to obtain explicit asymptotic approximations as $t \to \infty$, along distinct rays $x/t = $ const., as outlined below. For large time, the order of magnitude of the integrand is controlled at leading order by the real part of the exponent in (3.33), i.e. by the height of the surface $\rho_r(k;x/t)$ in (k_r, k_i, ρ_r)-space (figure 10). For a wide class of instabilities (but not for all of them!) the complex exponent $\rho(k;x/t)$ admits a stationary point k_* such that

$$\frac{\partial \rho}{\partial k}\left(k_*;\frac{x}{t}\right) = i\left[\frac{x}{t} - \frac{\partial \omega}{\partial k}(k_*)\right] = 0. \qquad (3.35)$$

In the vicinity of k_*, the complex function $\rho(k;x/t)$ is approximated by the Taylor expansion

$$\rho\left(k;\frac{x}{t}\right) \sim \rho\left(k_*;\frac{x}{t}\right) + \frac{1}{2}\frac{\partial^2 \rho}{\partial k^2}\left(k_*;\frac{x}{t}\right)(k-k_*)^2, \qquad (3.36)$$

and it is easily inferred that the local shape of the surface $\rho_r(k;x/t)$ around k_* is a saddle as sketched in figure 10 and k_* is a saddle point. The second step in the procedure consists of deforming the original contour F_k on the real k_r-axis into the steepest descent path F_ρ through the saddle point k_*, without crossing any 'hills' of the surface $\rho_r(k;x/t)$ at infinity (Bleistein & Handelsman 1986). By construction, the global maximum of $\rho_r(k;x/t)$ along the steepest descent path F_ρ is reached at the saddle point k_* and, for large time, one expects the dominant contribution to arise from the vicinity of the maximum at k_*. To leading order the integral along F_ρ may therefore be restricted to a small segment F_{ρ_1} of length 2ε around k_*. In classical steepest descent fashion, the function $\rho(k;x/t)$ is legitimately approximated by its Taylor expansion (3.36) along the steepest descent path around k_* and the function $f(k)$ by $f(k_*)$. In a final step, the integral F_{ρ_1} is extended to infinity

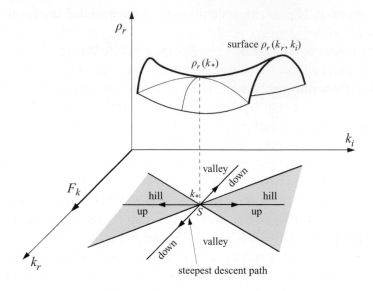

Figure 10. Local topology of surface $\rho_r(k_r, k_i)$ around saddle point k_*.

in both directions, the added contributions being only subdominant. The final result is

$$G(x,t) \sim \frac{f(k^*)}{\left[2\pi\dfrac{\partial^2\rho}{\partial k^2}\left(k_*;\dfrac{x}{t}\right)\right]^{1/2}} e^{\rho(k_*\,;x/t)\,t} . \qquad (3.37)$$

Upon reverting to the original variables via (3.34a, b), one reaches the following conclusion: the impulse response takes, along each spatio-temporal ray $x/t = $ const., the long-time asymptotic form

$$G(x,t) \sim \frac{\exp(\mathrm{i}\,[\pi/4 + k_*x - \omega(k_*)t])}{\dfrac{\partial D}{\partial \omega}\left[k_*, \omega(k_*)\right]\left[2\pi\dfrac{\partial^2\omega}{\partial k^2}(k_*)t\right]^{1/2}} , \qquad (3.38a)$$

where the complex wavenumber k_* is given by the saddle-point condition (3.35), i.e.

$$\frac{\partial\omega}{\partial k}(k_*) = \frac{x}{t} . \qquad (3.38b)$$

The asymptotic estimate (3.38a, b) neatly summarizes the spatio-temporal dynamics of instability waves governed by a dispersion relation $D(k,\omega) = 0$ with a single temporal mode $\omega(k)$. Many important concepts directly follow from the physical interpretation of these formulas.

Note first that the asymptotic impulse response takes the form of a wavepacket[1] as anticipated in the sketches of figure 8(a, d). According to (3.38b), an observer moving at the velocity V along the spatio-temporal ray $x/t = V$ in (x, t)-space, perceives a complex wavenumber k_* and frequency $\omega_* = \omega(k_*)$ such that the group velocity is precisely equal to V. The shape of the wave is then given by (3.38a) where k_* and ω_* depend on $x/t = V$. Thus relation (3.38b) readily extends the interpretation of the notion of group velocity to instability waves. Although $\partial\omega/\partial k$ is in general complex, it acquires physical significance only when it is real and equal to $x/t = V$. From the argument of the exponential in (3.38a), it is inferred that the temporal growth rate $\sigma(V)$ perceived by the moving observer is $\sigma = \omega_{*,i} - V k_{*,i}$. If $\sigma(V) > 0$ (resp. $\sigma(V) < 0$), perturbations increase (resp. decrease) exponentially along the ray $x/t = V$. The domain occupied by the unstable wavepacket in the (x, t)-diagrams of figure 8(b, d) is indeed rigorously defined by the condition $\sigma(x/t) > 0$. Typical shapes for the growth rate curve $\sigma(V)$ are illustrated in figures 11(a, b): the range of velocities V such that $\sigma(V) > 0$ provides a measure of the streamwise extent of the unstable wavepacket. The leading- and trailing-edge velocities $x/t = V^+$ and $x/t = V^-$ bounding the wavepacket are defined by $\sigma(V^+) = \sigma(V^-) = 0$, with $V^+ > V^-$. In typical cases, such as those illustrated in figures 6(a) and 7(a), the temporal growth rate $\omega_i(k)$ for k real displays a maximum $\omega_{i,max} = \omega_i(k_{max})$ at a real wavenumber k_{max}. Then

$$\frac{\partial\omega_i}{\partial k}(k_{max}) = 0 \tag{3.39}$$

and the corresponding group velocity $\partial\omega/\partial k(k_{max}) \equiv V_{max}$ is necessarily real. In view of (3.38b), the wavenumber k_{max} is therefore observed along the ray $x/t = V_{max}$ and it is readily verified that the associated growth rate $\sigma(V_{max}) = \omega_{i,max}$ is a global maximum over all V, as sketched in figure 11. The quantity V_{max} effectively provides a measure of the propagation velocity of the centre of the wavepacket.

A simple (and not surprising) criterion based on $\omega_{i,max}$ readily follows that distinguishes between linearly stable and unstable flows:

if $\omega_{i,max} < 0$, the temporal growth rate σ is negative for all $x/t = V$ and the flow is *linearly stable*;

if $\omega_{i,max} > 0$, the temporal growth rate σ is positive in a finite range of $x/t = V$ (figure 11a, b) and the flow is *linearly unstable*.

Of more crucial interest here is the so-called *absolute wavenumber* k_0 observed along the ray $x/t = V = 0$, i.e. at the source, for a stationary observer

---

[1] Expression (3.38a) looks deceptively like a plane wave: k^* and $\omega(k^*)$ are in fact functions of x/t via (3.38b)!

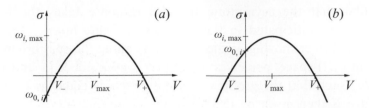

Figure 11. Temporal growth rate $\sigma \equiv (\omega_{*,i} - k_{*,i}V)$ as a function of observer velocity $x/t = V$ for the Ginzburg–Landau equation: (a) convective instability; (b) absolute instability.

in the laboratory frame. According to (3.38b), the complex wavenumber k_0 is defined by the zero-group-velocity condition

$$\frac{\partial \omega}{\partial k}(k_0) = 0, \qquad (3.40)$$

and the associated complex frequency $\omega_0 = \omega(k_0)$ is referred to as the *absolute frequency*. The *absolute growth rate* $\sigma(0) = \omega_{0,i}$ indicated in the $\sigma(V)$ diagrams of figure 11 measures the growth or decay of the wavepacket as observed in the laboratory frame. It is the quantity of interest to distinguish between convective and absolute instability as defined in (3.20) and (3.21). The following *Briggs (1964)–Bers (1983) criterion* ensues:

If $\omega_{0,i} < 0$, disturbances decay in the laboratory frame, ultimately leaving the source undisturbed and the flow is *convectively unstable*. As shown in figure 11(a), $0 < V_- < V_{\max} < V_+$ and the two edges of the impulse response are co-propagating away from the source.

If $\omega_{0,i} > 0$, disturbances grow exponentially in the laboratory frame, gradually contaminating the entire medium and the flow is *absolutely unstable*. As shown in figure 11(b), $V_- < 0 < V_{\max} < V_+$ and the two edges of the impulse response are counter-propagating[1].

Consider as a straightforward application the Ginzburg–Landau equation (3.15) and associated dispersion relation (3.16). Definition (3.40) immediately yields

$$k_0 = \frac{U}{2(\mathrm{i} - c_d)}, \qquad \omega_0 = -\frac{c_d U^2}{4(1 + c_d^2)} + \mathrm{i}\left[\mu - \frac{U^2}{4(1 + c_d^2)}\right]. \qquad (3.41)$$

The resulting domains of stability, convective instability and absolute instability are illustrated in μ, U parameter space in figure 12. At fixed values of U and c_d, increasing values of the parameter μ typically give rise

---

[1] The case $V_- < V_{\max} < 0 < V_+$ is naturally possible and it also leads to absolute instability.

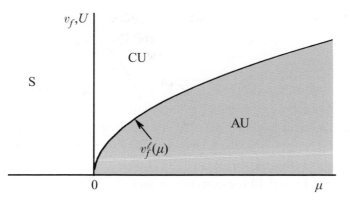

Figure 12. Stability diagram of the Ginzburg–Landau equation in the (μ, U)-plane. S: stable; CU: convectively unstable; AU: absolutely unstable. Front velocity $v_f(\mu)$ of supercritical Ginzburg–Landau equation (5.2) in advected frame $(U = 0)$; $v_f(\mu) = v_f^\ell(\mu)$ for all $\mu \geqslant 0$. Nonlinearly absolutely unstable domain (shaded region) in (μ, U)-space. A nonlinear global mode in a semi-infinite medium exists in the shaded region of linear absolute instability (Couairon & Chomaz 1997a).

to the successive transitions: stability $\rightarrow$ convective instability $\rightarrow$ absolute instability. Expressions (3.41) may be invoked to recast the Ginzburg–Landau dispersion relation (3.16) into the form

$$\omega = \omega_0 + \tfrac{1}{2}\omega_{kk}(k - k_0)^2, \qquad (3.42)$$

with $\omega_{kk} = 2(c_d - \mathrm{i})$. By applying the inverse transformations $k \rightarrow -\mathrm{i}\partial/\partial x$, $\omega \rightarrow \mathrm{i}\partial/\partial t$ to (3.42), the Ginzburg–Landau equation (3.15) is then rewritten equivalently as

$$\mathrm{i}\frac{\partial \psi}{\partial t} - \mathrm{i}\omega_{kk}k_0\frac{\partial \psi}{\partial x} - \left(\tfrac{1}{2}\omega_{kk}k_0^2 + \omega_0\right)\psi + \tfrac{1}{2}\omega_{kk}\frac{\partial^2 \psi}{\partial x^2} = 0. \qquad (3.43)$$

This convenient form, which explicitly displays the dependence of the coefficients on the absolute wavenumber k_0 and frequency ω_0, will be made use of later on in the chapter.

The absolute wavenumber k_0 and frequency ω_0 are defined by the zero-group-velocity condition (3.40). The dispersion relation $\omega(k) = \omega$ viewed as an equation for the complex unknown k therefore admits a double root k_0 at ω_0. In other words, the point k_0 must be a point of contact between two otherwise distinct spatial branches as ω approaches ω_0. For instance, in the case of the Ginzburg–Landau equation, the absolute wavenumber and frequency are given by (3.41) and the spatial branches $k^+(\omega)$ and $k^-(\omega)$, solutions of (3.42), are readily obtained as

$$k^{\pm}(\omega) = k_0 \pm \left(\frac{2}{\omega_{kk}}\right)^{1/2} (\omega - \omega_0)^{1/2}, \qquad (3.44)$$

thereby demonstrating the fact that k_0 is a double root at $\omega = \omega_0$. This observation has led Bers (1983) to devise an ingenious geometrical method to determine k_0 and ω_0. The scheme relies on monitoring the loci of $\omega(k)$ in the complex ω-plane and of $\{k^+(\omega), k^-(\omega)\}$ in the complex k-plane, as the integration contour L_ω is gradually lowered. A typical scenario is illustrated in figures 13 and 14 for convective and absolute instability respectively in the case of the complex Ginzburg–Landau equation. Figures 13(*a*) and 14(*a*) refer to the initial configuration prevailing in figure 9, where F_k lies on the real k_r-axis and L_ω is sufficiently high up to comply with causality. As L_ω is lowered, the branches $k^+(\omega)$ and $k^-(\omega)$ no longer remain confined to their respective upper and lower half-k-planes but they migrate towards one another (figures 13*b* and 14*b*). In this process, the spatial branch $k^+(\omega)$ crosses the k_r-axis into the lower half-k-plane. To preserve causality, the contour F_k must be deformed to maintain $k^+(\omega)$ and $k^-(\omega)$ on the same side of F_k. As L_ω is further lowered, there comes a stage (figures 13*c* and 14*c*) where F_k becomes pinched by $k^+(\omega)$ and $k^-(\omega)$ at a point k_0 necessarily associated with a double root of $\omega(k) = \omega$. Correspondingly, the temporal mode $\omega(k)$ develops a cusp at $\omega_0 = \omega(k_0)$ in the complex ω-plane (Kupfer, Bers & Ram 1987), which is the signature of the zero-group-velocity condition (3.40). If one attempts to push L_ω further down, the two spatial branches flip over and no contour F_k can be found that avoids intersecting them. Causality has therefore been violated and this last step is illegal! The only feature that distinguishes figures 13(*c*) and 14(*c*) is the level reached by L_ω when pinching takes place. In the convectively unstable case, $\omega_{0,i} < 0$ and L_ω is in the lower half-ω-plane. In the absolutely unstable case, $\omega_{0,i} > 0$ and L_ω is in the upper half-ω-plane.

As emphasized by Bers (1983), the only pertinent zero-group-velocity wavenumber is the first point encountered as L_ω is lowered, which involves pinching of F_k by two distinct k^+ and k^- branches initially located entirely in the upper and lower half-k-planes for high enough L_ω. Saddle points involving two k^+ or two k^- branches are irrelevant since they do not pinch F_k. The scheme unambiguously finds the absolute wavenumber k_0 and frequency ω_0.

The above geometrical analysis indicates that F_k and L_ω may be chosen to take a continuum of positions in the complex k- and ω-planes, provided

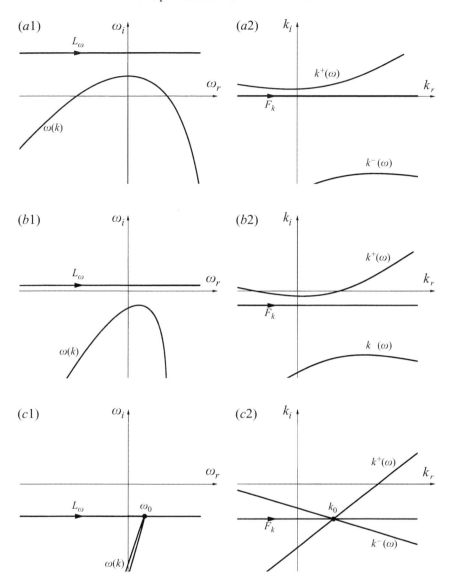

Figure 13. Locus of temporal and spatial branches as the L_ω contour is lowered in the complex ω-plane. Case of the Ginzburg–Landau equation in the convectively unstable range $0 \leqslant \mu \leqslant U^2/[4(1 + c_d^2)]$. See text for details.

that causality is preserved. If, in the course of deforming the contours, one of the k-branches has migrated away from its original half-k-plane, we unambiguously know on which side of F_k it must lie, as illustrated in figures 13(b) and 14(b). This information is essential in the evaluation of the k-integrals, as the next section demonstrates.

is triggered by the appearance of a pocket of absolute instability downstream of the nozzle exit, which acts as a source of instability waves and leads to the generation of synchronized oscillations.

As a second example of an application, consider the case of bluff body wakes discussed in § 2.2. Koch (1985) appears to have been the first to demonstrate the existence of a region of local absolute instability in the wake behind a blunt-edged plate and to suggest that it might be responsible for the onset of the Kármán vortex street. The circular cylinder wake also was shown by Triantafyllou, Triantafyllou & Chrysostomidis (1986) to display a region of absolute instability in the Kármán vortex shedding régime. The local analysis of Monkewitz (1988) is chosen here to illustrate the procedure and the main results for the case of circular cylinder wakes. The mean flow at each downstream station is modelled by the non-dimensional velocity profile

$$U(y; R, N) = 1 - R + 2R\, U_1(y; N),\tag{3.52}$$

where the shape function $U_1(y\,;\,N)$ is given by

$$U_1(y; N) = [1 + \sinh^{2N}\{y \sinh^{-1}(1)\}]^{-1},\tag{3.53}$$

and $R = (U_c - U_\infty)/(U_c + U_\infty)$ denotes the velocity ratio based on the free-stream velocity U_∞ and the centreline velocity U_c. The effect of varying R and the shape parameter N is illustrated in figure 18. Wake and jet profiles are obtained for $R < 0$ and $R > 0$ respectively. In the range $-1 < R < 0$, the wake is coflowing (figure 18a), whereas a finite core counterflow is present for $R < -1$ (figure 18c). The zero-centreline-velocity wake corresponds to the limiting value $R = -1$ (figure 18b). The shape parameter N allows a continuous tuning of the steepness of the shear layers (figure 18d): when $N = \infty$, $U_1(y; N)$ reduces to a top-hat profile. When $N = 1$, the $\mathrm{sech}^2\, y$ fully developed velocity distribution is recovered. Each downstream station in the wake effectively corresponds to specific settings of R and N.

The viscous dispersion relation is determined numerically by solving the Orr–Sommerfeld equation (3.6) for the family (3.52), subject to exponential-decay boundary conditions at $y = \pm\infty$. Note that the Reynolds number $\widehat{Re} = \overline{U}\delta/\nu$ based on the wake half-width δ and the average velocity $\overline{U} = (U_c + U_\infty)/2$ is different from its counterpart Re based on cylinder diameter d introduced in § 2.2. For wakes ($R < 0$), the varicose mode (odd $\phi(y)$) is found to be at most convectively unstable and the present discussion is limited to the case of the sinuous mode (even $\phi(y)$) which precisely displays the same symmetry properties as the Kármán vortex street. Final results are presented in figure 19. In the inviscid limit ($\widehat{Re} = \infty$), the onset of absolute instability first takes place at the shape parameter value $N \sim 2$

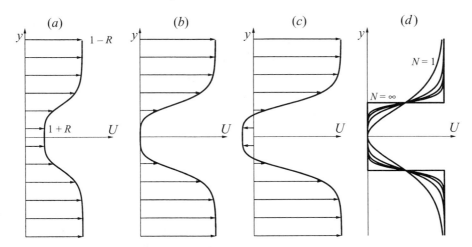

Figure 18. Effect of velocity ratio R and shape parameter N on basic wake velocity profile (3.52). (a) $-1 < R < 0$; (b) $R = -1$; (c) $R < -1$; (d) effect of increasing $N = 1, 2, 3, 4, \infty$ at $R = -1$.

for a coflowing wake $R = -0.85$. Thus, an appropriately shaped velocity profile may undergo a transition to absolute instability, *even in the absence of counterflow*. Viscous effects are seen to delay this onset: at finite Reynolds numbers, the transition curve is gradually displaced towards the counterflow side $R < -1$ with decreasing $\widehat{Re}$. Comparison with experiments indicates that the family (3.52), (3.53) faithfully reproduces the measured local velocity distributions in the near wake of circular cylinders in the range of cylinder Reynolds numbers $Re < Re_{Gc} = 48.5$. In this steady flow configuration, the streamwise station most susceptible to absolute instability is located one diameter dowstream of the cylinder axis at the point of maximum counterflow velocity, in the same cross-stream plane as the steady vortex centres within the recirculation bubble (see §2.2). Monkewitz's conclusions may then be summarized as follows: in the interval $5 < \widehat{Re} < 25$ corresponding to $2 < Re < 10$, the sinuous mode becomes locally convectively unstable in a gradually increasing streamwise domain around $x = d$. In the interval $25 < Re < 48.5$ corresponding to $10 < \widehat{Re} < 16$, a pocket of local absolute instability is nucleated around $x = d$, its streamwise extent increasing with Re to cover a larger and larger portion of the recirculation bubble. From the local instability point of view, the cylinder wake experiences the following transition scenario: locally stable everywhere $\mapsto$ region of local convective instability $\mapsto$ region of local absolute instability embedded within a convectively unstable domain. From the global instability point of view, the flow has undergone a single transition to an oscillatory régime at $Re = Re_{Gc}$ via

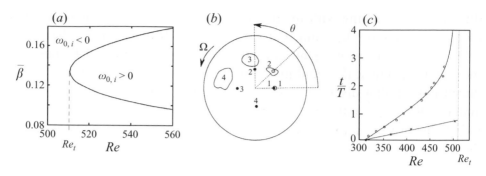

Figure 20. (*a*) Absolute/convective instability boundary ($\omega_{0,i} = 0$) in the ($Re, \bar{\beta}$)-plane for the boundary layer on a rotating disk. (*b*) Impulse response of the rotating disk boundary layer at $t = 0$, $T/4$, $T/2$, $3T/4$ corresponding to four source hole portions marked by a black circle, where T is the rotation period. (*c*) Wavepacket leading ($\times$) and trailing ($\circ$) edge trajectory in the ($Re, t/T$)-plane. The absolute/convective instability transition Reynolds number $Re_t = 510$ is indicated by a dotted line. After Lingwood (1995, 1996).

(r, θ, t)-space onto the (r-t)-plane. The resulting leading- and trailing-edge trajectories in the (r, t)-plane are depicted in figure 20(*c*). In the convectively unstable ring region defined by $Re_s < Re < Re_t$, the leading edge is seen to travel radially outward at a constant velocity, while its trailing edge gradually slows down to reach a complete stop for large time at the radial station $Re_t = 510$. The steepening of the trailing-edge trajectory merely reflects the proximity of the transition radius $Re_t = 510$ to absolute instability. As Re_t is approached, the local impulse response changes from that illustrated in figure 8(*b*) to that illustrated in figure 8(*c*). The predictions of the theory are therefore fully confirmed by experimental observations.

Even more striking is the fact that transition to turbulence, signalled by an abrupt order-of-magnitude increase in the background noise level, takes place at approximately the same critical radius $Re \sim 513$ in a variety of experiments including those of Lingwood (1996). The transition value is observed to be insensitive to low-level excitations. This feature seems to indicate that transition to turbulence is triggered by minute amounts of noise which are capable of being amplified in the region of local absolute instability, as soon as the radius $Re_t = 510$ is reached. This scenario is radically distinct from the one prevailing in non-rotating convectively unstable boundary layers and shear layers, where the local transition Reynolds number to turbulence is highly sensitive to the ambient disturbance level and spectral content.

Boundary layers on the rotating disk and on swept wings share common features: the radial velocity profile $U(z)$ of the former and the cross-flow velocity profile of the latter both display inflection points. This observation has led to the conjecture that transition to turbulence on swept wings might also be the result of an absolute instability. Unfortunately, this does not seem to be the case. According to the comprehensive search of Lingwood (1997) and Taylor & Peake (1998), swept-wing boundary layers do not exhibit transition to a genuine absolute instability: the zero-group-velocity condition (3.40) is never simultaneously satisfied along the chord and along the span.

4 Global instability analyses

The theoretical notions presented in §3 have been introduced in the framework of steady parallel shear flows that are strictly invariant with respect to translations in the streamwise x-direction. These techniques have then been applied to several spatially developing flows by assuming that they remain pertinent locally at each streamwise station, provided that x is 'frozen'. The objective of the present section is to justify this assumption and to demonstrate that parallel flow theory may be incorporated into a consistent global analysis of *slowly* spatially developing flows. The main underlying assumption is the following: the flows of interest, say the mixing layer generated downstream of a splitter plate illustrated in figure 1, should display two distinct well-separated length scales: a typical instability wavelength λ that has been shown in §3 to be of the same order of magnitude as the vorticity thickness δ_ω, and a length scale $L \sim (1/\delta_\omega)(\mathrm{d}\delta_\omega/\mathrm{d}x)^{-1}$ characterizing the streamwise non-uniformity of the basic flow. The small parameter

$$\varepsilon \sim \frac{\lambda}{L} \sim \frac{\delta_\omega}{L} \ll 1 \qquad (4.1)$$

is introduced to make this scale separation explicit. As discussed in Huerre & Rossi (1998), a consistent WKBJ asymptotic analysis (Bender & Orszag 1978) should then be implemented for the Navier–Stokes or Euler equations linearized around the streamwise developing basic flow $U = U(y; X)e_x + \varepsilon V(y; X)e_y$ of interest, where $X = \varepsilon x$ is a slow streamwise variable. As in §3, we choose instead to present the formulation in the idealized and less ponderous setting of the partial differential operator in one space dimension

$$D\left[-\mathrm{i}\frac{\partial}{\partial x}, \mathrm{i}\frac{\partial}{\partial t} ; X\right] \psi + \varepsilon D_\varepsilon\left[-\mathrm{i}\frac{\partial}{\partial x}, \mathrm{i}\frac{\partial}{\partial t} ; X\right] \psi = O(\varepsilon^2), \qquad (4.2)$$

where D_ε represents non-parallel flow corrections associated with the $O(\varepsilon)$ streamwise derivative $\varepsilon \partial U/\partial X$ of the basic flow $U(y; X)$ and with the cross-stream velocity $\varepsilon V(y; X)$. For brevity, the dependence on the global parameter frequency at each X defined by

$$\frac{\partial \omega}{\partial k}(k_0; X, R) = 0, \quad \omega_0(X, R) = \omega(k_0; X, R). \tag{4.12}$$

The local nature of the instability for a given flow is conveniently summarized by representing the growth rates $\omega_{i,\max}(X; R)$ and $\omega_{0,i}(X; R)$ as functions of streamwise distance X (Huerre & Monkewitz 1990). From a local instability standpoint, four broad classes of spatially developing flows may be distinguished, as sketched in figure 21. The first class (figure 21a) corresponds to flows that are *locally stable* everywhere ($\omega_{i;\max} < 0$, $\omega_{0,i} < 0$ for all X). If a region of convective instability is present (figure 21b) where $\omega_{i,\max} > 0$ but $\omega_{0,i} < 0$, the flow is said to be *locally convectively unstable*. The third class (figure 21c) pertains to flows that are *marginally absolutely unstable* with $\omega_{0,i}$ reaching almost zero at some X. In the fourth class (figure 21d), a region of absolute instability is present where $\omega_{i,\max} > 0$ and $\omega_{0,i} > 0$, and the flow is said to be *locally absolutely unstable*. Any given spatially developing flow may belong in succession to some or all of the four classes as an external control parameter R is increased through the ranges $R < R_c$ (figure 21a), $R_c < R < R_t$ (figure 21b), $R \sim R_t$ (figure 21c) and $R > R_t$ (figure 21d). Mixing layers have been seen to cover the last three classes as the velocity ratio increases through the value $R_t = 1.315$ whereas bluff-body wakes cover all possible cases as the Reynolds number Re increases through $Re_c = 5$ and $Re_t = 25$.

4.1 Spatial wave amplification in locally convectively unstable flows

The development of flows that are at most locally convectively unstable everywhere (figure 21b) is essentially controlled by the growth and decay of external perturbations. In the linear approximation, it is legitimate to follow separately the fate of each frequency ω_f as it evolves according to the local instability characteristics (4.4) prevailing at each streamwise station. In the context of the partial differential equation (4.2), the signalling problem associated with the real forcing frequency ω_f is governed by

$$\left\{ D\left[-i\frac{\partial}{\partial x}, i\frac{\partial}{\partial t}; X \right] + \varepsilon D_\varepsilon \left[-i\frac{\partial}{\partial x}, i\frac{\partial}{\partial t}; X \right] \right\} \psi_{\text{forcing}}(x, t) = \delta(x)\,e^{-i\omega_f t}. \tag{4.13}$$

The time-harmonic steady-state response is expected to be

$$\psi_{\text{forcing}}(x, t) \sim A^{\pm}(X)\exp\left(i\left[\frac{1}{\varepsilon}\int_0^X k^{\pm}(X'; \omega_f)\,dX' - \omega_f t \right] \right), \tag{4.14}$$

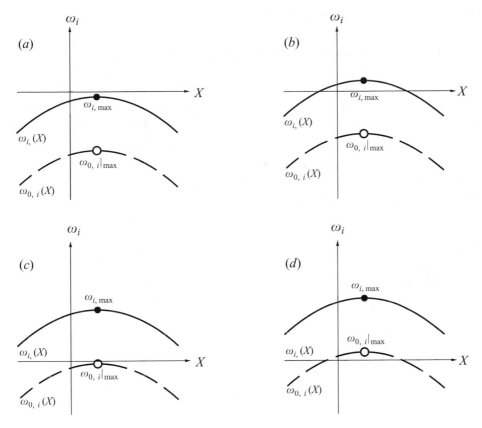

Figure 21. Classes of spatially developing flows according to the nature of the local instability. (*a*) Uniformly stable $R < R_c$; (*b*) convectively unstable $R_c < R < R_t$; (*c*) almost absolutely unstable $R \sim R_t$; (*d*) pocket of absolute instability $R > R_t$ (Huerre & Monkewitz 1990).

where the superscripts $+$ and $-$ refer to the domains $x > 0$ and $x < 0$ on either side of the source at $x = 0$. For simplicity, the discussion is restricted to the case where only one k^+-branch and one k^--branch exist. The amplitude functions $A^+(X)$ and $A^-(X)$ are given by (4.9) with $\omega = \omega_f$ and $k = k^+$ and $k = k^-$ respectively. The requirement that $\psi_{\text{forcing}}(x, t)$ be continuous across the source $x = 0$ results in the condition $A^+(0) = A^-(0) = 1$.

A similar formulation holds for the signalling problem associated with the Navier–Stokes or Euler equations linearized around a basic flow of the form $U = U(y; X) e_x + \varepsilon V(y; X) e_y$. If it is locally convectively unstable everywhere, the time-harmonic linear response to a perturbation of real frequency ω_f is given by (4.10) suitably adapted as in (4.14). The WKBJ formulation has been successfully applied in a variety of convectively unstable spatially developing shear flows: boundary layers (Bouthier 1972), jets (Crighton & Gaster 1976), mixing layers (Gaster, Wygnanski & Kit 1985), etc. For instance, the linear inviscid response predicted by (4.10) downstream of the source $(x > 0)$ has been compared with experimental observations in forced turbulent mixing layers by Gaster *et al.* (1985). The flow is forced mechanically at the trailing edge of the splitter plate separating the two streams by a small oscillating flap. Mean flow measurements confirm that velocity profiles are satisfactorily represented, in non-dimensional variables scaled with respect to the average velocity $\overline{U}$ and the initial thickness $\delta_\omega(0)/2$, by the self-similar family of hyperbolic tangent functions

$$U(y;X) = 1 + R \tanh \left[\frac{y}{\delta(X)} \right] \tag{4.15}$$

of varying non-dimensional thickness $\delta(X) \equiv \delta_\omega(X)/\delta_\omega(0)$. The self-similar nature of the local profiles (4.15) results in a local dispersion relation that is also self-similar: the local dispersion relation $D(k, \omega; X, R) = 0$ associated with $U(y;X) = 1 + R \tanh [y/\delta(X)]$ is simply deduced from the dispersion relation $D_{\text{th}}(k, \omega; R) = 0$ calculated in §3.1 for the parallel profile $U(y) = 1 + R \tanh y$ by the transformation

$$D(k, \omega; X, R) = D_{\text{th}}[k\,\delta(X), \omega\,\delta(X); R] = 0. \tag{4.16}$$

Since $\delta(X) = \delta_\omega(X)/\delta_\omega(0)$, $k\,\delta(X)$ and $\omega\,\delta(X)$ are then effectively scaled with respect to half the local vorticity thickness $\delta_\omega(x)/2$. The only part of the response that is observable is associated with the spatially amplified dowstream branch k_1^+ determined in §3.1. At leading order in ε, the local spatial growth rate is $-k_{1,i}^+(X, \omega_f)$. The streamwise development of a given forcing frequency ω_f is determined at leading order by the spatial growth rate curves depicted in figure 17, provided that $-k_{1,i}^+$ and ω_f are suitably scaled with respect to $\delta_\omega(x)/2$. As already discussed in §3.5, comparison with experiments confirms that the local dispersion relation (4.16) holds approximately. In order to pursue the evaluation of (4.10) to include terms of order unity in front of the exponential, one must calculate the complex eigenfunction solution $\Phi(y;X)$ of the Rayleigh equation (3.9) for the eigenvalue pair k_1^+, ω_f, at each station X. The amplitude function $A(X)$ is determined by solving the evolution equation (4.8) numerically. The theoretically predicted growth of the fluctuation amplitude is compared with experimental data in figure 22. The WKBJ approximation is seen to yield amplitude levels that exceed measured values by a significant margin. This discrepancy is not unexpected in view of the linear approximation that is enforced throughout.

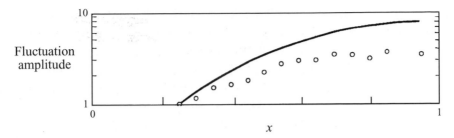

Figure 22. Overall amplification of disturbances with streamwise distance x from the trailing edge of the splitter plate for a spatially developing mixing layer of forcing Strouhal number $St \equiv f\delta_\omega(0)/\overline{U} = 0.96$. Open circles denote measured values and the continuous line is the WKBJ theoretical prediction. After Gaster *et al.* (1985).

4.2 Global intrinsic oscillations in locally absolutely unstable flows

According to the experimental observations presented in § 3.5, spatially developing flows that are locally absolutely unstable within some finite streamwise interval (figure 21*d*), for instance counterflow mixing layers of velocity ratio $R > R_t$, or bluff-body wakes of Reynolds number $Re > Re_{Gc} > Re_t$, display time-periodic oscillations that are relatively insensitive to low levels of external noise. The objective of this subsection is to outline a theoretical description of such synchronized states in terms of so-called *global modes*, i.e. extended wavepackets that live within the spatially developing flow of interest over a distance of the same order of magnitude as the non-uniformity length scale $L \gg \lambda$, and beat at a well-defined frequency. The formulation is restricted to the linear approximation and it is developed in the context of the one-space-dimensional partial differential equation (4.2) in an infinite medium.

Just as in the parallel flow case, it is appropriate to first introduce definitions of global stability and instability based on the long-time behaviour of the global Green function $G(x, t)$ governed by

$$\left\{ D \left[-i\frac{\partial}{\partial x}, i\frac{\partial}{\partial t}; X \right] + \varepsilon D_\varepsilon \left[-i\frac{\partial}{\partial x}, i\frac{\partial}{\partial t}; X \right] \right\} G(x, t) = \delta(x)\delta(t), \quad (4.17)$$

where $X = \varepsilon x$ is no longer a frozen parameter. The spatially developing flow is said to be *globally stable* if

$$\lim_{t \to \infty} G(x, t) = 0 \text{ for all } x. \quad (4.18)$$

It is said to be *globally unstable* if

$$\lim_{t \to \infty} G(x, t) = \infty \text{ for some } x. \quad (4.19)$$

When the flow is globally unstable, the long-time régime of G is expected to be dominated by a linear superposition of global mode solutions which, for steady spatially developing flows, take the form

$$\psi(x,t) = \phi(X)\,e^{-i\omega_G t}. \tag{4.20}$$

The main goal of the global analysis is the determination of the complex *global frequencies* ω_G and associated global eigenfunctions $\phi(X)$. Given the spectrum of global frequencies ω_G, criteria for global stability and instability readily follow: if $\omega_{G,i} < 0$ for all ω_G, the spatially developing flow is globally stable; if $\omega_{G,i} > 0$ for some ω_G, it is globally unstable. To illustrate the procedure involved in the calculation of global modes consider as usual the partial differential operator (4.2) in the infinite domain $-\infty < x < +\infty$, subjected to the boundary conditions

$$\lim_{x\to\pm\infty} \psi(x,t) = 0. \tag{4.21}$$

As a preliminary step and following reasoning presented in Chomaz, Huerre & Redekopp (1991) and Monkewitz, Huerre & Chomaz (1993), a *global frequency selection criterion* for ω_G is inferred by examining the long-time behaviour of the global Green function $G(x,t)$ defined in (4.17). In the same spirit as in §3.3, let $G(x,\omega)$ denote the time Fourier transform of $G(x,t)$ such that

$$G(x,t) = \frac{1}{2\pi}\int_{L_\omega} G(x,\omega)\,e^{-i\omega t}\mathrm{d}\omega, \tag{4.22}$$

where the straight line contour L_ω is chosen to be parallel to the real ω-axis and above all singularities to comply with causality (figure 23a). As in the signalling problem considered in §4.1, the form of $G(x,\omega)$ is immediately inferred from the general WKBJ approximation (4.3) for waves of arbitrary frequency ω. Thus,

$$G(x,\omega) \sim A^{\pm}(X)\exp\left(\frac{i}{\varepsilon}\int_0^X k^{\pm}(X';\omega)\,\mathrm{d}X'\right), \tag{4.23}$$

where the superscripts $+$ and $-$ refer to $x > 0$ and $x < 0$ respectively. The amplitude functions $A^{\pm}(X)$ are given by (4.9) with $k = k^+$ and $k = k^-$ respectively, and continuity across $x = 0$ requires that $A^+(0) = A^-(0) = 1$.

In order to proceed further, it is necessary to examine the behaviour of the WKJB solutions (4.23) in the complex X-plane. According to (4.9), the complex amplitude $A^{\pm}(X)$ is seen to exhibit a singularity at complex points X of zero group velocity:

$$\omega_k^{\pm} = \frac{\partial\omega}{\partial k}[k^{\pm}(X;\omega);X] = 0. \tag{4.24}$$

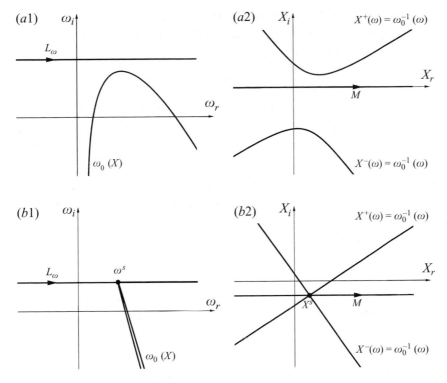

Figure 23. Global frequency selection criterion. (*a*) Integration contour L_ω in the complex ω-plane lies above the locus of singularity $\omega_0(X)$ as X travels on M; conversely, loci of $X^+(\omega)$ and $X^-(\omega)$ as ω travels on L_ω are well separated on either side of contour M in the complex X-plane. (*b*) The pinching process as L_ω is gradually lowered; global frequency is given by ω^s.

This zero-group-velocity condition effectively states that at such points, the spatial branches $k^+(X;\omega)$ and $k^-(X;\omega)$ coincide, i.e.

$$k^+(X;\omega) = k^-(X;\omega). \tag{4.25}$$

Equivalently, according to the definition (4.12) of the local absolute frequency $\omega_0(X)$, solutions of (4.24) or (4.25) are zeros of the relation

$$\omega_0(X) = \omega. \tag{4.26}$$

In the context of WKBJ theory (Bender & Orszag 1978), condition (4.25) defines so-called turning points in the vicinity of which the standard WKBJ solutions (4.23) break down. The fact that turning points are in general complex would seem at first sight to make them irrelevant on the 'physical' X_r-axis. It is, however, well established that immersion in the complex X-plane is necessary to compare the + and − spatial solutions. Such spatial solutions are in general exponentially small or large with respect to each other on the real X_r-axis, and it is precisely at complex turning points that they become comparable and may be connected.

In view of (4.26), $G(x, \omega)$ becomes singular at points $X = \omega_0^{-1}(\omega)$, where $\omega_0^{-1}(\omega)$ stands for the reciprocal function of $\omega_0(X)$. The dominant frequency ω_G emerging from the impulse response $G(x, t)$ is now qualitatively inferred by adapting the reasoning in §3.3. In the parallel flow case, the impulse response $G(x, t)$ was shown to be dominated, at large time in the laboratory frame $x/t = 0$, by the absolute frequency $\omega_0 = \omega(k_0)$ which involved pinching of the F_k-contour at k_0 by two spatial branches $k^+(\omega)$ and $k^-(\omega)$ originating from the upper and lower half-k-plane respectively (figures 13 and 14). The same result holds here, provided that the complex k-plane is replaced by the complex X-plane, as sketched in figure 23($a2$). Just as $k(\omega)$ is multiple-valued, with two branches $k^+(\omega)$ and $k^-(\omega)$, $\omega_0^{-1}(\omega)$ is also multiple-valued with, say, two branches $X^+(\omega)$ and $X^-(\omega)$ that are well separated on either side of the M-contour originally on the X_r-axis, provided that L_ω lies above all singularities $\omega_0(X)$ (figure 23$a1$). As L_ω is lowered, the M-contour may become pinched at a complex X^s by the two branches $X^+(\omega)$ and $X^-(\omega)$ (figure 23$b2$). Correspondingly, the absolute frequency locus $\omega_0(X)$ develops a cusp on the L_ω-contour at $\omega^s = \omega_0(X^s)$ in the complex ω-plane (figure 23$b1$). It is then no longer possible to lower L_ω and the dominant contribution to the long-time behaviour of $G(x, t)$ is expected to arise from the frequency ω^s. Typically, the pinching point X^s is a saddle point defined by

$$\frac{\mathrm{d}\omega_0}{\mathrm{d}X}(X^s) = 0. \tag{4.27}$$

The above argument makes it plausible that the dominant frequency emerging from the impulse response is given by $\omega^s = \omega_0(X^s)$, where the complex point X^s is determined by the saddle point condition (4.27). This global frequency selection criterion was originally derived by Chomaz *et al.* (1991) in the framework of the linearized complex Ginzburg–Landau equation with varying coefficients (4.5). Hunt & Crighton (1991) have devised an elegant procedure to calculate $G(x, t)$ exactly for the same model equation. For quadratic variations of $\omega_0(X)$, the long-time behaviour of $G(x, t)$ is then shown to reduce to a global mode of frequency ω^s. The frequency selection criterion (4.27) is therefore causal. In view of definition (4.12) for $\omega_0(X)$, it may be stated in the following alternative way: given the local dispersion relation $\omega(k; X)$, the global frequency ω_G is given, at leading order in ε, by $\omega_G \sim \omega^s = \omega(k^s; X^s)$, where the complex wavenumber k^s and spatial station

X^s satisfy the saddle point conditions

$$\frac{\partial \omega}{\partial k}(k^s; X^s) = \frac{\partial \omega}{\partial X}(k^s; X^s) = 0. \qquad (4.28a, b)$$

The global mode structure associated with $(4.28a, b)$, equivalently (4.27), may be determined by a systematic WKBJ approximation scheme. The procedure is now summarized in the context of the partial differential equation (4.2) and boundary conditions (4.21). A detailed derivation is given in Huerre & Rossi (1998). The absolute frequency $\omega_0(X)$ is assumed to display a saddle point at X^s defined by (4.27). The absolute growth rate $\omega_{0,i}(X)$ then exhibits a maximum $\omega_{0,i}|_{max}$ somewhere on the X_r-axis, as in the various classes of spatially developing flows depicted in figure 21. Substitution of global mode solutions of the form (4.20) into (4.2) and (4.21) leads to the eigenvalue problem in the streamwise direction

$$\left\{ D\left(-i\frac{d}{dx}, \omega_G; X\right) + \varepsilon D_\varepsilon \left(-i\frac{d}{dx}, \omega_G; X\right) \right\} \phi(X) = 0, \qquad (4.29a)$$

$$\lim_{X \to \pm\infty} \phi(X) = 0. \qquad (4.29b)$$

The extended wavepacket making up the global mode is expected to consist of k^+ and k^- spatial waves given by (4.3), i.e.

$$\phi^\pm(X) \sim A^\pm(X) \exp\left(\frac{i}{\varepsilon} \int_{X^s}^X k^\pm(X'; \omega^s) \, dX'\right), \qquad (4.30a)$$

and the global frequency is assumed, by construction, to admit an expansion of the form

$$\omega_G \sim \omega^s + \varepsilon\omega_2 + \dots. \qquad (4.30b)$$

The complex amplitude functions $A^\pm(X)$ are found to satisfy an amended version of evolution equation (4.8), namely

$$i\omega_k^\pm \frac{dA^\pm}{dX} + \left\{ \omega_2 - \delta\omega^\pm - \frac{1}{2}i\frac{D_{kk}^\pm}{D_\omega^\pm}k_X^\pm \right\} A^\pm = 0, \qquad (4.31)$$

where $\omega_k^\pm = (\partial\omega/\partial k)[k^\pm(\omega^s; X); X]$, $D_{kk}^\pm = D_{kk}[k^\pm(\omega^s; X), \omega^s; X]$, etc. Its solution is

$$A^\pm(X) = A^\pm(X^s) \exp\left\{ i\int_{X^s}^X \frac{\omega_2 - \delta\omega^\pm - \frac{1}{2}i(D_{kk}^\pm/D_\omega^\pm)k_{X'}^\pm}{\omega_k^\pm} \, dX' \right\}. \qquad (4.32)$$

To proceed further, it is essential to determine whether a given spatial branch is dominant (exponentially large) or subdominant (exponentially small) with respect to the other one in specific sectors of the complex X-plane. According

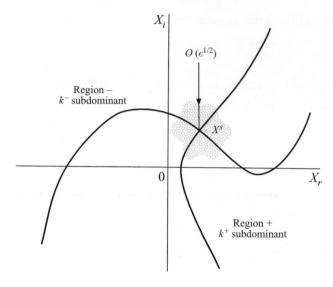

Figure 24. Sketch of outer WKBJ regions and inner turning point (or saddle point) region of size $O(\varepsilon^{1/2})$ in the complex X-plane. The curves represent Stokes lines around saddle point X^s.

to WKBJ theory (Bender & Orszag 1978), such domains are delineated by so-called Stokes lines emanating from turning points of the system. In the case of the second-order turning point X^s, there are four Stokes lines which delineate four sectors of angle $\pi/2$ around X^s, as depicted in figure 24. The $+$ and $-$ solutions are of equal order of magnitude on Stokes lines whereas, within a given Stokes sector, they each take on a specific dominant or subdominant character. Of particular interest are the two sectors marked $-$ and $+$ in figure 24, which include the boundary points $X_r = -\infty$ and $X_r = +\infty$ respectively. Let us assume that $k_i^+(X) > 0$ and $k_i^-(X) < 0$ for sufficiently large $|X|$ along the X_r-axis (the relationship between this assumption and causality is discussed in Le Dizès *et al.* 1996). Then $\phi^-(X)$ is exponentially small (resp. large) in the $-$ (resp. $+$) sectors, whereas $\phi^+(X)$ is exponentially small (resp. large) in the $+$ (resp. $-$) sectors. In order to comply with boundary conditions (4.29b), the global eigenfunction is necessarily solely composed of the $k^-(X)$ branch in the sector $-$ and of the $k^+(X)$ branch in the sector $+$. It must then be demonstrated that these solutions may be smoothly connected through the saddle point X^s by suitably choosing the eigenvalue ω_G, i.e. ω_2. Thus, the WKBJ approximations $\phi^\pm(X)$ specified in (4.30a), (4.32) appear as outer solutions valid in the $+$ and $-$ sectors of the complex X-plane respectively.

In order to study the inner shaded area around the saddle point X^s (figure

24), it is necessary to introduce an inner variable $\overline{X}$ based on the estimated size $O(\varepsilon^{1/2})$ of the inner region, defined by

$$\overline{X} = \frac{X - X^s}{\varepsilon^{1/2}}. \tag{4.33}$$

The inner global mode structure is taken to be of the form

$$\overline{\phi}(\overline{X}) \sim \mu(\varepsilon) \left[\overline{A}(\overline{X}) + \varepsilon^{1/2}\overline{\phi}_2(\overline{X}) + \varepsilon\overline{\phi}_3(\overline{X}) + \ldots\right] e^{ik^s(x - x^s)}, \tag{4.34}$$

where $\mu(\varepsilon)$ is an unknown gauge function and $\overline{A}(\overline{X})$ is the inner complex amplitude. Cumbersome but straightforward calculations up to $O(\varepsilon)$ lead to the inner evolution equation

$$\frac{1}{2}\omega_{kk}^s \frac{d^2\overline{A}}{d\overline{X}^2} - i\omega_{kk}^s k_{0X}^s \overline{X} \frac{d\overline{A}}{d\overline{X}} + \left[\omega_2 - \delta\omega^s - \frac{1}{2}\{\omega_{kk}(k_{0X}^s)^2 + \omega_{0XX}^s\}\overline{X}^2\right]\overline{A} = 0,$$
$$\tag{4.35}$$

where the various complex constants ω_{kk}^s, k_{0X}^s, $\delta\omega^s$ and ω_{0XX}^s denote the values taken by ω_{kk}, dk_0/dX, $\delta\omega$ and $d^2\omega_0/dX^2$ at X^s. If the substitution $-i\omega_2 \to \partial/\partial t$ is made, this equation effectively reduces to the linear Ginzburg–Landau equation with varying coefficients. The fact that this amplitude equation is encountered in a rational approximation scheme of the global mode spatial structure in the inner saddle point region, provides an *a posteriori* justification for its investigation as a model in itself. The inner evolution equation (4.35), together with matching conditions at large $|\overline{X}|$ between the inner solution and its outer counterparts (4.30a), constitute an eigenvalue problem. As derived in Huerre & Rossi (1998), the eigenvalue ω_2 may only take a discrete infinite set of values ω_{2n}. The most important result of this analysis then lies in the determination of the allowable global frequencies to $O(\varepsilon)$, which read

$$\omega_{Gn} \sim \omega^s + \varepsilon \left[\delta\omega^s - \frac{1}{2}i\omega_{kk}^s k_{0X}^s + (\omega_{kk}^s \omega_{0XX}^s)^{1/2}\left(n + \frac{1}{2}\right)\right], \tag{4.36}$$

where n is an integer and the complex frequency $\omega^s = \omega_0(X^s) = \omega(k^s; X^s, R)$ is given by the saddle point condition (4.27) or equivalently (4.28).

The above formulation was first implemented by Chomaz *et al.* (1991) and Le Dizès *et al.* (1996) in the framework of the linear Ginzburg–Landau equation with varying coefficients (4.5). The resulting linear global frequencies are given by (4.36) with $\delta\omega^s = 0$. A typical linear global eigenfunction is illustrated in figure 25 in the case where X_i^s is so small that the inner region of size $O(\varepsilon^{1/2})$ overlaps on the physical X_r-axis.

According to Monkewitz *et al.* (1993), the frequency selection criterion (4.36) still holds for two-dimensional spatially developing flows governed by

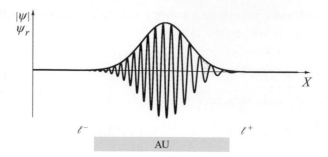

Figure 25. The linear global eigenfunction of complex Ginzburg–Landau equation (4.5) when X_i^s is small. Grey strip denotes locally absolutely unstable interval.

the linearized Navier–Stokes equations. The global mode spatial structure also remains valid, provided that the cross-stream eigenfunction $\Phi^\pm(y; X)$ is introduced in the manner (4.10) in all leading-order perturbation quantities. The global frequencies are seen to take infinitely many discrete values which are solely determined by the properties of the local dispersion relation at a single complex saddle point X^s. This location (in the complex X-plane!) acts as a smooth converter of k^+-waves into k^--waves. As demonstrated by Chomaz *et al.* (1991) and Le Dizès *et al.* (1996), global growth rates $\omega_{G,i}$ are necessarily lower than the maximum local absolute growth rate $\omega_{0,i}|_{\max}$ over all real X (figure 21), i.e.

$$\omega_{G,i} \leqslant \omega_{0,i}|_{\max} . \tag{4.37}$$

Thus, *a necessary condition for global instability is that a region of local absolute instability exist within the flow, as in figure 21(c, d)*. At most, convectively unstable flows (figure 21a, b) are necessary globally stable, i.e. they do not sustain amplified wavepackets of the form (4.20). This property provides an *a posteriori* justification for restricting the study of locally convectively unstable flows to the spatial response to external perturbations of given real frequency ω_f as analysed in §4.1. In order for a flow to become globally unstable, the region of absolute instability must be of sufficiently large streamwise extent. In the case of bluff-body wakes, there is a significant gap between the onset of absolute instability at $Re_t = 25$ and the onset of global instability at $Re_{Gc} = 48.5$, thereby indicating that the region of local absolute instability has to be fairly extensive to trigger a synchronized state. In the case of counterflow mixing layers, $R_t = 1.315$ and $R_{Gc} = 1.34$ almost coincide. In the case of the rotating-disk boundary layer, local absolute instability sets in at $Re_t = 510$ and transition to turbulence at $Re = 513$. There is no clearly observable global oscillation with a sharp frequency peak and it is speculated that a global wavepacket is indeed generated, but secondary instability mechanisms rapidly provoke transition to turbulence.

Is the linear selection criterion (4.27) consistent with the observed dynamics of spatially developing flows? This question has been addressed by using as a test case direct numerical simulations of the wake behind a blunt-edged plate aligned with the upstream flow. The computational studies of Karniadakis & Triantafyllou (1989), Oertel (1990) and Hammond & Redekopp (1997) among others, have established that this configuration undergoes a Hopf bifurcation to a global mode as the Reynolds number $Re = U_\infty h/v$ based on the free-stream velocity U_∞ and plate thickness h crosses a critical value Re_{Gc}. Hammond & Redekopp (1997) have made a detailed comparative study between global-mode theory and direct numerical simulations of the same flow, the main features of which are summarized below at the supercritical Reynolds number $Re = 160 > Re_{Gc}$, where $Re_{Gc} = 120$. All the results are made non-dimensional with respect to the free-stream velocity U_∞ and the plate thickness h. In the present framework, the Kármán vortex street displayed in figure 26(*a*) is considered as an extended wavepacket located within an underlying spatially developing mean flow. To illustrate this point of view, the streamwise distribution of the transverse velocity v is displayed in figure 26(*b*) at four different phases within a periodic cycle. The typical frequency spectrum presented in figure 26(*c*) confirms that the flow is periodic. The local instability characteristics are determined by solving the Orr–Sommerfeld equation (3.6) for the mean (i.e. time-averaged) velocity profile at each streamwise station. Application of the Briggs–Bers criterion (4.12) then yields the local absolute frequency $\omega_0(x)$. The saddle point criterion (4.27) may then be implemented by analytic continuation of the function $\omega_0(x)$ in the complex x-plane. It is found that $x^s = 0.79 + 0.078\mathrm{i}$ and $\omega_{0,r}(x^s) = 0.6321$. The predicted Strouhal frequency $St = fh/U_\infty$ is therefore $St = \omega_{0,r}(x^s)/2\pi = 0.1006$, to be compared with the 'observed' frequency (figure 26*c*) calculated in the numerical simulation $St = 0.1000$. Such a close agreement at a Reynolds number 25% above onset is somewhat unexpected: one would suspect nonlinearities to be strong enough to invalidate the linear frequency selection criterion. Note, however, that it has been applied here to the mean profile and not to the basic velocity profile in the absence of fluctuations. The mean profile includes the effect of Reynolds stresses, which partly takes into account nonlinearities.

It should be emphasized that the linear global frequency selection criterion (4.36) has been derived for flows that extend to infinity both upstream and downstream: the presence of the wake-producing body is simply ignored! In flows of semi-infinite extent say $0 \leqslant x < \infty$, for situations in which the

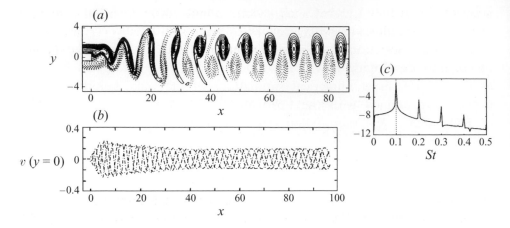

Figure 26. Direct numerical simulation of the wake behind a blunt-edged plate (Hammond & Redekopp 1997). (*a*) Vorticity contours (—, negative vorticity; $\cdots$, positive vorticity) at $Re = 320$; (*b*) global mode structure at $Re = 320$; (*c*) frequency spectrum at $x = 1$ and $y = 1.45$ for $Re = 160$.

absolute growth rate $\omega_{0,i}(x)$ is maximum at $x = 0$, the global frequency ω_G is, at leading order, given by the upstream-boundary absolute frequency $\omega_0(0)$ (Monkewitz *et al.* 1993). In this context, Woodley & Peake (1997) have demonstrated, through a rigorous asymptotic analysis, that airfoil wakes always support unstable global modes at large Reynolds numbers, provided they are sufficiently thick. By contrast, thin airfoil wakes only become globally unstable above a critical Reynolds number.

5 Absolute/convective and local/global instabilities in nonlinear systems

All the notions presented so far have been introduced in the framework of linear theory, whether one considers strictly parallel flows as in §3, or slowly spatially developing flows as in §4. In closing, we wish to highlight some of the recent advances made in extending absolute/convective and local/global instability concepts to the fully nonlinear régime. Only the essential results are presented and detailed derivations are omitted. Most studies have so far been limited to the analysis of nonlinear evolution equations in one space dimension for a perturbation field $\psi(x, t)$.

5.1 Nonlinear absolute/convective instability and front dynamics

Consider first the analogues of the strictly parallel flows studied in §3, namely nonlinear evolution equations in a doubly infinite domain, with constant coefficients. In general, such systems admit both a basic-state solution $\psi(x, t) = 0$ and finite-amplitude travelling wave solutions in certain parameter ranges. Popular examples include the *subcritical real Ginzburg–Landau equation*

$$\frac{\partial \psi}{\partial t} + U \frac{\partial \psi}{\partial x} - \mu \psi - \frac{\partial^2 \psi}{\partial x^2} = \psi^3 - \psi^5, \tag{5.1}$$

and the *supercritical complex Ginzburg–Landau equation*

$$\frac{\partial \psi}{\partial t} + U \frac{\partial \psi}{\partial x} - \mu \psi - (1 + \mathrm{i}c_d) \frac{\partial^2 \psi}{\partial x^2} = -(1 + \mathrm{i}c_n)|\psi|^2 \psi, \tag{5.2}$$

the properties of which are extensively discussed in Manneville (1990) and Cross & Hohenberg (1993). In the absence of nonlinear terms, both (5.1) and (5.2) reduce to the linear Ginzburg–Landau equation (3.15) considered in §3. In equation (5.2) the stabilizing cubic nonlinear term $|\psi|^2 \psi$ induces a supercritical Hopf bifurcation (Guckenheimer & Holmes 1983) to finite-amplitude travelling waves $\psi_2(x, t) = R \, \mathrm{e}^{\mathrm{i}(kx - \omega t)}$, k real, with $R = \sqrt{\mu - k^2}$ and $\omega = U k + c_d k^2 + c_n(\mu - k^2)$, as the control parameter μ crosses the linear instability threshold $\mu = 0$. In equation (5.1) the destabilizing cubic nonlinear term $+\psi^3$ induces a subcritical pitchfork bifurcation at $\mu = 0$ to unstable finite-amplitude states, but the stabilizing quintic nonlinear term ψ^5 ensures that stable steady spatially uniform states $\psi_2 = \pm[1/2 + (\mu + 1/4)^{1/2}]^{1/2}$ are recovered at large amplitude levels. In both systems, the stable finite-amplitude solutions ψ_2 are called *bifurcated states* to distinguish them from the basic state $\psi = 0$.

General definitions of nonlinear absolute/convective instability have been proposed in such a setting by Chomaz (1992). Consider an initial 'droplet' of bifurcated state introduced within a 'sea' of basic state as in figure 27. If all initial droplets of finite extent and amplitude shrink to zero, the system is said to be *stable* (figure 27a). If at least one droplet expands, it is *nonlinearly unstable* (figure 27b–d). The instability is *nonlinearly convective* if for all initial droplets of finite size and amplitude, the system relaxes to the basic state everywhere for large time (figure 27b). The instability is *nonlinearly absolute* if, for some initial droplet of finite size and amplitude, the system does not relax to the basic state everywhere for large time (figure 27d). These definitions should be compared with their linear counterparts in §3.2 (figure 8).

In the linear régime, the impulse response $G(x, t)$ was shown to take the form of a growing wavepacket delineated by leading and trailing edges of velocity V^+ and V^- such that $\sigma(V^\pm) = 0$, as illustrated in figures 8(b–d) and 11(a, b). According to these sketches, the linear absolute/convective nature

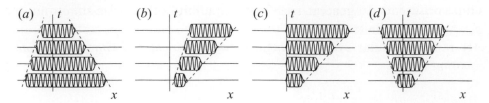

Figure 27. Nonlinear response to a localized 'droplet' of bifurcated state: (*a*) stable flow; (*b*) nonlinearly convectively unstable flow; (*c*) marginally nonlinearly convectively/absolutely unstable flow; (*d*) nonlinearly absolutely unstable flow. After Couairon & Chomaz (1997*a*).

of the instability may be determined by the direction of the trailing-edge velocity V^- defined by

$$\frac{\partial \omega}{\partial k}(k_*^-) = V^-, \qquad (5.3a)$$

$$\sigma(V^-) = \omega_i(k_*^-) - V^- k_{*i}^- = 0, \qquad (5.3b)$$

where k_*^- is the complex wavenumber observed along the ray $x/t = V^-$: if the trailing-edge velocity is such that $V^- > 0$ (resp. $V^- < 0$), the medium is convectively (resp. absolutely) unstable, as illustrated in figure 8(*b*, *d*).

In the nonlinear régime, the absolute/convective nature of the instability is similarly dictated by the direction of the front velocity V_f at the trailing edge of the droplet, as clearly seen from figure 27(*b*–*d*). The comprehensive analyses of Chomaz (1992) and Couairon & Chomaz (1996, 1997*a*) have firmly established the intimate connection between the notions of nonlinear absolute/convective instability and front velocity selection in pattern formation problems. In this context, it is essential to distinguish between the laboratory frame in which the advection velocity is U, as in (5.1) and (5.2), and the advected frame, where by definition $U = 0$. The velocity of the front separating the bifurcated state from the basic state is denoted V_f in the laboratory frame and v_f in the advected frame. To comply with the sign convention adopted in the front literature, *both V_f and v_f are counted positive when directed from the bifurcated region towards the basic-state region.* By definition, V_f and v_f are therefore connected through the relation $-V_f = U - v_f$, i.e.

$$v_f = U + V_f. \qquad (5.4)$$

Bearing in mind this sign convention, a front is then defined as a uniformly translating nonlinear solution of the form $\psi(x + v_f t)$ moving at the velocity v_f in the advected frame ($U = 0$). For example, in the case of the subcritical real

Ginzburg–Landau equation (5.1), the front solution satisfies the nonlinear eigenvalue problem

$$\frac{d^2\psi}{dx^2} - v_f \frac{d\psi}{dx} + \mu\psi + \psi^3 - \psi^5 = 0 \tag{5.5a}$$

with the boundary conditions

$$\psi(-\infty) = 0, \quad \psi(+\infty) = \psi_2. \tag{5.5b, c}$$

Equation (5.5a) may be viewed as a second-order dynamical system in x for $\psi(x)$ and $d\psi/dx(x)$. The existence or non-existence of front solutions satisfying (5.5b, c) is then directly related to the existence or non-existence of heteroclinic orbits linking the fixed point $\psi(-\infty) = 0$, $d\psi/dx(-\infty) = 0$ to the fixed point $\psi(+\infty) = \psi_2$, $d\psi/dx(+\infty) = 0$ in $(\psi, d\psi/dx)$-phase space. The analogue of the linear absolute/convective instability criterion based on the orientation of the trailing-edge velocity V^- may now be stated in the laboratory frame: if the trailing front velocity is such that $V_f < 0$ (resp. $V_f > 0$), the medium is nonlinearly convectively (resp. absolutely) unstable in the laboratory frame.

A brief introduction to front velocity selection criteria is now presented in the advected frame.[1] According to van Saarloos (1988, 1989), the velocity v_f of the front separating the bifurcated state from the basic state is, for a wide class of systems, given by the linear selection criterion

$$\frac{\partial\omega}{\partial k}(k_f^\ell) = -v_f^\ell, \tag{5.6a}$$

$$\sigma(v_f^\ell) = \omega_i(k_f^\ell) + v_f^\ell k_{fi}^\ell = 0. \tag{5.6b}$$

In other words, the front moves at a velocity v_f^ℓ such that in the co-moving frame associated with v_f^ℓ, the basic state is marginally absolutely/convectively unstable. Comparison between (5.3) rewritten in the advected frame ($U = 0$) and (5.6) immediately yields $v_f^\ell = -v^-$, the minus sign being due solely to different orientation rules for v_f^ℓ and v^-. Under such conditions, the trailing front velocity of the nonlinear droplet in figure 27(b–d) is identical to the trailing-edge velocity of the linear impulse response in figure 8(b–d). Thus, when the front velocity is linearly selected, the absolute/convective nature of the nonlinear instability is determined by the same criterion as in the linear régime. Under such circumstances, whenever the medium is linearly absolutely unstable it is also nonlinearly absolutely unstable. However van Saarloos (1988, 1989) has demonstrated that the front velocity does not

---

[1] Needless to say, identical results hold in the laboratory frame. The advected frame has been chosen to facilitate later reasoning.

always follow the linear criterion (5.6a, b). The following principle has been put forward by van Saarloos & Hohenberg (1992) to determine whether linear or nonlinear selection holds: if there exists a front with a velocity $v_f^{n\ell}$ and a spatial decay rate $k_{fi}^{n\ell}$ such that

$$v_f^{n\ell} > v_f^{\ell}, \quad |k_{fi}^{n\ell}| > |k_{fi}^{\ell}|, \qquad (5.7a, b)$$

then nonlinear selection prevails, i.e. the selected front is the fastest moving front of velocity $v_f^{n\ell}$ with the largest spatial decay rate $|k_{fi}^{n\ell}|$. Under such circumstances, the absolute/convective nature of the instability may not be the same in the linear and nonlinear régimes.

Consider as a first example the subcritical real Ginzburg–Landau equation (5.1) investigated by Couairon & Chomaz (1996, 1997a). The variations of the front velocity $v_f(\mu)$ in the advected frame ($U = 0$) are illustrated in figure 28. Application of the linear selection criterion (5.6a, b) to the dispersion relation $\omega = i(\mu - k^2)$ in the advected frame ($U = 0$) immediately yields $v_f^{\ell}(\mu) = 2\sqrt{\mu}$, represented by the inner parabola in figure 28. The fastest moving nonlinear front may be determined by analysing in detail all the possible configurations of heteroclinic orbits linking the basic state $\psi = 0$ to the stable finite-amplitude state ψ_2 in $(\psi, d\psi/dx)$-phase space. The final result in the advected frame ($U = 0$) is associated with the curve labelled $v_f^{n\ell}(\mu)$ in figure 28. In the range $\mu > 3/4$, the front velocity is seen to be linearly selected according to (5.6a, b), whereas in the interval $-3/16 < \mu < 3/4$, it is nonlinearly selected since $v_f^{n\ell} > v_f^{\ell}$.

The above results may now be expressed and interpreted in the laboratory frame of advection velocity U. The absolute/convective nature of the linear/nonlinear instability is specified in the various regions of the (μ, U)-control parameter plane in figure 28. The instability is nonlinearly convective (resp. absolute) if $V_f(\mu) = v_f(\mu) - U < 0$ (resp. $V_f(\mu) = v_f(\mu) - U > 0$), i.e. if $U > v_f(\mu)$ (resp. $U < v_f(\mu)$). In the former case, the trailing front is swept downstream (figure 27b) whereas in the latter it succeeds in making its way upstream against the incoming advection U (figure 27d). The nonlinear convective/absolute instability transition curve is given by $U = v_f(\mu)$ which is precisely the selected front velocity curve in figure 28. Advection is then exactly balanced by upstream propagation (figure 27c). In the range $\mu > 3/4$, the front velocity is linearly selected and the nonlinear absolute/convective instability threshold is given by $U = v_f^{\ell}(\mu)$. The nonlinear absolute instability domain then coincides with its linear counterpart (dark grey area of figure 28). In the range $-3/16 < \mu < 3/4$, the front velocity is nonlinearly selected and the nonlinear absolute/convective instability threshold is now

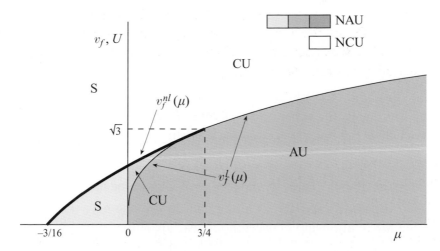

Figure 28. Front velocity $v_f(\mu)$ of the subcritical real Ginzburg–Landau equation (5.1) in advected frame ($U = 0$); $v_f(\mu) = v_f^{n\ell}(\mu)$ in the range $-3/16 \leqslant \mu \leqslant 3/4$, $v_f(\mu) = v_f^{\ell}(\mu)$ in the range $\mu \geqslant 3/4$. Nonlinearly absolutely unstable (NAU) domain (shaded region) in (μ, U)-space. The dark grey region represents the linearly absolutely unstable (AU) region which is totally embedded in the NAU region. The system is also NAU in the medium and light grey regions where it is linearly convectively unstable (CU) or linearly stable (S). The system is nonlinearly convectively unstable (NCU) in the unshaded region. A nonlinear global mode in a semi-infinite medium exists in the three shaded regions (Couairon & Chomaz 1997a).

given by $U = v_f^{n\ell}(\mu)$. In such a case, nonlinear absolute instability occurs in domains of parameter space where the medium is still linearly stable ($-3/16 < \mu < 0$, light grey area of figure 28) or linearly convectively unstable ($0 < \mu < 3/4$, medium grey area of figure 28). The linearly absolutely unstable domain $U < v_f^{\ell}(\mu)$ is seen to be embedded within a larger nonlinearly absolutely unstable domain $U < v_f^{n\ell}(\mu)$. This feature is not peculiar to the model under consideration, and the following important property holds: *linear absolute instability is a sufficient condition for nonlinear absolute instability*. This assertion is easily proved: linear absolute instability means by definition that $U < v_f^{\ell}(\mu)$ which according to (5.7a) also implies that $U < v_f^{n\ell}(\mu)$. *There may exist, in general, domains of parameter space where the medium is linearly stable or linearly convectively unstable, but where it is nonetheless nonlinearly absolutely unstable.*

Corresponding results (Couairon & Chomas 1997b) for the supercritical complex Ginzburg–Landau equation (5.2) are displayed in figure 12 at given settings of c_d and c_n. Application of the linear front selection criterion (5.6a, b) to $\omega = c_d k^2 + i(\mu - k^2)$ in the advected frame ($U = 0$) yields

$v_f^\ell(\mu) = 2\sqrt{\mu(1 + c_d^2)}$ represented by a parabola in figure 12. This curve is identical to the linear absolute/convective transition boundary in the laboratory frame ($U \neq 0$). In this example, the front velocity is always linearly selected and the threshold of linear absolute instability coincides with the threshold of nonlinear absolute instability for all μ. This feature has nothing to do with the supercritical nature of the bifurcation. It is the direct consequence of the linear selection mechanism governing the front velocity.

5.2 Nonlinear global modes in uniform semi-infinite media

Following the same approach as in the linear case, one may then proceed to determine the relationship between nonlinear absolute/convective instability concepts in doubly infinite uniform media and the onset of nonlinear global mode oscillations in systems that are no longer invariant under arbitrary space translations and Galilean transformations. Such a formulation has so far been attempted in only a few simple cases such as the subcritical real Ginzburg–Landau equation (5.1) with constant coefficients over the semi-infinite interval $x > 0$ studied by Couairon & Chomaz (1996, 1997a). A nonlinear global mode is then defined as a steady solution of (5.1) satisfying

$$\frac{d^2\psi}{dx^2} - U\frac{d\psi}{dx} + \mu\psi + \psi^3 - \psi^5 = 0, \qquad (5.8a)$$

and the boundary conditions

$$\psi(0) = 0, \quad \psi(\infty) = \psi_2 \; ;. \qquad (5.8b,c)$$

Note that (5.8a) is identical to (5.5a) provided one makes the substitution $U \mapsto v_f$: the only difference between the nonlinear front eigenvalue problem (5.5) and the nonlinear global mode problem (5.8) lies in the 'upstream' boundary conditions (5.5b) and (5.8b). According to (5.8b, c) a nonlinear global mode is possible if the stable manifold of the fixed point ψ_2 intersects the $d\psi/dx$-axis at a finite value of $d\psi/dx$ in phase space. The final results are summarized in figure 28: remarkably, the region of existence for nonlinear global modes in a semi-infinite medium is seen to coincide exactly with the region $U < v_f(\mu)$ of nonlinear absolute instability in a doubly infinite medium. Similar results hold for the supercritical complex Ginzburg–Landau equation (5.2), as shown in figure 12. All the model problems investigated by Couairon & Chomaz (1997a, b) support the conjecture that *the onset of nonlinear absolute instability coincides with the appearance of nonlinear global modes on the semi-infinite interval.*

   Let $\mu = \mu_f(U)$ be the reciprocal function of $U = v_f(\mu)$ and $\varepsilon = \mu - \mu_f \ll 1$

be the small parameter characterizing the departure from the nonlinear global mode threshold μ_f. The global spatial structure may be unfolded analytically in the vicinity of the nonlinear global mode boundary $U = v_f(\mu)$ by slightly perturbing the front solution prevailing at $\mu = \mu_f(U)$, so that it satisfies the boundary condition $\psi(0) = 0$. Explicit scaling laws may be derived for the characteristic healing length of the nonlinear global mode, defined as the distance Δx from the origin where $|\psi(\Delta x)| = 0.50\,|\psi_2|$. When the threshold for the appearance of a global mode coincides with the onset of linear absolute instability, i.e. when it is given by the *linear* front selection criterion $U = v_f^\ell(\mu)$, the healing length Δx varies as $\varepsilon^{-1/2}$. When the global mode threshold is determined by the *nonlinear* front selection criterion $U = v_f^{n\ell}(\mu)$, and this typically occurs while the system is still linearly stable or linearly convectively unstable, the healing length Δx varies as $\ln(1/\varepsilon)$.

The nonlinear global mode structure and healing length analytically predicted for the supercritical complex Ginzburg–Landau equation (5.2) have been compared by Couairon & Chomaz (1997b) to direct numerical simulations of Rayleigh–Bénard convection between two differentially heated horizontal plates with superimposed plane Poiseuille throughflow (Müller, Lücke & Kamps 1992). A typical nonlinear global mode resulting from computations in a finite box is illustrated in figure 29(a), which compares favourably with the analytically determined Ginzburg–Landau global mode in the vicinity of onset represented in figure 29(b). In the vicinity of onset, $\varepsilon = \mu - \mu_f \ll 1$, the healing length of the nonlinear global mode sustained by the Ginzburg–Landau model equation (5.2), is predicted to vary as

$$\Delta x \sim \frac{\pi}{\beta\,\varepsilon^{1/2}}, \tag{5.9}$$

where β is a numerically determined constant. When expressed in terms of the normalized length scale $L = \sqrt{\mu}\,\Delta x$ and normalized advection velocity $V_g = U\sqrt{\mu(1 + c_d^2)}$ the scaling law takes the form

$$L \sim \frac{\pi}{\beta}(1 - V_g^2/4)^{-1/2}. \tag{5.10}$$

According to figure 29(c), the direct nonlinear global-mode simulations of Rayleigh–Bénard convection by Müller *et al.* (1992) are seen to approach the scaling law (5.10) as the departure from onset $(1 - V_g^2/4)$ becomes small.

The distinct scaling law $\Delta x \sim \ln(1/\varepsilon)$, applicable to global-mode onset dictated by the nonlinear front selection criterion $U = v_f^{n\ell}(\mu)$, has recently been validated experimentally for the Kelvin–Helmholtz instability of an interface in a Hele-Shaw cell (Gondret *et al.* 1999).

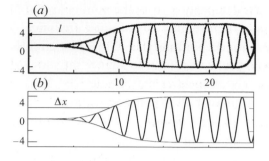
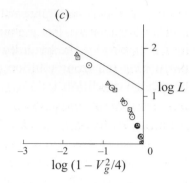

Figure 29. Comparison between predictions of the supercritical complex Ginzburg–Landau model (5.2) obtained by Couairon & Chomaz (1997*b*) and direct numerical simulations of Rayleigh–Bénard convection with through flow by Müller *et al.* (1992). (*a*) Vertical velocity field computed by Müller *et al.* (1992); (*b*) nonlinear global mode analytically determined near onset; (*c*) healing length L of nonlinear global mode predicted by (5.10) (continuous line) compared with numerical results of Müller *et al.* (1992).

5.3 Nonlinear global modes in spatially varying media

It is also possible to arrive at a consistent description of nonlinear global modes in doubly infinite spatially varying media by extending the linear WKBJ analyses of §4 to the fully nonlinear régime. Recent progress in this direction is summarized below in the context of the complex Ginzburg–Landau equation with varying coefficients:

$$\mathrm{i}\frac{\partial \psi}{\partial t} - \mathrm{i}\omega_{kk}(X)k_0(X)\frac{\partial \psi}{\partial x} - \left[\tfrac{1}{2}\omega_{kk}(X)k_0^2(X) + \omega_0(X)\right]\psi$$

$$+ \tfrac{1}{2}\omega_{kk}(X)\frac{\partial^2 \psi}{\partial x^2} = \gamma(X)|\psi|^2\psi, \qquad (5.11)$$

which is the nonlinear counterpart of (4.5). Following Pier & Huerre (1996) and Pier *et al.* (1998), time-periodic nonlinear global modes of the form (4.20) are sought in the doubly infinite domain $-\infty < X < +\infty$, subjected to the boundary conditions (4.21). The global frequency ω_G and the spatial eigenfunction $\phi(X)$ are then solutions of a nonlinear eigenvalue problem on $-\infty < X < +\infty$ obtained from (5.11), (4.21) by making the substitutions $\partial/\partial t \mapsto -\mathrm{i}\omega_G$, $\psi \to \phi$.

In contrast to the linear framework, the nonlinear global frequency ω_G is assumed to be real since it is associated with a fully nonlinear saturated solution. The method is closely patterned on the linear formulation presented in §4.2. In the vicinity of $X = \pm\infty$, the nonlinear global mode must be vanishingly small and it is governed by the linear equation (4.5). One may appeal to the known properties of the local linear dispersion relation (4.6) and its linear spatial branches (4.7). Assume, as in §4.2, that $k_i^{\ell+}(X;\omega) > 0$ and $k_i^{\ell-}(X;\omega) < 0$ for sufficiently large $|X|$ along the X_r-axis. The non-linear solution then involves solely the linear spatial branches $k^{\ell+}(X;\omega)$ and $k^{\ell-}(X;\omega)$ near $X = +\infty$ and $X = -\infty$ respectively: the eigenfunction is therefore approximated near $X = +\infty$ and $X = -\infty$ by the linear WKBJ solutions $\phi^+(X)$ and $\phi^-(X)$ defined in (4.30a), where ω^s is replaced by the unknown real frequency ω_G. Solving the nonlinear eigenvalue problem is tantamount to selecting the real frequency ω_G such that the $k^{\ell-}$ branch prevailing near $X = -\infty$ smoothly turns into the $k^{\ell+}$ branch prevailing near $X = +\infty$ as X increases. In locally linearly unstable X-intervals, the Ginzburg–Landau equation is known to support nonlinear travelling waves of the form

$$\psi(x,t) = \phi(X)\,\mathrm{e}^{-\mathrm{i}\omega t} \sim R(X)\exp\left(\mathrm{i}\left[\frac{1}{\varepsilon}\int^X k(X;\omega)\mathrm{d}X - \omega t\right]\right) \quad (5.12)$$

provided that the real wavenumber k, real frequency ω and real amplitude R satisfy the local nonlinear dispersion relation

$$\omega = \Omega^{n\ell}(k,R^2;X) = \omega_0(X) + \tfrac{1}{2}\omega_{kk}(X)[k-k_0(X)]^2 + \gamma(X)R^2. \quad (5.13)$$

The first part of this expression is simply the local linear dispersion relation $\omega = \Omega^{\ell}(k;X)$ defined in (4.6). The complex dispersion relation may also be recast into two real relations by elementary manipulations: taking the imaginary part of (5.13) yields $0 = \Omega_i^{n\ell}(k,R^2;X)$, which readily provides a relation between the real wavenumber k and amplitude R of the form

$$R^2 = \mathscr{R}_2(k;X). \quad (5.14)$$

Taking the real part of (5.13) yields $\omega = \Omega_r^{n\ell}(k,R^2;X)$, which, upon substituting (5.14), provides a relation between k and the real frequency ω of the form

$$\omega = \Omega^{NL}(k;X) \equiv \Omega_r^{n\ell}[k,\mathscr{R}_2(k;X);X]. \quad (5.15)$$

The above expression is referred to as the *reduced nonlinear dispersion relation*. Before proceeding to discuss the main results, it is essential to state some of its properties. Since (5.15) is quadratic in k, there exist two nonlinear spatial branches $k^{n\ell+}(X;\omega)$ and $k^{n\ell-}(X;\omega)$ which acquire physical significance only in spatial X-intervals where $R^2 = R_2(k;X) \geqslant 0$. Furthermore each linear spatial branch $k^{\ell\pm}$ may be continuously connected to a corresponding nonlinear branch $k^{n\ell\pm}$ through a locally neutrally stable station. Two types

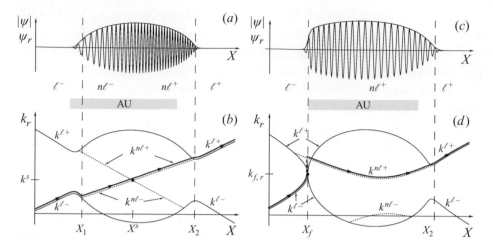

Figure 30. Nonlinear global eigenfunctions of the supercritical complex Ginzburg–Landau equation (5.11). (a,b) Soft global mode of frequency ω^s; (c,d) steep global mode of frequency ω_f. Upper graphs display the envelope $|\psi|(x,t)$ and the real part $\psi_r(x,t)$. Lower graphs display corresponding linear and nonlinear spatial branches in the (X,k_r)-plane. Linear spatial branches $k^{\ell\pm}$ are represented by solid lines and nonlinear spatial branches $k^{n\ell\pm}$ by dotted lines. The local wavenumber making up the nonlinear global mode in (a) and (c) follows the path indicated by a thick line. The wavenumber jump in the front at X_f in (d) is indicated by repeated arrows (Pier & Huerre 1996; Pier *et al.* 1998).

of fully nonlinear global modes, respectively referred to as soft (figure 30a,b) and steep (figure 30c,d), have so far been identified.

Soft global modes. According to Pier *et al.* (1998), the real frequency ω_G of soft global modes is given by $\omega_G \sim \omega^s \equiv \Omega^{NL}(k^s;X^s)$, where the real pair k^s, X^s satisfies the nonlinear saddle point conditions

$$\frac{\partial \Omega^{NL}}{\partial k}(k^s;X^s) = \frac{\partial \Omega^{NL}}{\partial X}(k^s;X^s) = 0. \tag{5.16}$$

The associated wavepacket spatial structure is illustrated in figure 30(a) by the functions $\psi_r(x,t)$ and $|\psi(x,t)|$, and in figure 30(b) by the variations of the real part of the local wavenumber $k_r(X;\omega^s)$. As a result of the imposed saddle point criterion (5.16), the nonlinear branch $k^{n\ell-}$ is observed to turn smoothly into the nonlinear branch $k^{n\ell+}$ at the station X^s. The following qualitative changes in the spatial structure take place as X varies from $X = -\infty$ to $X = +\infty$ (see figure 30a,b): near $X = -\infty$, only the complex spatially decaying $k^{\ell-}$-branch is allowed. According to the aforementioned properties of nonlinear spatial branches, $k^{\ell-}$ switches over to a real $k^{n\ell-}$ branch at a locally neutrally stable point X_1. As the saddle point X^s is reached, $k^{n\ell-}$ continues into $k^{n\ell+}$ which in turn switches over to a complex spatially decaying $k^{\ell+}$-branch at a second locally neutrally stable point X_2.

The soft nonlinear frequency selection criterion (5.16) is seen to be formally identical to its linear counterpart (4.28) provided that $\omega = \Omega^\ell$ is changed into $\omega = \Omega^{NL}$. There is however an important difference. The linear criterion pertains to complex values of k^s and X^s, whereas the nonlinear criterion is in a way simpler to apply: it is restricted to real values of k^s and X^s. The real saddle point X^s may be determined conveniently by monitoring the deformations of the linear and nonlinear spatial branches in the (X, k_r)-plane, as the real frequency ω is lowered. When ω decreases towards ω^s, the $k^{n\ell+}$- and $k^{n\ell-}$-branches typically approach each other and pinch at X^s.

Steep global modes. According to Pier *et al.* (1998), the real frequency ω_G of steep global modes is given by $\omega_G \sim \omega_f \equiv \Omega^\ell(k_f; X_f)$, where the complex wavenumber k_f and real station X_f satisfy

$$\frac{\partial \Omega^\ell}{\partial k}(k_f; X_f) = 0, \quad \Omega_i^\ell(k_f; X_f) = 0. \tag{5.17a, b}$$

This steep global mode criterion may readily be interpreted in terms of local linear instability properties. The zero-group-velocity condition (5.17a) singles out the complex absolute wavenumber k_f at X_f, and the neutral condition (5.17b) stipulates that the corresponding absolute growth rate ω_{fi} is zero at X_f. In other words, the steep global frequency ω_G is determined by the real absolute frequency ω_f prevailing at the station X_f where the linear instability changes from convective to absolute. The associated wavepacket structure is displayed in figure 30(c, d). As a result of the convective/absolute instability transition condition (5.17a, b), the linear branch $k^{\ell-}$ meets the linear branch $k^{\ell+}$ at X_f. More importantly, the locally neutrally stable condition prevailing at X_f allows a direct conversion of $k^{\ell-}$ into the nonlinear branch $k^{n\ell+}$. The following qualitative changes in the spatial structure take place as X varies from $X = -\infty$ to $X = +\infty$ (see figure 30c, d): near $X = -\infty$, the only allowable solution is, as usual, the complex spatially decaying $k^{\ell-}$-branch. At the convective/absolute instability transition station X_f, it is abruptly transformed into $k^{n\ell+}$ across a steep front. Finally $k^{n\ell+}$ switches over to a complex spatially decaying $k^{\ell+}$ branch at a locally neutrally stable point X_2 on the downstream side.

Although steep global modes are fully nonlinear, the determination of their frequency relies solely on the properties of the local linear dispersion relation $\omega = \Omega^\ell(k; X)$. Conditions (5.17a, b) are in essence identical to the linear front selection criterion (5.6a, b) provided that the front velocity v_f^ℓ is set equal to zero. Thus, a steep stationary front, precisely located at the transition station between linear convective and absolute instability, imposes its frequency on the entire medium. As in §5.1, the front separates the unstable state $\psi = 0$ on the linearly convectively unstable side to the left from the finite-amplitude travelling wave state on the linearly absolutely unstable side to the right.

As discussed in Pier, Huerre & Chomaz (2001), the nature (soft or steep) of the observed nonlinear global mode is dictated by the relative magnitudes of ω^s and ω_f, the mode of larger frequency being dominant. It can be demonstrated that a necessary condition for the existence of soft or steep global modes is the presence of a region of local linear absolute instability. More importantly, *steep global modes appear as soon as there exists a point of local linear absolute instability, even though the medium is still linearly globally stable.* Local linear absolute instability in a sense prevails over global linear stability and dictates the nature of the observed finite-amplitude state. By contrast soft global modes are only obtained further above the onset of steep global modes, provided that basic advection is small enough.

Couairon & Chomaz (1999*a*) have recently examined the asymptotic structure of fully nonlinear global modes in Ginzburg–Landau models that combine both a spatial origin at $x = 0$ as in §5.2, and slow spatial variations of the medium as in §5.3. Here again, predicted scaling laws for the modal amplitude and the streamwise position of its maximum are in good qualitative and quantitative agreement with the detailed experimental measurements of global modes in bluff-body wakes reported by Goujon-Durand, Jenffer & Wesfreid (1994).

6 Epilogue

The study of deceptively simple Ginzburg–Landau evolution models has been a recurring theme throughout the course of this chapter. Such an approach has allowed the introduction, with the minimum of algebraic complexity, of the theoretical notions underlying the distinction between flow amplifiers and oscillators. In the latter instance, explicit frequency selection criteria have been identified both in the linear and fully nonlinear régimes. In some cases, as in §5.2, quantitative scaling laws have been obtained that apply directly to real flows. But, for the most part, the theoretical formulations have had to be subsequently extended to the Navier–Stokes equations in order to precisely characterize the distinct spatio-temporal dynamics of a variety of shear flows: mixing layers, jets, wakes, boundary layers, etc. This strategy has been reasonably successful, both in terms of basic understanding and quantitative predictions. However, many areas remain relatively unexplored. The Briggs–Bers criterion has been generalized to three-dimensional instability waves (Brevdo 1991) but, with the exception of swept-wing boundary layers (Lingwood 1997; Taylor & Peake 1998), it has been applied to few shear flows. The study of nonlinear global modes has so far been restricted to one-dimensional evolution models and the generalization to real streamwise developing shear flows is likely to be non-trivial. Absolute/convective instability concepts have recently been extended to secondary instabilities which arise in spatially periodic model flows (Brevdo & Bridges 1996) or in the presence of fully nonlinear travelling waves (Couairon & Chomaz 1999*b*). In such cases, the onset of secondary absolute instability may result in spatio-temporal disorder, but it remains to determine whether the corresponding analysis in real shear flows will result in the identification of novel laminar–turbulent transition mechanisms.

It is in fact tempting to look for the possible onset of local absolute instability whenever a spatially developing flow undergoes a sudden transition to a qualitatively distinct dynamical régime, whether in the form of a synchronized global mode or a more exotic spatio-temporal state. For instance, vortex breakdown which is a ubiquitous phenomenon affecting, under certain conditions, concentrated vortices with axial flow, has recently been investigated in this light. The application of a moderate amount of swirl to jets and wakes effectively triggers an absolute instability in the absence of axial counterflow (Delbende, Chomaz & Huerre 1998; Loiseleux, Chomaz & Huerre 1998; Olendraru *et al.* 1999) but a definite connection with vortex breakdown remains to be established!

Acknowledgements

I am very grateful to Arnaud Couairon, Paul Manneville and Benoît Pier for their help and advice in putting this material together. The assistance and dedication of Brigitte Blanchard, Maryse Lecrinier, Martine Maguer and Dominique Toustou-Bouly at the Imprimerie de l'École Polytechnique has been essential to the completion of the manuscript.

References

ABLOWITZ, M. J. & FOKAS, S. 1997 *Complex Variables: Introduction and Applications.* Cambridge University Press.

BATCHELOR, G. K. 1967 *An Introduction to Fluid Dynamics.* Cambridge University Press.

BAYLY, B. J., ORSZAG, S. A. & HERBERT, T. 1988 Instability mechanisms in shear-flow transition. *Ann. Rev. Fluid Mech.* **20**, 359–391.

BENDER, C. M. & ORSZAG, S. A. 1978 *Advanced Mathematical Methods for Scientists and Engineers.* McGraw-Hill.

BERNAL, L. P. & ROSHKO, A. 1986 Streamwise vortex structure in plane mixing layers. *J. Fluid Mech.* **170**, 499–526.

BERS, A. 1983 Space-time evolution of plasma instabilities — absolute and convective. In *Handbook of Plasma Physics* (ed. M. N. Rosenbluth & R. Z. Sagdeev), pp. 451–517. North-Holland.

BLEISTEIN, N. & HANDELSMAN, R. A. 1986 *Asymptotic Expansions of Integrals*. Dover.

BOUTHIER, M. 1972 Stabilité linéaire des écoulements presque parallèles. *J. Méc.* **11**, 599–621.

BREVDO, L. 1991 Three-dimensional absolute and convective instabilities, and spatially amplifying waves in parallel shear-flows. *Z. Angew. Math. Phys.* **42**, 911–942.

BREVDO, L. & BRIDGES, T. 1996 Absolute and convective instabilities of spatially periodic flows. *Phil. Trans. R. Soc. Lond.* A **354**, 1027–1064.

BRIGGS, R. J. 1964 *Electron-Stream Interaction with Plasmas*. M.I.T. Press.

BROWAND, F. K. & WEIDMAN, P. D. 1976 Large scales in the developing mixing layer. *J. Fluid Mech.* **76**, 127–144.

BROWN, G. L. & ROSHKO, A. 1974 On density effects and large structures in turbulent mixing layers. *J. Fluid Mech.* **64**, 775–816.

CHOMAZ, J.-M. 1992 Absolute and convective instabilities in nonlinear systems. *Phys. Rev. Lett.* **69**, 1931–1934.

CHOMAZ, J.-M., HUERRE, P. & REDEKOPP, L. G. 1991 A frequency selection criterion in spatially developing flows. *Stud. Appl. Maths* **84**, 119–144.

COUAIRON, A. & CHOMAZ, J.-M. 1996 Global instabilities in fully nonlinear systems. *Phys. Rev. Lett.* **77**, 4015–4018.

COUAIRON, A. & CHOMAZ, J.-M. 1997a Absolute and convective instabilities, front velocities and global modes in nonlinear systems. *Physica* D **108**, 236–276.

COUAIRON, A. & CHOMAZ, J.-M. 1997b Pattern selection in the presence of a cross flow. *Phys. Rev. Lett.* **79**, 2666–2669.

COUAIRON, A. & CHOMAZ, J.-M. 1999a Fully nonlinear global modes in slowly varying flows. *Phys. Fluids* **11**, 3688–3703.

COUAIRON, A. & CHOMAZ, J.-M. 1999b Primary and secondary nonlinear global instability. *Physica* D **132**, 428–456.

CRIGHTON, D. G. & GASTER, M. 1976 Stability of slowly diverging jet flow. *J. Fluid Mech.* **77**, 397–413.

CROSS, M. C. & HOHENBERG, P. C. 1993 Pattern formation outside of equilibrium. *Rev. Mod. Phys.* **65**, 851–1112.

DELBENDE, I., CHOMAZ, J.-M. & HUERRE, P. 1998 Absolute/convective instabilities in the Batchelor vortex: a numerical study of the linear impulse response. *J. Fluid Mech.* **355**, 229–254.

DRAZIN, P. G. & REID, W. H. 1981 *Hydrodynamic Stability*. Cambridge University Press.

GASTER, M., WYGNANSKI, I. & KIT, E. 1985 Large scale structures in a forced turbulent mixing layer. *J. Fluid Mech.* **150**, 23–39.

GODRÈCHE, C. & MANNEVILLE, P. 1998 *Hydrodynamics and Nonlinear Instabilities*. Cambridge University Press.

GONDRET, P., ERN, P., MEIGNIN, L. & RABAUD, M. 1999 Experimental evidence of a nonlinear transition from convective to absolute instability. *Phys. Rev. Lett.* **82**, 1442–1445.

GOUJON-DURAND, S., JENFFER, P. & WESFREID, J. 1994 Downstream evolution of the Bénard–von Kármán instability. *Phys. Rev.* E **50**, 308–313.

GUCKENHEIMER, J. & HOLMES, P. 1983 *Nonlinear Oscillations, Dynamical Systems and Bifurcations of Vector Fields*. Springer.

HAMMOND, D. A. & REDEKOPP, L. G. 1997 Global dynamics of symmetric and asymmetric wakes. *J. Fluid Mech.* **331**, 231–260.

HERBERT, T. 1988 Secondary instability of boundary layers. *Ann. Rev. Fluid Mech.* **20**, 487–526.

HO, C. M. & HUANG, L. S. 1982 Subharmonics and vortex merging in mixing layers. *J. Fluid Mech.* **119**, 443–473.

HO, C. M. & HUERRE, P. 1984 Perturbed free shear layers. *Ann. Rev. Fluid Mech.* **16**, 365–424.

HUERRE, P. & MONKEWITZ, P. A. 1985 Absolute and convective instabilities in free shear layers. *J. Fluid Mech.* **159**, 151–168.

HUERRE, P. & MONKEWITZ, P. A. 1990 Local and global instabilities in spatially developing flows. *Ann. Rev. Fluid Mech.* **22**, 473–537.

HUERRE, P. & ROSSI, M. 1998 Hydrodynamic instabilities in open flows. In *Hydrodynamics and Nonlinear Instabilities* (ed. C. Godrèche & P. Manneville), pp. 81–294. Cambridge University Press.

HUNT, R. E. & CRIGHTON, D. G. 1991 Instability of flows in spatially developing media. *Proc. R. Soc. Lond.* A **435**, 109–128.

KÁRMÁN, T. VON 1921 Über laminare und turbulente Reibung. *Z. Angew. Math. Mech.* **I**, 233–252.

KARNIADAKIS, G. & TRIANTAFYLLOU, G. S. 1989 Frequency selection and asymptotic states in laminar wakes. *J. Fluid Mech.* **199**, 441–469.

KOCH, W. 1985 Local instability characteristics and frequency determination of self-excited wake flows. *J. Sound Vib.* **99**, 53–83.

KUPFER, K., BERS, A. & RAM, A. K. 1987 The cusp map in the complex-frequency plane for absolute instabilities. *Phys. Fluids* **30**, 3075–3082.

LASHERAS, J. C. & CHOI, H. 1988 Three-dimensional instability of a plane, free shear layer: an experimental study of the formation and evolution of streamwise vortices. *J. Fluid Mech.* **189**, 53–86.

LE DIZÈS, S., HUERRE, P., CHOMAZ, J.-M. & MONKEWITZ, P. A. 1996 Linear global modes in spatially developing media. *Phil. Trans. R. Soc. Lond.* A **354**, 169–212.

LIN, C. C. 1955 *The Theory of Hydrodynamic Stability*. Cambridge University Press.

LINGWOOD, R. J. 1995 Absolute instability of the boundary layer on a rotating disk. *J. Fluid Mech.* **299**, 17–33.

LINGWOOD, R. J. 1996 An experimental study of absolute instability of the rotating-disk boundary-layer flow. *J. Fluid Mech.* **314**, 373–405.

LINGWOOD, R. J. 1997 On the impulse response for swept boundary-layer flows. *J. Fluid Mech.* **344**, 317–334.

LOISELEUX, T., CHOMAZ, J.-M. & HUERRE, P. 1998 The effect of swirl on jets and wakes: Linear instability of the Rankine vortex with axial flow. *Phys. Fluids* **10**, 1120–1134.

MACK, L. M. 1976 A numerical study of the temporal eigenvalue spectrum of the Blasius boundary layer. *J. Fluid Mech.* **73**, 497–520.

MANNEVILLE, P. 1990 *Dissipative Structures and Weak Turbulence*. Academic.

MASLOWE, S. A. 1981 Shear flow instabilities and transition. In *Hydrodynamic Instabilities and the Transition to Turbulence* (ed. H. L. Swinney & J. P. Gollub), pp. 181–228. Springer.

MASLOWE, S. A. 1986 Critical layers in shear flows. *Ann. Rev. Fluid Mech.* **18**, 405–432.

MICHALKE, A. 1964 On the inviscid instability of the hyperbolic-tangent velocity profile. *J. Fluid Mech.* **19**, 543–556.

MICHALKE, A. 1965 On spatially growing disturbances in an inviscid shear layer. *J. Fluid Mech.* **23**, 521–544.

MONKEWITZ, P. A. 1988 The absolute and convective nature of instability in two-dimensional wakes at low Reynolds numbers. *Phys. Fluids* **31**, 999–1006.

MONKEWITZ, P. A. 1990 The role of absolute and convective instability in predicting the behavior of fluid systems. *Eur. J. Mech.* B/*Fluids* **9**, 395–413.

MONKEWITZ, P. A. & HUERRE, P. 1982 Influence of the velocity ratio on the spatial instability of mixing layers. *Phys. Fluids* **25**, 1137–1143.

MONKEWITZ, P. A., HUERRE, P. & CHOMAZ, J.-M. 1993 Global linear stability analysis of weakly non-parallel shear flows. *J. Fluid Mech.* **251**, 1–20.

MÜLLER, H. W., LÜCKE, M. & KAMPS, M. 1992 Transversal convection patterns in horizontal shear flow. *Phys. Rev.* A **45**, 3714–3726.

OERTEL, H. 1990 Wakes behind blunt bodies. *Ann. Rev. Fluid Mech.* **22**, 539–564.

OLENDRARU, C., SELLIER, A., ROSSI, M. & HUERRE, P. 1999 Inviscid instability of the Batchelor vortex: absolute/convective transition and spatial branches. *Phys. Fluids* **11**, 1805–1820.

ORSZAG, S. A. 1971 Accurate solution of the Orr–Sommerfeld stability equation. *J. Fluid Mech.* **50**, 689–703.

ORSZAG, S. A. & PATERA, A. T. 1981 Hydrodynamic stability of shear flows. In *Chaotic Behaviour of Deterministic Systems* (ed. G. Iooss, R. Helleman & R. Stora), pp. 624–662. North-Holland.

PERRY, A. E., CHONG, M. S. & LIM, T. T. 1982 The vortex-shedding process behind two-dimensional bluff bodies. *J. Fluid Mech.* **116**, 77–90.

PIER, B. & HUERRE, P. 1996 Fully nonlinear global modes in spatially developing media. *Physica* D **97**, 206–222.

PIER, B., HUERRE, P. & CHOMAZ, J.-M. 2001 Bifurcation to fully nonlinear synchronized structures in slowly varying media. *Physica* D. To appear.

PIER, B., HUERRE, P., CHOMAZ, J.-M. & COUAIRON, A. 1998 Steep nonlinear global modes in spatially developing media. *Phys. Fluids* **10**, 2433–2435.

PROVANSAL, M., MATHIS, C. & BOYER, L. 1987 Bénard–von Kármán instability: transient and forced regimes. *J. Fluid Mech.* **182**, 1–22.

SAARLOOS, W. VAN 1988 Front propagation into unstable states: marginal stability as a dynamical mechanism for velocity selection. *Phys. Rev.* A **37**, 211–229.

SAARLOOS, W. VAN 1989 Front propagation into unstable states. ii. linear versus nonlinear marginal stability and rate of convergence. *Phys. Rev.* A **39**, 6367–6390.

VAN SAARLOOS, W. & HOHENBERG, P. C. 1992 Fronts, pulses, sources and sinks in generalized complex Ginzburg–Landau equations. *Physica* D **56**, 303–367.

SOWARD, A. M. 2001 Thin aspect ratio $\alpha\omega$-dynamos in galactic discs and stellar shells. In *Advances in nonlinear dynamos* (ed. M. Núñez & A. F. Mas). Gordon and Breach. To appear.

STRYKOWSKI, P. J. & NICCUM, D. L. 1991 The stability of countercurrent mixing layers in circular jets. *J. Fluid Mech.* **227**, 309–343.

SWINNEY, H. L. & GOLLUB, J. P. 1981 *Hydrodynamic Instabilities and the Transition to Turbulence*. Springer.

TAYLOR, M. J. & PEAKE, N. 1998 The long-time behaviour of incompressible swept-wing boundary layers subject to impulsive forcing. *J. Fluid Mech.* **355**, 359–381.

TRIANTAFYLLOU, G. S., TRIANTAFYLLOU, M. S. & CHRYSOSTOMIDIS, C. 1986 On the formation of vortex streets behind stationary cylinders. *J. Fluid. Mech.* **170**, 461–477.

WILLIAMSON, C. H. K. 1989 Oblique and parallel modes of vortex shedding in the wake of a circular cylinder at low Reynolds numbers. *J. Fluid Mech.* **206**, 579–627.

WILLIAMSON, C. H. K. 1996 Vortex dynamics in the cylinder wake. *Ann. Rev. Fluid Mech.* **28**, 477–539.

WOODLEY, B. M. & PEAKE, N. 1997 Global linear stability analysis of thin aerofoil wakes. *J. Fluid. Mech.* **339**, 239–260.

Laboratoire d'Hydrodynamique (LadHyX), CNRS – École polytechnique,
F-91128 Palaiseau cedex, France

5

Turbulence

JAVIER JIMÉNEZ

1 Introduction

Any survey on turbulence is certain to be incomplete. There is too much that we do not know for us to give a definitive account on the subject but, as we shall see in the following pages, a lot has also been learned in the past century and many aspects of turbulence are now understood. Many of the problems are related to engineering applications, and reflect our inability to compute the mean behaviour of the flow with sufficient accuracy. But they betray a deeper lack of understanding. Although the mechanics of many turbulent flows are understood at a fundamental level, some important aspects are not, and turbulence remains an abundant source of basic physical problems. The aim of this article is to review what is known with some certainty, to describe briefly some areas of present research interest, and to point to some of the obvious gaps in our knowledge.

Such a wide programme will force us to neglect many important questions, and we will mostly restrict ourselves to constant-density flows at relatively high Reynolds numbers. We will say little about the influence of compressibility, even though it is crucial to many applications, or about subjects such as stratification, rotation, chemical reactions, combustion or the influence of electric and magnetic fields, although they all appear either in applications or in the large-scale flows of geophysics or astrophysics. Still less defensible will be our inattention to questions of flow stability and transition, which not only control the generation of turbulence and its behaviour at low Reynolds numbers, but whose effects persist in some cases into the high-Reynolds-number regime. Because of those residual effects we will not be able to avoid stability completely, but it certainly deserves a more detailed treatment than the one given here. Some of these subjects are discussed elsewhere in this volume, but others are excluded strictly because of lack of space, and the choice can only be described as personal.

Even so it will be seen that the subject matter of our discussion is exceptionally rich, and that it underlies most of the more complicated cases.

Historically, the disordered flow of water and air must have been familiar to mankind from very early, even if only by observation of the surface of rivers, and the engineers responsible for the large hydraulic works of the classical civilizations must have had some understanding of the effects of turbulent flow. This knowledge was not completely lost with the decline of the classical world, but the scientific study of turbulence only began towards the middle of the 19th century in Europe.

At that time the pressure drop in water pipes was a problem in practical hydraulics which, together with the drag on moving bodies, had been the subject of experimental and theoretical investigations for at least 150 years. It was known that it had two components, one linear and the other approximately quadratic in the fluid velocity, and that only the first one was dependent on the viscosity of the fluid. In 1854 Hagen and in 1857 Darcy published independently careful measurements in large pipes, and noted that the nonlinear component was associated with disordered motion in the fluid. They speculated that the increased drag was due to the energy spent in creating the fluctuating eddies.

Some time after that Boussinesq (1877) published a long paper clearly distinguishing between the two distinct kinds of flows, smooth and 'tumultuous', and introduced many of the themes which later became associated with turbulence research, such as an enhanced eddy viscosity and the idea that turbulent flows are too complicated for a deterministic description, and have to be treated statistically.

The transition between the two states was clarified by Reynolds (1883), who introduced dye at the inlet of a long glass pipe and observed the process by which the flow became disordered; this is the paper where he first used what became known as the 'Reynolds number' to describe the circumstances at which the transition takes place. In his second important paper on the subject (1894) he used for the first time the decomposition of the flow into mean and fluctuating parts, and introduced the concept of Reynolds stresses. It is from this paper that much of the modern statistical theory of turbulence derives. An account of this early period of hydrodynamics can be found in Rouse & Ince (1957). Most of the subsequent history is still the subject of current reference, and will be discussed in the appropriate places of the following sections.

There are many good books on turbulence, and this article cannot attempt to replace them. The three classic textbooks are those by Hinze (1975), Townsend (1976), and Tennekes & Lumley (1972), the first one encyclopaedic and the other two more personal. The last one, in particular, is still probably the best elementary introduction to the subject. The definitive treatise on the classical theory of turbulence and on its experimental support is the two-volume set by Monin & Yaglom (1975). There are few comparable modern general books on turbulence. The one that might best fit that description is that by Lesieur (1997) although somewhat biased towards the results of a particular modelling scheme. A book that may become classical is Frisch (1995), which deals almost exclusively with isotropic turbulence, and cannot therefore be considered as a general book on the subject. It contains one of the best modern descriptions of the fundamentals assumptions of turbulence theory, of its early historical development and, especially, on the subject of small-scale intermittency. Another excellent specialized book is McComb (1990), which leans towards the formalism of the statistical description of turbulence, but is also one of the few sources of information on the important technological field of non-Newtonian turbulence.

There are, finally, two books which, despite their age, are still indispensable. The short monograph by Batchelor (1953), although lacking experimental data, is still the reference work for the mathematical tools of isotropic turbulence, and a source of many analytical ideas which have not yet been superseded. The situation is similar with the book on boundary layers by Schlichting (1968) which, even though not dealing exclusively with turbulent layers, is still the best summary on their behaviour. A shorter modern book on that subject is due to Young (1989).

2 The small scales and the energy cascade

The first point to note about turbulent flows is that they are chaotic and disordered, and we will see later that they have many degrees of freedom. This, by itself, should not prevent us from computing them, since most macroscopic materials contain many more molecules than there are degrees of freedom in any turbulent flow, and statistical mechanics makes a virtue of the probabilistic aspects of these large numbers to obtain bulk properties which agree well with experiments. In turbulent flows we are also interested in mean quantities, and the fact that we cannot predict the fate of individual fluid particles is generally not important.

One of the reasons why turbulence theory is different from standard statistical mechanics is that it is dissipative. Even though it would appear that viscosity should have a weak effect at large Reynolds numbers, it cannot be neglected, and it influences the solutions drastically. Mathematically, this is because viscosity multiplies the terms with the highest derivatives, and cannot

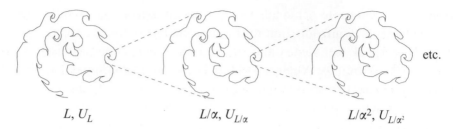

$$L, U_L \qquad\qquad L/\alpha, U_{L/\alpha} \qquad\qquad L/\alpha^2, U_{L/\alpha^2}$$

Figure 1. Schematic representation of the turbulent cascade. Different stages from left to right represent different magnifications of the same turbulent flow, showing eddies of decreasing characteristic lengths. The energy is transferred by instability processes towards the smaller scales, until it is dissipated by viscosity.

be removed without changing the character of the equations. Physically, what happens is that it changes the number of conserved quantities. Consider the momentum equation for an incompressible viscous fluid,

$$\partial_t \boldsymbol{u} + \boldsymbol{u} \cdot \nabla \boldsymbol{u} + \rho^{-1}\nabla p = v\nabla^2 \boldsymbol{u}, \qquad (2.1)$$

where p is the pressure, and ρ and v are the density and the kinematic viscosity of the fluid, both considered constant. If we multiply (2.1) by the vector velocity $\boldsymbol{u}$ and integrate by parts over a large volume, neglecting boundary terms, we obtain an evolution equation for the kinetic energy

$$\partial_t \int_V \tfrac{1}{2}|\boldsymbol{u}|^2 \, \mathrm{d}^3 x = -v \int_V |\nabla \boldsymbol{u}|^2 \, \mathrm{d}^3 x. \qquad (2.2)$$

It would seem that, as the viscosity v is made smaller, the right-hand side of (2.2) would also decrease and, in the limit of infinite Reynolds number, the kinetic energy would be conserved. That this is not so was the first quantitative experimental observation about turbulence, as we noted in the introduction in relation to the pressure loss in turbulent pipes. A review of the experimental evidence is provided by Sreenivasan (1998).

The paradox was clarified by Kolmogorov (1941), who introduced the concept of an energy cascade. He imagined that turbulence is formed by 'eddies' of many different sizes. Energy is injected into the largest ones, either at the boundaries or by the initial conditions, but they become unstable and transfer their energy to smaller eddies, which repeat the process to ever smaller sizes (figure 1). As long as the eddies are large, the transfer is inviscid and no energy is dissipated, but as they become smaller with successive instabilities, so do their Reynolds numbers. Eventually, energy reaches an eddy size (the 'Kolmogorov scale') for which viscosity cannot be neglected, and the right-hand side of (2.2) becomes appreciable. It is at this point that energy is dissipated.

The most successful prediction of this model is the distribution of energy between different scales, which we will derive here as a relation between the size and intensity of the different eddies. Consider homogeneous isotropic turbulence, in which the statistics are independent both of the location and of the orientation of the reference axes. Assume that the energy cascade is active at the same rate everywhere, and that energy is fed into the flow at some 'large' scale $\ell_0 = L$, generating eddies with a characteristic velocity difference u_L. The cascade hypothesis assumes that these eddies become unstable within their 'turnover' time scale $T_L = L/u_L$, and dump their energy into smaller eddies of size, say, $\ell_1 = L/2$. The rate of energy transfer per unit mass is $\varepsilon \sim u_L^2/T_L = u_L^3/L$. This process repeats itself for the smaller eddies that have just been created, and continues until the energy reaches a scale in which viscosity becomes important or, in the limit of zero viscosity, until the energy is broken into infinitesimally small eddies.

If the system is in equilibrium, the rate of energy transfer has to be the same for all eddy sizes, and the previous argument gives the characteristic velocity u_ℓ of eddies whose size is ℓ (say between ℓ and $\ell/2$),

$$\varepsilon \sim \frac{u_\ell^3}{\ell} \quad \Rightarrow \quad u_\ell \sim (\varepsilon\ell)^{1/3}. \tag{2.3}$$

We can use this equation to estimate the two length scales which bound the cascade. It is customary to characterize the velocity of the largest eddies by the root-mean-square value of the fluctuations of one velocity component, usually along the x_1-axis, $u' = \langle(u_1 - U_1)^2\rangle^{1/2}$, where $\langle\cdot\rangle$ stands for global spatial averaging, and $U_1 = \langle u_1 \rangle$. This is justified because it follows from (2.3) that the largest eddies have the largest velocity differences. For some large eddies it would be exactly true that

$$\varepsilon = \frac{u'^3}{L_\varepsilon}, \tag{2.4}$$

which can be used as a definition for the *integral* length L_ε at which energy is fed into the system.

To compute the smallest eddies for which the cascade can be considered inviscid, we estimate for each eddy size the time $T_\nu = \ell^2/\nu$ in which viscosity alone would damp it. If the inviscid instability time ℓ/u_ℓ is shorter than T_ν, instabilities break the eddy before viscosity has time to act, and the latter is not important. The opposite is true if the viscous time is shorter, and the crossover happens when both are of the same order. This is the Kolmogorov

Range	Length	Velocity	Velocity Gradient	Decay Time
Largest eddies	L_ε	u'	$u'/L_\varepsilon = \varepsilon/u'^2$	u'^2/ε
Inertial	ℓ	$u_\ell \approx 1.4(\varepsilon\ell)^{1/3}$	$u_\ell/\ell \sim (\varepsilon/\ell^2)^{1/3}$	$(\ell^2/\varepsilon)^{1/3}$
Kolmogorov	$\eta = (v^3/\varepsilon)^{1/4}$	$u_K = (\varepsilon v)^{1/4}$	$\omega' = (\varepsilon/v)^{1/2}$	$T_K = \omega'^{-1}$
Dissipative	ℓ	$\omega'\ell$	ω'	ℓ^2/v

Table 1. Characteristics of the turbulent eddies for the different ranges of isotropic turbulence.

dissipation length, usually called η, and such that

$$\frac{\eta^2}{v} = \frac{\eta}{(\varepsilon\eta)^{1/3}} \quad \Rightarrow \quad \eta = \left(\frac{v^3}{\varepsilon}\right)^{1/4}. \tag{2.5}$$

Eddies are separated in this way into two different classes. For the large ones, with $\ell \gg \eta$, instabilities act faster and viscosity is unimportant, while the opposite is true for the small eddies in which $\ell \ll \eta$. The former range is called *inertial*, although this term is sometimes restricted to the self-similar eddies for which it is also true that $\ell \ll L_\varepsilon$, while the latter is called *dissipative*. In the dissipative range the flow is smooth, and the velocity can be expanded as a Taylor series, with the result that the velocity differences are proportional to ℓ,

$$u_\ell \sim \omega'\ell, \tag{2.6}$$

where the root-mean-square vorticity magnitude ω' has been chosen as a characteristic value for the velocity gradients. Note that this is not true if we stay in the inertial range of scales, where a finite limit for $u_\ell/\ell \sim \ell^{-2/3}$ does not exist as ℓ decreases, and the flow cannot be described as a differentiable function.

Equating (2.3) and (2.6) at $\ell = \eta$ we can compute the order of magnitude of the characteristic velocities, gradients and time scales for the smallest eddies in the flow. The results for the different ranges are summarized in table 1.

It is clear from the table that the velocity and, therefore, the kinetic energy, is associated with large eddies, while the velocity gradients and, therefore, the viscous dissipation $v|\nabla u|^2$, are associated with eddies of the order of the Kolmogorov scale. This justifies the choice of scale for the gradients used in (2.6).

The time scale decreases monotonically with eddy size, and this has the important consequence that small scales tend to behave as if they were in equilibrium with the larger eddies which contain them. The small eddies evolve much faster than the large ones, and have time to adjust to local equilibrium before the large eddies change appreciably. This is the theoretical justification for studying homogeneous turbulence as a building block for other flows, as well as the reason for the equilibrium approximation used in (2.3). In most real flows turbulence is driven by large-scale forces, typically shear, which differ from one situation to another. These driving mechanisms are not universal, and neither are the large scales that they produce. The hope is that, since their energy is eventually passed to smaller eddies whose sizes and lifetimes are much shorter than theirs, those small scales would behave similarly in all cases, and would provide a common dissipation mechanism valid for all flows. Since these universal scales are too small to see the spatial variability of the large eddies, and too fast to sense any change in their characteristics during their own lifetime, they behave approximately as isotropic and homogeneous, and are controlled only by the local value of the energy dissipation. The cascade theory sketched in this section is only applicable to them.

Note next that, even if (2.4) and (2.5) are taken as definitions, they are only order-of-magnitude estimates. In actual turbulence there are eddies which are larger than the integral scale, while different effects of viscosity may become important for eddies which are somewhat larger or smaller than η. To go beyond estimates we have to define quantitative measures of the eddy behaviour. An obvious one is the root-mean-square velocity difference between two points spaced by a distance ℓ, which can be used as a precise definition for the otherwise somewhat nebulous concept of characteristic eddy velocity,

$$u_\ell = \langle [u_1(x + \ell) - u_1(x)]^2 \rangle^{1/2} \equiv \langle \Delta u^2 \rangle^{1/2}, \qquad (2.7)$$

where the average is taken over x, and ℓ is assumed to be along the same direction as the velocity component u_1. The square of this quantity is called the longitudinal second-order structure function, and it follows from the previous discussion of the cascade that it should be expressible either in terms of the large-scale quantities,

$$u_\ell = u' G_2(\ell / L_\varepsilon), \qquad (2.8)$$

if $\ell \gg \eta$, or of the Kolmogorov scale,

$$u_\ell = u_K F_2(\ell / \eta), \qquad (2.9)$$

if $\ell \ll L_\varepsilon$. The first representation holds only approximately, depending on the type of flow considered, but the second, which applies to scales which are removed from the non-universal driving mechanisms by several cascade

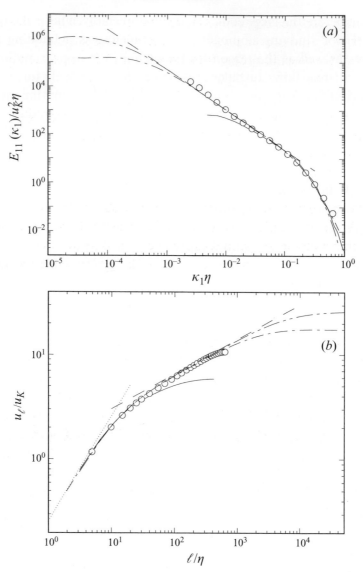

Figure 2. (*a*) One-dimensional spectra (see § 2.1 for definition), (*b*) root-mean-square velocity differences; both displayed in Kolmogorov scaling. ———, Decaying grid turbulence, $Re_\lambda = 60.5$, Comte-Bellot & Corrsin (1971); ∘, tidal channel. $Re_\lambda > 200$, Grant, Stewart & Moilliet (1962); —·—, —··—, wall region of boundary layer, $Re_\lambda = 600$ and 1450, Saddoughi & Veeravali (1994); ----, inertial theory (2.10) and (2.27), with $C_K = 1.5$; the dotted line in (*b*) is equation (2.15).

steps, should be essentially universal. In the inertial range, which links both regimes, the two representations hold, and it follows from (2.3) that

$$F_2 = C_1(\ell/\eta)^{1/3}, \qquad (2.10)$$

where C_1 is a universal empirical constant, $C_1 \approx 1.4$.

The main parameter characterizing the small-scale velocity distribution of different flows is the ratio between the length scales L_ε and η, which defines the extent of the inertial range. It follows from (2.4) and (2.5) that

$$L_\varepsilon/\eta = Re_L^{3/4}, \quad \text{where} \quad Re_L = u'L_\varepsilon/v \tag{2.11}$$

is a large-scale Reynolds number. It is customary in isotropic turbulence to write this relation in terms of a different Reynolds number, defined for historical reasons as

$$Re_\lambda = (15Re_L)^{1/2}. \tag{2.12}$$

This *microscale* Reynolds number can be written as $Re_\lambda = u'\lambda/v$, in terms of a length scale λ, called the Taylor microscale, intermediate between L_ε and η, but its interest is mainly historical and it is seldom used except as in (2.12). Note that (2.11) implies that the number of independent degrees of freedom in a volume L_ε^3 is

$$N_T \sim Re_L^{9/4} \sim Re_\lambda^{9/2}. \tag{2.13}$$

Fully turbulent flows require Re_λ to be larger than approximately 100, and one cannot speak of real turbulence below $Re_\lambda \approx 30$; the instabilities needed for the cascade do not develop in that regime. The highest Reynolds numbers have been measured in the atmospheric boundary layer, $Re_\lambda \approx 10^4$, although indications of still higher ones have been found in astrophysical observations. Typical industrial flows have $Re_\lambda \approx 100$–1000. In the wake of a person running $Re_\lambda \approx 500$ and, in the boundary layer of a large commercial aircraft, $Re_\lambda \approx 3000$. These are all large numbers, showing that most 'practical' flows are turbulent. They also imply that the number of degrees of freedom is large, potentially infinite in the high-Reynolds-number limit, and that a statistical theory is needed to handle them.

A transverse structure function can be defined as in (2.7) for the velocity components u_2 or u_3 that are perpendicular to the separation ℓ. For isotropic turbulence it can be shown (Batchelor, 1953, §3) that $u_{\ell,2}^2 = u_{\ell,3}^2 = 2u_{\ell,1}^2$, and that

$$\varepsilon = v\omega'^2 = 15v\langle(\partial u_1/\partial x_1)^2\rangle = \tfrac{15}{2}v\langle(\partial u_2/\partial x_1)^2\rangle, \tag{2.14}$$

from which it follows that the quantitative version of (2.6), for the velocity differences in an isotropic dissipative range, is

$$u_\ell = \omega'\ell/\sqrt{15}. \tag{2.15}$$

Experimental structure functions at several Reynolds numbers are shown in figure 2(b). The Kolmogorov scaling (2.9) works well at sufficiently small scales, and the extent of the inertial range increases with Re_λ, in agreement with (2.11). The fit to (2.10) in the inertial range and to (2.15) in the dissipation one are also good, although the transition between them, where the behaviour changes from linear to algebraic, is $\ell \approx 10\eta$, rather than the Kolmogorov scale itself.

We should finally mention a quantitative statement of the inertial cascade which is of interest as being one of the few analytic results in turbulence theory. If we write the Navier–Stokes equations (2.1) at two neighbouring points, multiply them by $\boldsymbol{u}$, and subtract one from the other, we obtain after some algebra an equation for u_ℓ^2, which can be interpreted as a measure of the kinetic energy at size ℓ. Because energy is conserved in the absence of viscosity, this equation takes, for homogeneous flows, the conservation form

$$\partial_t u_\ell^2 + \partial_{\ell_j} T_j(\ell) = \text{viscous terms}, \tag{2.16}$$

where T_j is an energy flux in the vector 'separation space' ℓ. This equation expresses that the energy at scale ℓ can only change by being transferred to other scales. The viscous terms are small as long as $\ell \gg \eta$, and can be neglected. The total energy transferred into or out of a given scale can be computed by integrating the flux T_j over the surface of the sphere $|\ell| = \ell$, and conservation implies that, in equilibrium flows for which $\partial_t = 0$, the energy flux ε is independent of ℓ. Even in non-equilibrium flows this is true if the scale is small enough, since u_ℓ tends to zero at small separations, and its time derivative becomes negligible.

That the energy transfer is independent of the separation was of course the original assumption of the energy cascade. What (2.16) provides is an analytic expression for ε, which can be written in terms of cubic combinations of the velocity differences. The expression is especially simple in the particular case of isotropic turbulence, and results in the famous Kolmogorov's '4/5th' law for the longitudinal third-order structure function,

$$S_3 = \langle \Delta u^3 \rangle = -\tfrac{4}{5}\varepsilon\ell. \tag{2.17}$$

This equation is often used to estimate ε in experimental flows, although there are comparatively large corrections at most practical Reynolds numbers, due either to the influence of the energy decay or of the large-scale forcing. It has sometimes been incorrectly used as a justification of the 1/3 exponent in (2.10). The velocity differences can either be positive or negative, and their mean value vanishes. Other odd moments, such as S_3, do not vanish because of a weak asymmetry between shortening and lengthening of fluid elements which is at the root of the energy transfer mechanism, but even moments, such as u_ℓ, simply measure the magnitude of the fluctuations. To deduce the latter from the former requires the additional assumption that the distribution of the velocity increments is identical at all scales, which would lead to (2.10) in any case, as was shown in the derivation of (2.3). Kolmogorov's equation, for example, applies only to the longitudinal velocity differences, for which ℓ and u are parallel, but symmetry implies that all the transverse structure functions of odd orders vanish identically in isotropic turbulence. A particularly careful derivation of (2.17) is given by Frisch (1995, §6), who also includes an interesting account of its history.

2.1 Spectra

The velocity distribution among eddies is often given in terms of spectra, instead of structure functions. Fourier theory is a well-developed field of classical analysis, and the present summary should only be seen as a qualitative introduction to its application in the representation of turbulent velocities. We will assume that the flow is spatially periodic in some large box of size L^3, and approximate the resulting series by integrals. A discussion independent of that artifact can be found, for example, in §2 of Batchelor (1953) or in §6 of Monin & Yaglom (1975).

Briefly, the wavenumber associated with a length scale ℓ is $\kappa = 2\pi/\ell$, and the velocity components, u_j, are expressed as Fourier series or integrals. From now on repeated indices will imply summation. For functions of a single spatial coordinate,

$$u_j(x) = 2\pi L^{-1} \sum \widehat{u}_{j,m} \exp(2\pi i m x/L) \approx \int_{-\infty}^{\infty} \widehat{u}_j(\kappa) \exp(i\kappa x)\, d\kappa. \qquad (2.18)$$

For functions of the three coordinates, (2.18) is trivially generalized by substituting κ by a three-dimensional wave-vector with components κ_n, and the argument of the exponential by the scalar product, $i\kappa_n x_n$. The Fourier integral becomes a triple integral over the three-dimensional wavenumber space.

The Fourier representation is only practical for homogeneous fields, since the exponentials on which it is based are essentially invariant to translations. If a flow is homogeneous or periodic in some directions but not in others, it makes sense to use mixed expansions which are only Fourier-like over some coordinates.

The practical importance of Fourier expansions in signal representation is due to the simple way in which they express the energy associated with a given range of scales. The Fourier expansion (2.18) can be interpreted as the expression of the infinite-dimensional vector $u_j(x)$ in terms of a particular set of orthogonal basis vectors. The usual representation of u_j by its value at each point can be seen as its expansion over the collocation basis formed by Dirac delta functions, such that $u_j(x)$ is the 'component' of u_j over the basis vector $\delta(\xi - x)$. The Fourier coefficients $\widehat{u}_j(\kappa)$ can be similarly interpreted as the components of u_j over the exponential basis vectors $v_\kappa = \exp(\mathrm{i}\kappa x)$, which are orthogonal with respect to the inner product

$$v_{\kappa'} \cdot v_{\kappa''} = \langle v_{\kappa'}(x) v_{\kappa''}^*(x) \rangle. \tag{2.19}$$

The asterisk represent complex conjugation, and the average is taken over x. The collocation basis is also orthogonal, so that the inner product of two vectors is in both cases a properly weighted sum of the products of their components. This leads to Parseval's theorem,

$$u \cdot w = \langle u(x)\, w^*(x) \rangle = 2\pi L^{-1} \int_{-\infty}^{\infty} \widehat{u}(\kappa)\, \widehat{w}^*(\kappa)\, \mathrm{d}\kappa. \tag{2.20}$$

In the particular case $u = w$, this formula represents the mean-square velocity, which can be expressed either as the sum of the squares of the velocity at each point, or as the sum of the squares of the absolute magnitudes of the Fourier transform. It is then natural to interpret the latter as an energy spectrum, proportional to the energy associated with the wavenumber κ. For real variables the energy spectrum is symmetric with respect to $\kappa = 0$, and is traditionally considered to be defined only for $\kappa \geqslant 0$. The proportionality constant in the definition of the spectrum is adjusted so that its integral is the mean-square value of the velocity. In the case of the one-dimensional spectrum of u_1 along x_1,

$$\langle u_1^2 \rangle = \int_0^{\infty} E_{11}(\kappa_1)\, \mathrm{d}\kappa_1. \tag{2.21}$$

Because u_ℓ, as defined in (2.7), is the contribution to the energy of the eddies up to size ℓ, the second-order structure function and the spectrum contain identical information. This gives a practical way of computing the latter, which can be expressed as the Fourier transform of the correlation function of the velocity,

$$R_{11}(\ell) = \langle u_1(x_1)\, u_1(x_1 + \ell) \rangle = \langle u_1^2 \rangle - \tfrac{1}{2} u_{\ell,1}^2, \tag{2.22}$$

$$E_{11}(\kappa_1) = \frac{2}{\pi} \int_0^{\infty} R_{11}(\ell) \cos(\kappa_1 \ell)\, \mathrm{d}\ell. \tag{2.23}$$

It is possible to define 'cospectra' between different velocity components as

$$E_{ij}(\kappa_1) = \frac{1}{\pi} \int_0^{\infty} \left[R_{ij}(\ell) + R_{ji}(\ell) \right] \cos(\kappa_1 \ell)\, \mathrm{d}\ell = 4\pi L^{-1} \operatorname{Re}(\widehat{u}_i \widehat{u}_j^*), \tag{2.24}$$

where Re is the real part. The proportionality constant is again chosen so that

$$\langle u_i u_j \rangle = \int_0^\infty E_{ij} \, d\kappa. \tag{2.25}$$

The spectrum then becomes a symmetric second-order tensor with components E_{ij}.

All the results that we have have obtained for the structure functions can be expressed in terms of spectra. The similarity form (2.9) can be rewritten in terms of the one-dimensional spectrum as

$$E_{11}(\kappa_1) = u_K^2 \eta F(\kappa_1 \eta), \tag{2.26}$$

where F is universal for $\kappa_1 L_\varepsilon \gg 1$. Its form in the self-similar inertial range can be found by arguments similar to those used in (2.3), but may also be obtained from simpler dimensional considerations. It in fact follows from the cascade argument that there is no characteristic length in that range, and that the spectrum may only depend on κ_1 and on the energy transfer rate ε. This implies that η should not be explicitly present in (2.26), which can only be true if

$$F = \tfrac{18}{55} C_K (\kappa_1 \eta)^{-5/3}. \tag{2.27}$$

The Kolmogorov constant C_K is related to C_1 in (2.10) by $C_1^2 \approx 1.3151 \, C_K$, and its experimental value is $C_K \approx 1.5$ (Monin & Yaglom 1975, 2: 351–355). The constant C_K was historically introduced as a premultiplier for the three-dimensional spectrum defined in §3. The numerical factor in front of (2.27) derives from isotropy considerations (Batchelor 1953, §3.4).

Figure 2(*a*) compares experimental spectra with (2.26) and (2.27). It is customary, as in that figure, to plot spectra as a function of the logarithm of the wavenumber. The reason is that the inviscid equations are invariant to geometric scaling, and the important relations are those between a given length scale and its multiples, rather than between scales which differ by a fixed amount. The logarithm converts scaling factors to increments along the axis. The log–log representation in figure 2 has the added advantage of displaying as straight lines the self-similar power laws but, in doing so, loses one of the useful graphic properties of the spectrum, which is to represent energies by integrals or areas. To remedy that, it is useful to plot the pre-multiplied spectrum, $\kappa_1 E_{11}(\kappa_1)$ versus $\log \kappa_1$. The extra factor in front of the spectrum compensates for the differential of the logarithm, and the integral property is restored,

$$\langle u_1^2 \rangle = \int_0^\infty \kappa_1 E_{11}(\kappa_1) \, d(\log \kappa_1). \tag{2.28}$$

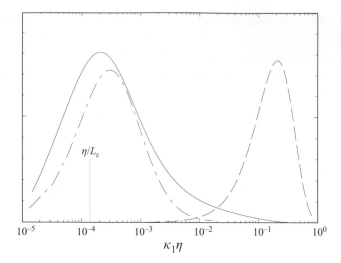

Figure 3. One-dimensional pre-multiplied spectra of: ———, $\kappa_1 E_{11}(\kappa_1)$; – – – – , $\kappa_1 G_{11}(\kappa_1)$; —·— , $\kappa_1 E_{12}(\kappa_1)$. Arbitrary units. The flow is a boundary layer at $Re_\lambda = 1450$, Saddoughi & Veeravali (1994). The vertical dotted line marks the location of the integral scale.

The areas beneath pre-multiplied spectra corresponds to energies, and spectral peaks show directly where the energy is concentrated. An illustration is given in figure 3, which displays pre-multiplied spectra of the velocity and of the velocity gradient $\partial u_1/\partial x_1$. The spectrum of the latter is $G_{11} = \kappa_1^2 E_{11}$, since differentiation of (2.18) is equivalent to multiplication by $i\kappa_1$ and, because of (2.2), is related to the energy dissipation. The two separate peaks show that dissipation is concentrated in scales of the order of the Kolmogorov length, while the energy is associated with the large scales of the order of L_ε. This separation of scales is what defines fully developed turbulence.

2.2 The statistical nature of turbulence

A word must be said about the nature of the averages used in turbulence. Up to now we have used the mean value over physical space but, in the non-uniform cases considered later, that will no longer be possible. Using averages implies randomness, although turbulence is the solution of the deterministic Navier–Stokes equations. The justification for using statistical methods, at least for the small scales, lies in the extreme sensitivity of the solution to small errors. It has been known for some time that even deterministic equations can be so sensitive to initial conditions that small differences grow exponentially with time, so that two solutions which originally differ only slightly become effectively uncorrelated after a short time (see Lorenz 1993).

To understand this intuitively consider two neighbouring particles advected by a smooth velocity field, $d_t x = u(x)$. If we write a linearized equation for the separation between the two particles, one at x and the other at $x + \Delta x$, we get a linear equation $d_t \Delta x = S \, \Delta x$, where $S = \nabla u$ is a matrix which is a smooth function of x, and therefore of t. The solution to this equation is $\Delta x = \Delta x_0 \exp[\int S \, dt]$. If at least one of the eigenvalues of the temporal average of S is positive, initially small differences between the two trajectories become exponentially amplified. This is for example always the case when u is incompressible, since the sum of the eigenvalues of S is its trace, which is the divergence of the velocity, and vanishes if the density is constant.

Gawedzki (1998) observed that the situation in turbulence is even worse and that, in some sense, turbulent solutions become uncorrelated even when their initial conditions are 'identical'. Consider either turbulence at infinite Reynolds number, so that $\eta = 0$, or at distances which are much larger than η. We have mentioned that the flow field is not analytic at those scales, and that gradients cannot be taken. Instead of being able to linearize the velocity differences we have to use (2.3), which gives $|\Delta u| \sim |\Delta x|^{1/3}$. The result is an equation of the type $d_t|\Delta x| \sim |\Delta x|^{1/3}$, whose solution is $|\Delta x|^2 \sim t^3$. This diffusion law was first observed by Richardson (1926), and implies that an infinitesimally small parcel of contaminant spreads algebraically with time. The interesting observation is that $|\Delta x| = 0$ when $t = 0$. Usually two trajectories only cross at singular points. What this result shows is that, at least in this approximation, all the points in the inertial cascade are singular, and that individual trajectories always become uncorrelated after a short time.

Whether the divergence of the trajectories is algebraic or exponential, it is possible to think of a turbulent solution as a particular realization of a class of random fields, each of them originating from different initial conditions, and to use statistics over those different solutions to study the properties of the class. This *ensemble* average is in general the right definition to use, but in practice, if the equations of motion and the boundary conditions are independent of time, the different instants of a single history can be used as a set of uncorrelated initial conditions, each of them generating a time-shifted version of the same history, and the temporal average at a given point can be used as a substitute for the ensemble average. In the following we will use whichever average makes sense for the problem at hand, without specifying it in detail.

3 Inhomogeneity and anisotropy

In the isotropic case considered up to now, all the components of the spectral tensor can be expressed in terms of a single 'three-dimensional' spectrum,

from the inhomogeneity of the mean flow, in what is essentially the first step of the energy cascade.

Consider the Reynolds decomposition (3.1). We can write an evolution equation for the kinetic energy of the turbulent fluctuations, $k = \langle \tilde{u}_j^2 \rangle / 2$,

$$(\partial_t + U_j \partial_j)k + \partial_j \phi_j = \Pi - \varepsilon. \tag{3.7}$$

The left-hand side contains the total derivative of the energy and the divergence of an energy flux, which is due to the velocity fluctuations and to the viscous diffusion,

$$\phi_j = \langle \tilde{u}_j(\tilde{p} + \tilde{u}_i^2/2) \rangle - \nu \partial_j k. \tag{3.8}$$

On the right-hand side we find the energy dissipation, as in isotropic turbulence,

$$\varepsilon = \nu \left\langle |\nabla \boldsymbol{u}|^2 \right\rangle, \tag{3.9}$$

and a new term which feeds energy into the fluctuations, and which is the work of the Reynolds stresses against the mean flow,

$$\Pi = \tau_{ij} \partial_j U_i. \tag{3.10}$$

It is this production term that is responsible for maintaining inhomogeneous turbulence. Since we have seen above that the Reynolds stresses are large-scale quantities, turbulent energy is fed into those large scales, and cascades towards dissipation through the usual Kolmogorov process. If the flow is not too far from isotropy or equilibrium, production and dissipation are the dominant processes and the system can still be approximately described by the isotropic mechanisms. Most of the spectra used in figure 2 to illustrate the universal character of the small scales correspond to inhomogeneous shear flows.

We may note at this stage that, even though turbulent flows dissipate energy more efficiently than laminar ones, turbulent dissipation is still a comparatively weak phenomenon. It follows from (2.10), and from the isotropy relations above (2.14), that the dissipation can be related to the transverse velocity differences by

$$\frac{k_\ell}{\varepsilon} \approx 10 \frac{\ell}{u_{\ell,2}}, \quad \text{where} \quad k_\ell = \tfrac{1}{2}(u_{\ell,1}^2 + u_{\ell,2}^2 + u_{\ell,3}^2) = \tfrac{5}{4}u_{\ell,2}^2 \tag{3.11}$$

is an estimate of the kinetic energy of the flow at size ℓ. The left-hand side of this equation is the time that it would take for the dissipation to damp the eddy, which is proportional to the eddy turnover time on its right-hand side. This agrees with our assumptions about the inertial range, but the proportionality coefficient is large, and the result is that an eddy can be sheared by a large amount, 10ℓ, before the subgrid dissipation has time to damp it.

We will see later that this 'quasi-linear' property of turbulent dissipation, although it has not led to a perturbation description of turbulence, simplifies considerably the treatment of shear-driven flows.

Energy considerations can be used to derive a rough estimate for the eddy viscosity coefficient introduced in the last section. If we use the energy production (3.10) as a surrogate for ε, express it in terms of the eddy viscosity as $\varepsilon \approx \Pi = 2v_\varepsilon S_{ij}S_{ij}$, and estimate the rate of strain as $S = u_{\ell,2}/\ell$, the relation (3.11) gives

$$v_\varepsilon \approx 0.125\,\ell\,u_{\ell,2}, \tag{3.12}$$

which can be written as

$$\frac{\ell\,u_{\ell,2}}{v_\varepsilon} \approx \frac{u_{\ell,2}^3}{\varepsilon\ell} \approx 8. \tag{3.13}$$

This implies that the 'effective' Reynolds number of *any* turbulent scale, based on the eddy viscosity due to still smaller scales, is constant. This argument, which is just a different way of expressing the self-similarity of the inviscid cascade, is the basis for a rule of thumb used sometimes in rough engineering estimates, which states that the effective Reynolds number of turbulent flows is $\Delta UL/v_\varepsilon \approx 10 - 50$. The present derivation shows that the rule is a lowest-order approximation, still coarser and even less justified than the eddy viscosity assumption, but nevertheless useful in preliminary evaluations.

Note also that v_ε is much bigger than the molecular viscosity. The ratio between the two is the (molecular) Reynolds number, and the molecular viscosity is always negligible. This is the main property of turbulence from the applications point of view, and extends to the diffusion of scalar contaminants. One example might help to bring it into focus. The molecular diffusion time of a drop of milk in a cup of coffee is about one day, while we are all familiar with how much faster is mixing with even moderate levels of turbulent stirring.

4 Intermittency

In our discussion of the cascade we have implicitly assumed that the dissipation is spatially homogeneous, or at least that the probability distributions of the various quantities are independent of the scale. It was found experimentally by Batchelor & Townsend (1949) that this is not true, and

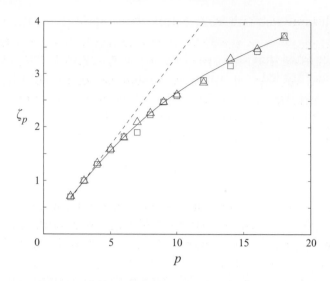

Figure 4. Scaling exponents of the longitudinal velocity structure functions. Symbols are two different experiments at $Re_\lambda \approx 500$–800. ———, Binomial model (4.2) with $c = 1.21$; – – – –, $\zeta_p = p/3$.

that the probability distribution (i.e. the normalized histogram) of the velocity differences becomes more intermittent (i.e. has a higher probability of extreme events) when it is measured at a smaller scale. Kolmogorov (1962, see also Oboukhov 1962) then proposed a modified cascade theory which included intermittency in the form of a spatial variation of the dissipation. The new assumption was that the probability of finding a velocity increment u_ℓ across a distance ℓ, near a point x, was a universal function of $u_\ell/(\varepsilon_\ell \ell)^{1/3}$, where ε_ℓ was the energy dissipation rate averaged over a ball of radius ℓ centred at x. The different statistics of the velocity differences at different scales would then be due to the increased granularity of the dissipation at small scales, and the problem of explaining the statistics of u_ℓ was translated to that of explaining the statistics of ε_ℓ. This 'refined similarity hypothesis' has been tested experimentally to good accuracy.

The theoretical and experimental work that followed these early observations form one of the most beautiful branches of turbulence theory. Even a summary is beyond the limited space of this chapter, but we will try to give the flavour of the different approaches. A more detailed account can be found in the book by Frisch (1995), and an expanded version of the points of view expressed in this section is given by Jiménez (1998a) or Jiménez & Wray (1998).

4.1 Pseudodynamics: multiplicative processes

A lot of the research in the field of intermittency has been based on the study of the longitudinal structure functions $S_p = \langle \Delta u^p \rangle$. Under fairly broad conditions, the full set of structure functions determines the probability density of a variable, so that this is equivalent to studying the histogram of the velocity differences. Under the assumption that the cascade is self-similar, S_p changes with separation as a power of ℓ,

$$S_p/u'^p \sim (\ell/L_\varepsilon)^{\zeta_p}. \tag{4.1}$$

The main experimental result on inertial-range intermittency is the set of exponents ζ_p (figure 4). In the original Kolmogorov hypothesis $\Delta u \sim \ell^{1/3}$ everywhere, and $\zeta_p = p/3$. Experiments show this not to be true. The structure functions decrease with separation more slowly than they should, implying that intense localized events become more common as the separation decreases.

There is a class of simple theories that reproduces this behaviour very well. Assume a cascade in which the scale of the eddies halves at every step, and assume that the probability distribution of the local dissipation in a given nth-generation eddy depends only on the value, $\varepsilon_{\ell,n-1}$, of the local dissipation in the eddy from which it originates. If the cascade is self-similar, the probability can only depend on the ratio $\varepsilon_{\ell,n}/\varepsilon_{\ell,n-1}$, and the cascade can be modelled as a process in which the dissipation is multiplied at every time step by a random factor with a given probability distribution. If we also assume that successive steps are independent, the scaling exponents ζ_p can be expressed in terms of the moments of the distribution of the generating factors. In the uniform cascade the multiplicative factor is the same for all eddies. As soon as more than one value of the factor is possible, the resulting dissipation distribution is intermittent. The intuitive reason is that some eddies are multiplied at every time step by the lowest possible factor, while others get multiplied by the highest possible one, and the disparity between the extremes grows at every cascade step.

Self-similar multiplicative cascades have been proposed in several contexts, and semi-popular but rigorous accounts of their properties and of their applications in other fields are given by Mandelbrot (1983) and Schroeder (1991). They were first used in turbulence by Gurvich & Yaglom (1967), although they were already implicit in the paper of Kolmogorov (1962).

There have been numerous models based on this idea, but few have been more successful in predicting ζ_p than the simple one of Meneveau & Sreenivasan (1991), who observed that the experimental exponents could be approximated by a binomial distribution of generating factors. Each cascade step consists of three bisections along the coordinate planes, in each of which the dissipation is distributed into unequal parts. One half of the eddy receives a fraction $c/2$ and the other one receives $1 - c/2$. The resulting exponents are

$$\zeta_p = \frac{p}{3} - 3 \log_2 \left[\frac{c^{p/3} + (2-c)^{p/3}}{2} \right]. \tag{4.2}$$

They are given as a solid line in figure 4, and fit the data almost perfectly. It should not be deduced from this that the cascade is as simple as the process described above. What the fit illustrates is how difficult it is to deduce information about statistical processes from their final results. Chhabra & Sreenivasan (1992) studied the distribution of multiplicative factors across cascade steps directly. They found a distribution which is tent-shaped, rather than binomial.

Parisi & Frisch (1985) gave an interesting geometric interpretation of the distributions which result from multiplicative processes. One can think of the law $\Delta u \sim \ell^{1/3}$ as meaning that, in the infinite-Reynolds-number limit, the velocity is singular with exponent $1/3$, so that its gradient diverges as $\ell^{-2/3}$ at small separations. Note that this does not mean that, for any finite Reynolds number, the flow is actually singular. What the cascade argument implies is that 'most' points in the flow have this singularity exponent, since the average energy transfer rate has to be independent of ℓ, but it says nothing about the possibility that some points may have different discontinuities. These sparse sets may contain gradients that diverge faster or slower than the mean, as long as their volume is small enough as $\ell \to 0$ that the integrated effect does not modify the global energy balance. Such zero-volume sets are called fractals, and are described in detail by Mandelbrot (1983) or Schroeder (1991).

This 'multifractal' interpretation can be obtained rigorously from the self-similar multiplicative model, and is therefore equivalent to it, but it provides a different representation which is useful in some cases. The reader should consult the book by Frisch (1995) for more details.

4.2 Structures

The main difficulty with the phenomenological theories discussed in the previous subsection is the lack of a definite dynamical model. Not only is the choice of the dissipation as the cascading variable arbitrary, but no process is suggested for its distribution among the descendent eddies at each cascade step. While these 'physics-independent' scaling arguments are not necessarily wrong, they are incomplete.

There is no lack of multiplicative processes in turbulence. The way in which the Navier–Stokes equations create higher velocity gradients and smaller scales in a three-dimensional incompressible fluid is well-known: the vorticity is amplified by being stretched by the velocity gradient tensor. The material derivative of the vorticity magnitude, $\omega = |\boldsymbol{\omega}|$, is

$$\frac{1}{2}\frac{D\omega^2}{Dt} = \omega_i s_{ij}\omega_j + v\omega_j\nabla^2\omega_j. \tag{4.3}$$

If we neglect the viscous terms and assume that the instantaneous strain $s_{ij} = (\partial_i u_j + \partial_j u_i)/2$ is random, isotropic and uncorrelated with the vorticity, (4.3) becomes a multiplicative diffusion process. It is actually possible to show from this equation that the predominant direction of the energy cascade is towards smaller scales, which we had assumed until now, but all attempts to use it to derive transition probabilities have failed, mainly because of the difficulty of representing the viscous term.

In the last decade the effect of vorticity amplification has been observed directly in numerical simulations and in laboratory turbulent flows. The observation is that vorticity concentrates into coherent filaments, essentially internally laminar, with diameters of the order of the Kolmogorov scale and lengths of the order of L_ε, which have lifetimes of the order of the large-scale turnover time of the flow. In spite of their small diameters, the velocity difference across their cores is of the order of the large-scale velocity u'. They have been shown by Belin *et al.* (1996) to be at least partly responsible for the anomalous scaling of the structure functions; if they are removed from the velocity signal, the scaling becomes much closer to that of a uniform cascade. Their properties, and related historical information, are summarized by Jiménez & Wray (1998).

These elongated coherent objects are incompatible with a strictly self-similar cascade. Not only are they difficult to classify into cascade steps, since they span the full range of length scales, but their long lifetimes suggest that they are themselves not cascading. There are reasons for that. Isolated vortex filaments are stable (Saffman 1992), and are only destabilized by sufficiently strong external perturbations. The salient feature of the observed vortex filaments is their intensity. Since they contain velocity differences $O(u')$ across distances which are only $O(\eta)$, their vorticity is much stronger than the average magnitude of the velocity gradients in the flow. They are therefore essentially unperturbed, and remain stable for long times.

In retrospect it is easy to see that this failure of self-similarity was an inevitable consequence of the cascade hypothesis, and that it should be a common feature of other self-similar intermittent processes. The key assumption that led us first to the hypothesis of a power-law cascade, and then to the multiplicative process, was that the only parameters that could influence the breakup of an eddy were its size ℓ, and its internal velocity difference u_ℓ. This is not completely true. Instabilities need initial perturbations, which are provided by the the neighbouring eddies. The original argument holds as long as the cascade is uniform and all the eddies have roughly the same intensity, since the neighbours have the same scaling as the eddy that we are considering. But as soon as the cascade becomes intermittent, some eddies find themselves in an environment that is much quieter than themselves, and their behaviour becomes different from the average. In essence, the transition probabilities, which we had originally assumed to depend only on the ratio $\varepsilon_{\ell,n}/\varepsilon_{\ell,n-1}$, really depend as well on $\varepsilon_{\ell,n}/\langle\varepsilon\rangle$. The cascade process is not the same for eddies in which this ratio is of order unity as for those in which it is large.

Self-similar multiplicative cascades therefore contain the seed of their own destruction, and themselves generate a coherent component which, in three-dimensional turbulence, takes the form of vortex filaments.

It should be stressed that neither the coherent filaments nor intermittency are important for most of the low-order mean quantities of turbulent flows. The scaling exponents shown in figure 4 are very close to Kolmogorov's below $p \approx 4$. The energy and the dissipation are controlled by the regular part of the cascade. The same was shown by Jiménez & Wray (1998) to be true for the coherent vortices. Their fractional volume, energy and dissipation are small, and decrease with increasing Reynolds number.

Intermittency is however not without theoretical and practical consequences. From the theoretical point of view it could be argued that the problem of turbulence is to characterize the set of singularities that is responsible for keeping the dissipation finite in the limit $v \to 0$. The original Kolmogorov answer is that it is a uniform distribution of $\ell^{1/3}$ singularities, but intermittency implies that the answer is more complicated.

From the practical point of view, there are processes which depend critically on the magnitude of the velocity gradients. At industrial scales, the probability of extinction in turbulent flames depends sharply on the local straining, and is responsible for most of the production of contaminants. A difference of a few percent in the probability of strong velocity gradients, even if it has no influence on the global averages, may make a big difference on the contaminant level. On a geophysical scale, tornadoes, storms and hurricanes are essentially the coherent products of intermittent atmos-

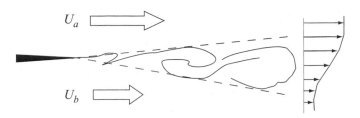

Figure 5. Sketch of a plane shear layer.

pheric turbulence (which is controlled by different processes from the ones described here). The probability of getting a tornado in a given city is at least as interesting as the mean wind velocity.

5 Free-shear flows

Most of the discussion has centred up to now on isotropic flows, on the grounds that they describe the universal behaviour of the small scales. In practice few flows are isotropic, and the rest of this chapter will be devoted to the study of shear-driven turbulence. Since we have seen that only the large scales are affected by shear, our discussion will be devoted mostly to them. The exception will be flows near walls, where the large scales are excluded by the geometry and the whole cascade is modified.

We saw in the introduction that it was in shear flows that turbulence was first recognized. They have continued to be the subject of technological attention, being closer than isotropic flows to problems of practical interest, and a lot is known about their behaviour. A lot of that information is found in engineering collections of experimental data, some of which, such as AGARD (1998), are designed with physics as well as engineering in mind.

The defining characteristic of these flows is the presence of a mean velocity gradient, which makes the large scales anisotropic and provides a continuous source of turbulent kinetic energy. The simplest are those away from walls, including shear layers, jets and wakes behind bodies. The plane shear layer is archetypical. Two parallel streams of different velocities are separated by a thin plate and come together as the plate ends. The layer grows from the resulting velocity discontinuity and thickens downstream (figure 5). Shear layers appear in the initial regions of many other flows, such as jets and separation regions. They are technologically important because they generate noise and dissipate energy, and are therefore a source of drag, but also because, if the two streams contain different fluids, it is in the shear layer that the mixing occurs. Most industrial combustors, and a lot of

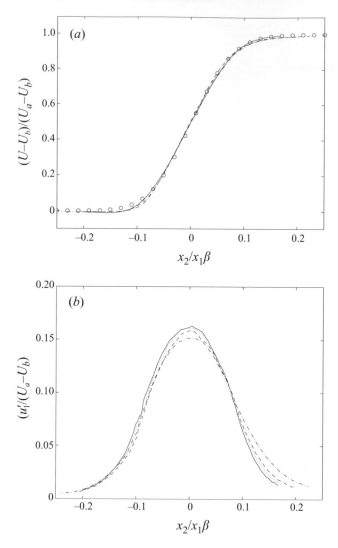

Figure 6. Plane shear layer. (*a*) Mean streamwise velocity, (*b*) root-mean-square streamwise velocity. $\beta = 0.3$, $(U_a - U_b)\theta_0/\nu = 2900$, where θ_0 is the initial momentum thickness. —·—, $\beta x/\theta_0 = 40$; ----, 60; ········, 90; ——, 135 (from Delville, Bellin & Bonnet 1988); ○ in (*a*) is the constant-eddy-viscosity approximation.

the mixers in the chemical industry, are designed around shear layers. It is probably because of this that they were one of the first turbulent flows to be understood in detail, and also one of the first to be controlled.

A lot can be said about the mixing layer from dimensional considerations. An ideal layer grows from a discontinuity of zero thickness and, in the absence of viscosity, has no length scale except for the distance to the origin.

The only other parameters are the velocity of the two free streams, $U_a > U_b$, and the mean velocity, or the r.m.s. fluctuations, must therefore take the form

$$\frac{U}{U_a - U_b}, \frac{u'}{U_a - U_b}, \ldots = f(x_2/x_1, \beta), \tag{5.1}$$

where $\beta = (U_a - U_b)/(U_a + U_b)$ is a parameter characterizing the velocity ratio, and x_1, x_2 are the streamwise and transverse coordinates (see figure 6). In real flows the thickness of the boundary layers on both sides of the splitter plate provides a length scale for the mixing region. A useful definition is the momentum thickness,

$$\theta = \int_{-\infty}^{\infty} \frac{U - U_b}{U_a - U_b} \left(1 - \frac{U - U_b}{U_a - U_b}\right) dx_2, \tag{5.2}$$

which is related to the momentum deficit with respect to a sharp discontinuity. Similarity in shear layers is typically achieved at distances from the plate of the order of several hundred times the initial momentum thickness, although the precise value depends on the details of the boundary layers. The origin of coordinates in (5.1) should for this reason be taken at some distance from the end of the splitter plate, to account for the initial development of the layer.

It can be seen from the small values of x_2/x_1 in figure 6 that the layer spreads slowly. This is a common property of shear flows which could have been anticipated from the observation in §3.1 that turbulent dissipation is a weak process. The turbulent region cannot spread without dissipating energy, and the result is that the flow is slender, much longer than it is wide. This allows it to be approximated as quasi-parallel. For shear layers, for example, we can think of the flow as spreading laterally while being advected downstream with a mean velocity $U_c = (U_b + U_a)/2$. The downstream coordinate is thus converted to an evolution time $t = x/U_c$ and, in that moving frame of reference, the only velocity scale is the difference $U_a - U_b$. The lateral spreading rate must then be proportional to

$$\frac{d\theta}{dx} \approx \frac{1}{U_c} \frac{d\theta}{dt} \approx C_L \frac{U_a - U_b}{U_a + U_b}. \tag{5.3}$$

This analysis is only approximate, but the prediction that the spreading rate should be proportional to the parameter β is satisfied well (Brown & Roshko 1974), with $C_L \approx 0.037$–0.046. The large scatter is not experimental error, and will be discussed later.

The shape of the profile is also represented well by a crude eddy-viscosity approximation. Assume an eddy viscosity which is constant across the layer

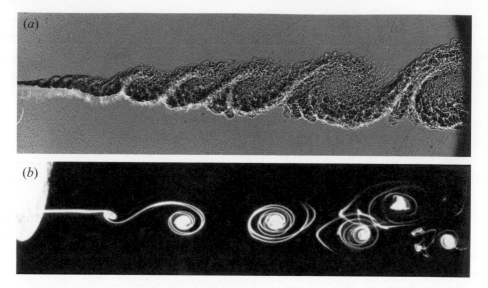

Figure 7. (*a*) Shear layer at high Reynolds number between two stream of different gases, Brown & Roshko (1974). Reynolds number based on maximum visual
thickness, $Re \approx 2 \times 10^5$. (*b*) Initial development of a low-Reynolds-number velocity
discontinuity, Freymuth (1966). $Re \approx 7500$.

and that, on dimensional grounds, has the form $v_\varepsilon = C'(U_a - U_b)\theta$. Neglecting
longitudinal gradients and transforming the problem to a temporal one in
the advective frame of reference, the mean velocity satisfies a diffusion
equation whose solution is an error function. It agrees quantitatively with
the measured velocity profiles if $C' \approx 0.063$ (figure 6*a*). The 'Reynolds
number' of the turbulent shear layer, based on its momentum thickness, on
the velocity difference, and on that eddy viscosity, is about 15, which is of the
same order as the effective Reynolds numbers obtained in §3.1. Note that
this is independent of the actual molecular Reynolds number, underscoring
again that molecular viscosity is negligible for the large scales of turbulent
flows, and therefore for their Reynolds stresses and velocity profiles.

Given the success of these crude approximations, which essentially substitute turbulence by a homogeneous fluid with modified viscosity, it was a
surprise to find that the largest scales of the plane shear layer are anything
but homogeneous, and that the flow can be understood, in large part, in
terms of linear stability theory.

The key observation was made by Brown & Roshko (1974), although
indications had been accumulating for several years. They found that the
interface between the two streams takes the form of large, organized, quasi-
two-dimensional structures, which span essentially the whole width of the
layer (figure 7*a*). Those structures are strikingly reminiscent of the linear

instabilities of a smoothed velocity discontinuity, which had been computed earlier by Michalke (1964), and it soon became clear that not only the shape of the structures, but also their wavelengths and internal organization, agreed remarkably well with the two-dimensional stability results which one would expect only to apply for laminar mixing layers (figure 7*b*).

5.1 Linear stability of shear flows

We have invoked flow instabilities at several points in our discussion, but never actually discussed them in detail. The theory of hydrodynamic stability is a broad and well-developed field. Chapter 4 deals with instabilities in open shear flow, such as mixing layers, and contains material relevant to the present discussion. Several of the books mentioned in the introduction devote chapters to that subject, especially those by Monin & Yaglom (1975) and Lesieur (1997). Classic textbooks on flow stability are those by Betchov & Criminale (1967) and Drazin & Reid (1981). The one by Chandrasekhar (1961) deals little with shear flows, and is mostly oriented towards geophysical and astrophysical applications.

The following paragraphs contain the minimum necessary to understand the dynamics of free shear flows, and do not attempt to justify the results that they quote.

Hydrodynamic stability is similar to that of other mechanical systems. Small perturbations to stable equilibrium states remain small, and die in the presence of friction. In unstable cases, they grow until the system moves far from its original state. The concept can be extended to more complicated dynamical structures, such as periodic orbits or complex attractors, but needs some modification when applied to turbulent flows. Since the basic flow is not steady, the instabilities have to compete with it to be able to grow, and only instabilities with sufficiently fast growth rates are relevant.

Flow instabilities are studied by linearizing the equations of motion about a laminar equilibrium solution. The procedure is similar to the Reynolds decomposition introduced in § 3, but the perturbations are assumed to be small, and all the quadratic terms are neglected, including the Reynolds stresses. The resulting equations are linear, and can be solved by superposition of elementary solutions, which are harmonic functions along the directions in which the basic flow is homogeneous. Moreover, if the basic flow is steady, the temporal behaviour of each individual mode can be expanded in terms of exponentials. Thus, if the base flow depends only on the coordinate x_2, the perturbations can be expanded in terms of

$$\tilde{u} = \hat{u}(\kappa_1, \kappa_3, x_2) \exp[i(\kappa_1 x_1 + \kappa_3 x_3) + \sigma t], \qquad (5.4)$$

which are only solutions if a certain eigencondition,

$$\sigma = \sigma(\kappa_1, \kappa_3), \tag{5.5}$$

is satisfied. For the (simpler) temporal problem that we will discuss here, the spatial wavenumbers κ_1 and κ_3 are real, while the temporal eigenvalue σ can be complex. Eigenfunctions whose eigenvalues have positive real parts grow exponentially in time, while those with negative real parts decay. Any initial condition has to be expanded in terms of all the spatial wavenumbers but, after a while, only those with unstable eigenvalues grow, while the stable ones disappear. From the point of view of long-term behaviour, only the former have to be studied.

Neutral modes, whose eigenvalues are purely imaginary or zero, or even those which are just slightly decaying, are special cases. In general they do not decay or grow in the relevant time scales and, in the presence of unstable modes, are swamped by them. They can however grow linearly under certain conditions of symmetry and, if no exponential mode exists, they may become dominant. We will find them in wall turbulence, but they are not important in free shear flows, which tend to be exponentially unstable.

For parallel flows depending on a single coordinate, $U = [U_1(x_2), 0, 0]$, the most unstable eigenmodes are two-dimensional in the spanwise direction, with $\kappa_3 = 0$, and the perturbations only contain spanwise vorticity. We will restrict ourselves to that case, which includes the plane shear layer, although eigenfunctions are in general three-dimensional and include all the vorticity components.

The instabilities of parallel flows can be classified into two groups. The 'inertial' ones are essentially independent of viscosity, and are unstable as long as the Reynolds number is above a generally low threshold. Their eigenvalues are of the same order as the shear of the base flow, and they therefore grow in times comparable to a single turnover time. The instabilities of the second group appear in flows which are stable to perturbations of the first type, and depend on viscosity for their destabilization mechanism. These flows are typically stable for very low and for very high Reynolds numbers, and become unstable only in some intermediate range. The growth rates of this second class of instabilities tend to be much slower than those of the inertial type, and are therefore not important in turbulent flows. On the other hand, they control transition in some wall-bounded cases.

In the absence of rotation and body forces, the basic inertial instability is that of an isolated vorticity layer, which breaks down, essentially by wrinkling, in times which are of the order of its internal shear time. The mechanism, known as Kelvin–Helmholtz, is sketched in figure 8. The layer

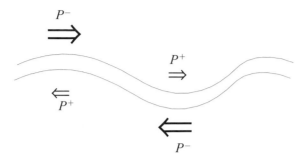

Figure 8. Sketch of the instability mechanism of a free vorticity layer. See text for details.

is equivalent to a smeared velocity jump which, in the advective frame of reference, reduces to two counterflowing streams. If the layer is locally deformed upwards, the velocity of the upper stream increases, and its pressure drops, and the opposite is true of the lower stream. If the interface were stationary this would result in a pressure difference between the two streams. In the real case that difference is substituted by a vertical acceleration that drives the fluid upwards and increases the deformation. Note that this argument implies that the instability waves travel approximately with a velocity which is the average of both streams. This mechanism is robust and works in spite of possible imperfections of the vorticity layers, which are therefore almost always unstable. It is also essentially independent of the Reynolds number, but the requirement that the layer should deform as a whole limits the instability to relatively long wavelengths. As an example, a shear layer with a hyperbolic tangent velocity profile is only unstable to perturbations whose wavelengths are longer than $19\,\theta$, and the maximum growth rate occurs for wavelengths approximately equal to $30\,\theta$ (Michalke 1964).

5.2 Nonlinear development and reduced models

As the instability progresses, the vorticity tends to concentrate at the inflection points of the wave and, in the nonlinear regime, collapses into discrete vortex blobs. The subsequent evolution can be approximately analysed as an infinite row of point vortices (actually vortex lines), which are subject to a new instability. The vortices wander away from the centre of the row and orbit in pairs, in what can be seen as a discrete Kelvin–Helmholtz instability, leading to the formation of a new row of quasi-particles, each twice as strong as the original ones, and with a wavelength that is also twice as long.

growth rate, across its maximum, and finally towards the neutral point. At this point it stops growing and begins to damp. The maximum intensity is reached there, which is not where the wave is growing faster, but where it has had more time to grow, and these most mature structures are the ones that are dominant at each location.

Two simple models based on this idea were analysed by Liou & Morris (1992), where references can be found to previous attempts. In them the large scales are linear eigenfunctions of the mean velocity profile, corrected for spatial growth and non-parallel effects, and grow from a broad initial spectrum of perturbations. This part of the model is linear, but it does not lead to unbounded growth. Since the thickness of the profile increases downstream, each wavelength grows for a finite time, and its total amplification is finite. The nonlinear closure of the model is done through the Reynolds stresses, which are generated by the amplified waves, and which are responsible for the growth of θ. The generic form of these models is

$$\widehat{u}^{-1} \partial_{x_1} \widehat{u}(\kappa_1) = \theta^{-1} \sigma(\kappa_1 \theta), \tag{5.6}$$

$$\partial_{x_1} \theta = \int |\widehat{u}|^2 \, d\kappa_1, \tag{5.7}$$

where the first equation represents the growth of the different modes due to the linear instability, and the second one is the thickening of the layer due the Reynolds stresses. It is easy to check that these equations result in algebraic, rather than exponential, growth, but the details depend on the initial spectrum. The main ingredient missing in (5.6) and (5.7) is a mechanism to relax the effect of the initial condition, which implies redistributing energy among spectral modes. One possibility was discussed by Goldstein & Lieb (1988), who studied the nonlinear interaction of a mode with its subharmonic, once its linear growth rate drops to zero, and which is equivalent to reintroducing eddy pairing in (5.6). A different approach was taken by Jiménez (1980), who noted that downstream pairings produce pressure disturbances, which are felt upstream as subharmonic perturbations and influence the initial spectrum. Which mechanism is dominant in real flows is unclear.

5.3 Mixing and control

The presence of coherent structures has important consequences. Perhaps the most obvious is the possibility of control. It is clear that, if the large structures are responsible for the growth of the layer, and if they in turn originate from the amplification of small initial perturbations, it should be possible to control the characteristics of the layer by manipulating the initial conditions. This was shown to be true by the early experiments of Oster & Wygnanski (1982), and has been used since then in practical manipulations of flows which include shear layers, jets, wakes and separating flows. This sensitivity of the flow to the initial conditions, which can persist for very long distances downstream, also explains the large scatter in the experimental values of the growth rate, which was mentioned above.

The second important effect of the coherent structures is their influence on turbulent mixing which, as we noted at the beginning of this section, is one of the important applications of this particular flow. Before the discovery of the structures it was known that the mean concentration profile of two mixing streams was similar to that of the velocity, and it was believed that it reflected the molecular mixing of the two species. This turns out not to be true in the presence of well-defined coherent structures, which move large parcels of unmixed fluid across the layer without actually mixing them, thus decreasing, or at least modifying, the rate of chemical reaction. Since we have seen that the formation of the structures is not necessarily a residual effect of transition, and that they can re-form at large Reynolds numbers as instabilities of the mean profile, this effect is not restricted to the initial part of the mixing layer.

It is however modified by the amount of three-dimensionality in the small-scale structure, which controls the effective eddy viscosity and the details of the amalgamation mechanism. Several mixing control strategies have therefore been based on introducing three-dimensional vorticity at the splitter plate, thus promoting the secondary instabilities and preventing the coherence. This has been used, for example, to increase the level of molecular mixing and to decrease noise.

In general, and although aspects such as the influence of compressibility, of density differences, and of the three-dimensionality of the mean flow, are still subjects of active research, it can be said that free-shear flows are amongst the best understood, and most easily controlled, turbulent flows.

6 Wall-bounded flows

In contrast to the isotropic turbulent cascade, and to free-shear flows, wall-bounded turbulence is the subject of much of current research, and the cause of many outstanding controversies. These flows, which include boundary layers, pipes and channels, are one of the main open problems in turbulence.

Few aspects underscore this more clearly than the uncertainty that surrounds their scaling properties. Consider a circular pipe of radius R, whose mean velocity profile is shown in figure 9(a) as a function of the distance x_2

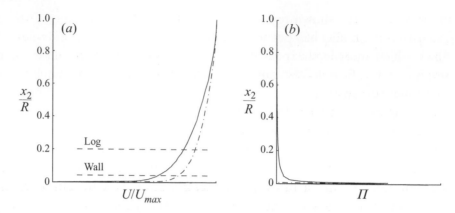

Figure 9. Profiles from a turbulent pipe. (*a*) Mean velocity. (*b*) Turbulent energy production. ———, $Re_\tau \approx 2000$; —·— , 5×10^5. Zagarola & Smits (1997). The dashed lines in (*a*) mark the classical upper limits for the logarithmic and wall regions; the latter for $Re_\tau \approx 2000$.

from the wall. It is tempting to normalize the profile with the velocity U_{max} at the centreline, which is also the maximum velocity difference across the flow, but it is clear from the figure that most of that difference is concentrated in a thin layer near the wall, and that it may not be relevant to the central part of the pipe. This is even clearer when the two Reynolds numbers in the figure are compared. They suggest that the scaling in the core of the pipe should apply to relative, rather than absolute, velocities, and that the latter are isolated from the wall by what can almost be considered a velocity discontinuity across a thin layer.

The classical velocity scaling is based on the turbulent shear stress. It follows from the momentum equation that the total shear stress is $\tau_{12} = \tau_w(1 - x_2/R)$, where τ_w is the stress at the wall. If the Reynolds number is high enough, the viscous contribution is negligible except in the immediate neighbourhood of the wall, and $\tau_{12} \approx -\langle \tilde{u}_1 \tilde{u}_2 \rangle$. It then makes sense to use the friction velocity, defined by $u_\tau^2 = \tau_w$, as a scale for the velocity fluctuations and for the differences in the mean velocity. The latter is based on the idea that the mean velocity gradient is due to large turbulent eddies whose intensity is $O(u_\tau)$. The same scaling applies very near the wall, where the Reynolds stresses go to zero, and the viscous expression of the wall stress, $\tau_w = v\partial U/\partial x_2$, is equivalent to

$$U/u_\tau = x_2 u_\tau/v. \qquad (6.1)$$

The right-hand side of this equation is a Reynolds number based on the distance to the wall, and introduces a viscous length scale $\delta_\tau = v/u_\tau$. Quantities normalized with u_τ and δ_τ are said to be in 'wall units', and are denoted by a $^+$ superscript, so that (6.1) is written $U^+ = x_2^+$. Wall scaling is essentially the same as Kolmogorov's, since $\delta_\tau \approx \eta$ at the wall. The friction Reynolds number, $Re_\tau = u_\tau R / \nu$, is equivalent to the large-scale Reynolds number Re_L used for the isotropic cascade.

The presence of two length scales, δ_τ and R, makes the velocity profile inhomogeneous. Although they are the same lengths as the Kolmogorov and integral scales in homogeneous turbulence, the wall segregates them in space as well as in size. We can think of the core of the pipe as 'normal' turbulence, with a full cascade and a full range of length scales. As we approach the wall, however, the large scales no longer 'fit', and only the Kolmogorov range is left.

In between these two extremes there is an overlap region, where $x_2^+ \gg 1$ but $x_2 / R \ll 1$, in which some of the large eddies are present, while others are excluded by the wall. In the classical picture the only possible length scale in this region is x_2, and the mean velocity profile is logarithmic. There are several arguments that lead to this conclusion, but the following, due to Townsend (1976), is specially instructive. Assume as a first approximation that the local energy dissipation, given by the Kolmogorov argument as $u_\tau^3 / K x_2$, is in equilibrium with the production,

$$\Pi = u_\tau^2 \frac{\partial U}{\partial x_2} = \frac{u_\tau^3}{K x_2}, \qquad \Rightarrow \qquad U^+ = \frac{1}{K} \log x_2^+ + A. \qquad (6.2)$$

The Kármán constant K is universal, essentially related to the dynamics of the cascade, and has been measured in different wall flows as $K \approx 0.4$. The additive constant A depends on the details of the flow near the wall, and is not directly determined by the overlap region. Its experimental value for smooth walls is $A \approx 5$, but changes by orders of magnitude if the wall is rough, or through the effect of certain manipulations of the near-wall region.

An alternative derivation of (6.2) is based on using an eddy viscosity $\nu_\varepsilon = K u_\tau x_2$. The contrast with free-shear flows is clear; in that case we were able to use a single length scale for the eddy viscosity because the same (large) structures were dominant across the whole flow, while here, in a more inhomogeneous situation, that is no longer possible.

Figure 10 shows that the logarithmic profile is very well satisfied experimentally, approximately between $x_2^+ > 100$ and $x_2 / R < 0.2$. Those limits have been overlaid on figure 9, showing that most of the velocity drop occurs in the near-wall layer, and that the core of the pipe is in reality a very weak shear flow. A rough estimate of u_τ, obtained by extrapolating (6.2) to the centre of the pipe, is $U_{max} / u_\tau \approx 20$–$30$, which varies little with Reynolds

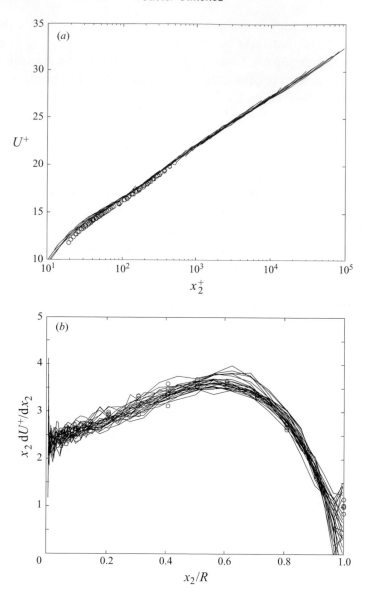

Figure 10. The logarithmic velocity profile in smooth pipes. (*a*) Mean velocity in wall scaling, $x_2/R < 0.2$; (*b*) inverse of the Kármán constant in outer scaling. ——, $Re_\tau = 1700$–10^6, Zagarola & Smits (1997); ∘ , $Re_\tau = 1600$–3900, Perry *et al.* (1986).

number because of the logarithmic dependence; the velocity at $x_2^+ = 100$ is $U/u_\tau = 16$.

In this classical view the near-wall viscous layer corresponds to the dissipative range of scales of isotropic turbulence, the core plays the role of the energy-containing eddies, and the logarithmic region that of the inertial

	Scales	Profile	Limits
Near-wall	u_τ, δ_τ	$U^+ = x_2^+$	$x_2^+ < 100$
Overlap	$u_\tau, x_2; R\,(?)$	$U^+ = K^{-1} \log x_2^+ + A$	$x_2 \ll R$ and $x_2^+ \gg 1$
Core	$R; u_\tau\,(?)$	$U_{max}^+ - U^+ = f(x_2/R)$	$x_2/R > 0.1 - 0.2$

Table 2. Classical scaling of wall flows.

range, but the analogy is not complete. In the isotropic case the small scales are net recipients of energy, which they dissipate. The wall region is different. Because the mean velocity gradient is highest there, the production of turbulent energy has a very sharp maximum there (figure 9b). Even when the dissipation, which is also maximum near the wall, is taken into account, the region below $x_2^+ \approx 30$ is a net source of turbulent energy, which it exports to the rest of the flow as a spatial energy flux. Other parts of the flow, especially the centre of the pipe, where $\Pi = 0$ because $\partial U / \partial x_2 = 0$, are energy sinks, and the global balance is between energy generated near the wall at the Kolmogorov scale, transported to large-scale eddies in the centre of the pipe. This spatial inverse cascade is superimposed on the normal, quasi-isotropic, Kolmogorov local energy decay from large to small scales (Jiménez 1999).

A summary of the classical scalings and velocity laws for the different layers is given in table 2, and it should be emphasized again that they account for most of the available data within experimental accuracy. They have however been challenged by different groups on theoretical grounds, as noted by the question marks in the table.

The complicating factors are several. The weakest argument is using u_τ as a velocity scale for the outer flow. This is particularly true for external boundary layers, since the shear stress tends to zero near the edge of the layer, and the argument used at the beginning of the section cannot be applied. Oberlack (1999) has proposed a scaling for this region which is equivalent to assuming no velocity scale, and which leads to an exponential velocity profile. George & Castillo (1997) have proposed scaling the velocity with U_{max}, which leads to power laws.

The problem in the overlap region is the possible influence of the outer scales, which would be similar to the influence of the Reynolds number discussed in §4 for the inertial range. The strict self-similarity needed for the derivation of (6.2) implies that everything in that region scales in wall variables, which is equivalent to assuming strict Kolmogorov scaling in the dissipation range. This is probably not true. While some variables, like U and u_1' scale well, others like u_2' may not, although it is not clear whether the problem is insufficiently high experimental Reynolds numbers, or incomplete

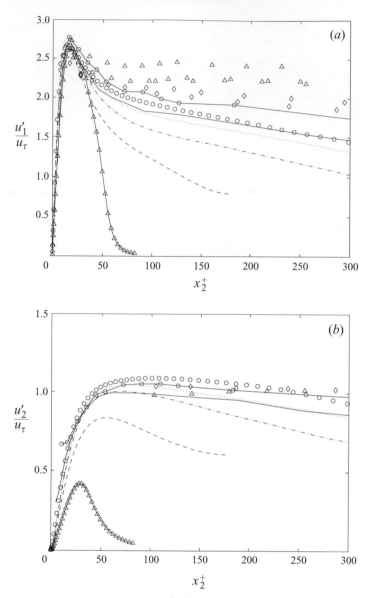

Figure 11. (*a*) Root-mean-square longitudinal, and (*b*) wall-normal velocity fluctuations. Wall scaling; various sources. Lines are channels, $Re_\tau = 180\text{--}10^3$; symbols are boundary layers, $Re_\theta = U_\infty\theta/\nu = 1400\text{--}13000$. —△—, $Re_\tau \approx 60$, modified numerical channel by Jiménez & Pinelli (1989). To a rough approximation, $Re_\tau \approx Re_\theta/2$.

similarity (see figure 11). Barenblatt (1993) has argued for the latter, and pointed that it could lead to power laws for the velocity in the overlap region. Zagarola, Perry & Smits (1997) have challenged his analysis of the experimental data, and argued for the classical result.

The theoretical arguments on both sides are inconclusive and, in the absence of a full theoretical solution of the Navier–Stokes equations at infinite Reynolds number, both scalings should be considered possible. The experimental consequences for the velocity profiles seem to be too small for practical measurement. Moreover, as the Reynolds number is increased in a given wind tunnel the viscous length becomes smaller, and wall roughness begins to be important. The Reynolds number becomes irrelevant in this limit and should be substituted by the ratio between the boundary layer thickness and the roughness. While there are numerous studies of flows over rough walls, they are complicated by variations in the geometry of the roughness, and do not approach the level of quality of those over smooth ones. They are however potentially easier to control than the latter, and would certainly be useful in clarifying the theoretical controversies.

The question of the interaction of the inner and outer scales is more general than the form of the velocity profile. It has been realized for a long time that large scales can exist near the wall as long as their velocities are parallel to it. These so-called 'inactive' motions are traditionally assumed not to influence the mean profile, since they do not include wall-normal velocities and should therefore not contribute to the Reynolds stresses, although Saddoughi & Veeravali (1994) give some evidence to the contrary. They are experimentally observed, and near-wall spectra of the tangential velocity components contain longer wavelengths than the wall-normal ones. This would suggest that the wall-normal fluctuations, whose size is limited by the presence of the wall, should scale well in wall units, while the tangential ones may see the effect of the outer flow, and of the Reynolds number. In fact, the opposite seems to be true (figure 11), underscoring again our lack of understanding.

6.1 Flow structures

In parallel to the statistical information discussed above, a lot is known about the dynamical structures of wall flows.

The best understood region is that very near the wall, below $x_2^+ \approx 100$, where viscosity is important and the flow, if not laminar, is at least relatively smooth. As we have said above, it is roughly equivalent to the dissipative range in the cascade, but acts as a source, rather than as a sink, of turbulent energy. It is dominated by two types of structures: streamwise velocity streaks, and quasi-streamwise vortices (figure 12). The former were first recognized by Kim, Kline & Reynolds (1971), and are long ($x_1^+ \approx 1000$) sinuous arrays of alternating streamwise jets superimposed on the mean

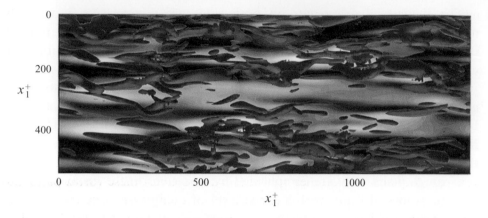

Figure 12. Structures in the wall region of a plane turbulent channel. The view is from the interior of the channel into the wall, with the flow from left to right. The shades of grey are the $\partial_{x_2} u_1$ at the wall, and may be used as a measure of the streamwise velocity near the wall. Dark regions label low shear, and therefore the low-velocity streaks. Coloured objects are isosurfaces of streamwise vorticity $\omega_1 = \pm 0.2$ (purple is positive). Only the region below $x_2 = 80$ is represented. Axes are labelled in wall units. $Re_\tau = 180$. Numerical simulation by A. Pinelli and M. Uhlmann.

shear, with an average spanwise separation $x_3^+ \approx 100$. At the spanwise locations where the jets point forward, the wall shear is higher than the average, while the opposite is true for the 'low-velocity' streaks where the jets point backwards. The quasi-streamwise vortices are slightly tilted away from the wall, and each one stays in the near-wall region only for $x_1^+ \approx 200$. Several vortices are associated with each streak, with a longitudinal spacing of the order of $x_1^+ \approx 400$ (Jiménez & Moin 1991). Some of the vortices are connected to the trailing legs of coherent-vortex arches (hairpins) which are present in the outer part of the boundary layer, but most merge into disorganized vorticity after leaving the immediate neighbourhood of the wall (Robinson 1991).

It is known that the vortices cause the streaks by deforming the mean velocity gradient, moving high-speed fluid towards the wall and low-speed fluid away from it (Blackwelder & Eckelmann 1979). There is less agreement on the mechanism by which the vortices are produced, but it is generally believed that they are the result of the instability of the streaks, thus creating a closed turbulence generation cycle near the wall (Kim *et al.* 1971). Swearingen & Blackwelder (1987) suggested that the instability involved was that of the wall-normal vorticity layers which separate the low- from the high-velocity streaks, a suggestion later modified and extended by Jeong *et al.* (1997), Waleffe (1997) and Kawahara *et al.* (1998). The details of the near-wall cycle itself have been increasingly clarified by the first two of these groups, and recent numerical experiments by Jiménez & Pinelli (1999) strongly suggest that it is autonomous, able to generate turbulent fluctuations without the assistance of the core region.

Waleffe (1997) has proposed a weakly nonlinear reduced model that is, however, more complicated than that for free-shear flows. The main step is the exponential instability of a three-dimensional velocity field which includes the mean shear and the velocity streaks. The eigenfunctions of this instability are three-dimensional and include pairs of quasi-streamwise vortices. A quadratic nonlinearity is needed to convert these vortex pairs into a mean pumping flow ('rolls'), two-dimensional in the cross-plane, (x_2, x_3), which regenerates the streak. The last step is interesting, because it involves the amplification of a weak second-order effect by a neutral instability of the mean flow, whose growth is only linear in time. We mentioned in the previous section that those modes should not be important in the presence of the faster shear, but it is easy to check that two-dimensional flows in the cross-plane behave independently from the shear, while they advect it as a passive scalar. In essence all the energy of the instability is extracted from the shear, instead of from the cross-flow, and the latter is never damped. The mean shear thus plays two roles in this model: it provides energy for the exponential instability of the streaks, and it is deformed by the rolls to regenerate the streaks. While this model is very suggestive, it should be mentioned that it is nowhere near as effective as those of free-shear flows in generating quantitatively correct results.

The other interesting region is the logarithmic layer. We have mentioned that it contains 'inactive' motions in which the wall-normal scales are compressed, while the wall-parallel ones are not. It is clear that these modes may be very anisotropic, and that the normal Kolmogorov cascade may not work for them. Since they are longer than x_2, while we have seen that the logarithmic profile can be written as a balance between the instability of the eddies of size x_2 and the local energy production, these eddies are too slow to participate in that process. They reside *above* the normal energy cascade, which is only carried by eddies which are smaller than x_2. On the other hand, being very anisotropic, they carry Reynolds stresses, and they are the seat of the spatial inverse cascade mentioned above. This can be used to estimate their energy spectrum, which is different from Kolmogorov's. The dimensional argument is the following. Near the wall the conserved quantity is the Reynolds stress u_τ^2, and the spectrum can only depend on it, on the wavenumber κ_1, and possibly on x_2 and R. But we are considering scales which are much longer than the former, so that $\kappa_1 x_2 \ll 1$, and much shorter than the latter, so that $\kappa_1 R \gg 1$. Neither of the two parameters x_2 and R may then enter the spectrum, and the only possible combination is

$$E_{11}(\kappa_1) \sim u_\tau^2 \kappa_1^{-1}. \tag{6.3}$$

This spectrum is observed in most wall flows for $\kappa_1 x_2 \ll 1$, and has been carefully documented by Perry, Henbest & Chong (1986). Note that the argument only works if $x_2/R \ll 1$, and coincides in space with the logarithmic region. It has been argued by Jiménez (1999) that the κ_1^{-1} spectrum, the logarithmic similarity profile, and the inverse spatial cascade, are different aspects of the same phenomenon. Note also that, since κ^{-1} is not integrable near $\kappa = 0$, this spectral range contains most of the fluctuating turbulent energy.

These eddies are responsible for a significant difference between wall-bounded and other turbulent flows. In most other cases the integral scale is uniform across the flow, and the eddies not subject to Kolmogorov scaling are only those which are larger than a given fraction of L_ε. The number of non-Kolmogorov modes per integral volume is therefore $O(1)$, and it could be argued that the only role of a theory of those flows is to understand the isotropic cascade, which contains a potentially infinite number of degrees of freedom as $Re \to \infty$. The finite number of large-scale modes could in principle be handled by a sufficiently powerful numerical simulation.

The situation near walls is different. The integral scale is $L_\varepsilon \sim x_2$, and all the modes larger than x_2 are anisotropic. The number of those modes in a volume $R^2 \, \mathrm{d}x_2$ is $\mathrm{d}N \sim R^2 \, \mathrm{d}x_2/x_2^3$. Integrating between $x_2 = \delta_\tau$ and R, the total number of non-Kolmogorov modes in a volume R^3 is

$$N \sim Re_\tau^2, \tag{6.4}$$

which is infinite in the high-Reynolds-number limit, and only slightly smaller than the total number of turbulent degrees of freedom, $N_T \sim Re_L^{9/4}$. This implies that a statistical theory is needed for the large-scale structures of wall flows, as much as for the small-scale ones, and that this theory is fundamentally different from Kolmogorov's.

It is interesting to observe that the dynamics of the κ_1^{-1} eddies is essentially linear. Since these large eddies are slow with respect to the shear, which from (6.2) is $O(u_\tau/x_2)$, they are passively deformed by it. Hunt (1984) has exploited a different aspect of this idea, studying the effect of the neighbourhood of a wall on pre-existing isotropic turbulence. The orders of magnitude are the same as in the case of shear, and the eddies are linearly deformed by the potential flow induced by the impermeability condition at the wall. The prediction is that the large-scale spectra of both tangential velocity components are enhanced, while the normal component is blocked by the wall and has little energy at sizes longer than x_2. The κ_1^{-1} behaviour does not appear spontaneously in this case. In most shear flows, only the streamwise velocity has a κ_1^{-1} spectrum, clearly because the shear provides a source of energy for u_1 but not for u_3, and the structures are much longer in the stream direction than along the span. It may be significant that κ_1^{-1} spectra of comparable extent have been observed for both tangential components by Hoxey & Richards (1992) in the atmospheric boundary layer, where shear may not be the dominant energy source, and very near rough walls, where energy is injected directly by the roughness elements (S. G. Saddoughi, private communication).

There is an interesting suggestion here. Large structures whose energy is in the streamwise direction and in the streamwise velocity component can only be described as jets, and visualizations by Jiménez (1998b) confirm that the κ_1^{-1} structures of the logarithmic region look like large streaks, similar to those in the viscous layer, although turbulent and at a much larger scale. We have seen that their time scales are such that they are linearly deformed by the mean shear, and it is known that the asymptotic result of the linear deformation of isotropic turbulence by shear is a series of jets (Townsend 1976, pp. 80–87; Lee, Kim & Moin 1990). The theory of linear deformation of turbulence, known as rapid distortion theory (RDT), is nothing more than the solution of the initial value problem for the same linearized perturbation equations used to study hydrodynamic instability, in the case in which only neutral modes are present. The suggestion is that, in the same way that all the small scales of turbulent flows are universal because they are fast enough to be controlled by themselves, the largest scales of free and wall-bounded shear flows, although apparently very different from each other, are also controlled by the same process of linear interaction with the mean shear. In the free-shear case, in which there is at least one unstable eigenvalue of the linear equations, the dominant structure is the corresponding eigenfunction. In the wall-bounded case, where only neutral eigenvalues are available, we find the elongated structures of RDT. In both cases a weakly nonlinear coupling is needed to prevent unbounded growth. We saw that in free-shear flows it is the two-dimensional deformation of the mean velocity profile by the quadratic Reynolds stresses. No complete model is available for wall flows, but it is tempting to speculate about a turbulent version of the near-wall model described above (see for example Townsend 1976, pp. 328–332).

7 Computing turbulence

7.1 Direct numerical simulations

We saw in §2 that one of the defining characteristics of high-Reynolds-number turbulence is that the kinetic energy and the Reynolds stresses are associated with length scales which are different from those of the energy dissipation, and that both are linked by a quasi-equilibrium inertial range which is isotropic and universal. The different strategies for computing turbulent flows differ in which scales they compute explicitly and which ones they model.

At one end of the spectrum of computational methods is direct numerical simulation (DNS), which explicitly computes everything up to, and including, the energy dissipating scales. The velocity field is smooth at those sizes, so that derivatives and Taylor series expansions can be used, and standard numerical analysis applies. High-order numerical simulations at resolutions of a few Kolmogorov scales, or low-order ones at somewhat higher resolutions, provide results which differ from the true ones by as little as required. The quality of DNS is only limited by how much one is willing to spend in terms of resolution, domain size, and running time to collect statistics. When they can be obtained, the results of DNS are indistinguishable from laboratory experiments, and the scatter among careful simulations is generally smaller than among comparable experiments. For a review of recent simulations, see Moin & Mahesh (1998).

Unfortunately the price of direct simulations is high. Since the ratio between the integral and the dissipation lengths is $O(Re_\lambda^{3/2})$, the number of grid points needed to simulate a cube whose size is of the order of the integral scale is $O(Re_\lambda^{9/2})$. Sufficient statistics require that the simulation should be run for a few turnover times, L_ε/u', while considerations of numerical accuracy limit the time step to be shorter than the time it takes for a fluid particle to cross one grid element, $\Delta x/u'$. The number of time steps needed is then of the order of $L_\varepsilon/\Delta x = O(Re_\lambda^{3/2})$, and the total number of operations is

$$N = O(Re_\lambda^6). \tag{7.1}$$

Even for very moderate Reynolds numbers, $Re_\lambda \approx 100$, this means 10^{12} operations, while other considerations, such as the need for several variables per grid point and several operations per variable, add one or two orders of magnitude.

Modern commercial computers process 10^9 floating-point operations per second (flops), and 10^{12} flops will soon be possible on demonstration machines. Petaflops (10^{15}) commercial machines are expected by the year 2020.

Simple direct simulations therefore run now in a few hours on commercial computers, and in a few seconds on prototypes. They are expected to run in milliseconds in twenty years. These are reasonable times, and DNS is a very useful tool for studying simple flows at low and moderate Reynolds numbers.

It has several advantages over laboratory experiments. An obvious one is that, once a flow has been simulated, it is completely accessible to observation, including three-dimensional views and variables which are difficult to obtain in any other way. Even more important is the possibility of simulating 'imaginary' flows, using equations and boundary conditions which differ from the real ones in almost arbitrary aspects, and allowing us to check partial processes or hypothetical mechanisms, or to test proposed control strategies. Direct simulations have already made important contributions to turbulence research, and will undoubtedly be used increasingly in the future, as improved hardware and algorithms extend their capabilities.

They are however limited in scope, and are likely to remain that way for some time. The exponent in (7.1) is high, and a moderate increase in Reynolds number implies a large one in operation count. Industrial turbulent flows have Reynolds numbers in the range of $Re_\lambda \approx 10^3$, and atmospheric turbulence, of 10^4. Computer speed has historically increased by a factor of 10 every seven years (Brenner 1996) and, assuming that the same trend continues to hold, it will take 50 years until direct numerical simulations of *simple* industrial flows are possible, and a century before we can tackle geophysical ones.

7.2 Large-eddy simulations

Fortunately, other strategies are available. The isotropic inertial range is more or less universal, and can be parameterized only by the energy transfer rate. If we can estimate that rate, and use it to model the effect of the inertial range, we should be able to avoid computing not only the dissipation scales, but all those which are approximately isotropic and in equilibrium. This is the principle of large-eddy simulations (LES). The large scales are computed directly, and the dissipation scales, and most of the inertial cascade, are substituted by a 'subgrid' model.

Large-eddy simulations are implemented in terms of filtered variables, defined as

$$\bar{u}_i = \int K(\boldsymbol{x} - \boldsymbol{y}) u_i(\boldsymbol{y}) \, \mathrm{d}\boldsymbol{y}, \qquad \text{etc.,} \tag{7.2}$$

where $K(\boldsymbol{x})$ is a smoothing kernel. These are essentially local averages, and the result is a local version of the Reynolds decomposition (3.1). Note that one effect of smoothing is to allow us to differentiate the filtered field, which we would not have been able to do otherwise in the inertial range of scales. Assuming that the filter commutes with differentiation, and applying it to the Navier–Stokes equations, we obtain

$$\left.\begin{array}{c} \partial_t \bar{u}_i + \partial_j(\bar{u}_i \bar{u}_j) + \rho^{-1}\partial_i p^* = \partial_j(\nu \partial_j \bar{u}_i + \bar{\tau}_{ij}), \\[2mm] \partial_i \bar{u}_i = 0, \end{array}\right\} \tag{7.3}$$

which are similar to the Reynolds-averaged equations (3.2) and (3.3) with a modified subgrid stress tensor. The latter has been separated into a trace-free component

$$\bar{\tau}_{ij} = T_{ij} - (T_{kk}/3)\delta_{ij}, \quad \text{where} \quad T_{ij} = \bar{u}_i \bar{u}_j - \overline{u_i u_j}, \tag{7.4}$$

and an isotropic one that has been absorbed into a modified pressure

$$p^* = \bar{p} - \rho\, T_{kk}/3. \tag{7.5}$$

If we think of the filter as an average over a local 'box', and of the velocity as separated into a smoothed part and a subgrid fluctuation, the subgrid stresses are the flux of momentum due to the fluctuation across the 'walls' of the filter, while the correction to the pressure is the subgrid kinetic energy. The latter needs no explicit modelling, since the only role of pressure in incompressible flows is to ensure continuity. The correction term in (7.5) is due to the difference between enforcing continuity for the true velocity field or for the filtered one and, if the latter is done correctly, the effective pressure is computed automatically. The true pressure only has to be computed if it is needed for some reason, such as in acoustics or in flows involving cavitation.

The divergence of the subgrid shear stresses, however, has to be modelled. One of the most popular models was introduced by Smagorinsky (1963), and is an extension of the eddy-viscosity idea (3.6),

$$\bar{\tau}_{ij} = 2\nu_\varepsilon \bar{S}_{ij}, \qquad \text{with} \qquad \nu_\varepsilon = C_S \Delta^2 |\bar{S}|, \tag{7.6}$$

where Δ is a measure of the width of the filter, $\bar{S}_{ij}$ is the rate-of-strain tensor computed with the filtered velocity $\bar{u}$, and $|\bar{S}|^2 = \bar{S}_{ij}\bar{S}_{ij}$. In this equation Δ has been used as the length scale, and $|\bar{S}|^{-1}$ as a time scale in the eddy viscosity. As with other eddy-viscosity formulations, there is little reason to believe that (7.6) is true, and in particular that $\bar{\tau}_{ij}$ and $\bar{S}_{ij}$ are parallel tensors. In fact, checks on real flows show that both tensors are only weakly correlated, in spite of which (7.6) works well in many situations. The reason seems to be that all that is needed by the large scales is to have a mechanism that dissipates the correct amount of energy at the end of the cascade.

If an equation for the kinetic energy of the filtered velocities is obtained from (7.3), the product $-\Pi_S = -\overline{\tau}_{ij}\overline{S}_{ij}$ appears on the right-hand side as an energy sink. If the Reynolds number is large, and if the filter is far enough from the Kolmogorov scale, the large-scale viscous dissipation $v|\nabla\overline{u}|^2$ is negligible, and Π_S acts as a substitute for ε. The energy is transferred to the subgrid scales, where viscosity would normally get rid of it. In the simulation, the dissipation is done by the model.

This interpretation can be used to estimate the Smagorinsky constant C_S. The dissipation is written as

$$\varepsilon = \Pi_S = 2C_S\Delta^2|\overline{S}|^3 \qquad (7.7)$$

and the magnitude $|\overline{S}|$ is expressed in terms of the energy spectrum. Any filtering operation (7.2) is equivalent to a multiplication of the spectral tensor by a window in Fourier space which, for smoothing filters, damps the high wavenumbers. The norm $|\overline{S}|^2$ becomes a weighted integral over a neighbourhood of the origin in Fourier space which, because the spectrum of the velocity gradients increases with increasing wavenumbers, is dominated by scales whose sizes are $O(\Delta)$. If the filter is narrow enough for those scales to be isotropic, the result is that C_S is a universal constant, which depends only on the filter. For the particular case of a local average over a cubic box of side Δ, $C_S \approx 0.03$. In practice the filter is not necessarily cubical but, because we are essentially integrating a spherically symmetric spectrum over a volume in Fourier space, anisotropic filters can be used with the same value of C_S as long as we take $\Delta^3 = \Delta_1\Delta_2\Delta_3$.

In shear flows, which is where LES is really needed, the corrections due to anisotropy are fairly large, and a different strategy has proved more useful. In those 'dynamic' models the Smagorinsky 'constant', or some equivalent formulation, is determined locally so that the smallest scales of the resolved flow satisfy a self-similarity constraint. This usually involves the comparison of two filtering operations of different widths, but the details depend on the formulation. The reader should consult recent reviews by Lesieur & Metais (1996) and Moin (1997) for details, as well as for successful applications.

The robust mechanism of the energy cascade ensures that even fairly large errors in the subgrid model have only minor effects on the integral scales of the flow. In numerical experiments reported by Jiménez & Moser (2000), simulations of decaying isotropic turbulence were carried using Smagorinsky constants that were on purpose chosen too high or too low by a factor of 2. After a short transient, which had the effect of modifying the initial conditions of the flow, the decay rate returned to the correct experimental value. What happens is that the largest-scale stages of the cascade work correctly, independently of what happens underneath. If the subgrid model is not sufficiently dissipative, the energy accumulates in the smaller scales near the filter, and $|S|$ increases until (7.7) reaches the right level. The opposite happens if the model is too dissipative. In both cases, if the range of directly simulated scales is long enough, the largest scales, which contain the energy and the Reynolds stresses, are not modified.

This is actually a demonstration of the idea, which we have implicitly used up to now, of a one-directional cascade. Causality is from large to small scales, and how the energy is dissipated in the latter does not influence the former, as long as the amount is correct. An even more dramatic demonstration was given by Brachet *et al.* (1998), who experimentally generated isotropic turbulence in superfluid helium. Dissipation in that system is not due to viscosity, but to the random reconnection of quantized vortex lines, in spite of which the energy spectrum in the inertial range was indistinguishable from the normal one, including the value of the Kolmogorov constant.

It is interesting to note that, in spite of this independence, the cascade is not really unidirectional. The subgrid transfer Π_S provides a definition for the pointwise energy transfer, which can be measured by explicitly filtering fully resolved simulations. The result is that the cascade is only direct (large to small) in about 60% of the points, and inverse in the rest (Piomelli *et al.* 1991). The overall direct energy transfer is the difference between these two large opposing fluxes. This is the reason for the main limitation of eddy-viscosity models, including Smagorinsky's and the dynamic ones, which is that, as long as $\nu_\varepsilon > 0$, the cascade is direct everywhere. Locally negative eddy viscosities are generally forbidden by numerical considerations, and the result is that eddy-viscosity models need too little stress to produce a given dissipation (Jiménez & Moser 1998). It is therefore generally impossible to produce with them the right dissipation and the right stresses at the same time, and LES using these models only works if the filter is chosen narrow enough that just the dissipation has to be modelled. The stresses have to be carried by the resolved field.

This still makes LES a practical alternative for the computation of free-shear flows, even at very high Reynolds numbers. We saw in §3 that the stresses are carried by scales which are at most one order of magnitude smaller than L_ε. Including them requires only that $\Delta \approx L_\varepsilon/10$, so that a large-eddy simulation does not need more that a few thousand grid points per cubic integral scale. While this is a large number, it is independent of *Re*. The situation with wall-bounded flows is different, since we have seen that the number of anisotropic modes in them is potentially infinite. This is the reason why they have resisted up to now simulation at large Reynolds

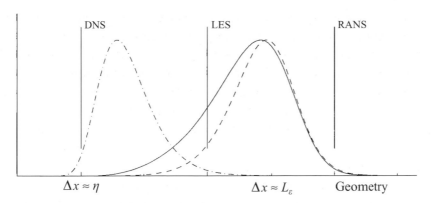

Figure 13. Sketch of the resolution requirements for the different simulation schemes for turbulent flows. The three curves are pre-multiplied spectra for: ———, turbulent kinetic energy; – – – – , shear stress; —·— , energy dissipation. Note that the abscissae, $\Delta x \sim \kappa^{-1}$, are plotted in the opposite direction as in figure 3.

numbers, and why this is one of the most active areas of current research in large-eddy simulation.

7.3 Reynolds-averaged Navier–Stokes simulations

The oldest and least general scheme for computing turbulent flows is to solve directly the Reynolds-averaged Navier–Stokes (RANS) equations (3.2) and (3.3). These equations are not 'closed', because of the presence of the unknown Reynolds stresses. It is possible to write evolution equation for those terms, but they contain triple products, which cannot be expressed in term of simpler quantities, etc. This process leads to an infinite hierarchy of equations of higher moments, which has to be closed at some point with a model. We have used several times in this chapter 'one-point' closures, in the form of eddy viscosities, which we have estimated in *ad hoc* ways. It should be remembered that the result was different for free-shear flows and for the wall layer, although we were able to give *a posteriori* physical reasons for our choice in each case. This is in general the problem with RANS. Since they try to model *all* the turbulent scales, including the non-universal energy-containing ones, RANS models cannot be universal, and have to be adapted to the different cases.

The development of RANS models, either in the form of eddy-viscosity formulas, or in the more sophisticated one of evolution equations for the Reynolds stresses, is an industry in itself, and has been quite successful in providing approximations for flows of practical interest. Good introductions can be found in the books by Launder & Spalding (1972) and Wilcox (1993).

A typical example is the popular k–ε method. It solves two 'evolution' equations: one for the fluctuating kinetic energy k, and another for the dissipation ε, which is written in an *ad hoc* way, using the same structure as for the first one. The diffusion fluxes in both equations are assumed to be proportional to the gradients of the respective variables. The kinetic energy provides a velocity scale, and the dissipation a time scale, for the eddy viscosity $v_\varepsilon \sim k^2/\varepsilon$. There are various empirical coefficients which are adjusted to give the right results for canonical flows, such as decaying turbulence, the logarithmic velocity profile, etc. Given the crudeness of this procedure, this model is extraordinarily successful in computing the mean velocity profiles of industrial flows, and most aeroplanes, cars, or air-conditioners have been computed, at best, using a k–ε model.

Part of the reason is that the way of adjusting the empirical coefficients is designed to give the right results for flows which are not too different from the ones used for calibration. It is also an accepted procedure in RANS to use slightly different sets of coefficients for different flows, if the results are better that way. In this sense RANS models are sophisticated interpolation tables, but the best ones are based on sound physical reasoning, paying attention to such requirements as conservation properties in the non-dissipative limit, tensor invariance, etc., which go a long way towards defining the correct physics. Even if RANS is mostly seen today as an industrial device, it has a distinguished history in turbulence research, and many of the intuitive concepts of time and length scales have been honed by their application to it.

The resolution requirements for RANS are not different, in principle, from LES, but they are quite a bit cheaper in practice. Since all the turbulence fluctuations are included in the model, there is usually no need to consider the flow unsteady, and it is enough to look for steady equilibrium solutions. Neither is it necessary to represent the integral turbulent scales, and the grid is controlled solely by the geometry of the flow. The geometry also controls the integral scales, so that the requirements for RANS and LES are proportional to one another, independent of the Reynolds number. But the proportionality constant may be $O(10-100)$ in each direction, for a total saving of three to six orders of magnitude. On the other hand, RANS methods are *ad hoc*, work poorly in non-equilibrium situations, and cannot be trusted in situations for which calibration experiments are not available.

The resolution requirements for the three simulation methods are sketched in figure 13, with respect to the turbulence spectrum. DNS has to resolve the dissipation peak, LES the energy and stress peaks, while RANS only has to resolve the geometry.

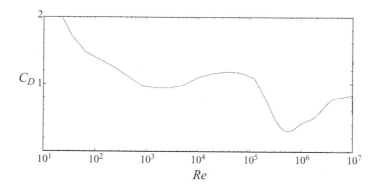

Figure 14. Drag coefficient of a circular cylinder, as a function of $Re = U_\infty D/\nu$. From Oertel (1990).

8 Conclusions

In the preceding pages we have tried to give an overview of our present knowledge of simple turbulent flows. The intended message is that turbulence is not a 'mystery', but a complex physical phenomenon whose essential mechanisms are well understood, at least in the incompressible limit, and with the possible exception of wall-bounded flows. This does not mean that computing turbulent flows is easy, or even possible in many cases, in the same way that playing billiards is hard although the dynamics of billiard balls are understood in great detail. Most turbulent flows are geometrically complex, and involve widely different scales even without taking turbulence into account.

Consider the evolution with Reynolds number of the drag of a circular cylinder (figure 14), which has often been used as an example of the complexity of turbulence. It certainly seems that there has to be something mysterious about a sudden drop of anything in the range $Re = 10^5–10^6$, which are huge numbers. The reality is simpler. This particular bifurcation has been studied in great detail, for example by Schwebe (1983), and is associated with the transition to turbulence of the laminar boundary layers ahead of the separation point, whose Reynolds number is more modest, $Re_\theta \approx 100$. Many other seemingly 'unexplained' phenomena become easier to understand when the right parameter is identified.

It may be that turbulence will always remain complex in this sense, even after we completely understand all its pieces, but that would be no more surprising than the example of billiards, or of chess.

Still, challenges remain. We have neglected many topics in our review, some of which, like the influence of compressibility, are very poorly understood.

parcels of fluid that have different densities from their surroundings. Thus the study of thermals and turbulent plumes has been an active area of research for more than 50 years. Despite this history many outstanding questions remain, such as the influence of background flow (turbulence and mean shear), interactions between thermals, and the detailed nature of the entrainment process.

Another area of active research is atmospheric convection in complex topography and with time-dependent heating. Critical times for atmospheric pollution in urban conurbations are the mornings and evenings when the atmosphere is warming up or cooling down. This is the time when traffic and the associated emissions are the heaviest. Many cities are built in valleys and the transitional flows are often dominated by drainage flows down hillsides. These flows combine aspects of convection above heated surfaces with gravity-driven flows in ways that are still poorly understood.

The object of this chapter is to describe the fundamental aspects of convective flows, so that the underlying physics is revealed and explained. I will discuss topics related to the environment in the individual sections, and point out the challenges that remain.

Perhaps the most universal experience of convection comes from cooking. Liquid in a pot placed over a flame moves as it heats, and this motion transfers the heat from the base of the pot to the interior of the liquid so that a fairly uniform temperature is maintained throughout the volume. When a fluid is heated, the added energy increases the kinetic energy of the vibrating molecules and their mean free path increases. This additional vibrational energy increases the volume of the fluid and, since mass is conserved, the density ρ of the fluid decreases. Under the action of gravity, less-dense fluid rises and convection ensues.

The addition of heat increases the internal energy of the fluid and is measured as an increase of temperature T. The thermodynamic measure of this internal energy is the enthalpy, where the change in enthalpy associated with a temperature change ΔT is given by $C_p \Delta T$, where C_p is the specific heat at constant pressure. Variations in the internal energy of a liquid may also be produced by different concentrations of a solute. Spatial increase in ionic concentration causes a reduction in the molecular mean free path and a consequent increase in density.

These spatial variations of internal energy are modified in time within a stationary fluid by the transmission of enthalpy or ionic concentration to neighbouring regions. This process is called conduction (of heat) or diffusion (of a solute) and the flux of internal energy is assumed to be proportional to the gradient of enthalpy or concentration (Fourier's law).

Fluids with one-dimensional spatial variations in density caused by variations in internal energy are said to be stratified. Common experience suggests that convection occurs in fluids in which the density is lower at the bottom than the top, as in the case of the heated cooking pot. This density variation is called *unstable stratification*. Similarly, fluid where the density decreases with height has *stable stratification*. Although this is loose terminology I will use it here. The other stratification of interest is where the density varies in the horizontal direction; I will call this *horizontal stratification*.

The effect of stratification on the motion of the fluid takes two forms. The first is the fact that, even for fluid in uniform motion, variations in density produce variations in momentum. The second, which is the subject of this chapter, results from variations in the body force ρg, when the fluid is in a gravitational field g. In most flows of environmental relevance the latter is the most significant and variations in the fluid momentum are neglected – this is the Boussinesq approximation.

If the fluid is at rest in a gravitational field, the pressure at any location is given by the weight of fluid above. The fluid will be accelerated from rest if there are unbalanced pressure variations that can exert a force on a fluid volume. For fluid to remain at rest it is necessary that the pressure is uniform in the horizontal and, since the weight is given by the mass of fluid above a point, the density must also be horizontally uniform. Therefore, motion will always be generated by gravity when the fluid is horizontally stratified. If the density is horizontally uniform then the fluid can be at rest with the pressure p given by the hydrostatic relation

$$\nabla p = \rho \boldsymbol{g}. \qquad (1.1)$$

This state is one of equilibrium for a fluid and can be examined to determine whether the equilibrium is stable or unstable, by considering a small departure such as that shown in figure 1. Consider a small vertical displacement of fluid as shown and suppose that the time taken for the displacement is small so that the displaced element conserves its internal energy. The vertical displacement produces horizontal density variations and consequently motion in the fluid. For the case of stable stratification (figure 1a), the raised parcel is denser than fluid surrounding it and a downwards restoring force is applied to the parcel. Thus the equilibrium is stable. For the case of unstable stratification (figure 1b) the parcel is buoyant and the force is upwards and the equilibrium is unstable.

In both cases the flow is produced by buoyancy forces associated with horizontal variations in density. When the fluid is displaced, gradients in the pressure field are no longer parallel to density and vorticity is produced

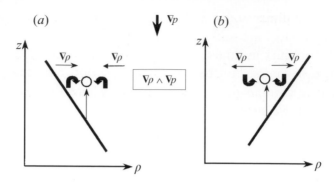

Figure 1. Small vertical perturbations lead to horizontal density gradients shown by the horizontal arrows. Horizontal vorticity normal to the page is generated by the baroclinic torque $\nabla\rho \wedge \nabla p$, as indicated by the curved arrows. The vectors are shown for an upward displacement of the parcel. In (a) the torque produces a vortex pair that moves the parcel downwards and the stratification is statically stable, and in (b) the induced vortex pair moves the parcel further upwards and the stratification is statically unstable.

(see (1.4)). This is called 'baroclinic generation of vorticity'. For convection to occur it is necessary for vertical density variations to be converted into horizontal variations by perturbations to the fluid. The horizontal density gradients depend on the horizontal scale of the perturbations and so there is a tendency for the flow to be collected into convective elements – or convection cells – and for the properties such as the vertical heat transfer to be determined by these. Sometimes the scales of the cells are determined internally, such as when a plane layer of fluid is heated uniformly from below. In other cases variations in the heating or geometrical effects, such as topography, determine their size and form.

In a gravitational field, variations in density set the potential energy of the fluid. When buoyant fluid moves upwards, on average the potential energy of the fluid decreases. The release of this energy provides the kinetic energy for the flow and also energy that is dissipated and added to the internal energy of the fluid. The motion redistributes the fluid so that less-dense fluid moves upwards and hence the tendency is for stable stratification to be produced. Consider stratified fluid in a closed domain with no inputs of internal energy. In the absence of diffusion, the density of each fluid element is conserved and the ultimate state is one where the fluid is stably stratified with no horizontal density gradients. The potential energy of this final state is the minimum that can be achieved and the difference between this potential energy and the initial state is called the available potential energy. Convection in this case will convert all the available potential energy into kinetic energy which will ultimately be dissipated by viscosity to heat.

The effect of viscous dissipation and diffusion of density is to inhibit the convective flow and the ratio of the driving buoyancy forces and these dissipative forces is given by a dimensionless parameter, the Rayleigh number *Ra*, defined by

$$Ra = \frac{g \Delta \rho l^3}{\rho \nu \kappa}, \tag{1.2}$$

where $\Delta \rho$ is a measure of the density difference, l is a typical length scale, and ν and κ are the kinematic viscosity and molecular diffusivity, respectively. When $Ra \gg 1$, the buoyancy forces dominate and, when the stratification is unstable, convection will ensue. For environmental flows Ra is large (or if it is not then convection is not of interest), but near any boundary the scale of the motion is small and heat enters the fluid by conduction, through a thin boundary layer.

The equations governing fluid flow are the Navier–Stokes equations

$$\boldsymbol{u}_t + \boldsymbol{u} \cdot \nabla \boldsymbol{u} = -\frac{1}{\rho} \nabla p + \boldsymbol{g} + \nu \nabla^2 \boldsymbol{u}, \tag{1.3}$$

where the usual notation is employed. Noting that $\boldsymbol{u} \cdot \nabla \boldsymbol{u} = \nabla \wedge \boldsymbol{\omega} - \frac{1}{2} \nabla |\boldsymbol{u}|^2$, the curl of (1.3) yields the vorticity equation for $\boldsymbol{\omega} = \nabla \wedge \boldsymbol{u}$ in the form

$$\boldsymbol{\omega}_t + \boldsymbol{u} \cdot \nabla \boldsymbol{\omega} = \boldsymbol{\omega} \cdot \nabla \boldsymbol{u} + \frac{1}{\rho^2} \nabla \rho \wedge \nabla p + \nu \nabla^2 \boldsymbol{\omega}. \tag{1.4}$$

The effect of buoyancy is shown in the middle term on the left-hand side of (1.4), $(1/\rho^2) \nabla \rho \wedge \nabla p$. This term is non-zero whenever surfaces of constant pressure and density are non-parallel. In a stationary fluid under gravity the pressure is hydrostatic and constant-pressure surfaces are horizontal. Hence, as discussed above, if the density field contains any horizontal variations, vorticity will be generated, and flow will occur. When the background density field is such that the density increases in the direction of gravity, buoyancy forces provide a restoring force and damped oscillations (internal gravity waves) occur. Such a density field is said to be statically stable. On the other hand, if the density decreases in the direction of gravity the motion is amplified and convection ensues. Such a stratification is said to be statically unstable.

The term $(1/\rho^2) \nabla \rho \wedge \nabla p$ is zero if $p = p(\rho)$ and such a fluid is called 'barotropic'. A trivial example is an unstratified fluid with constant density. If $(1/\rho^2) \nabla \rho \wedge \nabla p \neq 0$ the fluid is called 'baroclinic' and the flow results from the baroclinic generation of vorticity. In non-stationary fluids, such as those in a rotating frame of reference (see §6), surfaces of constant pressure are not necessarily perpendicular to gravity, and so the terminology has a more general usage.

In a stationary flow it is clear that the primary baroclinic generation is concerned with horizontal gradients of density. Since convection is usually thought of in terms of vertical gradients, e.g. warm air under cold air, or heating the base of a saucepan, one of the aims of this chapter is to explain how vertical gradients can be changed into horizontal gradients.

In §2, the classical problem of the flow generated by a temperature difference imposed across an infinite horizontal layer of fluid is discussed. This discussion shows how stability analysis reveals the central role played by the Rayleigh number and examines the flow at various values of Ra. Section 3 describes the convective plume that is the predominant element of high-Rayleigh-number convection. A formal derivation of plume equations from the equations of motion is given. Solutions of the plume equations are discussed for a number of environmental flows. Section 4 describes double-diffusive convection that occurs when two components such as a solute and heat stratify a fluid. The effects of horizontally imposed temperature variations are discussed in §5, and the influence of the rotation of the Earth is discussed in §6.

2 Rayleigh–Bénard convection

The simplest case to consider, as a paradigm of convection, is that of a layer of fluid of depth H contained between two horizontal boundaries with the lower boundary maintained at a higher temperature $T_0 + \Delta T$ than the upper boundary at temperature T_0 (figure 2). A possible steady state is for the fluid to be at rest and for the temperature gradient $T_z = \Delta T / H$ to be constant. Heat is conducted from the lower boundary to the upper boundary and the heat flux $-kT_z$ (where $k = \rho C_p \kappa$ is the thermal conductivity) is constant in space and time. However, this configuration has heavy fluid on top of light fluid and is potentially unstable. The instability was first discussed by Rayleigh (1916) and has been the subject of a very extensive literature since.

2.1 The linear stability problem

The stability problem may be formulated as follows. The governing equations are the Boussinesq, energy and continuity equations, which, for an

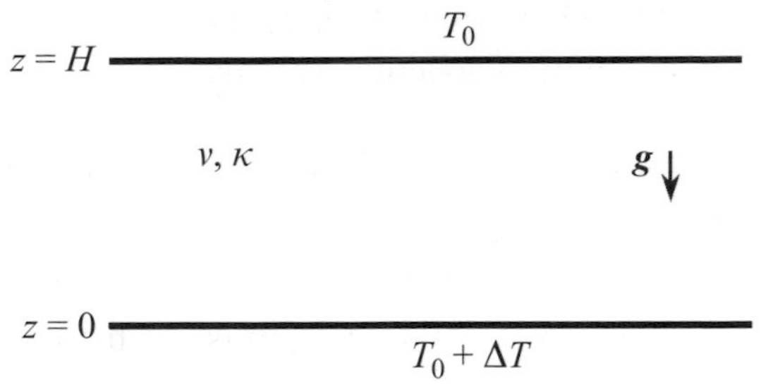

$$z = H \rule{6cm}{0.4pt}$$
$$T_0$$

$$\nu, \kappa \qquad\qquad g \downarrow$$

$$z = 0 \rule{6cm}{0.4pt}$$
$$T_0 + \Delta T$$

Figure 2. A schematic of the Rayleigh–Bénard problem of convection between two infinite horizontal boundaries.

incompressible fluid, take the form

$$u_t + u \cdot \nabla u = -\frac{1}{\rho_0} \nabla p + g \frac{\rho}{\rho_0} + \nu \nabla^2 u, \qquad (2.1)$$

$$T_t + u \cdot \nabla T = \kappa \nabla^2 T, \qquad (2.2)$$

$$\nabla \cdot u = 0, \qquad (2.3)$$

and

$$\rho = \rho_0 [1 - \alpha(T - T_0)]. \qquad (2.4)$$

Here, ρ represents the density, and (2.4) is a linearization of the equation of state $\rho = \rho(T)$ about the mean temperature T_0 at which the density is ρ_0. The Boussinesq assumption requires that departures of density from the mean, ρ_0, are small and that the molecular properties of the fluid, the viscosity ν and the diffusivity of heat κ, are constant. This approximation is valid for many geophysical flows, as discussed by Spiegel & Veronis (1960). Application of the Boussinesq assumption means that $\Delta T \ll T_0$ and is consistent with the simplified equation of state (2.4), where α, the coefficient of expansion, is constant. For a perfect gas $\alpha = T_0^{-1}$, where T_0 is measured in Kelvin.

Noting that in the undisturbed state the temperature is given by the diffusion profile $T_z = \Delta T / H$, it is convenient to make (2.1)–(2.4) dimensionless using the following (starred) non-dimensional variables

$$x = x^* H, \quad t = t^* H^2 \kappa^{-1}, \quad u = u^* H^{-1} \kappa, \quad p = p^* \rho_0 \kappa^2 H^{-2}, \quad T = T^* \Delta T.$$

In terms of these dimensionless variables, and on dropping the stars, the equations become

$$u_t + u \cdot \nabla u = -\nabla p + \sigma Ra T k + \sigma \nabla^2 u, \qquad (2.5)$$

$$T_t + u \cdot \nabla T = \nabla^2 T, \qquad (2.6)$$

$$\nabla \cdot u = 0, \qquad (2.7)$$

where k is a unit vector in the upward vertical direction. The strength of the convection is governed by the Rayleigh number

$$Ra = \frac{g\alpha\Delta T H^3}{\nu\kappa} \tag{2.8}$$

and the properties of the fluid are described by the Prandtl number

$$\sigma = \frac{\nu}{\kappa}. \tag{2.9}$$

For natural flows the Prandtl number $\sigma \geqslant O(1)$: for heat in air $\sigma = 0.7$ and for heat in water $\sigma = 7$.

The behaviour of small departures from the rest state, with the constant (dimensional) temperature gradient $T_z = \Delta T/H$ is governed by linearized forms of (2.5)–(2.7) given by

$$u_t = -\nabla p + \sigma Ra\,Tk + \sigma\nabla^2 u, \tag{2.10}$$
$$T_t = -w + \nabla^2 T \tag{2.11}$$

and

$$\nabla \cdot u = 0. \tag{2.12}$$

Equations (2.10)–(2.12) are five equations in five unknowns. Eliminate the pressure from (2.10) and use (2.12) to obtain

$$\left(\sigma^{-1}\partial_t - \nabla^2\right)\nabla^2 w - Ra\,\nabla_H^2 T = 0,$$

where $\nabla_H^2 \equiv \partial_x^2 + \partial_y^2$ is the horizontal Laplacian.

Then, by (2.11), we obtain a single equation for the vertical velocity

$$\left(\partial_t - \nabla^2\right)\left(\sigma^{-1}\partial_t - \nabla^2\right)\nabla^2 w - Ra\,\nabla_H^2 w = 0. \tag{2.13}$$

This linear equation has constant coefficients (for a Boussinesq fluid) and admits normal-mode solutions of the form

$$w = \text{Re}\left[\widehat{w}(z)e^{i(kx+ly)+st}\right]. \tag{2.14}$$

Substitution of (2.14) in (2.13) gives

$$\left[\left(s + K^2 - \partial_z^2\right)\left(s\sigma^{-1} + K^2 - \partial_z^2\right)\left(\partial_z^2 - K^2\right) + Ra\,K^2\right]\widehat{w} = 0, \tag{2.15}$$

where $K = \sqrt{k^2 + l^2}$ is the horizontal wavenumber of the perturbation.

The vertical structure of $\widehat{w}(z)$ is determined by the boundary conditions on the top and bottom boundaries.

2.1.1 *Boundary conditions*

If the temperature is fixed at each boundary, the temperature perturbation

$$T = 0 \quad \text{on} \quad z = 0, 1. \tag{2.16}$$

If, instead, the heat flux is held fixed, these boundary conditions are replaced by

$$T_z = 0 \quad \text{on} \quad z = 0, 1.$$

The vertical velocity is zero at each boundary and hence

$$\widehat{w} = 0 \quad \text{on} \quad z = 0, 1. \tag{2.17}$$

Traditionally two forms for the velocity boundary conditions are applied.

(i) *Rigid (no-slip) boundaries*

The most realistic case, at least for the lower boundary, is the case of no-slip boundaries where $\boldsymbol{u} = \boldsymbol{0}$. We then have

$$u_x + v_y = 0 \quad \text{on} \quad z = 0, 1$$

and hence from (2.3),

$$\widehat{w}_z = 0 \quad \text{on} \quad z = 0, 1. \tag{2.18}$$

(ii) *Free (stress-free) boundaries*

It is mathematically more convenient to consider stress-free boundaries on the two horizontal surfaces. Since $w_x = w_y = 0$, then $u_z = v_z = 0$ at $z = 0, 1$. Again, from continuity (2.3)

$$\widehat{w}_{zz} = 0 \quad \text{on} \quad z = 0, 1. \tag{2.19}$$

Since the equations are reduced to a single equation for the vertical velocity, it is convenient to replace (2.16) by a condition on $\widehat{w}$. This is achieved by forming the horizontal divergence of the horizontal momentum equations (2.10) and using (2.11) to obtain

$$\sigma Ra \nabla_H^2 T = \nabla^2 w_t - \sigma \nabla^4 w. \tag{2.20}$$

Then (2.16), (2.18) and (2.19) imply that

$$\widehat{w}_{zzzz} = 0 \quad \text{on} \quad z = 0, 1. \tag{2.21}$$

For stress-free, conducting boundaries $\widehat{w} = \sin n\pi z$, where n is an integer, and (2.15) becomes a quadratic for s

$$\left(s + K_n^2\right)\left(s\sigma^{-1} + K_n^2\right) K_n^2 - Ra K^2 = 0, \tag{2.22}$$

where $K_n^2 = K^2 + n^2\pi^2$ is the wavenumber of the disturbance. The solution of (2.22) is

$$s = -\tfrac{1}{2}(\sigma + 1)K_n^2 \pm \left(\tfrac{1}{4}(\sigma - 1)^2 K_n^4 + \sigma Ra\, K^2 K_n^{-2}\right)^{1/2}. \qquad (2.23)$$

For all values of the Prandtl number σ, s is real for $Ra > 0$. Thus marginal stability is determined by 'exchange of stabilities' and is given by $s = 0$. The smallest value of the Rayleigh number occurs when $n = 1$ and is given by

$$Ra = \frac{(K^2 + \pi^2)^3}{K^2}. \qquad (2.24)$$

Remarkably, this result is independent of σ. The minimum value of the Rayleigh number occurs for $K_c = \pi/\sqrt{2}$ and takes the value

$$Ra_c = \frac{27\pi^4}{4} \approx 657.5. \qquad (2.25)$$

The above analysis reveals a number of interesting features of relevance to natural flows. First the critical value of the Rayleigh number is $O(10^3)$. The critical value depends on the boundary conditions, and other (more realistic) boundary conditions (see Drazin & Reid 1981) give larger values but still of the same order of magnitude. For a layer of water 0.1 m deep, a Rayleigh number of $O(10^3)$ is achieved with a temperature difference 10^{-4} K, and so almost any realistic unstable temperature distribution will produce Rayleigh numbers far in excess of critical values. And, since Ra increases rapidly with H, deeper layers produce even higher values of Ra. Consequently, natural flows are usually unstable whenever the density distribution is statically unstable.

The second fact that emerges from the linear analysis is that the most unstable mode has the least vertical structure (i.e. $n = 1$). The convection consists of cells where the buoyant fluid is carried upwards from the lower boundary and negatively buoyant fluid is carried downwards from the upper boundary (for this problem the flow is completely symmetrical about the mid-plane) by flow that is undirectional. The vertical velocity changes sign for higher vertical modes and only for the lowest vertical mode is the buoyant fluid carried continuously upwards. This strong correlation between vertical velocity and buoyant fluid results from the momentum generated by the buoyancy forces and persists at high Rayleigh numbers, even when the convection is turbulent.

The third feature of the linear analysis is that only the horizontal *scale* of the marginal mode is determined. Any combination of horizontal wavenumbers (k, l) with the correct length K_c satisfies (2.25), and so the horizontal shape of the convection cells is not determined by the analysis. Further analysis, such as a consideration of nonlinearity must be carried out to determine the shape. Considerable work has been done on this problem and a good place to start reading is Drazin & Reid (1981, p. 435). Bénard (1900) observed a hexagonal planform in his original experiments with a free upper surface, and these are now known to be caused by variations in surface tension. (Hexagons are also observed in convection without surface tension, as a result of other effects.) The value of K_c implies that the horizontal half-wavelength of the marginal mode is $\sqrt{2}$ and so the horizontal scale of each convection cell is larger than, but of the same order as, the depth of the layer. This feature also persists at high Rayleigh number.

2.2 Heat transport

When convection occurs heat is transferred between the boundaries by advection; hot fluid is moved upwards by the flow. The heat transfer is greater than that by conduction, and this increase is measured by the Nusselt number

$$Nu = \frac{F}{k\Delta T H^{-1}}. \tag{2.26}$$

The denominator of (2.26) is the heat flux that would result from steady conduction, and Nu is a dimensionless value of the actual heat flux F. For the conductive regime $Nu = 1$ and, in general, $Nu > 1$ when convection is present. For fixed fluid properties (i.e. fixed Prandtl number), dimensional analysis suggests that

$$Nu = Nu(Ra), \tag{2.27}$$

and much attention has been given to the form of this functional relationship, particularly at high Ra. A prediction of (2.27) allows the heat flux to be determined in terms of the external parameters of the flow (since Ra depends on the temperature difference and the fluid depth) and so is of considerable practical, as well as conceptual, value. There are two general scaling arguments which I will discuss.

Before doing so, it is of interest to consider another interpretation of Nu. Near the boundaries heat is transferred by conduction since the vertical velocities are zero at the boundaries. Conduction boundary layers of thickness δ are found, and the heat flux F_δ across each layer is

$$F_\delta = kT_\delta \delta^{-1}, \tag{2.28}$$

Authors	Turbulent/ mean flux	Ratio of heat and velocity radii, λ	Entrainment coefficient, α
Rouse, Yin & Humphreys (1952)	0.03	1.16	0.12
George, Alpert & Tamanini (1977)	0.19	0.92	0.16
Nakagome & Hirita (1977)	−0.26	1.14	0.14
Chen & Rodi (1980)	0.14	0.92	0.16
Papanicolaou & List (1988)	−0.02	1.06	0.13
Shabbir & George (1994)	0.20	0.92	0.16

Table 1. Values of the ratio of turbulent to mean buoyancy fluxes, the ratio of the buoyancy to velocity radii and the entrainment coefficient (for top-hat variables) obtained in experiments. The variations in the values of the ratio of the turbulent fluxes to the mean fluxes give an indication of the precision of the experiments.

3.2 Equations of motion

Axisymmetric steady flow of an inviscid incompressible fluid with no diffusion of mass and with no swirl, with a velocity field $\boldsymbol{u} = (u, 0, w)$ in cylindrical coordinates (r, θ, z) with z vertical, is governed by the equations

$$uu_r + wu_z = -\frac{1}{\rho_0}p_r, \tag{3.8}$$

$$uw_r + ww_z = -\frac{1}{\rho_0}p_z - g\frac{\rho}{\rho_0}, \tag{3.9}$$

$$u\rho_r + w\rho_z = 0, \tag{3.10}$$

$$\frac{1}{r}(ru)_r + w_z = 0. \tag{3.11}$$

Here the vertical velocity $w \to 0$ as $r \to \infty$, so that the vertical motion is confined to the plume.

In a turbulent plume, each variable may be expressed as the sum of a mean $\bar{\rho}$ and fluctuating component ρ' in the form

$$\rho = \bar{\rho} + \rho'. \tag{3.12}$$

Here the overbar represents a time mean such as would be recorded by an instrument at a fixed location. Measurements show that the fluctuations are comparable with the mean, and that the mean Gaussian profile (3.5) is more a measure of the intermittency (i.e. the fraction of time the position is occupied by buoyant fluid) of the plume than the reduction in the fluctuation magnitude. The vertical density flux in the plume is proportional to

$$\int_0^\infty \overline{w\rho}\,r\,\mathrm{d}r = \int_0^\infty \overline{w}\,\overline{\rho}\,r\,\mathrm{d}r + \int_0^\infty \overline{w'\rho'}\,r\,\mathrm{d}r. \tag{3.13}$$

Since the fluctuating components scale on the mean values, the two terms on the right-hand-side of (3.13) are roughly proportional to one another. This relationship is found in experimental studies. Table 1 shows measured values of the turbulent buoyancy flux as a fraction of the mean flux. Generally, the turbulent flux is a small fraction of the mean flux (the scatter in the values gives an indication of the precision of the experiments). Similar remarks apply to the turbulent volume and momentum fluxes. Thus in (3.8)–(3.11) the variables may be replaced by their mean values on the understanding that there may be some $O(1)$ scaling constants in the equations. This replacement will be made here, and the overbar will be dropped for simplicity.

3.3 The entrainment assumption

On integrating (3.11) across the plume, we obtain

$$\int_0^\infty rw_z\,\mathrm{d}r = -\int_0^\infty (ru)_r\,\mathrm{d}r,$$

and hence

$$\frac{\mathrm{d}}{\mathrm{d}z}\int_0^\infty rw\,\mathrm{d}r = \frac{1}{2\pi}\frac{\mathrm{d}Q}{\mathrm{d}z} = -\left[ru\right]_0^\infty = -ru|_\infty. \tag{3.14}$$

where Q is the volume flux carried by the plume (3.6). Hence the rate of increase in volume flux in the plume is compensated by an inflow from infinity. The plume acts like a line sink to the exterior flow, and potential theory can be used to calculate the flow in the environment (Taylor 1958).

The inflow from infinity is driven by the entrainment into the turbulent plume since, by conservation of volume in the ambient fluid, $ru|_\infty = ru|_{\text{plume edge}}$. It is not possible to calculate this inflow directly, and it is necessary to represent it in an approximate form. Morton *et al.* (1956) introduced the 'entrainment assumption' that the inflow velocity at the plume edge, the entrainment velocity u_e defined by

$$bu_e = -ru|_{\text{plume edge}}, \tag{3.15}$$

is proportional to the mean vertical velocity w in the plume at any height.

Since, from (3.2) $w \propto z^{-1/3}$ and $b \propto z$, then (3.14) and (3.15) imply that $Q \propto z^{5/3}$, which is the result (3.7) obtained on the basis of similarity theory for an unstratified ambient fluid. Hence, the assumption that the entrainment

3.5 Plume equations

The governing equations lead naturally to variables integrated across the plume. The simplest form of these equations is to represent the plume by 'top-hat' profiles for velocity and buoyancy, in which these quantities are taken to be constant within the plume and zero outside. The top-hat velocity $\overline{w}$ is defined as

$$\pi \overline{w} b^2 = 2\pi \int_0^\infty wr \, dr = Q, \tag{3.26}$$

$$\pi \overline{w^2} b^2 = 2\pi \int_0^\infty w^2 r \, dr = M. \tag{3.27}$$

In terms of the volume and momentum fluxes, the top-hat variables are

$$\overline{w} = \frac{M}{Q} \quad \text{and} \quad b = \frac{Q}{\pi^{1/2} M^{1/2}}. \tag{3.28}$$

Write (3.14) in the form

$$\frac{dQ}{dz} = -2\pi r u \big|_\infty = 2\pi b u_e, \tag{3.29}$$

where the entrainment velocity

$$u_e = \alpha \overline{w} \tag{3.30}$$

is taken to be proportional to the top-hat velocity, adopting the entrainment assumption of Morton *et al.* (1956). The constant α is the 'entrainment constant' and is determined from experiment (see § 3.7). Using the conditions (3.28) and (3.30) the volume flux equation is

$$\frac{dQ}{dz} = 2\pi^{1/2} \alpha M^{1/2}. \tag{3.31}$$

To obtain an equation for the momentum flux, traditionally (e.g. Turner 1979, § 6.1.2) the right-hand side of (3.24) is equated with $g'b$, i.e.

$$\int_0^\infty g'r \, dr = g'b, \tag{3.32}$$

and the buoyancy flux is taken proportional to $b^2 \overline{w} g'$, where $\overline{w}$ is the top-hat velocity (3.26). Comparison with (3.24) shows that this assumption is equivalent to

$$\int_0^\infty g'r \, dr = \frac{\displaystyle\int_0^\infty g'wr \, dr \displaystyle\int_0^\infty wr \, dr}{\displaystyle\int_0^\infty w^2 r \, dr}. \tag{3.33}$$

If all variables are self-similar, such that

$$w(r, z) = F_w(r/b) \, f_w(z),$$ (3.34)

$$g'(r, z) = F_{g'}(r/b) \, f_{g'}(z)$$ (3.35)

and

$$b = b(z),$$ (3.36)

then (3.33) implies that

$$\frac{\displaystyle\int_0^\infty F_w F_{g'} r \, dr \int_0^\infty F_w r \, dr}{\displaystyle\int_0^\infty F_{g'} r \, dr \int_0^\infty F_w^2 r \, dr} = 1.$$ (3.37)

For top-hat profiles with $F_w = \overline{w} H(1 - r^2/b^2)$ and $F_{g'} = \overline{g'} H(1 - r^2/b^2)$, where H is the Heaviside function, (3.37) is satisfied identically. For Gaussian radial distributions of velocity and density given by (3.5) the left-hand side of (3.37) is $2/(1 + \lambda^2)$. Equality is achieved when the radial distributions of velocity and density are the same ($\lambda = 1$).

With these definitions the above conservation equations can be written as

$$\frac{dQ}{dz} = 2\pi^{3/2} \alpha M^{1/2},$$ (3.38)

$$\frac{dM}{dz} = \frac{1 + \lambda^2}{2} \frac{BQ}{M},$$ (3.39)

$$\frac{dB}{dz} = -N^2 Q.$$ (3.40)

3.6 A plume in an unstratified environment

In an unstratified fluid $N^2 = 0$ and (3.40) shows that the buoyancy flux B is conserved. Then (3.38) and (3.39) may be combined to give

$$\frac{dM}{dz} = a \frac{BQ}{M^{3/2}},$$ (3.41)

where $a = (1 + \lambda^2)/(4\pi^{3/2}\alpha)$. Integration gives

$$Q = \left(\frac{4}{5aB}\right)^{1/2} M^{5/4} + cQ_s,$$ (3.42)

where Q_s is the volume flux at the source and c is a dimensionless constant defined as

$$c = 1 - \left(\frac{4}{5aB} \right)^{1/2} \frac{M_s^{5/4}}{Q_s},$$ (3.43)

where M_s is the momentum flux at the source.

A pure plume has zero volume and momentum fluxes at the source and so (3.42) becomes

$$Q = \left(\frac{4}{5aB} \right)^{1/2} M^{5/4},$$ (3.44)

and using (3.41) we get

$$Q = \left(\frac{1 + \lambda^2}{2} \right)^{1/3} CB^{1/3} z^{5/3},$$ (3.45)

where the constant $C = \frac{6}{5}\alpha \left(\frac{9}{10}\alpha \right)^{1/3} \pi^{2/3}$ is related to the entrainment constant. Thus the scaling relations, e.g. (3.7), are recovered but with the constants now given explicitly in terms of the entrainment constant α.

In practice a plume, e.g. that rising from a chimney, will have non-zero volume and momentum fluxes at the source. In such cases the plume is said to be 'forced' (Morton 1959) and these initial conditions influence the behaviour of the plume close to the source. Since the volume flux (3.42) and the momentum flux (3.27) increase with height, the former by entrainment and the latter by buoyancy-generated momentum, eventually the initial values become inconsequential and the 'pure plume' solution (3.44) holds asymptotically. On dimensional grounds, the heights over which the initial volume and momentum fluxes are important are given by the length scales

$$L_Q = \frac{Q_S^{5/3}}{B^{1/5}} \quad \text{and} \quad L_M = \frac{M_s^{3/4}}{B^{1/2}},$$ (3.46)

respectively. Details of these effects are discussed in detail in Fischer *et al.* (1979).

Since (3.41) applies at heights greater than $O(L_Q, L_M)$ it is possible to define a virtual origin such that the solution (3.42) applies for $Q_s \neq 0$, $M_s \neq 0$. The position of the virtual origin depends on the source properties, and extensive discussion may be found in Hunt & Kaye (2000).

3.7 Values of the entrainment constant

The entrainment constant α in (3.31) must be obtained from measurement. Several methods have been used, such as measuring the profile of the plume and applying a similarity hypothesis. The most accurate way to determine α

is to measure the plume volume flow rate directly. A technique discovered by Baines (1983) involves the release of a plume in an enclosure through which a known volume flow rate is imposed on the ambient. An interface forms at the height at which the plume volume flux equals the imposed flux and α is determined from (3.45). Values found by various authors are given in table 1. Recent experiments suggest that a value of $\alpha = 0.13$ should be used when top-hat variables are employed.

3.8 Plume rise in a stratified fluid

Many environmental flows involve the motion of convective elements through stably stratified ambient fluid characterized by a buoyancy frequency N. Turbulent elements entrain ambient fluid and become less buoyant as they rise. Since the density of the ambient decreases with height, in many circumstances the densities of the element and the ambient become equal at some height z_s. At heights $z > z_s$ the element is negatively buoyant and further rise is inhibited.

For a buoyant plume with buoyancy flux B_s at the source rising through fluid with constant N, dimensional analysis shows that

$$z_s = c B_s^{1/4} N^{-3/4}, \tag{3.47}$$

where c is a dimensionless constant. This result has famously been verified over four decades of vertical scale from the laboratory to plumes from fires in the atmosphere (Morton *et al.* 1956), with $c = 2.5$. This is a remarkable result showing that the force balances at large geophysical scales occur in small-scale laboratory experiments at lower Reynolds numbers, and is an indication of the robustness of plume theory. The formula has been applied to plumes rising into the stratosphere from volcanic eruptions and is used to calculate the strength of the eruption (Woods 1995).

The rise of a plume in a stratified environment can also be calculated using the top-hat plume equations (3.38)–(3.40). The results of these calculations are shown in figure 4 taken from Morton *et al.* (1956). From (3.37) we see that the buoyancy flux is not conserved with height, and at height $\bar{z}_s$ the buoyancy flux $B(z) = 0$. The plume is still rising at this height ($w > 0$) and for $z > \bar{z}_s$ buoyancy forces decelerate the plume and, since the volume flux increases, the radius $b(z)$ increases rapidly. Observations show that the plume fluid then falls downwards and spreads out at a level $z \approx \bar{z}_s$. The calculations show that

$$\bar{z}_s = \bar{c} B_s^{1/4} N^{-3/4}, \tag{3.48}$$

where $\bar{c} = 2.2$, in good agreement with the dimensional argument.

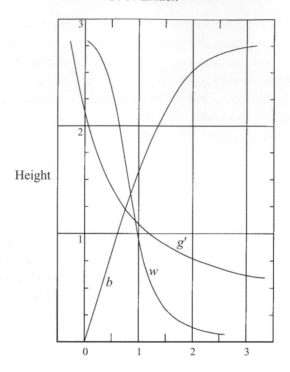

Figure 4. The (dimensionless) vertical velocity w, buoyancy g' and plume radius b as functions of height for an axisymmetric plume rising through a stably stratified fluid with constant N, from Morton *et al.* (1956).

It is worth considering the reasons for this agreement since there are a number of implicit assumptions in both calculations. In the dimensional argument the use of the source buoyancy flux B_s implies that B is conserved, while it clearly is not. In the top-hat calculations it is assumed that the entrainment assumption is valid when $N \neq 0$. From the relations (3.2) and (3.4) we see that the vertical velocity $w(z)$ depends weakly on B and $b(z)$ is independent of B. Hence the momentum and volume fluxes in the plume are not sensitive to the decrease in $B(z)$. This is also apparent in the top-hat solutions which show behaviour very similar to rise through an unstratified ambient fluid until $z \approx \bar{z}_s$. The agreement with observations is a strong verification of the appropriateness of the entrainment assumption.

If the buoyancy frequency N is not constant with height, then under some circumstances the plume may rise indefinitely. One obvious case is when $N = 0$ for $z > z_1$, where $z_1 < \bar{z}_s$. Caulfield & Woods (1998) have shown that even if $N \neq 0$ everywhere, top-hat theory predicts the plume can still 'escape' provided N decreases sufficiently rapidly with height.

3.9 Convection in confined spaces

The preceding discussion has assumed that the environment is infinite. In an ensemble of plumes with mean horizontal spacing R, each plume can effectively entrain the fluid in an area order R^2. If, in addition, the vertical rise of the plume is limited by, say, rise through stable stratification, the effects of the finite environment may be modelled by considering a single plume in an enclosed space.

Consider a plume rising from a point source on the floor of an enclosure with insulating boundaries of height H and horizontal area R^2 initially filled with uniform fluid. Fluid rises in the plume and on reaching the ceiling, spreads out horizontally, as a gravity current, forming a buoyant region at the top of the enclosure. A circulation is established in the fluid surrounding the plume, with ambient fluid entrained horizontally into the plume being replaced by downward advection. The buoyant fluid at the top is advected downwards and is replaced by plume fluid spreading out at the ceiling. As time increases more of this buoyant layer is re-entrained and, at any time, the plume fluid arriving at the ceiling is the most buoyant fluid in the enclosure and a stable stratification is established.

This process, known as a 'filling box', was first described by Baines & Turner (1969). They noted that, since the buoyancy flux is constant, the mean density in the enclosure decreases linearly with time and they assumed that the environment established a self-similar stratification $\rho_e(z) - kt$ for some constant k. The solution may be determined from the plume equations (3.38)–(3.40) together with equations for volume and mass conservation in the environment:

$$\overline{w}b^2 = WR^2 \tag{3.49}$$

and

$$k - W\frac{\partial \rho_e}{\partial z} = 0, \tag{3.50}$$

where $W(z)$ is the vertical velocity in the environment.

The motion of the initial front, i.e. the leading edge of the fluid first to descend after the plume has been switched on, is calculated from volume conservation

$$Q(z_F) = W_F R^2, \tag{3.51}$$

where W_F is the velocity of descent of the front at z_F. The volume flux $Q(z_F)$ at that level is given by (3.45) which, appropriate to a plume in unstratified

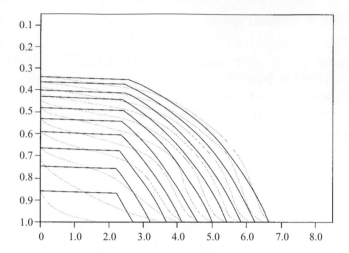

Figure 5. Theoretical (solid lines) and experimental (dashed lines) results for a single-plume filling box. The ordinate is the dimensionless height within the box and the abscissa is a dimensionless measure of the density, for a dense plume with source at the top of the box. The theory is derived using the plume equations (3.38)–(3.40) together with (3.49) and (3.50), and the experimental results are from Kaye (1998).

surroundings, remains valid for $z < z_F$. Hence

$$z_F = \frac{H}{\left(1 + t/t_1\right)^{3/2}}, \tag{3.52}$$

where $t_1 = 5R^2/(3cH^{2/3}B^{1/3})$ is a characteristic time scale for the space. This time scale may be written as

$$t_1 = \frac{5}{3} \frac{R^2 H}{cB^{1/3}H^{5/3}} \tag{3.53}$$

and is proportional to the volume R^2H of the space divided by the volume flux $cB^{1/3}H^{5/3}$ in the plume at the height of the ceiling. Thus the initial front descends on a time scale given by the time it takes for all the ambient fluid in the space to pass through the plume.

After this time a stable density stratification is established within the enclosure and the asymptotic form of this stratification may be obtained by solving (3.38)–(3.40) with (3.49) and (3.50) and assuming that $\rho_e(z,t) = \rho_e(z) - kt$. The solution is shown in figure 5. The form of the stratification closely resembles $z^{-1/3}$ because it is primarily determined by entrainment into the plume. Worster & Huppert (1983) calculated the time-dependent aspects of the filling box explicitly and their results confirm that the similarity solution is achieved. This behaviour has been confirmed in laboratory experiments by Kaye (1998), the results of which are also shown in figure 5. The presence of stable stratification observed in the interior of high-*Ra* Rayleigh–Bénard convection (Gille 1967) is due to the filling-box process. Plume dynamics explain the otherwise puzzling observation that the transport of heat is up the mean temperature gradient.

The flow extends in a straightforward way to any number of equally spaced plumes of equal buoyancy flux, provided the plumes do not interact before reaching the top of the enclosure. However, departures from symmetry can significantly alter the flow. For example, if the plumes are not equally spaced such that the effective values of R are different and/or if they have different buoyancy fluxes B, then (3.53) shows that the characteristic time scale t_1 will vary in space and horizontal density gradients will be established. The resulting flow will develop mean horizontal circulations caused by the fact that the buoyancy flux is not constant in space under these conditions. In fact, Wong & Griffiths (2000) show that the stratification can support shear-mode internal waves and lead to complex flow structures.

3.10 Convection in ventilated spaces

An application of plume theory concerns flow and stratification in ventilated spaces. This topic has application to the natural ventilation of buildings (Linden 1999) and recent trends in building design have raised new questions concerning convective flows. Modern commercial and industrial buildings tend to favour tall and open-plan spaces with significant internal gains (heat sources) from equipment, personnel and space heating and from solar insolation. In order to reduce costs there has been a move from fully air-conditioned buildings towards natural ventilation. In the extreme geometries of modern buildings, traditional design guidelines are inappropriate and research has been directed at understanding the flows generated by heat sources within these spaces.

The simplest problem, the steady flow resulting from a single heat source within a single ventilated space, is discussed by Linden, Lane-Serff & Smeed (1990). The flow depends on the locations of the openings to the space, and we restrict attention here to the case of two openings one at the top and one at the bottom with areas a_t and a_b, respectively. The heat source with buoyancy flux B is located at the floor of the enclosure of height H. The plume rising above the heat source spreads out along the ceiling and establishes a buoyant layer in the upper part of the space. Since the density of the air within the space is less than or equal to that outside, the hydrostatic pressure gradient inside is less than that outside (see figure 6).

upper layer which is given by

$$g' = C^{-1}B^{2/3}h^{-5/3}. \tag{3.60}$$

The interface height depends on the areas of the top and bottom openings and is controlled by the smaller of the two openings as can be seen by the limit $a_t \ll a_b$ of (3.58) which becomes

$$A^* \approx \sqrt{2}c_t a_t. \tag{3.61}$$

Therefore, it is possible to control the ventilation of a space by constricting a top opening, allowing occupants to alter the lower openings at will, independently of the strength of the internal gains. Similar results may be obtained for multiple sources of buoyancy, although the resulting stratification is more complex (Cooper & Linden 1996; Linden & Cooper 1996).

3.11 Concluding remarks

The study of isolated convective elements has many applications to the environment. Plumes form naturally above chimneys, volcanoes, and other isolated buoyancy sources such as 'black smokers' found at the bottom of the ocean. This section has therefore concentrated on a careful description of the basic properties of a plume and some extensions and applications. Plumes also occur naturally when a surface is heated, as buoyant fluid collects in coherent rising elements such as occurs when the ground is heated by the sun or, conversely, when the surface of the ocean is cooled. The scalings discussed here provide estimates of the vertical velocities associated with convection in the lower atmosphere and also in the upper ocean.

Under strong convective conditions the coherent rise and fall of fluid in the atmosphere can have surprising effects, such as increasing the maximum *ground-level* concentration of a pollutant released from a chimney. This is because pollutant is carried both up and down by convective flows and so higher concentrations are obtained than if the pollutant is smeared throughout the convecting layer.

There are many other effects of practical importance that are beyond the scope of this section. Traffic pollution is worst during rush hours, which are times (early morning and evening) when the convection is beginning or stopping. Filling-box processes are responsible for bringing pollutant, initially raised to high levels in plumes, back down to the source levels. Transient processes such as the erosion of the night-time inversion and the change in convection once the heat comes from within the ground – which causes a transition from plumes to thermals (Hunt 1999) – are important areas of current research. Thermals may be studied in a similar manner to plumes, using both similarity arguments and the entrainment assumption. The ideas in this section can be applied to line sources of buoyancy and to plumes rising in a moving environment, and the reader is referred to Turner (1979) for details.

4 Double-diffusive convection

Convective flows can occur in a fluid even when the density decreases with height. For convection to occur in this case it is necessary that the density field is produced by two or more components which diffuse at different rates. The classical example is water stratified with temperature and salinity and the notation in this section follows this case. (Convection in this case is often referred to as thermohaline convection.)

When one of these components is unstably distributed within the fluid, circumstances can occur where the potential energy of this component is released and drives convection. As discussed in § 2, a layer of water heated from below is unstable when a critical Rayleigh number is exceeded. Suppose that, in addition to the unstable temperature gradient, there is a gradient of salinity with the most-saline water at the bottom. Provided the salinity gradient is small enough, the density of the layer will still increase with height and it seems reasonable to expect that convection will still occur. If the salinity gradient is increased sufficiently it can overcome the temperature gradient so that the density gradient changes sign and the density decreases with height. Even in this case, the potentially unstable temperature field exists, and it is possible for the energy associated with it to be released against the restoring forces associated with the stable density field.

For convection to occur it is necessary that motion results in a reduction of the potential energy of the density field. Such a redistribution of density requires that the centre of mass of the fluid layer is lowered, as is the case when buoyant fluid moves upwards and negatively buoyant fluid moves downwards. Opposing this motion are dissipative processes caused by diffusion and viscosity. Since, in the example above, the energy in the temperature field is undiminished the question arises as to whether it can be released against the additional dissipative processes (e.g. internal gravity waves) associated with the overall stable density gradient provided by the salt field. Clearly a simple spatial redistribution of the density field cannot work. Somehow the temperature and salinity fields must act independently so that the redistribution of density can be such as to lower the centre of mass.

For a Boussinesq liquid the effects of temperature T and salinity S on the density field can be approximated by an equation of state in the form

$$\rho = \rho_0 \left[1 - \alpha(T - T_0) + \beta(S - S_0) \right], \tag{4.1}$$

where (ρ, T, S) depart only slightly from a reference state (ρ_0, T_0, S_0), and α and β are coefficients that give the change in density per unit change in temperature and salinity, respectively, and are defined by

$$\alpha = -\frac{1}{\rho}\frac{\partial \rho}{\partial T} \quad \text{and} \quad \beta = \frac{1}{\rho}\frac{\partial \rho}{\partial S}. \tag{4.2}$$

Salinity and temperature are advected by the flow according to the conservation equations

$$T_t + \boldsymbol{u} \cdot \nabla T = \kappa_T \nabla^2 T \tag{4.3}$$

and

$$S_t + \boldsymbol{u} \cdot \nabla S = \kappa_S \nabla^2 S, \tag{4.4}$$

where κ_T and κ_S are the molecular diffusion coefficients for heat and salt, respectively. Since (4.1) implies that the effects of salinity and temperature are effectively the same and (4.3) and (4.4) are identical unless $\kappa_T \neq \kappa_S$, the temperature and salinity fields can only be uncoupled if their molecular diffusivities are different. In that case it is possible that the energy stored in the temperature field can be released and convection may occur.

Consequently, provided

(i) there are two components contributing to the density field and one is unstably distributed and

(ii) the two components have different diffusivities

convection may be possible even if the overall density field is statically stable. Convection in this case is called double-diffusive convection.

As in the Rayleigh–Bénard problem (§2) it is necessary for dissipative processes to be overcome and so a form of critical Rayleigh number will exist. The stability analysis will be given after the physical mechanisms are discussed.

The components are distinguished by their rates of diffusion and there is an asymmetry in the problem depending on whether the faster component (as is the case discussed above) or the slower component is unstably distributed. The instability mechanism is different in the two cases and these are now discussed.

4.1 Finger instability

We consider first the case where the slower diffusing component (S) is unstably distributed. This configuration can be readily produced at home by taking a tall glass and half-filling it with cold water, placing a small sponge on the surface and then gently pouring warm water on top of the sponge so that it soaks through and forms a layer on top of the cold water. A two-layer stable temperature (T) stratification is set up. If a little food colouring is first added to the warm water this will help the visualization. Dissolve a very small amount of salt (just a few grains) in warm water and add this to the glass just before it is full, and watch what happens.

This configuration can be idealized as a two-layer stratification with the upper layer warmer and saltier, but less dense, than the lower layer. Consider a small perturbation that causes a small parcel of upper-layer fluid to cross the interface into the lower layer. Heat and salt diffuse from the parcel but, because heat is lost from the parcel more rapidly than salt, it reaches thermal equilibrium with its surroundings while still retaining some excess salt. The parcel is therefore denser than the fluid in the lower layer and continues to descend, and convection ensues. Similarly a parcel raised into the upper layer will gain heat but remain fresh and so continue to rise. The argument applies equally well to continuous vertical gradients of temperature and salinity.

Convection occurs because horizontal variations in density are produced by lateral diffusion processes. These horizontal density variations produce vorticity by baroclinic generation through the $\nabla \rho \wedge \nabla p$ term in (1.4), and are quite consistent with the motion taking place within a mean vertical density gradient that is statically stable. Thus there is no contradiction with convection occurring in a stably stratified fluid.

The above argument also shows the importance of lateral diffusion, and for this to occur efficiently it is necessary that the horizontal scale of the motions be small, although not so small that viscosity prevents them. Thus the convection cells tend to be thin structures and to look like vertical fingers. Observations in the glass will show that the ends of the descending fingers have small mushroom-like structures on them. These are caused by the locally unstable vertical density fields at the ends of the fingers.

4.2 Diffusive convection

In the case where the faster diffusing component is unstably distributed – in the thermohaline case this is when warm salty water lies beneath cold fresh water – the instability is harder to visualize and explain! Imagine the case

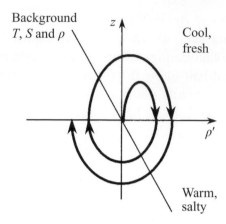

Figure 7. The motion in ρ', z phase space (where ρ' is its perturbation density) of the trajectory of a parcel displaced upwards in a fluid stratified with two components in a diffusive sense. The parcel rises from its neutral position ($\rho' = 0$, $z = 0$) and loses heat and salt and increases in density until it is restored to $z = 0$. It then gains heat and salt but loses density when $z < 0$. Since the contrasts in T and S are greater during the $z < 0$ trajectory than during the previous $z > 0$ trajectory the parcel returns to its original position $z = 0$ less dense than when it left. This process repeats and sets up growing oscillations as shown.

where the vertical gradients of salinity and temperature are uniform and consider the vertical displacement of a particle given a small upward vertical velocity from its original (equilibrium) position. If there were no diffusion or viscosity the parcel would exhibit simple harmonic motion with a frequency related to the buoyancy frequency of the background stratification.

Diffusion of heat and salt modifies this picture. While rising from its original position the parcel is surrounded by a cooler and less saline environment and so loses heat and salt, but since $\kappa_T > \kappa_S$ its density will increase. As the parcel rises, the background density decreases and the parcel is subject to a restoring force that decelerates the parcel to rest and then reverses its motion. On the downward trajectory the parcel continues to lose heat and salt and its density increases further and it arrives back at its original level cooler, fresher and denser than it started.

As the parcel descends below its original level it now loses density and is subject to an upward buoyancy force. Because it was cooler and fresher when it returned to its original level the contrasts of T and S with the ambient are greater, and the parcel gains more buoyancy by diffusion than it lost during the upward motion. Consequently, the parcel arrives at its original level again, now warmer, saltier and less dense than when it first rose from there. This process repeats and leads to a set of growing oscillations as illustrated in figure 7. This form of instability is called an overstable oscillation, and is common in systems with positive feedback and a time lag. As will be seen below, this behaviour occurs for a range of parameters that are hard to produce in the laboratory.

4.3 Linear stability analysis

The linear stability analysis given in §2.1 extends to the double-diffusive case by considering a layer of fluid with horizontal boundaries maintained at constant temperature and salinity. The governing equations are, after manipulations as in §2.1,

$$(\sigma^{-1}\partial_t - \nabla^2)\nabla^2 w = Ra\nabla_H^2 T - Rs\nabla_H^2 S, \tag{4.5}$$
$$(\partial_t - \nabla^2)\nabla^2 T = w \tag{4.6}$$

and

$$(\partial_t - \tau\nabla^2)\nabla^2 S = w. \tag{4.7}$$

Parameters are rendered dimensionless as in §2.1 and those governing the stability are the Prandtl number $\sigma = \nu/\kappa_T$ and the Rayleigh number $Ra = g\alpha\Delta T H^3/(\nu\kappa_T)$, found in the thermal problem, and two additional parameters. These parameters are the diffusivity ratio,

$$\tau = \frac{\kappa_S}{\kappa_T}, \tag{4.8}$$

and a measure of the salinity field, a saline Rayleigh number,

$$Rs = \frac{g\beta\Delta S H^3}{\nu\kappa_T}. \tag{4.9}$$

As in the Rayleigh–Bénard case there is a conductive solution with the fluid at rest and with linear variations of temperature and salinity with height between the boundaries. The stability of that solution can be examined by a perturbation of the form

$$w = \widehat{w}(z)e^{iKx+st}$$

as, in this case also, only the horizontal scale of the perturbation is determined. For free boundaries with constant temperature and salinity

$$\widehat{w}(z) = \sin n\pi z, \quad n \text{ an integer},$$

and on substituting into (4.5)–(4.7) we obtain the eigenvalue equation for the growth rate s, which is now a cubic

$$s^3 + (\sigma + \tau + 1)K_n^2 s^2 + \left[(\sigma + \sigma\tau + \tau)K_n^4 - (Ra - Rs)\sigma K^2 K_n^{-2}\right]s$$
$$+ \sigma\tau K_n^6 + (Rs - \tau Ra)\sigma K^2 = 0, \tag{4.10}$$

where $K_n^2 = K^2 + n^2\pi^2$. In the case when $\tau = 1$, (4.10) reduces to $(s + K_n^2)$ times the quadratic equation (2.22), with $Ra - Rs$ in place of Ra. This shows that when $\tau = 1$, the fluid is only unstable when the density is statically unstable, and therefore that unequal diffusion coefficients are essential for the double-diffusive character of the instability.

4.3.1 Properties of the eigenvalue equation (4.10)

The cubic eigenvalue equation may be written symbolically as

$$f(s) = s^3 + as^2 + bs + c = 0,$$

where

$$a \equiv (\sigma + \tau + 1)K_n^2, \tag{4.11a}$$
$$b \equiv (\sigma + \sigma\tau + \tau)K_n^4 + (Rs - Ra)\sigma K^2 K_n^{-2}, \tag{4.11b}$$
$$c \equiv \sigma(\tau K_n^6 + (Rs - \tau Ra)K^2). \tag{4.11c}$$

The signs of b and c depend on the signs of the heat and salt Rayleigh numbers and their magnitude with respect to the viscous and diffusive processes as represented by σ and τ. For $\tau < 1$, (4.10) has three roots, one of which must be real, while the other two are either real or complex conjugates and have stationary values (if they exist) at

$$s_{1,2} = \frac{-a \pm \sqrt{a^2 - 3b}}{3}.$$

It is convenient to consider the properties of (4.10) in four quadrants of the (Rs, Ra)-plane and the four regions shown in figure 8. These regions are defined by the lines $b = 0$ and $c = 0$ which, in the limits $|Ra| \gg 1$, $|Rs| \gg 1$ (i.e. when molecular effects are unimportant), are equivalent to

$$Ra = Rs$$

and

$$Ra = \frac{1}{\tau}Rs,$$

respectively.

While we are primarily concerned with the cases where the fluid is stably stratified, i.e. where $b > 0$, we will consider the full (Rs, Ra)-plane.

Quadrant IV: Warm fresh over cool salty

In this case $Rs > 0$, $Ra < 0$ so that both components are stably distributed and $b > 0$, $c > 0$ over the entire quadrant Thus $s_{1,2}$ are both real and negative if $b < \frac{1}{3}a^2$ and both are imaginary if $b > \frac{1}{3}a^2$. In either case, since

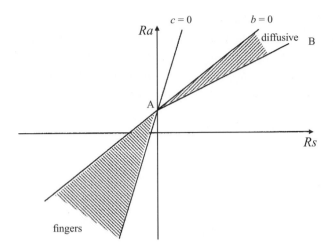

Figure 8. Linear stability of the double-diffusive system in the (Rs, Ra)-plane. The regions shown correspond to the signs of the coefficients of the cubic characteristic equation (4.10). Regions where fingers and diffusive convection occur in statically stable fluid are shown shaded.

$c > 0$, all roots are negative or have negative real parts (see equation (4.17) below). Therefore, this configuration is stable.

Quadrant II: Cool salty over warm fresh

This is the inverse of the previous case, with $Rs < 0$ and $Ra > 0$. Provided the individual values of the Rayleigh numbers are large enough, $b < 0$ and $c < 0$. Then $s_{1,2}$ are both real, one positive and one negative, implying that $f(s)$ has one positive real root. In this case both components are unstably distributed and the net density stratification is statically unstable. Instability sets in with a direct mode in a form analogous to thermal convection. It is stabilized only by viscous and diffusive effects. When these are large enough, either b or c become positive and the system may be stabilized.

Quadrant III: Warm salty over cool fresh

This is the finger configuration with $Ra < 0$, $Rs < 0$ and static stability requires $|Ra| > |Rs|$ i.e. the system lies to the right of the line $b = 0$. If $b < \frac{1}{3}a^2$ there are two real, stationary values $s_{1,2} < 0$ and for $b > \frac{1}{3}a^2$ the curve is monotonic. In either case, when $c < 0$ there is one positive real root of $f(s)$ and there is a direct mode of instability, namely, fingers. When $c > 0$, all roots are negative and the system is stable. The region of finger instability therefore lies between the lines $b = 0$ and $c = 0$, and is shown shaded in figure 8.

Quadrant I: Cool fresh over warm salty

This final case is the 'diffusive' configuration: $Ra > 0$, $Rs > 0$. We require $Rs > Ra$ for static stability, and hence $b > 0$ and $c > 0$ as in Quadrant IV. Thus the real root has negative real part but, as shown below in (4.17), the complex roots may have positive real parts. Hence instability, if it occurs, must set in at marginal stability as an oscillation. This is the diffusive case.

For fixed fluid properties (fixed σ and τ) the real root $s = 0$ in (4.10) gives

$$Ra - \tau^{-1}Rs = K_n^6 K^{-2}, \tag{4.12}$$

which is equivalent to (2.34) for thermal convection. As there, the minimum value of the right-hand side occurs when $n = 1$ and $K = \pi/\sqrt{2}$, and marginal stability is given by

$$Ra - \tau^{-1}Rs = \frac{27\pi^4}{4}. \tag{4.13}$$

Thus $R = Ra - \tau^{-1}Rs$ acts as an effective Rayleigh number and the form of the convection is analogous to thermal convection. Instability sets in as an exchange of stabilities and the form of the marginal mode is roughly square. When Ra is positive and Rs is negative both gradients are destabilizing and they combine in R to give a destabilizing density gradient: this case is gravitational convection of an unstable density field and is not double diffusion.

On the other hand when Ra is negative (stabilizing) and Rs is negative (destabilizing) then the stratification is in the finger configuration. In this case again instability sets in by exchange of stabilities. As has already been pointed out, in most natural circumstances Ra (and Rs) are likely to be much greater than $O(10^3)$, and the stability boundary is approximated well by

$$Rs = \tau Ra, \tag{4.14}$$

or, in terms of the temperature and salinity differences across the layer,

$$\beta \Delta S = \tau \alpha \Delta T. \tag{4.15}$$

For heat and salt in water $\tau \approx 10^{-2}$, so that a small salinity difference can destabilize the layer even in the presence of a strong stable temperature field. This is why only a few grains of salt are needed to produce salt fingers in a glass.

Another implication of large Ra and Rs is that any imbalance from (4.14) means that $R \gg 27\pi^4/4$, and the instability is far from the marginal state. Stern (1960) showed that the fastest growing mode in this case has a horizontal wavenumber

$$K \approx \left(\frac{g\alpha\Delta T}{H\nu\kappa_T} \right)^{1/4}, \qquad (4.16)$$

which, for $Ra \gg 1$, implies a much smaller horizontal scale than the marginal mode. This reflects that maximum growth occurs when horizontal diffusion is most efficient and is why the observed convection takes the form of long, thin fingers.

Diffusive instability (Quadrant IV) occurs when the complex roots of (4.10) have positive real part. Writing the roots as s_1 and $p \pm iq$ then the coefficients of the cubic are

$$a = -s_1 - 2p, \qquad b = p^2 + q^2 + 2s_1 p, \qquad c = -s_1(p^2 + q^2).$$

For the diffusive case $a > 0$, $b > 0$ and $c > 0$. Hence $s_1 < 0$. Oscillatory instability occurs when $p = 0$, and so $ab = c$. From (4.11) we obtain the marginal condition

$$Ra - \frac{\sigma + \tau}{\sigma + 1} Rs = \frac{(\sigma + \tau)(1 + \tau)}{\sigma} \frac{K_n^6}{K^2}. \qquad (4.17)$$

This condition is only satisfied in the first quadrant of the (Rs, Ra)-plane, and so only occurs under the conditions appropriate for diffusive stratification (cool fresh over warm salty).

Again the minimum of the right-hand side of (4.17) occurs when $n = 1$ and $K = \pi/\sqrt{2}$, and (4.17) becomes

$$Ra - \frac{\sigma + \tau}{\sigma + 1} Rs = \frac{(\sigma + \tau)(1 + \tau)}{\sigma} \frac{27\pi^4}{4}. \qquad (4.18)$$

This line is shown as AB on figure 8 in the limit $\tau \to 0$ appropriate to heat and salt in water. In this limit the lines $b = 0$, $c = 0$ and AB intersect on the Ra-axis at the value $Ra = 27\pi^4/4$, the critical value for thermal convection between free, conducting boundaries.

As before the practical case of interest is when both Ra and Rs are large compared with 10^3 when (4.18) may be approximated by

$$Ra = (\sigma + \tau)(\sigma + 1)^{-1} Rs \qquad \text{or} \qquad \alpha\Delta T = (\sigma + \tau)(\sigma + 1)^{-1}\beta\Delta S. \qquad (4.19)$$

For heat and salt ($\tau \simeq 10^{-2}, \sigma \simeq 7$) in water this approximates to $\alpha\Delta T \approx 7\beta\Delta S/8$, while for two solutes in water ($\tau \approx O(1)$), instability occurs for $\alpha\Delta T \approx \beta\Delta S$. In both these cases instability of the diffusive mode sets in close to the case where the density gradients of the two components almost compensate, and this is one reason why it is hard to observe (Shirtcliffe 1967).

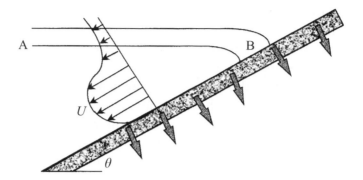

Figure 10. A schematic showing flow down a slope resulting from surface cooling. The isopycnals sketched are those resulting from cooling by the boundary in the absence of motion.

It has already been noted that, at high Rayleigh number, convection takes the form of plumes or thermals that collect fluid from the heated boundary layer and carry it vertically away from the boundary. As fluid moves towards the base of the plume it continues to be heated from below, so that within the boundary layer horizontal temperature differences occur, with the warmest fluid at the base of the plume. This horizontal density gradient drives the fluid horizontally. The question then arises as to how quickly this fluid travels in these circumstances.

Consider a layer of fluid of density ρ with a lower boundary of slope θ maintained at a density $\rho + \Delta\rho$. There is a buoyancy force along the slope of magnitude $g' \sin\theta$, which is balanced by dissipative forces. For large Prandtl number laminar flow buoyancy is balanced by viscosity giving

$$\frac{\nu U}{\delta^2} \approx g' \sin\theta, \tag{5.1}$$

where δ is the boundary layer thickness. Since heat enters the boundary layer by conduction, $\delta \approx \sqrt{\kappa t} \approx \sqrt{\kappa x U^{-1}}$, where x is the distance along the boundary, and so

$$U \approx \left(\frac{g' x \sin\theta}{\sigma}\right)^{1/2}. \tag{5.2}$$

Since both the velocity and the boundary-layer thickness increase with distance x along the boundary there is an increase of the volume flux along the boundary. This increase is caused by entrainment of ambient fluid.

For turbulent flows, the buoyancy force is balanced by the stress exerted by the flow at the ground and by entrainment of ambient fluid. For a slope flow of depth δ this balance is

$$\frac{U^2}{\delta} \approx g' \sin \theta, \tag{5.3}$$

and the depth of the flow is related to the distance x along the slope by

$$\delta = Ex, \tag{5.4}$$

where E is an entrainment parameter. Substituting (5.4) into (5.3) gives

$$U \approx (Eg'x \sin \theta)^{1/2}, \tag{5.5}$$

which is similar to (5.2).

The entrainment coefficient E is a function of slope and is clearly different in the two cases of a heated or cooled slope. For a heated slope, convective elements are produced at the ground and these rise through the slope flow promoting vertical mixing and entertainment. In this case E is fairly insensitive to the slope angle θ. Plume theory (§ 3.1) gives a velocity scale $w \approx (B\delta)^{1/3}$, where B is the buoyancy flux per unit area. Hence the up-slope flow scales as $U \approx (B\delta)^{1/3} \sin \theta$, leading to a weak deep flow.

When the slope is cooled the down-slope flow is stable and takes the form of a gravity current travelling down the slope. The entrainment now is strongly dependent on the slope with $E \propto \sin \theta$ (Turner 1979). The flow is strongly confined within a thin layer and large velocities are achieved.

For small slopes and fixed σ, (5.2) may be replaced by

$$U \approx (g'h)^{1/2}, \tag{5.6}$$

where $h(x)$ is the vertical rise of the boundary. This scaling is the classical gravity-current scaling (Simpson 1997) for a horizontal buoyancy difference g' over a height h, and shows the equivalence between the two flows.

6 Effects of rotation

Geophysical flows are influenced by the rotation of the Earth, and convective flows are no exception. Both the atmosphere and the oceans are thin layers of fluid and the relevant component of the Earth's rotation is the local vertical component described by the Coriolis parameter $f = 2\Omega \sin \theta$, where Ω is the Earth's rotation rate and θ is the latitude. For motions on a sufficiently small horizontal scale that variations in latitude are unimportant, the curvature of the Earth's surface can be ignored and the Coriolis parameter can be taken as a constant. The atmosphere or ocean can be treated as a plane layer rotating about a vertical axis with rotation rate $\frac{1}{2}f$.

The equations governing convection in this rotating layer are given by the generalization of (2.1)

$$u_t + u \cdot \nabla u + f \wedge u = -\frac{1}{\rho_0}\nabla p + g\frac{\rho}{\rho_0} + \nu \nabla^2 u, \tag{6.1}$$

and (2.2)–(2.3). The centrifugal force can be incorporated into the pressure field and plays only a minor role, except that for a stratified fluid with diffusion there is no state of rest since the parabolic density surfaces are not solutions of the steady diffusion equation (see Acheson 1990). In general, the departure from the rest state is small and for geophysical purposes may be ignored. However, in some astrophysical flows this circulation, called the Eddington–Sweet circulation, may be significant.

The first effect of rotation may be seen by considering slow steady motion of an inviscid unstratified fluid. Then (6.1) reduces to the geostrophic balance

$$f \wedge u = -\nabla\frac{p}{\rho_0}. \tag{6.2}$$

Taking the curl of (6.2), with f constant, gives

$$f \cdot \nabla u - f \nabla \cdot u = 0, \tag{6.3}$$

and since, for an incompressible fluid, $\nabla \cdot u = 0$, we have

$$f \cdot \nabla u = 0. \tag{6.4}$$

Thus the fluid velocity cannot vary in the direction of the rotation vector; (6.4) is the Taylor–Proudman theorem. Rotating flows have a 'stiffness' in the direction parallel to the rotation axis, which has significant implications for both the onset of convection and also on the form of the convection cells at supercritical Rayleigh numbers. It is expected that the constraint imposed by the Taylor–Proudman theorem should inhibit convective flows that involve vertical velocities that increase away from one boundary and then decrease again approaching the opposite boundary.

The second effect of rotation is that it provides a restoring force to disturbances and so can support waves. This restoring force may be illustrated by considering the outward radial displacement of a horizontal ring of fluid. Conservation of angular momentum implies that it rotates more slowly and in the rotating frame the velocity is negative, i.e. anticyclonic – see figure 11. The Coriolis force on this ring acts to the right as shown, i.e. inward, and provides a restoring force. Similarly an inward perturbation leads to cyclonic motion and an outward restoring force. This leads to inertial waves which have frequencies less than $|f|$. In general, fluid systems that support waves

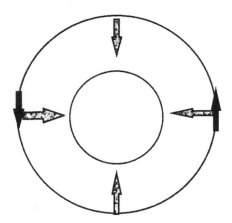

Figure 11. A sketch illustrating the formation of anticyclonic motion and the associated Coriolis restoring force when a ring of fluid expands in a rotating system.

may also have instabilities that are oscillatory, and this is the case for rotating flows.

Making (6.1) dimensionless as in §2 gives the following equations:

$$u_t + u \cdot \nabla u + \sigma T a^{1/2} k \wedge u = -\nabla p + \sigma R a T k + \sigma \nabla^2 u, \qquad (6.5)$$

$$T_t + u \cdot \nabla T = \nabla^2 T, \qquad (6.6)$$

$$\nabla \cdot u = 0, \qquad (6.7)$$

where the Taylor number, defined as

$$Ta = \frac{f^2 H^4}{\nu^2}, \qquad (6.8)$$

is a measure of the strength of the rotation.

The linear analysis analogous to that performed for the Rayleigh–Bénard problem can be done using the method for the latter given in §2. Equations (6.5)–(6.8) may be linearized, reduced to a single equation for the vertical velocity and solved in terms of normal modes. Details may be found in Chandrasekhar (1961, chap. 3). For free, constant-temperature boundaries, instability sets in by exchange of stabilities and the critical Rayleigh number is given by the minimum of

$$Ra = \frac{K_n^6 + n^2 \pi^2 Ta}{K^2}, \qquad (6.9)$$

where, as in §2, n is the vertical mode number. The minimum occurs for

$n = 1$, in which case

$$Ra = \frac{(K^2 + \pi^2)^3 + \pi^2 Ta}{K^2}. \tag{6.10}$$

The critical value is obtained when K^2 is the solution of

$$(K^2 + \pi^2)^2(2K^2 - \pi^2) + \pi^2 Ta = 0. \tag{6.11}$$

From (6.11) we note that the critical wavenumber K_c increases with Ta, and from (6.10) that Ra_c also increases with increased rotation. For large values of the Taylor number Ta, (6.10) and (6.11) give the asymptotic results

$$K_c \to Ta^{1/6} \quad \text{and} \quad Ra_c \to Ta^{2/3} \quad \text{as} \quad Ta \to \infty. \tag{6.12}$$

Thus the convection is inhibited by rotation and the scale of the marginal mode decreases. Qualitatively similar results are found for rigid boundaries, and the stability curve was checked experimentally by Nagawa & Frenzen (1955). Extensive experiments were also carried out by Rossby (1969) for a range of relatively low Ra and Ta, and these investigations found that the convection cells took the form of vortices. The spacing d between the vortices for $Ra_c < Ra < 30Ra$ was found to be $d/H \approx 5.6Ta^{-1/6}$ where $Ra_c \approx 8.7Ta^{2/3}$. At higher Rayleigh numbers $Ra \geqslant 10^3 Ra_c$, Boubnov & Golitsyn (1986) found $d/H \approx Ra^{1/9} Ta^{-1/4}$ in experiments with a heated lower boundary and a free upper surface, while Maxworthy & Narimousa (1994) found $d/H \approx Ra^{1/3} Ta^{-3/8}$ for a cooled rigid upper boundary, suggesting that the boundaries may play an important role.

The form of the convection is strongly affected by rotation. Fluid leaving the boundary layer causes convergence near the boundary and induces cyclonic vorticity. As this fluid approaches the opposite boundary the divergence induces anticyclonic vorticity. This flow is sketched in figure 12, adapted from Sakai (1997). This schematic is confirmed by his experiments in which the convection was visualized using thermochromatic liquid crystals. Sakai found that the scaling supports the form found by Maxworthy & Narimousa (1994), and provides a theoretical argument based on the assumption that the flow outside the strong vortices is geostrophic (see (6.2)).

Griffiths (2000) describes similar experiments where strong vortices are found, but no longer associated with convection cells. Instead the vortices appear as instabilities that occur on line plumes that originate from the thermal boundary layers. Similar instabilities were found by Bush & Woods (1999).

The above discussion is not immediately relevant to natural flows since both the Rayleigh and Taylor numbers involve molecular properties. Both

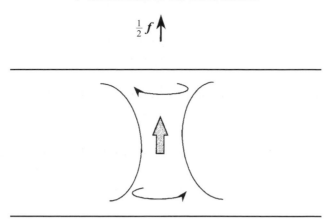

Figure 12. A sketch showing the induced rotation in the convection cells by (positive) background rotation (adapted from Sakai 1997).

Ra and *Ta* are large in most circumstances, and molecular effects only play a significant role near marginal stability. A more natural parameter for geophysical flows comes from a consideration of the balance between buoyancy and Coriolis forces (Griffiths 1987). This balance may be represented in a number of ways, but the underlying parameter is a Rossby number $Ro = U/fL$, where U and L are representative velocity and length scales, respectively. When the convection is driven by a buoyancy flux B in a layer of depth H, $U \approx (BH)^{1/3}$ and $Ro = (B/H^2)^{1/3}f^{-1}$. For $Ro \gg 1$, the convection is not affected by the rotation, while for $Ro \ll 1$, rotation has a controlling influence. Experiments suggest that the transition occurs around $Ro \approx 0.1$. Below this value the convection takes the form of intense vortices with a horizontal spacing that decreases with decreasing Ro.

In addition to the global Rossby number based on the layer depth, above a heated boundary there is a local Rossby number $Ro_l = (Bz^{-2})^{1/3}f^{-1}$, where z is the height measured from the boundary. Near the boundary $Ro_l \gg 1$ and rotation is unimportant. At greater heights the Rossby number decreases and eventually rotation changes the structure of the convection. This effect has been verified in laboratory experiments and is believed to apply to convection in the ocean mixed layer caused by surface cooling (Fernando, Chen & Boyer 1991; Helfrich 1994; Jones & Marshall 1993).

In some circumstances the effects of rotation can be very subtle. Even in flows in which $Ro \gg 1$, rotation can be important if the convection has large horizontal scales associated with it. One example is that produced by a local source of heat of radius R in a region of horizontal scale L. If $Ro \gg 1$, the convection above the source itself is unaffected by rotation and buoyant fluid rises as a collection of plumes to the top of the region where it spreads horizontally. The buoyancy of the fluid at the top of this region scales on $B^{1/3}H^{-1/3}$, and this buoyant fluid spreads out one Rossby deformation radius $R_D = (BH)^{1/3}f^{-1} = RoH$. Hence if the horizontal scale $L > HRo$, rotation will inhibit the horizontal spreading of the plume and baroclinic instability results. Recent numerical work (Julien *et al.* 1999) confirms laboratory observations that rotation inhibits entrainment in plumes as a result of the Taylor–Proudman constraint. This inhibition occurs when the plume Rossby number decreases to about 0.1. As a result plumes remain more coherent and are responsible for most of the vertical heat transport.

Acknowledgements

I would like to thank Colin Caulfield for extensive discussions about plumes and particularly with the formulation in § 3.5.

References

ACHESON, D. J. 1990 *Elementary Fluid Dynamics*. Oxford University Press.

BAINES, W. D. 1983 A technique for the direct measurement of the volume flux of a plume. *J. Fluid Mech.* **132**, 247–256.

BAINES, W. D. & TURNER, J. S. 1969 Turbulent convection from a source in a confined region. *J. Fluid Mech.* **37**, 51–80.

BARENBLATT, G. I., BERTSCH, M., DAL PASSO, R., PROSTOKISHIN, V. M. & VGHI, M. 1993 A mathematical model of turbulent heat and mass transfer in stably stratified shear flow. *J. Fluid Mech.* **253**, 341–358.

BÉNARD, H. 1900 Les tourbillons cellulaires dans une nappe liquide. *Rev. Gen. Sci. Pure Appl.* **11**, 1261–1271 and 1309–1328.

BOUBNOV, B. M. & GOLITSYN, G. S. 1986 Experimental study of convective structures in a rotating fluid. *J. Fluid Mech.* **167**, 503–531.

BUSH, J. W. M. & WOODS, A. W. 1999 Vortex generation by line plumes in a rotating stratified fluid. *J. Fluid Mech.* **388**, 289–313.

CASTAING, B., GUNARATNE, G., HESLOT, F., KADANOFF, L., LIBCHABER, A., THOMAS, S., WU., X.-Z., ZALESKI, S. & ZANETTI, G. 1989 Scaling of hard turbulence in Rayleigh–Benard convection. *J. Fluid Mech.* **204**, 1–39.

CAULFIELD, C. P. & WOODS A. W. 1998 Turbulent gravitational convection from a point source in a non-uniformly stratified environment. *J. Fluid Mech.* **360**, 229–248.

CHANDRASEKHAR, S. 1961 *Hydrodynamic and Hydromagnetic Stability*. Clarendon.

CHAVANNE, X. 1997 Observations of the ultimate regime in Rayleigh–Benard convection. *Phys. Rev. Lett.* **79**, 3648–3651.

CHEN, C. J. & RODI, W. 1980 *Vertical Turbulent Buoyant Jets*. Pergamon.

COOPER, P. & LINDEN, P. F. 1996 Natural ventilation of an enclosure containing two buoyancy sources. *J. Fluid Mech.* **311**, 153–176.

DEARDORFF, J. W. & WILLIS, G. E. 1967 Investigation of turbulent convection between horizontal plates. *J. Fluid Mech.* **28**, 675–704.

DRAZIN, P. G. & REID, W. R. 1981 *Hydrodynamic Stability*. Cambridge University Press.

FERNANDO, H. J. S., CHEN, R.-R. & BOYER, D. L. 1991 Effects of rotation on convective turbulence. *J. Fluid Mech.* **228**, 513–547.

FISCHER, H. B., LIST, E. J., KOH, R. C. Y., IMBERGER, J. &. BROOKS, N. H. 1979 *Mixing in Inland and Coastal Waters*. Academic.

GEORGE, W. K., ALPERT, R. L. & TAMANINI, F. 1977 Turbulence measurements in an axisymmetric buoyant plume. *Intl J. Heat Mass Transfer* **20**, 1145–1154.

GILLE, J. 1967 Interferometric measurement of temperature gradient reversal in a layer of convecting air. *J. Fluid Mech.* **30**, 371–384.

GOLDSTEIN, R. J. CHIANG, H. D. & SEE, D. L. 1990 High Rayleigh number convection in a horizontal enclosure. *J. Fluid Mech.* **213**, 111–126.

GRIFFITHS, R. W. 1987 Effect of Earth's rotation on convection in magma chambers. *Earth Planet. Sci. Lett.* **85** 525–536.

GRIFFITHS, R. W. 2000 Developments in high Rayleigh number convection. In *Developments in Geophysical Turbulence* (ed. R. M. Kerr & Y. Kimura), Kluwer.

HELFRICH, K. R. 1994 Thermals with background rotation and stratification. *J. Fluid Mech.* **259**, 265–280.

HOWARD, L. N. 1964 Convection at high Rayleigh number. *Proc. 11th Intl Congr. Appl. Mech.* pp. 1109–1115. Springer.

HUNT, G. R. & KAYE, N. 2000 Virtual origin correction for lazy turbulent plumes. *J. Fluid Mech.* (submitted).

HUNT, J. C. R. 1999 Eddy dynamics and kinematics of convective turbulence. In *Convection in the Atmosphere*, (edited by E. J. Plate, E. E. Federovich, D. X. Viegas & J. C. Wyngaard), pp. 41–82. Kluwer.

HUPPERT, H, F. & LINDEN, P. F. 1979 On heating a stable salinity gradient from below. *J. Fluid Mech.* **95**, 431–464.

HUPPERT, H. E. & TURNER, J. S. 1981 Double-diffusive convection. *J. Fluid Mech.* **106**, 299–329.

JONES, H. & MARSHALL, J. M. 1993 Convection in a neutral ocean; a study of open-ocean deep convection. *J. Phys. Oceanogr.* **23**, 1009–1039.

JULIEN, K., LEGG, S., MCWILLIAMS, J. & WERNE, J. 1999 Plumes in rotating convection. Part 1. Ensemble statistics and dynamical balances. *J. Fluid Mech.* **391**, 151–187.

KAYE, N. 1998 Interacting Turbulent Plumes. PhD thesis, University of Cambridge.

KRAICHNAN, R. H. 1962 Turbulent thermal convection at arbitrary Prandtl number. *Phys. Fluids* **5**, 1374–1389.

LIBCHABER, A., SANO, M. & WU, X. 1990 About thermal turbulence. *Physica* A **163**, 258–264.

LINDEN, P. F. 1974 A note on the transport across a diffusive interface. *Deep-Sea Res.* **21**, 283–287.

LINDEN, P. F. 1999 The fluid mechanics of natural ventilation. *Ann. Rev. Fluid Mech.* **31**, 201–238.

LINDEN, P. F. & COOPER, P. 1996 Multiple sources of buoyancy in a naturally ventilated enclosure. *J. Fluid Mech.* **311**, 177–192.

LINDEN, P. F., LANE-SERFF, G. F. & SMEED, D. A. 1990 Emptying filling boxes: the fluid mechanics of natural ventilation. *J. Fluid Mech.* **212**, 309–335.

LINDEN, P. F. & SHIRTCLIFFE, T. G. L. 1978 The diffusive interface in double-diffusive convection. *J. Fluid Mech.* **87**, 417–432.

MAXWORTHY, T. & NARIMOUSA, S. 1994 Unsteady turbulent convection into a homogeneous rotating fluid, with oceanographic applications. *J. Phys. Oceanogr.* **17**, 865–887.

MORTON, B. R. 1959 Forced plumes. *J. Fluid Mech.* **5**, 151–163.

MORTON, B. R., TAYLOR, G. I. & TURNER, J. S. 1956 Turbulent gravitational convection from maintained and instantaneous sources. *Proc. R. Soc. Lond.* **234**, 1–23.

NAGAWA, Y. & FRENZEN, P. 1995 A theoretical and experimental study of cellular convection in rotating fluids. *Tellus* **7**, 1–21.

NAKAGOME, H. & HIRIITA, M. 1977 The structure of turbulent diffusion in an axisymmetric turbulent plume. *Proc. 1976 ICHMT Seminar on Turbulent Buoyant Convection*, pp. 361–372.

PAPANICOLAOU, P. N. & LIST, E. J. 1988 Investigations of round turbulent buoyant jets. *J. Fluid Mech.* **195**, 341–391.

PHILLIPS, O. M. 1972 Turbulence in a stratified fluid: is it unstable? *Deep-Sea Res.* **19**, 79–81.

RAYLEIGH, LORD 1916 On convection currents in a horizontal layer of fluid when the higher temperature is on the under side. *Phil. Mag.* **32**, 59–546.

RICOU, F. P. & SPALDING, D. B. 1961 Measurements of entrainment by axisymmetric turbulent jets. *J. Fluid Mech.* **11**, 21–32.

ROONEY, G. G. & LINDEN, P. F. 1996 Similarity considerations for non-Boussinesq plumes in an unstratified environment. *J. Fluid Mech.* **318**, 237–250.

ROSSBY, H. T. 1969 A study of Benard convection with and without rotation. *J. Fluid Mech.* **36**, 309–335.

ROUSE, H., YIH, C. S. & HUMPHREYS, H. W. 1952 Gravitational convection from a boundary source. *Tellus* **4**, 210–210.

SAKAI, S. 1997 The horizontal scale of convection in the geostrophic regime. *J. Fluid Mech.* **333**, 85–95.

SCHMITT, R. W. 1979 The growth rate of supercritical salt fingers. *J. Mar. Res.* **37**, 419–436.

SHABBIR, A. & GEORGE, W. K. 1994 Experiments in a round turbulent buoyant plume. *J. Fluid Mech.* **275**, 1–32.

SHIRTCLIFFE, T. G. L. 1967 Thermosolutal convection: observation of the overstable mode. *Nature* **213**, 489–490.

SHIRTCLIFFE, T. G. L. 1973 Transport and profiles measurements of the diffusive interface in double-diffusive convection with similar diffusivities. *J. Fluid Mech.* **213**, 111–126.

SIGGIA E. D. 1994 High Reynolds number convection. *Ann. Rev. Fluid Mech.* **26**, 137–168.

SIMPSON, J. E. 1997 *Gravity Currents in the Environment and the Laboratory*, 2nd Edn. Cambridge University Press.

SPIEGEL, E. A. & VERONIS, G. 1960 On the Boussinesq approximation for a compressible fluid. *Astrophys. J.* **131**, 442–447.

STERN, M. E. 1960 The 'salt fountain' and thermal convection. *Tellus* **12**, 172–175.

TAYLOR, G. I. 1958 Flow induced by jets. *J. Aero. Space. Sci.* **25**, 464–465.

TURNER, J. S. 1967 Salt fingers across a density interface. *Deep-Sea Res.* **14**, 599–611.

TURNER, J. S. 1979 *Buoyancy Effects in Fluids*, 2nd Edn. Cambridge University Press.

WONG, A. B. D., GRIFFITHS R. W. & HUGHES, G. O. 2000 Shear layers driven by turbulent plumes. *J. Fluid Mech.* (submitted).

Woods, A. W. 1995 The dynamics of explosive volcanic eruptions. *Rev. Geophys.* **33**, 405–530.

Worster, M. G. & Huppert, H. E. 1983 Time dependent density profiles in a filling box. *J. Fluid Mech.* **132**, 457–466.

Wu, X. & Libchaber, A. 1990 Scaling relations in thermal turbulence: the aspect ratio dependence. *Phys. Rev. A* **45**, 842–845.

Department of Mechanical and Aerospace Engineering, University of California, San Diego, 9500 Gilman Drive, La Jolla, CA, 92130-0411, USA

7

Reflections on Magnetohydrodynamics

H. K. MOFFATT

1 Introduction

Magnetohydrodynamics (MHD) is concerned with the dynamics of fluids that are good conductors of electricity, and specifically with those effects that arise through the interaction of the motion of the fluid and any ambient magnetic field $B(x, t)$ that may be present. Such a field is produced by electric current sources which may be either external to the fluid (in which case we may talk of an 'applied' magnetic field), or induced within the fluid itself. The induction of a current distribution $j(x, t)$ by flow across the field B is the result of Faraday's 'law of induction'. The resulting Lorentz force distribution $F(x, t) = j \wedge B$ is generally rotational, i.e. $\nabla \wedge F \neq 0$, and therefore generates vorticity in the fluid. There is thus a fundamental interaction between the velocity field v and the magnetic field B, an interaction which not only leads to modification of well-understood flows of 'conventional' fluid dynamics, but also is responsible for completely new phenomena that simply do not exist in non-conducting fluids.

There are three major fields of application of magnetohydrodynamics, which will be discussed in this survey in the following order.

1.1 Liquid-metal magnetohydrodynamics

This is a branch of the subject which has attracted increasing attention over the last 30 years, and which has been much stimulated by the experimental programmes of such laboratories as MADYLAM in Grenoble, and the Laboratory of Magnetohydrodynamics in Riga, Latvia. These programmes have been motivated by the possibility of using electromagnetic fields in the processing of liquid metals (and their alloys) in conditions that frequently require high temperatures and high purity. These fields can be used to levitate samples of liquid metal, to control their shape, and to induce internal

347

stirring for the purpose of homogenization of the finished product, all these being effects that are completely unique to magnetohydrodynamics. Electromagnetic stirring (exploiting the rotationality of the Lorentz force) is used in the process of continuous casting of steel and other metals; and magnetic fields have an important potential use in controlling the interface instabilities that currently limit the efficiency of the industrial process of extracting aluminium from the raw material cryolite. The industrial exploitation of MHD in such contexts is of relatively recent origin, and great developments in this area are to be expected over the next few decades.

1.2 Magnetic fields in planetary physics and astrophysics

Nearly all large rotating cosmic bodies, which are either partly or wholly fluid in composition, exhibit magnetic fields of predominantly internal origin, i.e. fields associated with internal rather than external currents. Such fields may be generated by amplification of a very weak applied field, a process that may be limited by the conductivity of the fluid. More dramatically however, such fields may be the result of a 'dynamo instability' which is entirely of internal origin; this can occur (as in the liquid core of the Earth) if buoyancy-driven convection interacts with Coriolis forces to produce 'helicity' in the flow field. When the dynamo instability occurs, the field energy grows exponentially until the Lorentz forces are strong enough to modify the flow; at this stage, the magnetic energy is at least of the same order as the kinetic energy of the flow that generates it, and may even (as in the Earth context) be much greater. A fundamental understanding of this process in the context of turbulent flow has developed slowly over the last 50 years; and our present understanding of turbulent dynamo action, although incomplete, must surely be regarded as one of the great achievements of research in turbulence of this last half-century.

1.3 Magnetostatic equilibrium, structure and stability

This large area of MHD is also important in astrophysical contexts; but it has received even greater stimulus from the field of fusion physics and the need to design fusion reactors containing hot fully ionized gas (or plasma) isolated from the containing solid boundaries by a suitably engineered magnetic field – such an arrangement being conventionally described as a 'magnetic bottle'. In ideal circumstances, the gas in such a bottle is at rest, in equilibrium under the mutual action of Lorentz and pressure forces. Pressure forces of course tend to make the gas expand; the Lorentz force must thus be such as to prevent this expansion. Moreover the equilibrium, if it is to be effective, must be stable: the energy of the system must be minimal with respect to variations associated with the vast family of perturbations available to such a system. It is this problem of stability, first recognized and analysed in the 1950s, which has bedevilled the subject ever since, and which still stands in the way of development of an energy-producing thermonuclear reactor. Progress has nevertheless been sustained, and there is still a degree of optimism that the full stability problem can be eventually, if not solved, at least sufficiently tamed to allow design of commercially viable reactors. This is truly one of the immense scientific challenges that continues to face mankind at the dawn of the new millennium. Its solution would provide a clean source of low-cost energy at least until the next reversal of the Earth's magnetic field; by then we may have other problems to worry about!

2 Fundamental principles

The electromagnetic field is governed by Maxwell's equations, and it is legitimate in all the above contexts to adopt the 'magnetohydrodynamic approximation' in which displacement current and all associated relativistic effects are neglected. The magnetic field is then related to current by Ampère's equation

$$\nabla \wedge \boldsymbol{B} = \mu_0 \boldsymbol{j} \qquad \text{with} \qquad \nabla \cdot \boldsymbol{B} = 0, \tag{2.1}$$

where μ_0 is constant ($4\pi \times 10^{-7}$ in SI units). Moreover $\boldsymbol{B}$ evolves according to Faraday's law of induction which may be expressed in the form

$$\partial \boldsymbol{B}/\partial t = -\nabla \wedge \boldsymbol{E}, \tag{2.2}$$

where $\boldsymbol{E}(\boldsymbol{x}, t)$ is the electric field in the 'laboratory' frame of reference. The current in this frame is related to $\boldsymbol{E}$ and $\boldsymbol{B}$ by Ohm's law:

$$\boldsymbol{j} = \sigma(\boldsymbol{E} + \boldsymbol{v} \wedge \boldsymbol{B}), \tag{2.3}$$

where $\boldsymbol{v}(\boldsymbol{x}, t)$ is the velocity field and σ is the electrical conductivity of the fluid. Like viscosity, σ is temperature dependent, and will therefore in general be a function of $\boldsymbol{x}$ and t in the fluid. We shall, however, neglect such variations, and treat σ as a given constant fluid property. Note that the field $\boldsymbol{E}' = \boldsymbol{E} + \boldsymbol{v} \wedge \boldsymbol{B}$ appearing in (2.3) is the electric field in a frame of reference moving with the local fluid velocity $\boldsymbol{v}$.

From the above equations, we have immediately

$$\frac{\partial \boldsymbol{B}}{\partial t} = \nabla \wedge (\boldsymbol{v} \wedge \boldsymbol{B}) + \eta \nabla^2 \boldsymbol{B}, \tag{2.4}$$

where $\eta = (\mu_0\sigma)^{-1}$, the 'magnetic diffusivity' (or 'resistivity') of the fluid. Equation (2.4) is the famous 'induction equation' of magnetohydrodynamics, describing the evolution of B if $v(x, t)$ is known. The equation has a marvellous generality: it holds quite independently of the particular dynamical forces generating the flow (e.g. whether these are of thermal or compositional origin, whether the Lorentz force is or is not important, whether Coriolis forces are present or not); it holds also whether v is incompressible ($\nabla \cdot v = 0$) or not. Equation (2.4) may be regarded as the vector analogue of the scalar advection–diffusion equation

$$\frac{\partial \theta}{\partial t} + v \cdot \nabla\theta = \kappa\nabla^2\theta, \tag{2.5}$$

which describes the evolution of a scalar contaminant $\theta(x, t)$ subject to molecular diffusivity κ. Clearly, it is desirable to extract from (2.4) as much information as we can, before specializing to any particular dynamical context.

Note first the striking analogy, first pointed out by Batchelor (1950), between equation (2.4) and the equation for vorticity $\omega = \nabla \wedge u$ (the use of u rather than v here is deliberate) in a non-conducting barotropic fluid of kinematic viscosity v:

$$\frac{\partial \omega}{\partial t} = \nabla \wedge (u \wedge \omega) + v\nabla^2\omega. \tag{2.6}$$

The analogy is incomplete in that ω is constrained by the relationship $\omega = \nabla \wedge u$, whereas B and v in (2.4) suffer no such constraint: there is in effect far more freedom in (2.4) than there is in (2.6)! Nevertheless a number of results familiar in the context of (2.6) do carry over to the context of (2.4).

2.1 The magnetic Reynolds number

First, suppose that the system considered is characterized by a length scale l_0 and a velocity scale v_0. Naïve comparison of the two terms on the right of (2.4) gives

$$\frac{|\nabla \wedge (v \wedge B)|}{|\eta\nabla^2 B|} \sim \frac{v_0 l_0}{\eta} = R_m, \tag{2.7}$$

the magnetic Reynolds number, defined by obvious analogy with the Reynolds number of classical fluid dynamics. The three major areas of application of magnetohydrodynamics specified in the introduction may be discriminated in terms of this single all-important dimensionless number:

(a) $R_m \ll 1$: this is the domain of liquid-metal magnetohydrodynamics in most (but not quite all) circumstances of potential practical importance. In this situation, diffusion dominates induction by fluid motion, and the magnetic field in the fluid is determined, at least at leading order, by geometrical considerations (the geometry of the fluid domain and of the external current-carrying coils or magnets). The additional field induced by the fluid motion is weak compared with the applied field.

(b) $R_m = O(1)$: this is the regime in which dynamo instability, analogous in some respects to the dynamical instabilities of a flow that may lead to turbulence, may occur. Such instability requires that, in some sense, induction represented by the term $\nabla \wedge (v \wedge B)$ in (2.4) dominates the diffusion term $\eta \nabla^2 B$, which tends to eliminate field in the absence of external sources. Thus dynamo instability may be expected to be associated with an instability criterion of the form

$$R_m > R_{mc}, \qquad (2.8)$$

where R_{mc} is a critical magnetic Reynolds number, presumably of order unity; it must immediately be added that a condition such as (2.8) may be necessary but by no means sufficient for dynamo instability, which may require additional much more subtle conditions on the field v.

(c) $R_m \gg 1$: here we are into the domain of 'nearly perfect conductivity', in which inductive effects dominate diffusion. The limit $R_m \to \infty$ (or $\sigma \to \infty$ or $\eta \to 0$) may be described as the perfect-conductivity limit. In this formal limit, B satisfies what is known as the 'frozen-field equation'

$$\frac{\partial B}{\partial t} = \nabla \wedge (v \wedge B), \qquad (2.9)$$

which implies that the flux Φ of B across any material (Lagrangian) surface S is conserved:

$$\frac{d\Phi}{dt} = 0 \qquad \text{where} \qquad \Phi = \int_S B \cdot n \, dS. \qquad (2.10)$$

This is Alfvén's theorem, the analogue of Kelvin's circulation theorem, and one of the results for which the incomplete analogy between (2.4) and (2.6) is reliable. The fact that, in the perfect-conductivity limit, magnetic lines of force are 'frozen in the fluid' provides an appealing picture of how a magnetic field may develop in time. By analogy with vorticity in an incompressible fluid, any motion that tends to stretch magnetic field lines (or 'B-lines' for short), will also tend to intensify magnetic field. Is this a manifestation of dynamo action? Sometimes it is, but by no means invariably, as we shall see.

2.2 Magnetic helicity conservation

Since $\nabla \cdot \boldsymbol{B} = 0$, we may always introduce a vector potential $\boldsymbol{A}(\boldsymbol{x}, t)$ with the properties

$$\boldsymbol{B} = \nabla \wedge \boldsymbol{A}, \qquad \nabla \cdot \boldsymbol{A} = 0. \tag{2.11}$$

The 'uncurled' version of (2.9) is then

$$\frac{\partial \boldsymbol{A}}{\partial t} = \boldsymbol{v} \wedge \boldsymbol{B} - \nabla \varphi, \tag{2.12}$$

where $\varphi(\boldsymbol{x}, t)$ is a scalar field satisfying

$$\nabla^2 \varphi = \nabla \cdot (\boldsymbol{v} \wedge \boldsymbol{B}). \tag{2.13}$$

Note that the 'gauge' of $\boldsymbol{A}$ may be changed by the replacement $\boldsymbol{A} \to \boldsymbol{A} + \nabla \chi$ for arbitrary scalar χ. The equation $\nabla \cdot \boldsymbol{A} = 0$ implies a particular choice of gauge. Equations (2.9) and (2.12) may be written in equivalent Lagrangian form:

$$\frac{\mathrm{D}}{\mathrm{D}t} \left(\frac{\boldsymbol{B}}{\rho} \right) = \frac{\boldsymbol{B}}{\rho} \cdot \nabla \boldsymbol{v}, \qquad \frac{\mathrm{D}A_i}{\mathrm{D}t} = v_j \frac{\partial A_j}{\partial x_i} - \frac{\partial \varphi}{\partial x_i}. \tag{2.14}$$

This form is most convenient for the proof of conservations of magnetic helicity, defined as follows: let S be any closed 'magnetic surface', i.e. a surface on which $\boldsymbol{B} \cdot \boldsymbol{n} = 0$, and let V be the (material) volume inside S. Then the magnetic helicity $\mathcal{H}_M$ in V is defined by

$$\mathcal{H}_M = \int \boldsymbol{A} \cdot \boldsymbol{B} \, \mathrm{d}V, \tag{2.15}$$

a pseudo-scalar quantity that may easily be shown to be independent of the gauge of $\boldsymbol{A}$. In the limit $\eta = 0$, this quantity is conserved (Woltjer 1958); for

$$\frac{\mathrm{d}}{\mathrm{d}t} \int \boldsymbol{A} \cdot \boldsymbol{B} \, \mathrm{d}V = \int \frac{\mathrm{D}\boldsymbol{A}}{\mathrm{D}t} \cdot \boldsymbol{B} \, \mathrm{d}V + \int \boldsymbol{A} \cdot \frac{\mathrm{D}}{\mathrm{D}t} \left(\frac{\boldsymbol{B}}{\rho} \right) \rho \, \mathrm{d}V, \tag{2.16}$$

and, using (2.14), this readily gives

$$\frac{\mathrm{d}}{\mathrm{d}t} \int \boldsymbol{A} \cdot \boldsymbol{B} \mathrm{d}V = \int \nabla \cdot [\boldsymbol{B}(\boldsymbol{v} \cdot \boldsymbol{A} - \varphi)] \, \mathrm{d}V$$
$$= \int_S (\boldsymbol{n} \cdot \boldsymbol{B})(\boldsymbol{v} \cdot \boldsymbol{A} - \varphi) \mathrm{d}S = 0. \tag{2.17}$$

Note that this result holds whether the fluid is incompressible or not; it merely requires that the fluid be perfectly conducting.

This invariant admits interpretation in terms of linkage of the $\boldsymbol{B}$-lines (which are frozen in the fluid) (Moffatt 1969). To see this, consider the simplest 'prototype' linkage (figure 1) for which $\boldsymbol{B}$ is zero except in two flux

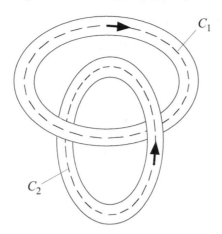

Figure 1. Linked flux tubes; here the linkage is right-handed and the linking number $n = 1$.

tubes of small cross-section centred on (unknotted) curves C_1 and C_2; the field lines within each tube are supposed to be unlinked, one with another. The sole linkage is then that between the two tubes, or equivalently between C_1 and C_2. Let Φ_1 and Φ_2 be the fluxes of $\boldsymbol{B}$ within these two tubes. The integral (2.16) may then be simply evaluated; it degenerates to

$$\mathscr{H}_M = \Phi_1 \oint_{C_1} \boldsymbol{A} \cdot \mathrm{d}\boldsymbol{x} + \Phi_2 \oint_{C_2} \boldsymbol{A} \cdot \mathrm{d}\boldsymbol{x}. \qquad (2.18)$$

The two line integrals are zero if C_1 and C_2 are unlinked (more strictly if the flux of $\boldsymbol{B}$ across any surface spanning either C_1 or C_2 is zero); if C_1 and C_2 are linked, with linking number n, then, from (2.18)

$$\mathscr{H}_M = \pm 2n\Phi_1\Phi_2, \qquad (2.19)$$

where the $+$ or $-$ is chosen according to whether the linkage is positive or negative; thus, for example, it is positive in figure 1, in which the sense of relative rotation is right-handed; if either arrow is reversed then the sign changes.

In general, $\boldsymbol{B}$-lines are not closed curves, and the above simple interpretation of $\mathscr{H}_M$ is not available. Nevertheless, $\mathscr{H}_M$ does always carry some limited information about the topology of the magnetic field (Arnold 1974; see also Arnold & Khesin 1998). This will be further considered in §6 below.

3 The Lorentz force and the equation of motion

As indicated in the introduction, the Lorentz force in the fluid (per unit volume) is $F = j \wedge B$. With $j = \mu_0^{-1} \nabla \wedge B$, this admits alternative expression in terms of the Maxwell stress tensor

$$F_i = \frac{1}{\mu_0} \frac{\partial}{\partial x_j} \left(B_i B_j - \tfrac{1}{2} B^2 \delta_{ij} \right) . \tag{3.1}$$

The first contribution, which may equally be written $\mu_0^{-1}(B \cdot \nabla)B$, represents a contribution to the force associated with curvature of B-lines directed towards the centre of curvature; the second contribution, which may be written $-(2\mu_0)^{-1} \nabla B^2$, represents (minus) the gradient of 'magnetic pressure'

$$p_M = (2\mu_0)^{-1} B^2 . \tag{3.2}$$

In an incompressible fluid of constant density ρ, the equation of motion including these contributions to the Lorentz force may be written in the form

$$\frac{\partial v}{\partial t} + v \cdot \nabla v = -\nabla \chi + \frac{1}{\mu_0 \rho} B \cdot \nabla B + \nu \nabla^2 v , \tag{3.3}$$

where

$$\chi = (p + p_M)/\rho , \tag{3.4}$$

and ν is the kinematic viscosity of the fluid. Coupled with the induction equation (2.4) and appropriate boundary conditions, this determines the evolution of the fields $\{v(x, t), B(x, t)\}$.

We may illustrate this with reference to a phenomenon of central importance in MHD, namely the ability of the medium to support transverse wave motions in the presence of a magnetic field.

3.1 Alfvén waves

Suppose that the medium is of infinite extent, and that we perturb about a state of rest in which the fluid is permeated by a uniform magnetic field B_0. Let $v(x, t)$ and $B = B_0 + b(x, t)$ be the perturbed velocity and magnetic fields. Then with the notation

$$V = (\mu_0 \rho)^{-1/2} B_0 , \qquad h = (\mu_0 \rho)^{-1/2} b , \tag{3.5}$$

the linearized forms of equations (2.4) and (3.3) (neglecting squares and products of v and b) are

$$\left. \begin{array}{l} \partial v/\partial t = -\nabla \chi + V \cdot \nabla h + \nu \nabla^2 v , \\ \partial h/\partial t = (V \cdot \nabla)v + \eta \nabla^2 h . \end{array} \right\} \tag{3.6}$$

These equations admit wave-like solutions with $\chi = \text{const.}$ and

$$\{\boldsymbol{v},\ \boldsymbol{h}\} \propto e^{i(\boldsymbol{k}\cdot\boldsymbol{x}-\omega t)} \tag{3.7}$$

provided

$$\left.\begin{array}{l} (-i\omega + vk^2)\boldsymbol{v} = i(\boldsymbol{k}\cdot\boldsymbol{V})\boldsymbol{h}, \\ (-i\omega + \eta k^2)\boldsymbol{h} = i(\boldsymbol{k}\cdot\boldsymbol{V})\boldsymbol{v}. \end{array}\right\} \tag{3.8}$$

Hence, for a non-trivial solution,

$$(i\omega - vk^2)(i\omega - \eta k^2) = -(\boldsymbol{k}\cdot\boldsymbol{V})^2, \tag{3.9}$$

which is a quadratic equation for ω with roots

$$\omega = -\tfrac{1}{2}i(\eta + v)k^2 \pm \tfrac{1}{2}\left\{4(\boldsymbol{k}\cdot\boldsymbol{V})^2 - (\eta - v)^2 k^4\right\}^{1/2}. \tag{3.10}$$

In the ideal-fluid limit ($\eta = 0$, $v = 0$), the roots are real:

$$\omega = \pm(\boldsymbol{k}\cdot\boldsymbol{V}), \tag{3.11}$$

a dispersion relationship of remarkable simplicity. Note that the group velocity associated with these waves is

$$\boldsymbol{c}_g = \nabla_k \omega = \pm\boldsymbol{V}, \tag{3.12}$$

a result evidently independent of the wave-vector $\boldsymbol{k}$. The associated non-dispersive waves are known, after their discoverer, as Alfvén waves, and the velocity $\boldsymbol{V}$ is the Alfvén velocity (Alfvén 1950).

When $\eta + v \neq 0$, these waves are invariably damped. In the liquid-metal context, the damping is predominantly due to magnetic resistivity ($\eta \gg v$); if, moreover, the field $\boldsymbol{B}_0$ is strong in the sense that $|\boldsymbol{k}\cdot\boldsymbol{V}| \gg \eta^2 k^4$, then (3.8) approximates to

$$\omega \approx -\tfrac{1}{2}i\eta k^2 \pm \boldsymbol{k}\cdot\boldsymbol{V}, \tag{3.13}$$

and the nature of the damping is clear.

If, on the other hand, $|\boldsymbol{k}\cdot\boldsymbol{V}| \ll \eta^2 k^4$, then the two roots (3.8) approximate to

$$\omega \approx -i\eta k^2,\ -i(\boldsymbol{k}\cdot\boldsymbol{V})^2/\eta k^2. \tag{3.14}$$

Now waves do not propagate, both modes are damped, and the first mode is damped much more rapidly than the second. Filtering of the first mode from equations (3.6) may be achieved by simply dropping the term $\partial\boldsymbol{h}/\partial t$; in this approximation, the field perturbation $\boldsymbol{h}$ is instantaneously determined by the flow $\boldsymbol{v}$.

4 Electromagnetic shaping and stirring

4.1 Two historic experiments

Two early papers provide examples of the manner in which applied magnetic fields and/or currents may contribute to the shaping of the region occupied by a conducting fluid and to the stirring of the fluid within this region. The first is that of Northrup (1907) who described an experiment in which a steady current is passed through a layer of conducting liquid (sodium–potassium alloy NaK) covered by a layer of (non-conducting) oil. The configuration described by Northrup is shown, in plan and elevation, in figure 2. The steady current $j(x)$ passes through a constriction between two electrodes on the boundary; the increased current density in this region gives rise to an increased magnetic field encircling the current lines (via Ampere's Law); the resulting Lorentz force distribution causes depression of the oil/NaK interface. Northrup attributes this observation to his friend Carl Hering who, he says, "jocosely called it the 'pinch phenomenon' ". The famous 'pinch effect' is precisely the effect of contraction of a (compressible) cylindrical column of fluid carrying an axial current due to the self-induced radial magnetic pressure gradient, a behaviour of fundamental importance in plasma containment devices (figure 3); as Northrup and Hering recognized, a similar effect occurs even in liquid metals under experimentally realizable laboratory conditions. In the complex geometry of figure 2, the nonlinear deformation (i.e. shaping) of the interface is determined by a balance between Lorentz forces, gravity and surface tension; this is too complex to be calculated analytically, and numerical techniques would be required to solve this type of three-dimensional problem.

Northrup also commented that the liquid on the inclined surface of the interface 'showed great agitation'. This is a manifestation of the stirring effect, which, as will be shown below, inevitably accompanies the shaping influence of the magnetic field.

The second historic experiment is that described by Braunbeck (1932) and illustrated in figure 4. Here, a cylindrical capsule containing liquid metal is suspended on a torsion wire, and is subjected to a *rotating* magnetic field, as indicated. Such a field may be regarded as the superposition of two alternating fields in quadrature and at right angles, and may be easily produced using a three-phase power supply. Braunbeck observed that the capsule rotated in the direction of rotation of the field, achieving an equilibrium angle of rotation depending on the conductivity of the liquid; the device may thus in principle be calibrated to determine the conductivity of small samples of liquid metal. What is happening in equilibrium is that the liquid

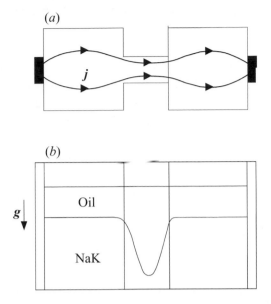

Figure 2. Sketch of Northrup's (1907) experiment: current flows through the NaK under a layer of oil; the constriction leads to a Lorentz force distribution which depresses the oil/NaK interface. (*a*) Plan; (*b*) elevation.

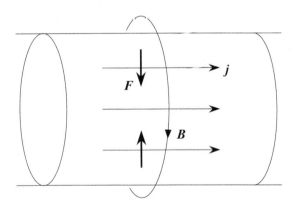

Figure 3. The pinch effect: axial current in the cylinder generates an azimuthal magnetic field; the resulting magnetic pressure gradient causes radial contraction of the cylinder.

metal rotates within the stationary container, exerting on it a viscous torque which is balanced by the restoring torque transmitted by the torsion wire. We shall analyse the details of this flow, and describe certain limitations of the description, below. Here, we may simply say that the stirring effect of an alternating or rotating magnetic field is a fundamental mechanism that is

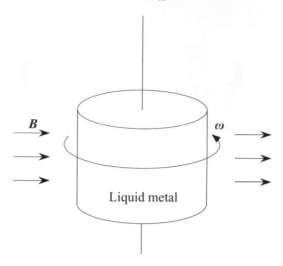

Figure 4. Sketch of the experiment of Braunbeck (1932): a capsule, containing liquid metal and suspended on a torsion wire, is subjected to a rotating magnetic field. This induces rotation in the fluid; the viscous torque on the capsule is then in equilibrium with the torque transmitted by the wire.

widely exploited in liquid-metal technology; thus, for example, it is used in the continuous casting of steel to stir the melt before solidification in order to produce a more homogeneous end-product (see, for example, Moffatt & Proctor 1984).

4.2 Electrically induced stirring

Northrup's problem as described above is one of a general class of problems in which a steady current distribution $j(x)$ is established in the fluid in a domain $\mathscr{D}$ through prescription of an electrostatic potential distribution $\varphi_S(x)$ on the boundary S of this domain. This current, together with any external current closing the circuit, is the source for the magnetic field $B(x)$, and flow is driven by the force $F = j \wedge B$. If $R_m \ll 1$, then additional (induced) currents are weak, i.e. $j(x)$ is effectively the same as if the conductor were at rest. Hence j is determined through solution of a Dirichlet problem:

$$j = -\sigma \nabla \varphi, \tag{4.1}$$

where

$$\nabla^2 \varphi = 0 \text{ in } \mathscr{D}, \qquad \varphi = \varphi_S(x) \quad \text{on} \quad S, \tag{4.2}$$

and $B(x)$ is then given by (2.1).

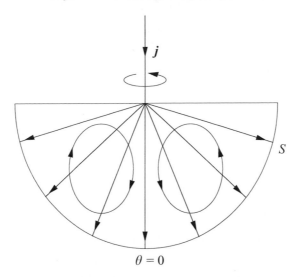

Figure 5. The weldpool problem: the Lorentz force drives the fluid towards the axis $\theta = 0$ and a strong jet flow develops along this axis; the flow is subject to instability involving swirl about this axis (Bojarevičs *et al.* 1989).

If the boundary S is entirely rigid, then the electromagnetic problem thus defined is conveniently decoupled from the fluid dynamical problem, and may be solved, either analytically or numerically, as a preliminary to determination of the force field F and the resulting flow field v. If, on the other hand, as in Northrup's problem, part of the surface S is an interface with another fluid (usually non-conducting) then the dynamical problem of determining the shape of S is coupled with the electromagnetic problem of determining j and B, an altogether more complex situation. Two problems of practical importance, described in the following subsections, have attracted much interest.

4.2.1 The weldpool problem

Here, current is injected through a point electrode at the surface S; the current spreads out radially (figure 5), and the configuration is locally axisymmetric about the direction of injection. Both j and B are singular at the point of injection, but of course these singularities are removed if the finite size of any real electrode is allowed for. Even so, the solution to this problem has some curious features that deserve comment here; for extended discussion, see Bojarevičs *et al.* (1989).

With local spherical polar coordinates (r, θ, φ) as indicated in figure 5, the current in the fluid is given by

$$j = (J/2\pi r^2, 0, 0) \qquad (4.3)$$

and the corresponding field B is then given by

$$B = \left(0, 0, \frac{\mu_0 J \sin\theta}{2\pi r(1+\cos\theta)}\right). \qquad (4.4)$$

Note the singularities at $r = 0$. The Lorentz force is given by

$$F = j \wedge B = \left(0, \frac{-\mu_0 J^2}{4\pi^2 r^3} \frac{\sin\theta}{1+\cos\theta}, 0\right). \qquad (4.5)$$

This force, directed towards the axis $\theta = 0$, tends to drive the fluid towards this axis; being incompressible, the fluid has no alternative but to flow out along the axis in the form of a strong axisymmetric jet, which becomes increasingly 'focused' for decreasing values of the fluid viscosity.

Experimental realization of this flow indicates a behaviour that is not yet fully understood: the flow is subject to a strong symmetry-breaking instability in which the fluid spontaneously rotates, in one direction or the other, about the axial direction $\theta = 0$. It is believed that this rotation controls the singularity of axial velocity that otherwise occurs if the dimensionless parameter

$$K = \mu_0 J^2 / \rho v^2 \qquad (4.6)$$

exceeds a critical value K_c of order 300; a recent discussion of this perplexing phenomenon is given by Davidson *et al.* (1999).

4.2.2 Aluminium smelting

The essential ingredients of the industrial processes by which aluminium is extracted from the raw material cryolite (an electrolytic salt, sodium aluminium fluoride) are shown in figure 6. Current passes from the anodes to the cathode through the cryolite and a layer of molten aluminium grows from the cathode. There is now a jump of conductivity across the fluid/fluid interface so that the current lines are refracted across this interface. The magnetic field B is again that due to the current distribution in the fluids and the external circuit. The Lorentz force drives a flow in both fluids, the shape of the interface being controlled by gravity (surface tension effects being weak on the scale of this industrial process). The process is further complicated by instabilities of the interface to which the steady configuration may be subject (for detailed discussion, see Moreau 1990). This is a problem of enormous industrial importance: in effect the efficiency of the process is

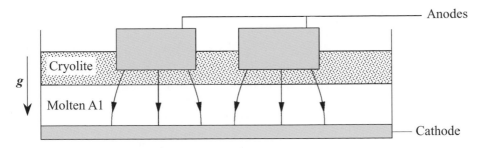

Figure 6. Sketch of the aluminium smelting process (Moreau 1990). The interface between the liquid cryolite and the molten aluminium is subject to instabilities of magnetohydrodynamic origin, which limit the efficiency of the process.

limited by instabilities that may lead to contact between the aluminium and the anodes, a 'short-circuiting' that would terminate the process. This is a billion-dollar industry for which an understanding of the fundamentals of magnetohydrodynamics would appear to be a first essential.

4.3 Inductive stirring

An equally important stirring mechanism is that associated with the application of an alternating (AC) magnetic field

$$B(x,t) = \mathrm{Re}\left[\hat{B}(x)\mathrm{e}^{-\mathrm{i}\omega t}\right], \tag{4.7}$$

such a field being produced by AC currents in external circuits (figure 7). The field diffuses into the conductor and a Lorentz force distribution $F(x,t) = j \wedge B$ is established. This force distribution has a time-averaged part $\bar{F}(x)$ and an additional periodic part with period 2ω and zero mean. The resulting flow consists of a steady part driven by $\bar{F}(x)$, and a time-periodic part whose amplitude is controlled by the inertia of the fluid. We shall focus on the steady part in the present discussion, bearing in mind that this steady part may be unstable, in which case a turbulent response to the force $\bar{F}(x)$ may be envisaged. We suppose that $R_m \ll 1$, so that again $B(x,t)$ may be calculated as if the conductor were at rest.

4.3.1 The high-frequency limit

The qualitative nature of the force field $\bar{F}(x)$ is best appreciated by first considering the high-frequency situation, in which $B(x,t)$ is confined to a thin magnetic boundary layer (the skin effect) just inside the surface S. The

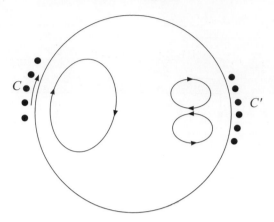

Figure 7. Stirring induced by AC fields; the sketch indicates the flow pattern that would be driven by a travelling magnetic field from the source coils C and a stationary AC field from the coils C; a wide range of patterns of stirring may be generated by appropriate engineering of the external coils and appropriate choice of field frequency ω.

thickness of this boundary layer is

$$\delta_M = (2\eta/\omega)^{1/2},$$
(4.8)

and is small when ω is large. Outside the conductor, the field $\hat{B}(x)$ is given by

$$\hat{B} = \nabla\Phi, \qquad \nabla^2\Phi = 0,$$
(4.9)

and a boundary condition, which in the high-frequency limit is just

$$\partial\Phi/\partial n = 0 \quad \text{on} \quad S.$$
(4.10)

In effect, the field is perfectly excluded from the conductor in this limit. Of course, the potential Φ has 'prescribed' singularities at the external coils. Let us regard this 'external' problem as solved; then $B_S = (\nabla\Phi)_S$ is known, as a tangential field on the surface S. Note that the surface divergence of B_S is non-zero in general; in fact

$$\nabla_S \cdot B_S = -\frac{\partial B_n}{\partial n} = -\frac{\partial^2\Phi}{\partial n^2}\bigg|_S.$$
(4.11)

Within the skin inside the conductor, the field $\hat{B}$ and hence $\hat{j}$ can be calculated by standard methods; the mean force $\hat{F} = \frac{1}{2}\text{Re}(\hat{j}^* \wedge \hat{B})$ may then be found; here, the $*$ denotes the complex conjugate. Details may be found in Moffatt (1985*b*); the result is

$$F = \frac{1}{2\mu_0\delta_M} \left[|B_S|^2 n - \mathrm{Im}B_S^*(\nabla \cdot B_S) \right] e^{-2\zeta/\delta_M}, \qquad (4.12)$$

where ζ is a normal coordinate directed into the conductor. The first term here is a strong inwardly directed normal component, which is responsible for the *shaping* of any part of S that is free to move in the normal direction. By contrast, *stirring* of the fluid is associated with the curl of F, given by

$$\nabla \wedge F = - \left(\frac{\rho}{\delta_M} \right) (n \wedge Q_S) e^{-2\zeta/\delta_M}, \qquad (4.13)$$

where

$$Q_S = \frac{1}{\mu_0\rho} \left[\nabla_S \tfrac{1}{2}|B_S|^2 - \mathrm{Im}B_S^*\nabla_S \cdot B_S \right], \qquad (4.14)$$

a vector field defined on the surface S, and having the dimensions of an acceleration.

This force distribution will clearly generate a highly sheared flow within the boundary layer (thickness $O(\delta_M)$). We may estimate the net effect, on the reasonable assumption that fluid inertia is negligible compared with viscous effects within the layer. On this 'lubrication' assumption the velocity field within the layer, assuming a no-slip condition at $\zeta = 0$, is

$$v = -\tfrac{1}{8}\delta_M^2 v^{-1} Q_S(x) \left(e^{-2\zeta/\delta_M} - 1 \right), \qquad (4.15)$$

so that asymptotically

$$v \sim \left(\delta_M^2/8v \right) Q_S(x) \quad \text{for} \quad \zeta \gg \delta_M. \qquad (4.16)$$

What this means is that, under the influence of the magnetic field, the no-slip condition must be replaced by a condition of 'prescribed tangential velocity' as given by (4.16) for the driven flow in the interior of the conductor.

A corresponding analysis using a free-surface condition $\partial u/\partial \zeta = 0$ on $\zeta = 0$ leads to an effective 'stress' condition

$$\tau_S = -\rho v \left(\partial u/\partial \zeta \right)\big|_{\zeta=\infty} = \tfrac{1}{4}\rho\delta_M Q_S(x) \qquad (4.17)$$

as a boundary condition for the interior flow. Note that this effective stress is independent of kinematic viscosity v.

Thus, in the high-frequency limit, the influence of the magnetic field is confined to the magnetic boundary layer, or skin, in such a way as to simply replace the no-slip (or zero-stress) conditions on S by an effective slip (or effective stress) distribution on S, which is then responsible for generating an internal flow, a flow whose topology will clearly be influenced, and indeed controlled, by the function $Q_S(x)$, which is determined through (4.14) by the surface field $B_S(x)$.

As the frequency is decreased, the skin depth δ_M increases, and the above simple description ceases to be valid; the force distribution penetrates more and more into the interior of the fluid, and extends throughout the fluid when δ_M increases to the scale L of the fluid domain.

4.3.2 The case of a circular cylinder

The prototype example is just as conceived by Braunbeck (1932): a horizontal rotating field applied to a circular cylinder with axis vertical and containing conducting fluid. Here,

$$B_S = 2B_0 e^{i\theta} e_\theta, \qquad (4.18)$$

where B_0 is the value of field magnitude far from the cylinder, and θ is the angular coordinate. Hence, in this case, $|B_S|^2$ is uniform on S, and

$$\nabla_S \cdot B_S = \frac{2i}{a} B_0 e^{i\theta} \quad \text{on} \quad S. \qquad (4.19)$$

Hence, from (4.14) and (4.15), the effective slip velocity just inside the skin is

$$v \approx \left(\eta B_0^2 / \mu_0 \rho \omega v a \right) e_\theta. \qquad (4.20)$$

The resulting motion is a rigid body rotation of the fluid inside this skin, with angular velocity

$$\Omega = \eta B_0^2 / \mu_0 \rho \omega v a^2, \qquad (4.21)$$

a result obtained originally by Moffatt (1965). The viscous stress and resulting couple G (per unit length of cylinder) acting on the cylinder are easily calculated, with the result

$$G = 2\pi \left(\frac{2\eta}{\omega} \right)^{1/2} \frac{B_0^2}{\mu_0 \rho}. \qquad (4.22)$$

Remarkably, this couple, although of viscous origin, is independent of v; this is because the core angular velocity, given by (4.21), is proportional to v^{-1}. In Braunbeck's experiment, it is the torque (4.22) that is ultimately in equilibrium with the torque transmitted by the torsion wire, which is proportional to the net angle of rotation; thus measurement of this angle provides a means of determination of η, and hence of the conductivity σ.

There are of course a number of limitations of the above type of analysis that should be borne in mind. First, the high-frequency approximation is valid only if $\delta_M \ll a$, i.e. $\omega \gg \eta/a^2$. This is not however a serious restriction;

the problem can be solved exactly for arbitrary ω in terms of Bessel functions. More seriously however, the analysis fails if the field strength B_0 becomes so strong that Ω in (4.21) becomes comparable with ω; for it is in fact not the *absolute* angular velocity ω of the field that is relevant to the induction of currents, but rather the *relative* angular velocity $\omega - \Omega$ between field and fluid. This strong field situation requires a major modification of approach (see, for example, Moreau 1990, where problems of this type are extensively treated).

4.3.3 Magnetic levitation

A high-frequency field also offers the possibility of levitating a volume (usually a small droplet) of conducting liquid in the complete absence of any rigid boundary support (figure 8); here, it may be more appropriate to talk of a magnetic basket rather than a magnetic bottle! The field is again expelled from the conductor (except in a thin skin); the first contribution to the force F in (4.12) corresponds to the effect of magnetic pressure p_M on the surface S; if S can so adjust itself that each vertical column of liquid of length Δz is support by the magnetic pressure difference Δp_M between bottom and top, i.e.

$$\Delta p_M - \rho g \Delta z, \tag{4.23}$$

then levitation is possible. This can be achieved by placing the main current sources of the field B *below* the liquid sample as in the configuration envisaged in figure 8(a). In the axisymmetric configuration of figure 8(b), the magnetic pressure vanishes on the axis of symmetry, and (4.23) cannot be satisfied; one must rely on surface tension to compensate gravity for the column of fluid immediately adjacent to this axis.

Configurations of this kind have been studied in some detail by Mestel (1982) and by Sneyd & Moffatt (1982). As pointed out in the latter paper, the effective stress given by (4.17) drives an interior flow with closed streamlines, whose intensity is limited only by viscosity. This is because integration of the steady equation of motion in the form

$$0 = v \wedge \omega - \nabla \left(\frac{p}{\rho} + \tfrac{1}{2} v^2 \right) + F - \nu \nabla \wedge \omega \tag{4.24}$$

round a closed streamline C gives

$$\oint_C F \cdot dx = \nu \oint_C dx \cdot (\nabla \wedge \omega), \tag{4.25}$$

i.e. inertia and pressure forces play no part in the equilibrium that is established. In practice, the viscosity in liquid metals is small; put another way,

(a)

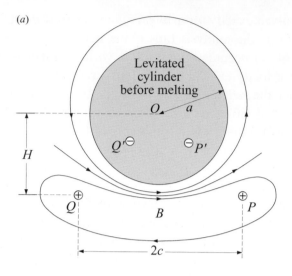

(b)

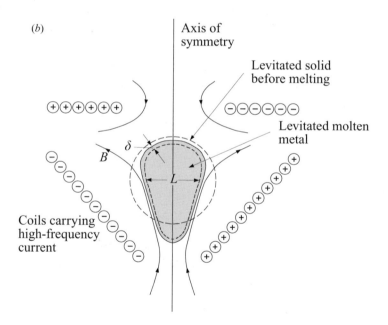

Figure 8. Magnetic levitation by high-frequency magnetic field, providing support via magnetic pressure distribution on the surface (Sneyd & Moffatt 1982).

the Reynolds number of the flow that is necessarily driven in the interior of a levitated droplet of radius of the order of 10 mm or greater is very large. It seems likely therefore that this interior flow will in these circumstances be turbulent. The 'great agitation' referred to by Northrup (see §4.1 above) also presumably indicated a state of turbulent flow just inside the liquid metal – and for similar reasons.

5 Dynamo theory

Dynamo theory is concerned with explaining the origin of magnetic fields in stars, planets and galaxies. Such fields are produced by currents in the interior regions, which would in the normal course of events be subject to ohmic decay, in the same way that current in an electric circuit decays if not maintained by a battery. This decay can however be arrested, and indeed reversed, through inductive effects associated with fluid motion; when this happens, the fluid system acts as a self-exciting dynamo. The magnetic field grows spontaneously from an arbitrary weak initial level, in much the same way as any other perturbation of an intrinsically unstable situation.

The possibility of such 'dynamo' instabilities may be understood with reference to the induction equation (2.4), which governs field evolution if the velocity field $v(x,t)$ in the fluid is regarded as 'given'. Let us suppose for simplicity that the fluid fills all space, and that the velocity field is steady, i.e. $v = v(x)$. We must suppose also that B has no 'sources at infinity' which would mitigate against the concept of an internally generated dynamo.

We may seek 'normal mode' solutions of (2.4) of the form

$$B(x,t) = \mathrm{Re}\left[\hat{B}(x)\,e^{pt}\right], \tag{5.1}$$

where, by substitution,

$$p\hat{B} = \nabla \wedge (v \wedge \hat{B}) + \eta \nabla^2 \hat{B}. \tag{5.2}$$

When coupled with the requirement that $\hat{B}(x)$ should be either a localized field of finite energy or, for example, a space-periodic field if v is space-periodic, this constitutes an eigenvalue problem which (in principle) determines a sequence of possibly complex eigenvalues $p_1, p_2, \dots$, which may be ordered so that

$$\mathrm{Re}\,p_1 \geqslant \mathrm{Re}\,p_2 \geqslant \mathrm{Re}\,p_3 \geqslant \dots, \tag{5.3}$$

and corresponding 'eigenfields' $\hat{B}_1(x), \hat{B}_2(x), \dots$. If $\mathrm{Re}\,p_1 > 0$, then the corresponding field

$$B(x,t) = \mathrm{Re}\left[\hat{B}_1(x)\,e^{p_1 t}\right] \tag{5.4}$$

exhibits dynamo behaviour: it grows exponentially in intensity, the growth being oscillatory or non-oscillatory according to whether $\text{Im}\, p_1 \neq 0$ or $= 0$. The mode of maximum growth-rate is clearly the one that will emerge from an arbitrary initial condition in which all modes may be present.

If this exponential growth occurs, then it can persist only for as long as the velocity field $v(x)$ remains unaffected by the Lorentz force. This is the 'kinematic phase' of the dynamo process. Obviously, however, the Lorentz force increases exponentially with growth rate $2\text{Re}\, p_1$ (considering only the mode (5.4)) and so ultimately the back-reaction of the Lorentz force on the fluid motion must be taken into account. This is the 'dynamic phase' of dynamo action in which the nature of the supply of energy to the system (via the dynamic equation of motion) must be considered. While great progress has been made over the past 50 years towards a full understanding of the kinematic phase, the highly nonlinear dynamic phase has proved far more intractable from an analytical point of view, and is likely to remain a focus of much research effort, both computational and analytical, over the next few decades.

5.1 Fast and slow dynamos

During the kinematic phase, the growth rate (p say) of the most unstable mode is determined in principle by the velocity field $v(x)$ and the parameter η which appears in equation (5.2). If $v(x)$ is characterized by velocity scale v_0 and length scale l_0, then on dimensional grounds,

$$p = (v_0/l_0)f(R_m),\qquad(5.5)$$

where, as in §2.1, $R_m = v_0 l_0/\eta$. An important distinction between dynamos described as 'fast' or 'slow' has been introduced by Vainshtein & Zel'dovich (1972). A fast dynamo is one for which $f(R_m) = O(1)$ as $R_m \to \infty$, i.e. the growth rate p scales on the dynamic time scale l_0/v_0. A slow dynamo, by contrast, is one for which $f(R_m) \to 0$ as $R_m \to \infty$; for example, if $f(R_m) \sim R_m^{-1/2}$ as $R_m \to \infty$, then $p \sim (v_0/l_0)R_m^{-1/2}$, and diffusivity η continues to influence the growth rate even in the limit $\eta \to 0$. The distinction is an important one because, on the galactic scale, R_m is extremely large, and a slow dynamo is likely to have little relevance in such contexts. This has led to an intensive search for dynamos that can legitimately be described as fast (see Childress & Gilbert 1995); however, in the strict sense indicated above, no such dynamo has yet been found! All known dynamos are slow; diffusivity remains important no matter how small η may be. The situation is again somewhat analogous to that governing the vorticity equation in turbulent flow: viscous effects remain important (in providing the mechanism for dissipation of kinetic energy) no matter how small the kinematic viscosity v may be.

When a dynamo enters the dynamic phase (assuming that sufficient time is available for it to do so) the distinction between fast and slow behaviour disappears; in either case, the growth rate must decrease, ultimately to zero when an equilibrium between generation of magnetic field by the (modified) velocity field and ohmic dissipation of magnetic field is established.

5.2 A little historical digression

The history of dynamo theory up to 1957, when Cowling's seminal monograph *Magnetohydrodynamics* was published, was characterized by the gravest uncertainty as to whether any form of self-exciting dynamo action in a spherical body of fluid of uniform conductivity was possible at all. One of the main firm results in this regard was negative: this was Cowling's (1934) theorem, which stated in its simplest form that steady axisymmetric dynamo action is impossible. This reinforced the view prevalent at the time, and originally stated in the geomagnetic context by Schuster (1912) that 'the difficulties which stand in the way of basing terrestrial magnetism on electric currents inside the earth are insurmountable'. Fortunately this pessimistic conclusion has been eroded and now completely reversed with the passage of time. For example, Cook (1980) writes that 'there is no theory other than a dynamo theory that shows any sign of accounting for the magnetic fields of the planets'. This view is echoed by Jacobs (1984) who writes 'There has been much speculation on the origin of the Earth's magnetic field ... The only possible means seems to be some form of electromagnetic induction, electric currents flowing in the Earth's core'.

If we try to identify one single development over the last half-century that has revolutionized our view of the subject, that development must surely be the 'mean-field electrodynamics' proposed in the seminal paper of Steenbeck, Krause & Rädler (1966), and foreshadowed in earlier 'pre-seminal' papers of Parker (1955) and Braginski (1964). This theory, which led in the fully turbulent context to the discovery of the famous α-effect – the appearance of a mean electromotive force parallel to the mean magnetic field – lies at the heart of the dynamo process as currently understood. In effect, it provides a starting point for any modern approach to dynamo theory; as such, it merits the closest study. A simplified account is presented in the following sub-sections. A systematic treatment may be found in the research monographs of Moffatt (1978) and Krause & Rädler (1980). Extensive treatment, with particular reference to astrophysical application, may also be found in the books of Parker (1979) and Zeldovich, Ruzmaikin & Sokoloff (1983).

5.3 Mean-field electrodynamics

Let us suppose that the velocity $v(x, t)$ appearing in equation (2.4) is turbulent – i.e. random in both space and time (see Chapter 5). We may suppose that there is some source of energy for this turbulence (e.g. through some random stirring mechanism) so that v is statistically stationary in time. It is then natural to adopt the notation $\langle \ldots \rangle$ for a time average, over any interval long compared with the time scale $t_0 = l_0/v_0$ characteristic of the energy-containing eddies of the turbulence; here l_0 is the scale of these eddies and v_0 the r.m.s. value of v. For simplicity, we may suppose that $\langle v \rangle = 0$.

The field $B(x, t)$ may be decomposed into mean and fluctuating parts:

$$B(x, t) = B_0(x, t) + b(x, t), \tag{5.6}$$

where $B_0(x, t) = \langle B(x, t) \rangle$ and $\langle b \rangle \equiv 0$. Here we allow B_0 to depend on t; this must be interpreted as allowing for slow evolution on a time scale T much greater than t_0. Separation of the time scales t_0 and T ($t_0 \ll T$) is the key to the solution of the problem. The theory may equally be developed in terms of separation of *spatial* scales l_0 and the scale $L(\gg l_0)$ on which the mean field develops.

The mean of equation (2.4) gives an evolution equation for $B_0(x, t)$:

$$\frac{\partial B_0}{\partial t} = \nabla \wedge \mathscr{E} + \eta \nabla^2 B_0, \tag{5.7}$$

where $\mathscr{E} = \langle v \wedge b \rangle$, the mean electromotive force arising through interaction of the fluctuating fields. Now the problem is like the 'closure' problem of turbulence: we need to find a relationship between $\mathscr{E}$ and B_0 in order to solve (5.7). But now, in contrast to the intractable problem of turbulence, we have the separation of scales to help us.

Subtracting (5.7) from (2.4) gives an equation for the fluctuating field b:

$$\frac{\partial b}{\partial t} = \nabla \wedge (v \wedge B_0) + \nabla \wedge (v \wedge b - \mathscr{E}) + \eta \nabla^2 b. \tag{5.8}$$

Without further approximation, it is difficult, if not impossible, to solve for b in terms of B_0 and v. However, we may make progress by noting that, for given $v(x, t)$, equation (5.8) establishes a linear relationship between b and B_0; and hence, since $\mathscr{E}$ is linearly related to b, between $\mathscr{E}$ and B_0.

Now both $\mathscr{E}$ and B_0 are average fields varying on the slow time scale, and on the large length scale L; hence this linear relationship between $\mathscr{E}$ and $\boldsymbol{B}_0$ may be represented by a series of the form

$$\mathscr{E}_i = \alpha_{ij} B_{0j} + \beta_{ijk} \frac{\partial B_{0j}}{\partial x_k} + \dots , \qquad (5.9)$$

where $\alpha_{ij}, \beta_{ijk}, \dots$ are tensor (actually pseudo-tensor) coefficients which are in principle determined by the statistical properties of the turbulent field $\boldsymbol{v}(\boldsymbol{x}, t)$ and the parameter η which intervenes in the solution of (5.8). Note that successive terms of (5.9) decrease in magnitude by a factor of order l_0/L; and that any terms involving time derivatives, i.e. $\partial \boldsymbol{B}_0/\partial t$, may be eliminated in favour of space derivatives through recursive appeal to equation (5.7). That the coefficients $\alpha_{ij}, \beta_{ijk}, \dots$ are pseudo-tensors (rather than tensors) should be evident from the fact that $\mathscr{E}$ is a polar vector (like velocity) whereas $\boldsymbol{B}_0$ is an axial vector (like angular velocity).

We may go further on the simplifying assumption that the turbulence is homogeneous and isotropic. In this case, α_{ij} and β_{ijk} share these properties, i.e. they are isotropic pseudo-tensors invariant under translation, i.e. independent of $\boldsymbol{x}$ (they are already independent of t from the assumption that $\boldsymbol{v}$ is statistically stationary). Isotropy is a very strong constraint; it implies that

$$\alpha_{ij} = \alpha \delta_{ij} , \qquad \beta_{ijk} = \beta \epsilon_{ijk} \dots , \qquad (5.10)$$

where now α is a pseudo-scalar quantity and β is a pure scalar. Hence (5.9) simplifies to

$$\mathscr{E} = \alpha \boldsymbol{B}_0 - \beta (\nabla \wedge \boldsymbol{B}_0) + \dots ; \qquad (5.11)$$

the next term in this series involves $\nabla \wedge (\nabla \wedge \boldsymbol{B}_0)$ and so on. This is the required relationship between $\mathscr{E}$ and $\boldsymbol{B}_0$.

Substituting this back into (5.7) gives the mean-field equation in its simplest form:

$$\frac{\partial \boldsymbol{B}_0}{\partial t} = \alpha (\nabla \wedge \boldsymbol{B}_0) + (\eta + \beta) \nabla^2 \boldsymbol{B}_0 + \dots , \qquad (5.12)$$

where the terms indicated by $+\dots$ involve higher derivatives of $\boldsymbol{B}_0$, and may presumably be neglected when the scale of variation of $\boldsymbol{B}_0$ is large. It is clear from the structure of (5.12) that β must be interpreted as an 'eddy diffusivity' associated with the turbulence (although there is no guarantee from the above treatment that β must invariably be positive!). The first term on the right of (5.12) will however always dominate the evolution, provided $\alpha \neq 0$ and the scale of $\boldsymbol{B}_0$ is sufficiently large. Before going further it is therefore essential to find a means of calculating α explicitly and determining the conditions under which this key parameter is definitely non-zero.

5.4 First-order smoothing

To do this, it is legitimate to consider the situation in which B_0 is constant; the reason for this is that α is independent of the field $B_0(x, t)$, and we are free to make any assumption about this field that simplifies calculation of α. The assumption that B_0 is constant is equivalent to considering the conceptual limit $L \to \infty$, $T \to \infty$.

Under this condition, the fluctuation equation (5.8) becomes

$$\frac{\partial b}{\partial t} = (B_0 \cdot \nabla)v + \nabla \wedge (v \wedge b - \mathscr{E}) + \eta \nabla^2 b. \tag{5.13}$$

Now we know that, if the magnetic Reynolds number $R_m = v_0 l_0 / \eta$ is *small*, then the induced field b is weak compared with B_0 (actually $b = O(R_m) B_0$; see below). In this circumstance, the awkward term $\nabla \wedge (v \wedge b - \mathscr{E})$ in (5.8) is negligible compared with the other terms in the equation and may be neglected. We then have a linear equation with constant coefficients which may be solved by elementary Fourier techniques. Let us then make this assumption and explore the consequences.

It is illuminating to consider first the situation in which v is a circularly polarized wave of the form

$$v(x, t) = v_0(\sin(kz - \omega t), \cos(kz - \omega t), 0), \tag{5.14}$$

and $B_0 = (0, 0, B_0)$ (see figure 9). Note that the vorticity associated with (5.14) is

$$\omega = \nabla \wedge v = kv, \tag{5.15}$$

and the associated helicity density,

$$\mathscr{H} = v \cdot \omega = kv^2 = kv_0^2, \tag{5.16}$$

is maximal (since ω is parallel to v) and constant. Then we easily find that (5.13) is satisfied provided

$$b = \frac{B_0 v_0 k}{\omega^2 + \eta^2 k^4} \left(\eta k^2 \cos \varphi - \omega \sin \varphi, -\omega \cos \varphi - \eta k^2 \sin \varphi, 0 \right) \tag{5.17}$$

where $\varphi = kz - \omega t$. We then obtain

$$v \wedge b = \frac{-B_0 v_0^2 k^3 \eta}{\omega^2 + \eta^2 k^4} (0, 0, 1). \tag{5.18}$$

For this case of a single 'helicity' wave (5.14), $v \wedge b$ turns out to be uniform, so the awkward term in (5.13) is identically zero! It follows from (5.18) that

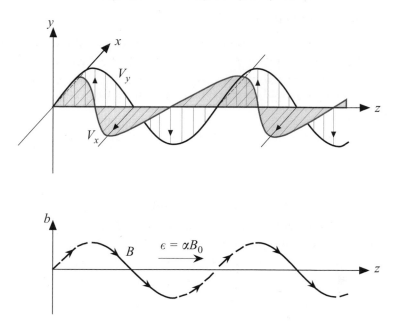

Figure 9. A circularly polarized travelling wave $v(x, t)$ of the form (5.14) induces and interacts with the field b which is also circularly polarized (equation (5.17)), but phase-shifted relative to v. The resulting $v \wedge b$ is uniform and parallel to B_0.

$\mathscr{E} = \alpha B_0$, where

$$\alpha = -\frac{\eta k^2}{\omega^2 + \eta^2 k^4} \mathscr{H} . \tag{5.19}$$

It is obviously desirable to write the result in this form, both left- and right-hand sides of the equation being pseudo-scalars.

It is important to note the origin of the 'α-effect' contained in (5.19): it is the phase-shift between b and v caused by molecular diffusivity η which leads to a non-zero value for α; it is a strange fact that, although diffusion is responsible for the decay of magnetic field in the absence of fluid motion, it is also responsible for the appearance of an α-effect which, as shown below, is a vital ingredient of the self-exciting dynamo process.

The velocity field (5.14) is of course rather special; it should be clear however that if we consider a random superposition of such waves with wave-vectors k, frequencies ω, and amplitudes $\hat{v}(k, \omega)$ isotropically distributed, then, provided the awkward term of (5.13) can be neglected, a result generalizing (5.19) can be obtained; all the contributions from different wave modes are additive, and the final result is

$$\alpha = -\tfrac{1}{3}\eta \iint \frac{k^2 \mathscr{H}(\boldsymbol{k},\omega)\,\mathrm{d}^3\boldsymbol{k}\,\mathrm{d}\omega}{\omega^2 + \eta^2 k^4}, \tag{5.20}$$

where $\mathscr{H}(\boldsymbol{k},\omega)$ is the helicity spectrum function of the velocity field $\boldsymbol{v}$, with the property

$$\langle \boldsymbol{v} \cdot \boldsymbol{\omega} \rangle = \iint \mathscr{H}(\boldsymbol{k},\omega)\,\mathrm{d}^3\boldsymbol{k}\,\mathrm{d}\omega. \tag{5.21}$$

The factor $\tfrac{1}{3}$ appears in (5.20) from averaging over all directions. The main thing to note again is the direct relationship between α and the helicity of the turbulent field.

This theory, usually described as 'first-order smoothing' theory, is limited to circumstances in which, as indicated above, $|\boldsymbol{b}| \ll |\boldsymbol{B}_0|$. This condition is satisfied if $R_m \ll 1$; it is also satisfied under the alternative condition

$$|v_0 k/\omega| \ll 1 \tag{5.22}$$

for values of k and ω making the dominant contributions to the field of turbulence. The condition (5.22) is relevant in a rapidly rotating system in which the turbulence is more akin to a field of weakly interacting inertial waves whose frequencies ω are of the order of the angular velocity Ω of the system; in this context, the condition (5.22) is one of small Rossby number.

If neither of the above conditions is satisfied, then it is not legitimate to neglect the awkward term in (5.13); no fully satisfactory theory is as yet available for the determination of α in these circumstances. Nevertheless, one may assert that the key property of turbulence required to provide a non-zero value for α is that its statistical properties should *lack reflectional symmetry*, i.e. should be non-invariant under change from a right-handed to a left-handed frame of reference; it is this same property that is necessary to provide non-zero mean helicity $\mathscr{H} = \langle \boldsymbol{v} \cdot \boldsymbol{\omega} \rangle$, which is indeed a simple measure of 'lack of reflectional symmetry', and so a loose connexion between α and $\mathscr{H}$ is to be expected.

The concept of isotropic turbulence that lacks reflectional symmetry is quite novel! One often thinks of a turbulent vorticity field in pictorial terms as like a random field of spaghetti; to picture turbulence lacking reflectional symmetry, think rather of a pasta in which each pasta element is twisted with the same sense of twist – say right-handed: if the vorticity field follows the sense of these elements, it remains statistically isotropic but acquires positive helicity. For a more realistic example, it suffices to consider Bénard convection in a rotating system (see, for example, Chandrasekhar 1961); it turns out that the average of $\boldsymbol{v} \cdot \boldsymbol{\omega}$ over horizontal planes is non-zero and antisymmetric about the centreplane (Moffatt 1978, chap. 10).

5.5 Dynamo action associated with the α-effect

Let us return now to the mean field equation (5.12) in the form

$$\frac{\partial \boldsymbol{B}}{\partial t} = \alpha \nabla \wedge \boldsymbol{B} + \eta_e \nabla^2 \boldsymbol{B}, \tag{5.23}$$

where now $\eta_e = \eta + \beta$ is the 'effective diffusivity', and we drop the suffix 0 on $\boldsymbol{B}_0$, to simplify notation. The simplest way to treat (5.23) is to seek solutions which have the Beltrami property

$$\nabla \wedge \boldsymbol{B} = K \boldsymbol{B}, \tag{5.24}$$

where K is a constant. For example, the field

$$\boldsymbol{B} = (c \sin Kz + b \cos Ky, a \sin Kx + c \cos Kz, b \sin Ky + a \cos Kx) \tag{5.25}$$

(the 'abc' field) has this property, as may be easily verified. We shall suppose that K is chosen to have the same sign as α, i.e. $\alpha K > 0$, and that $|K|$ is small, so that the scale $L \sim K^{-1}$ of $\boldsymbol{B}$ is large compared with the scale l_0 of the underlying turbulence that gives rise to the α-effect. Of course (5.23) implies that

$$\nabla^2 \boldsymbol{B} = -\nabla \wedge (\nabla \wedge \boldsymbol{B}) = -K^2 \boldsymbol{B}, \tag{5.26}$$

and so (5.23) becomes

$$\frac{\partial \boldsymbol{B}}{\partial t} = \left(\alpha K - \eta_e K^2\right) \boldsymbol{B}, \tag{5.27}$$

so that

$$\boldsymbol{B}(\boldsymbol{x}, t) = \boldsymbol{B}(\boldsymbol{x}, 0) e^{pt}, \tag{5.28}$$

where

$$p = \alpha K - \eta_e K^2. \tag{5.29}$$

Obviously we have exponential growth of the mean field $\boldsymbol{B}$ provided $\alpha K > \eta_e K^2$, i.e. provided

$$|K| < |\alpha|/\eta_e. \tag{5.30}$$

Thus we have a dynamo effect, in which the mean field (i.e. the field on the large scale L) grows exponentially provided L is large enough.

If we adopt the low-R_m estimate for α (based on (5.20)),

$$\alpha \sim \frac{1}{\eta} l_0^2 \mathcal{H} \sim R_m v_0 \tag{5.31}$$

assuming $\mathcal{H} \sim v_0^2/l_0$, and note that in this low-R_m situation, $\beta \ll \eta$ (actually $\beta = O(R_m^2)\eta$), then the scale of maximum growth rate is given (from (5.29)) by

$$L \sim K^{-1} \sim R_m^{-2} l_0 . \tag{5.32}$$

Hence a new magnetic Reynolds number $\tilde{R}_m$ defined in terms of L rather than l_0 is given by

$$\tilde{R}_m = \frac{L v_0}{\eta} \sim R_m^{-2} R_m \sim R_m^{-1} . \tag{5.33}$$

The condition $R_m \ll 1$ implies that $\tilde{R}_m \gg 1$: the field grows on a scale L at which the corresponding $\tilde{R}_m$ is *large*.

We can now summarize the implications of the above discussion. If a conducting fluid is in turbulent motion, the turbulence being homogeneous and isotropic but having the crucial property of 'lack of reflectional symmetry' (a property that, as observed above, can be induced in a rotating fluid through interaction of buoyancy-induced convection and Coriolis forces), then a magnetic field will in general grow from an arbitrarily weak initial level, on a scale L large compared with the scale l_0 of the turbulence. This is one of these remarkable situations in which 'order arises out of chaos', the order being evident in the large-scale magnetic field. It may be appropriate to follow the example of Richardson (1922) by summarizing the situation in rhyme:

Convection and diffusion,
In turb'lence with helicity,
Yields order from confusion
In cosmic electricity!

This totally general principle applies no matter what the physical context may be, whether on the planetary, stellar, galactic or even super-galactic scale. It is this generality that makes the approach described above, which derives from that pioneered by Steenbeck *et al.* (1966), so intensely appealing.

5.6 The back-reaction of Lorentz forces

The exponential growth associated with the dynamo action described above must ultimately be arrested by the action of Lorentz forces which may have a two-fold effect: (i) the generation of motion on the scale L of the growing field, and (ii) the suppression of the turbulence (or at least severe modification) on the scale l_0 which provides the α-effect. The latter effect is the easier to analyse, since one has as a starting point the Alfvén-wave type of analysis described in §3 above: the effect of a strong locally uniform magnetic field is to cause the constituent eddies of the turbulence either to propagate as Alfvén waves, or, if resistive damping is strong (as it is when R_m is small), to decay without oscillation.

There are two aspects of the behaviour that are worth noting. First, a locally uniform magnetic field induces strong local anisotropy in the turbulent field, modes with vorticity perpendicular to the magnetic field being most strongly damped. The dynamic effect of the field may be estimated in terms of a (dimensionless) *magnetic interaction parameter* N defined by

$$N = \frac{B_0^2 l_0}{\mu_0 \rho \eta v_0}, \tag{5.34}$$

and becomes strong when the local mean field B_0 grows strong so that $N \gg 1$. It has been shown (Moffatt 1967) that under this condition, and provided $NR_m^3 \ll 1$, turbulence that is initially isotropic becomes locally two-dimensional (invariant along the direction of B_0), with kinetic energy ultimately equally partitioned between the components (u, v) perpendicular to B_0 and the component w parallel to B_0, i.e.

$$\langle w^2 \rangle = \langle u^2 \rangle + \langle v^2 \rangle. \tag{5.35}$$

This obviously has implications for the α-effect, which is suppressed in magnitude and which no longer remains isotropic. The effects can be exceedingly complex, and it would be inappropriate to attempt to describe them here; there can be no question however that, if the turbulence is maintained by a prescribed force distribution $f(x, t)$, then the large-scale magnetic field must ultimately saturate at a level determined in principle by the statistics of this forcing (in particular by the mean square values $\langle f^2 \rangle$ and $\langle f \cdot \nabla \wedge f \rangle$) in conjunction with the relevant physical properties of the fluid, namely v and η.

The second feature to note is that on scales of order l_0 and smaller, the spectrum $\Gamma(k)$ of field fluctuations b is related to the spectrum $E(k)$ of v in a way that can be derived from the fluctuation equation (5.8). This calculation was first carried out by Golitsin (1960) (see also Moffatt 1961); it shows that if there is an inertial range of wavenumbers in which $E(k) \sim k^{-5/3}$ (see Chapter 5), then in this range, $\Gamma(k) \sim k^{-11/3}$. This is one prediction where theory is now corroborated by experiment: Odier, Pinton & Fauve (1998) found, in experiments on liquid gallium with values of R_m

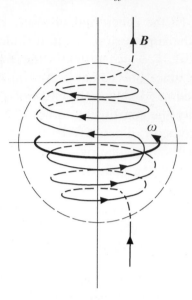

Figure 10. Generation of toroidal fields by differential rotation: the poloidal field is 'gripped' by the flow, and 'cranked' around the axis of symmetry (see, for example, Moffatt 1978).

in the range 1.3 to 15, that $\Gamma(k)$ scales like $k^{-\mu}$, with $\mu = 3.7 \pm 0.2$, neatly embracing the value $11/3 = 3.66$. In a branch of magnetohydrodynamics in which experimental results are sparse, this result stands out, and gives hope that more central aspects of dynamo theory may also soon be subject to experimental investigation and verification.

5.7 The $\alpha\omega$-dynamo

It would be misleading to leave the subject of dynamo theory without mentioning a mechanism of field generation that is just as important as the α-effect in the context of stellar and planetary magnetism, namely the generation of toroidal (or zonal) magnetic field from poloidal (or meridional) field by differential rotation. This mechanism, which operates in an axisymmetric system, is easily understood with reference to figure 10: if the angular velocity $\omega(r, \theta)$ is non-uniform along a field line of the poloidal field $\boldsymbol{B}_P(r, \theta)$, i.e. if $\boldsymbol{B}_P \cdot \nabla\omega \neq 0$, then the field line will be 'cranked' around the axis of symmetry, and a toroidal component $B_T(r, \theta)$ will be generated. This process is ultimately limited by field diffusion; detailed analysis (see, for example, Moffatt 1978) shows that, if $R_{m\omega}$ is a magnetic Reynolds number based on the strength and scale of the differential rotation, then B_T saturates at a level $O(R_{m\omega})|\boldsymbol{B}_P|$.

This of course is on the assumption that, through some other mechanism, the field $\boldsymbol{B}_P$ is itself maintained at a steady level. There is no axisymmetric mechanism that can achieve this (Cowling 1934), and it is of course here that the α-effect is required. The αω-dynamo incorporates both effects in the dynamo cycle:

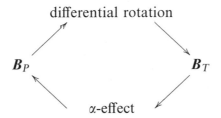

The α-effect here operates through the action of helical convection (i.e. convection influenced by Coriolis forces), which acts on the toroidal field in much the same manner as described in §5.4 above. The non-axisymmetric character of such a convective process is what allows the αω-dynamo to escape the strait-jacket of Cowling's theorem.

Many variants of the αω-dynamo, involving particular distributions of $\alpha(r,\theta)$ and $\omega(r,\theta)$, have been investigated (see particularly Krause & Rädler 1980), and much effort has been applied over the last 25 years to incorporate the difficult back-reaction of Lorentz forces in these models. Many difficulties still remain to be overcome, particularly in the two most prominent spheres of application, the Earth and the Sun. But it seems likely that, in the ultimate theories of the origins of both geomagnetism and heliomagnetism which may be expected to evolve over the next 50 years, both α-effect and ω-effect will survive as essential ingredients.

6 Relaxation to magnetostatic equilibrium

We turn now to a problem that is in some respects the opposite of the dynamo problem. As we have repeatedly observed, the Lorentz force $\boldsymbol{F} = \boldsymbol{j} \wedge \boldsymbol{B}$ is in general rotational and so drives a fluid motion; through this mechanism, magnetic energy can be converted to kinetic energy of motion, and if the fluid is viscous this energy is dissipated into heat. Thus, even if the fluid can be regarded as perfectly conducting (i.e. $\eta = 0$ or $R_m = \infty$) so that Joule dissipation is negligible, there is an alternative indirect route by which magnetic energy may dissipate, and this dissipation mechanism will persist for so long as $\nabla \wedge \boldsymbol{F} \neq 0$.

Insofar as the fluid is perfectly conducting however, the topology of the magnetic field is conserved during this 'relaxation' process; in particular, all links and knots present in the magnetic flux tubes at the initial instant $t = 0$ are present at least for all finite times $t > 0$. It is physically obvious that the magnetic energy cannot decay to zero under such circumstances; the only alternative possibility is that the field relaxes to an equilibrium compatible with its initial topology and for which $\nabla \wedge F = 0$. In such a state

$$F = j \wedge B = \nabla p, \tag{6.1}$$

i.e. the Lorentz force is balanced by a pressure gradient and the fluid is at rest: the field is in magnetostatic equilibrium (see, for example, Biskamp 1993).

We shall look more closely at the details of this process in the following subsections; for the moment it need merely be noted that we are faced here with an intriguing class of problems of variational type: to minimize a positive functional of the field B (here the magnetic energy) subject to conservation of the field topology; this topological constraint is conveniently captured by the requirement that only frozen-field distortions of B, i.e. those governed by the frozen-field equation

$$\frac{\partial B}{\partial t} = \nabla \wedge (v \wedge B), \tag{6.2}$$

are to be considered. We know that this equation guarantees conservation of magnetic helicity $\mathscr{H}_M$ (§ 2.2 above); but since (6.2) tells us that the field B in effect deforms with the flow, *all* its topological properties (i.e. those properties that are invariant under continuous deformation) are automatically conserved under this evolution.

We may look at this also from a Lagrangian point of view. Let the Lagrangian particle path associated with the flow $v(x, t)$ be given by

$$x \rightarrow X(x, t), \quad \partial X / \partial t = v, \quad X(x, 0) = x. \tag{6.3}$$

Then the Lagrangian statement equivalent to (6.2) is

$$\frac{B_i(X)}{\rho(X)} = \frac{B_j(x)}{\rho(x)} \frac{\partial X_i}{\partial x_j}, \tag{6.4}$$

where ρ is the density field. The deformation tensor $\partial X_i / \partial x_j$ encapsulates both the rotation and stretching of the field element of B as it is transported from x to X in time t. If the fluid is incompressible, then of course $\rho(X) = \rho(x)$. Note that (6.3) is a family of mappings continuously dependent on the parameter t, and being the identity mapping at $t = 0$; in the language of topology, it is an 'isotopy'.

6.1 The structure of magnetostatic equilibrium states

Equation (6.1) places a strong constraint on the structure of possible magnetostatic equilibria; it implies that

$$\boldsymbol{j} \cdot \nabla p = 0 \qquad \text{and} \qquad \boldsymbol{B} \cdot \nabla p = 0, \tag{6.5}$$

so that, in any region where $\nabla p \neq 0$, both the current lines ($\boldsymbol{j}$-lines) and field lines ($\boldsymbol{B}$-lines) must lie on surfaces $p = \text{const.}$

In any region where $\nabla p \equiv 0$, it follows from (6.1) that $\boldsymbol{j}$ is parallel to $\boldsymbol{B}$, i.e.

$$\boldsymbol{j} = \gamma(\boldsymbol{x})\boldsymbol{B} \tag{6.6}$$

for some scalar field $\gamma(\boldsymbol{x})$. The field $\boldsymbol{B}$ is then described as 'force-free' in this region. There is a considerable literature devoted to the subject of force-free fields (see particularly Marsh 1996); here we simply note that, on taking the divergence of (6.6) and using $\nabla \cdot \boldsymbol{j} = \nabla \cdot \boldsymbol{B} = 0$, we obtain

$$\boldsymbol{B} \cdot \nabla \gamma = 0, \tag{6.7}$$

so that now the $\boldsymbol{B}$-lines lie on surfaces $\gamma = \text{const.}$ The only possible escape from this (topological) constraint is when $\nabla \gamma \equiv 0$ and $\gamma = \text{const.}$ The field $\boldsymbol{B}$ is then a 'Beltrami' field. We have seen one example in (5.25) above; a more general field satisfying $\nabla \wedge \boldsymbol{B} = \gamma \boldsymbol{B}$ may be constructed as a superposition of circularly polarized Fourier modes in the form

$$\boldsymbol{B}(\boldsymbol{x}) = \int_{|\boldsymbol{k}|=\gamma} \left(\hat{\boldsymbol{b}}(\boldsymbol{k}) + i\hat{\boldsymbol{k}} \wedge \hat{\boldsymbol{b}}(\boldsymbol{k}) \right) e^{i\boldsymbol{k}\cdot\boldsymbol{x}}, \tag{6.8}$$

where $\hat{\boldsymbol{k}} = \boldsymbol{k}/|\boldsymbol{k}|$ and $\boldsymbol{k} \cdot \hat{\boldsymbol{b}}(\boldsymbol{k}) = 0$. With such a field, the $\boldsymbol{B}$-lines are no longer constrained to lie on surfaces; the particular field (5.25) exhibits the property of chaotic wandering of $\boldsymbol{B}$-lines (Hénon 1966; Arnold 1966; Dombre *et al.* 1986), and it may be conjectured that this is a generic property of force-free fields of the general form (6.8).

6.2 Magnetic relaxation

Suppose now that incompressible fluid is contained in a domain $\mathcal{D}$ with fixed rigid boundary $\partial\mathcal{D}$. For the reasons already indicated, let us suppose that this fluid is viscous and perfectly conducting. Such a combination of fluid properties is physically artificial; but no matter – this is a situation where the end justifies the means! We suppose that at $t = 0$ the fluid is at rest and that it supports a magnetic field $\boldsymbol{B}_0(\boldsymbol{x})$ where $\boldsymbol{n} \cdot \boldsymbol{B}_0 = 0$ on $\partial\mathcal{D}$. Clearly this condition of tangency of the field at the surface $\partial\mathcal{D}$ persists under the be analytic, even if $\boldsymbol{B}_0$ is analytic. Second, and more subtly, it is known that there are always 'islands of regularity' within any sea of chaos; during relaxation, the boundaries of such islands can become infinitely deformed so that the region of chaos within which the field $\boldsymbol{B}^E$ satisfies (6.20) may exhibit an arbitrarily complex geometry; the simplicity of (6.20) is gained at the cost of geometrical complexity of the region within which it is valid.

Magnetic relaxation is a process that can in principle be realized numerically; however, no satisfactory way has yet been found to provide accurate simulation in three dimensions of the frozen-field equation (6.2). Field relaxation in two dimensions *has* been accomplished by Linardatos (1993), who shows a clear tendency for current sheets to form due to collapse of the separatrices passing through saddle points of the field; but the problem of chaos does not arise for two-dimensional fields.

We note that the formation of current sheets, which may be recognized as a natural concomitant of magnetic relaxation to a minimum-energy state, is of absolutely central importance in solar physics: it is a prime mechanism for the explosive activity associated with solar flares and for consequential heating of the solar corona (Parker 1994). Whereas dynamo theory is central to an understanding of the internal origins of solar magnetism, magnetic relaxation provides the key to a proper understanding of observed external magnetic activity. The two process, both fundamental, are complementary in more senses than one (Parker 1979; see also Priest 1982).

7 Concluding remarks

In this essay, I have attempted to convey the flavour, rather than the detail, of three overlapping areas of magnetohydrodynamics in which I have been successively involved over the last 40 years. In 1960, the subject was relatively young and it attracted a huge cohort of researchers, active particularly in the contexts of astrophysics and fusion (plasma) physics. Great progress was made, notably in dynamo theory in the 1960s, the 1970s being more a period of consolidation. In more recent years, as in other branches of fluid mechanics, the computer has played an increasingly important role in allowing the investigation of nonlinear effects frequently beyond the reach of analytical study. Some fascinating new areas have emerged, most notably the area treated in §6 above in which the global topology of the magnetic field plays a central role. It may be noted that there is an exact analogy between the magnetostatic equation (6.13) and the steady Euler equation of ideal hydrodynamics, so that the magnetic relaxation technique provides an indirect means of determining solutions of the steady Euler equations of arbitrary streamline topology (Moffatt 1985a). There is a rich interplay between MHD and 'ordinary' fluid dynamics that stems from this analogy, in subtle conjunction with the (different) analogy between (2.4) and (2.6). Magnetohydrodynamics is a richly rewarding field of study, not only in its own right, but also through the illumination of more classical areas of fluid dynamics that this interplay provides.

References

ALFVÉN, H. 1950 *Cosmical Electrodynamics.* Oxford University Press.

ARNOLD, V. I. 1966 Sur la topologie des écoulements stationnaires des fluides parfaits. *C. R. Acad. Sci. Paris* **261**, 17–20.

ARNOLD, V. I. 1974 The asymptotic Hopf invariant and its applications. In *Proc. Summer School in Differential Equations, Erevan 1974.* Armenian SSR Acad. Sci. [English transl: *Sel. Math. Sov.* **5** (1986), 327–345].

ARNOLD, V. I. & KHESIN, B. A. 1998 *Topological Methods in Hydrodynamics.* Springer.

BAJER, K. & MOFFATT, H. K. 1990 On a class of steady confined Stokes flows with chaotic streamlines. *J. Fluid Mech.* **212**, 337–363.

BATCHELOR, G. K. 1950 On the spontaneous magnetic field in a conducting liquid in turbulent motion. *Proc. R. Soc. Lond.* A **201**, 405–416.

BERGER, M. A. & FIELD, G. B. 1984 The topological properties of magnetic helicity. *J. Fluid Mech.* **147**, 133–148.

BISKAMP, D. 1993 *Nonlinear Magnetohydrodynamics.* Cambridge University Press.

BOJAREVIČS, V., FREIBERGS, YA., SHILOVA, E. I. & SHCHERBININ, E. V. 1989 *Electrically Induced Vortical Flows.* Kluwer.

BRAGINSKII, S. I. 1964 Theory of the hydromagnetic dynamo. *Sov. Phys. JETP* **20**, 1462–1471.

BRAUNBEK, W. 1932 Eine neue Methode electrodenloser Leitfähigkeitsmessung. *Z. Phys.* **73**, 312–334.

CHANDRASEKHAR, S. 1961 *Hydrodynamic and Hydromagnetic Stability.* Clarendon.

CHILDRESS, S. & GILBERT, A. D. 1995 *Stretch, Twist, Fold: the Fast Dynamo.* Springer.

CHUI, A. Y. K. & MOFFATT, H. K. 1995 The energy and helicity of knotted magnetic flux tubes. *Proc. R. Soc. Lond.* A **451**, 609–629.

COOK, A. H. 1980 *Interiors of the Planets.* Cambridge University Press.

COWLING, T. G. 1934 The magnetic field of sunspots. *Mon. Not. R. AstrON. Soc.* **94**, 39–48.

COWLING, T. G. 1957 *Magnetohydrodynamics.* Interscience; 2nd Edn 1975, Adam Hilger Ltd., Bristol.

DOMBRE, T., FRISCH, U., GREENE, J. M., HÉNON, M., MEHR, A. & SOWARD, A. M. 1986 Chaotic streamlines in the ABC flows. *J. Fluid Mech.* **167**, 353–391.

DAVIDSON, P. A., KINNEAR, D., LINGWOOD, R. J., SHORT, D. J. & HE, X. 1999 The role of Ekman pumping and the dominance of swirl in confined flow driven by Lorentz forces. *Eur. J. Mech.* B/*Fluids* **18**, 693–711.

GOLITSIN, G. 1960 Fluctuations of magnetic field and current density in turbulent flows of weakly conducting fluids. *Sov. Phys. Dokl.* **132**, 315–318.

HÉNON, M. 1966 Sur la topologie des lignes de courant dans un cas particulier. *C. R. Acad. Sci. Paris* A **262**, 312–314.

JACOBS, J. A. 1984 *Reversals of the Earth's Magnetic Field.* Adam Hilger Ltd., Bristol.

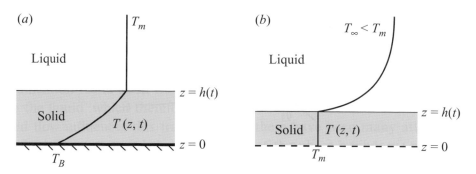

Figure 2. (*a*) Solidification of a melt from a cooled boundary at $z = 0$. The latent heat is conducted back through the solid to the boundary. (*b*) In contrast, when a crystal nucleates and grows into a supercooled melt, the latent heat is conducted into the liquid. Here, the line $z = 0$ is a symmetry plane of the one-dimensional crystal.

Equations (2.3)–(2.5) admit a similarity solution

$$T - T_B = (T_m - T_B)\frac{\mathrm{erf}\,\eta}{\mathrm{erf}\,\lambda}, \qquad h(t) = 2\lambda\sqrt{\kappa t}, \qquad (2.6a, b)$$

where

$$\mathrm{erf}\,x \equiv \frac{2}{\sqrt{\pi}} \int_0^x \mathrm{e}^{-u^2}\,\mathrm{d}u$$

is the error function and

$$\eta = \frac{z}{2\sqrt{\kappa t}} \qquad (2.7)$$

is a similarity variable. The coefficient λ, which is a measure of the solidification rate, satisfies the transcendental equation

$$G(\lambda) \equiv \sqrt{\pi}\lambda \mathrm{e}^{\lambda^2}\mathrm{erf}\,\lambda = S^{-1}, \qquad (2.8)$$

where the Stefan number

$$S = \frac{L}{C_p(T_m - T_B)} \qquad (2.9)$$

is the ratio of the latent heat of solidification to the sensible heat required to cool the newly formed solid to the boundary temperature. The solution to equation (2.8) is displayed in figure 3(*a*). We see that λ decreases as S increases; the solidification rate is slow when the latent heat is large.

A useful approximation can be made when the Stefan number is large. Because the solidification rate is then small, the temperature field in the solid

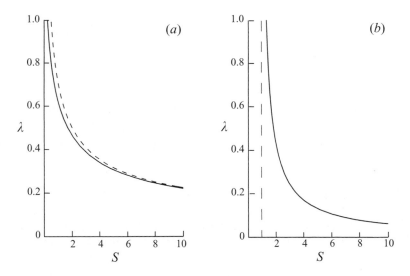

Figure 3. The scaled growth rates λ as functions of the Stefan number S predicted from the models illustrated in figure 2. The dashed curve in (a) is the result of the quasi-stationary model. The asymptote in (b) is at $S = 1$.

region is quasi-steady, hence linear, and the Stefan condition (2.5) can be approximated by

$$\rho L \dot{h} \simeq k \frac{(T_m - T_B)}{h}. \tag{2.10}$$

This equation is readily solved to give

$$h \simeq \sqrt{\frac{2\kappa t}{S}}, \qquad \text{i.e.} \qquad \lambda \simeq \frac{1}{\sqrt{2S}}. \tag{2.11}$$

The dashed curve in figure 3(a) shows that this gives a good approximation, with less than 7% relative error for $S > 2$.

A similar, one-dimensional solution can be found when the liquid is initially supercooled with temperature $T_\infty < T_m$. In this case (depicted in figure 2b), the solid is isothermal, with temperature T_m, the temperature in the liquid satisfies the diffusion equation (2.3), and the Stefan condition gives

$$\rho L \dot{h} = -k \frac{\partial T}{\partial z}\bigg|_{z=h+}. \tag{2.12}$$

There is again a similarity solution

$$T - T_\infty = (T_m - T_\infty) \frac{\text{erfc}\,\eta}{\text{erfc}\,\lambda}, \qquad h = 2\lambda\sqrt{\kappa t} \tag{2.13}$$

with, now,

$$F(\lambda) \equiv \sqrt{\pi}\lambda e^{\lambda^2} \operatorname{erfc}\lambda = S^{-1}, \qquad \text{where} \qquad S = \frac{L}{C_p(T_m - T_\infty)} \qquad (2.14)$$

and the complementary error function $\operatorname{erfc}x \equiv 1 - \operatorname{erf}x$.

The solution for λ is displayed in figure 3(b), where we see that no solution exists for $S < 1$. It is certainly possible to devise an experiment with $S < 1$ so the question remains: what happens in this case and why is the present mathematical formulation inadequate?

2.3 Kinetic undercooling

The problem lies in the assumption that the solid–liquid interface has temperature T_m. The freezing temperature T_m is an equilibrium temperature at which solid and liquid can coexist without change of phase. For solidification to proceed, the phase boundary must have a temperature below T_m, which is found to be related to the rate of solidification. The simplest case, appropriate to interfaces that are molecularly rough (unfacetted), is that

$$V_n = \mathscr{G}(T_m - T), \qquad (2.15)$$

where the kinetic coefficient $\mathscr{G}$ is constant (Kirkpatrick 1975). With this new boundary condition replacing (2.4b), there is no closed-form solution of the equations for general values of the Stefan number but it is straightforward to show that

$$h \sim \mathscr{G}(T_m - T_\infty)t, \qquad t \ll 1. \qquad (2.16)$$

Further, it can be shown that

$$h \sim 2\lambda\sqrt{\kappa t} \quad \text{as} \quad t \to \infty \quad \text{when} \quad S > 1 \qquad (2.17)$$

and that

$$h \sim (1 - S)\mathscr{G}(T_m - T_\infty)t \quad \text{as} \quad t \to \infty \quad \text{when} \quad S < 1 \qquad (2.18)$$

(Umantsev 1985). Kinetic undercooling gives a finite solidification rate at time zero, solidification rates remain finite for all time, and the similarity solution is recovered asymptotically where it exists ($S > 1$). In general terms we can say that when $S > 1$ the rate of solidification is controlled by the rate of removal of latent heat, while for $S < 1$ the kinetics of molecular attachment is rate controlling.

In the first case described above, of solidification from a cooled boundary, kinetics plays a minor role, simply regularizing the solution at early times.

We shall see later, however, that kinetics can play a major role when coupled with convection of the liquid.

2.4 The Gibbs–Thomson effect

Solidification is intrinsically concerned with interfaces, and one of the fundamental properties of interfaces is that intermolecular forces acting between the materials either side of them result in a surface energy. This manifests itself at a fluid–fluid interface as a surface tension, which creates a pressure jump across the interface when it is curved (see Chapter 1). The pressure variations generated at a corrugated fluid–fluid interface drive flows in both fluids that tend to restore the interface to a planar state. Excess surface energy associated with a corrugated solid–melt interface similarly drives the interface towards a planar state but the mechanism is different. Although there is a pressure jump across a curved solid–melt interface, it is taken up by elastic stresses in the solid and does not drive any flow. However, the same intermolecular forces that give rise to surface tension change the phase equilibria causing, in particular, the equilibrium freezing temperature to be depressed when the interface is convex towards the liquid (at a crest). This is expressed mathematically by the Gibbs–Thomson relationship

$$T_e = T_m - \Gamma \nabla \cdot \boldsymbol{n}, \tag{2.19}$$

where T_e is the equilibrium freezing temperature, $\boldsymbol{n}$ is the normal to the interface pointing into the liquid, $\Gamma = \gamma T_m / \rho_s L$, and γ is the surface energy. Note that $\nabla \cdot \boldsymbol{n}$ is the curvature of the interface. Thus heat flows from troughs to crests, promoting solidification at the troughs and melting at the crests, and the interface is driven towards a planar state.

2.5 Morphological instability

The ideas of diffusion-limited solidification and surface energy combine in determining the stability of the one-dimensional solutions presented earlier. As a simple illustration of the principal ideas,[1] consider a temperature field given by

$$T(x, z, t) = T_m + Gz + \hat{\theta} e^{i\alpha x - \alpha z + \sigma t} \tag{2.20}$$

---

[1] The analysis is presented here rather informally but can be derived rigorously from an asymptotic analysis of the full governing equations when the characteristic wavelength of the disturbance is much shorter than the diffusion length, i.e. in the limit $\alpha \gg \Delta T / G = V / \kappa$. This can be justified *a posteriori* as shown below.

in the liquid region

$$z > \zeta(x,t) = \hat{\zeta}e^{i\alpha x + \sigma t} . \tag{2.21}$$

These represent a sinusoidal perturbation to a planar solid–liquid interface at $z = 0$ in a frame of reference fixed to the undisturbed interface and the corresponding perturbation to the temperature field satisfying the quasi-steady diffusion equation. The perturbation coefficients $\hat{\theta}$ and $\hat{\zeta}$ are related by the Stefan condition and linearized Gibbs–Thomson relationship,

$$\rho L(V + \zeta_t) = -kT_z \quad \text{and} \quad T = T_m + \Gamma \zeta_{xx} \quad (z = \zeta). \tag{2.22a,b}$$

By substituting expressions (2.20) and (2.21) into (2.22), and keeping only those terms that are linear in $\hat{\theta}$ and $\hat{\zeta}$, we obtain

$$\rho L \sigma \hat{\zeta} = k\alpha \hat{\theta} \quad \text{and} \quad \hat{\theta} + G\hat{\zeta} = -\Gamma \alpha^2 \hat{\zeta}, \tag{2.23}$$

which combine to give

$$\sigma(\alpha) = \frac{k}{\rho L}\alpha(-G - \Gamma \alpha^2) . \tag{2.24}$$

There is a close mathematical analogy between this instability and the Saffman–Taylor instability that occurs when a viscous fluid displaces a more-viscous fluid in a porous medium or Hele-Shaw cell. The dispersion relation (2.24) is identical in form to equation (2.26) of Chapter 2 and is represented by a graph similar to figure 5 of that chapter.

We see that if G is positive (the temperature increases into the liquid) then the growth rate of the perturbations σ is negative; all disturbances decay and the interface is stable. On the other hand, if G is negative then there is a range of wavenumbers $0 < \alpha < \alpha_c \equiv \sqrt{-G/\Gamma}$ in which the growth rate σ is positive. Disturbances with wavenumbers less than α_c (wavelengths greater than $2\pi/\alpha_c$) will grow to form corrugations of the solid–liquid interface: the interface is said to be morphologically unstable.

The instability arises because diffusion of heat between the interface and the liquid is enhanced where the interface protrudes into the liquid, as described in Chapter 2, figure 2, and in a review by Langer (1980). When the liquid is supercooled ($G < 0$), heat is diffused away from the interface into the liquid, preferentially at protrusions, which causes them to grow. Opposing this tendency is the Gibbs–Thomson effect that causes perturbations to a planar interface to decay, as we saw in the previous section.

The competition between these two effects selects a length scale for the most unstable perturbations: diffusion is more rapid on small length scales, which causes σ to increase with α; surface energy acts even more strongly on smaller length scales, which causes σ ultimately to decrease with α. The maximum growth rate is readily determined from (2.24) to occur at $\alpha = \alpha_c/\sqrt{3}$. This corresponds to a wavelength

$$\lambda \propto \sqrt{\Gamma/(-G)} = \sqrt{l_\Gamma l_T}, \qquad (2.25)$$

where $l_T = \Delta T/(-G)$ is a characteristic length scale for temperature variations and $l_\Gamma = \Gamma/\Delta T$ is the so-called capillary length corresponding to an undercooling of ΔT. Typical values of l_T and l_Γ are a few centimetres and a few nanometres respectively, so $l_T \gg l_\Gamma$. Equation (2.25) expresses the rule of thumb that the length scale characteristic of morphological instabilities is the geometric mean of the diffusion length and the capillary length.

In this section we have seen how the rate of solidification of a pure melt is usually constrained by the rate at which latent heat can be transported, since the rate of molecular attachment is usually much faster than thermal transport rates. In consequence, we have seen that when the Stefan number is large, as is true in many practical applications, thermal diffusion fields can be approximated as being quasi-steady. Finally, we have seen that planar solidification fronts are morphologically unstable when the liquid is supercooled. The rest of this chapter will build on these ideas to examine the influence of fluid flow and to study the effects of dissolved solutes.

3 Convective heat transfer

The Stefan condition (2.1) expresses a balance between conductive heat transfer through the solid phase, heat transfer through the melt and the latent heat released at a solidification front. The heat flux carried by the melt is enhanced by fluid motions, which may be forced by external agents (e.g. stirring) or induced by buoyancy forces arising as the melt is cooled prior to being solidified. In this section we shall explore the role of convective heat transfer in determining the rate of solidification.

3.1 Flow near a stagnation point

One of the fundamental flows near a rigid boundary is the flow near a stagnation point, which is a ubiquitous feature of bounded flows and is therefore important to understand in its own right. It is, in addition, convenient for our present purpose since the heat transfer it induces is independent of position parallel to the boundary and so its influence on solidification is one-dimensional.

Consider a solid of uniform temperature T_m growing into a supercooled

These equations, together with the far-field condition, $C \to C_0$ as $z \to \infty$, have a steady one-dimensional solution with $\zeta = 0$ and

$$C = C_0 + C_0 \left(\frac{1 - k_D}{k_D} \right) e^{-Vz/D} . \tag{5.4}$$

Morphological stability of this system can be examined by considering the perturbed solutions

$$C = C_0 + C_0 \left(\frac{1 - k_D}{k_D} \right) e^{-Vz/D} + \hat{\phi} e^{-\mu Vz/D} e^{i\alpha x' + \sigma t'}, \qquad \zeta = \hat{\zeta} e^{i\alpha x' + \sigma t'} \tag{5.5}$$

where $x' = Vx/D$, $t' = V^2 t/D$, α and σ are all dimensionless and the constant μ is determined from (5.1) to be

$$\mu = \tfrac{1}{2} \left[1 + \sqrt{1 + 4(\alpha^2 + \sigma)} \right] . \tag{5.6}$$

Equations (5.2) and (5.3) are linearized to give

$$-G_c \sigma \hat{\zeta} = k_D G_c \hat{\zeta} + (\mu - 1 + k_D) \hat{\phi} \tag{5.7}$$

and

$$G \hat{\zeta} = -m G_c \hat{\zeta} - m \hat{\phi} - \alpha^2 \frac{\Gamma V^2}{D^2} \hat{\zeta} , \tag{5.8}$$

where

$$G_c = -C_0 \frac{1 - k_D}{k_D} \frac{V}{D} < 0 \tag{5.9}$$

is the concentration gradient in the melt at the undisturbed solid–liquid interface. These equations combine to give the growth rate implicitly (since $\mu = \mu(\sigma)$) by

$$\sigma = -k_D + (\mu - 1 + k_D) \left(1 - M^{-1} - \frac{l_\Gamma}{l_C} \alpha^2 \right) , \tag{5.10}$$

where $M = -mG_c/G$ is the morphological number, $l_C = D/V$ is the solutal diffusion length and

$$l_\Gamma = \frac{\Gamma k_D}{m C_0 (1 - k_D)} \tag{5.11}$$

is a capillary length. Equation (5.10) gives a rather complicated relationship for the growth rate σ. However, it can be shown that its key features occur when $\alpha \gg 1$, with $\sigma = O(\alpha)$, whence $\mu \sim \alpha$ (from (5.6)) and

$$\sigma \sim -k_D + \alpha \left(1 - M^{-1} - \frac{l_\Gamma}{l_C} \alpha^2 \right) . \tag{5.12}$$

Note that the first term on the right-hand side is retained even though $\alpha \gg 1$, since the term in brackets can be small near marginal stability. From this simpler expression it is readily determined that morphological instability occurs ($\sigma > 0$) when

$$M^{-1} < 1 - \frac{l_\Gamma}{l_C}\alpha^2 - \frac{k_D}{\alpha} . \tag{5.13}$$

The minimum value of M for which morphological instability can occur is therefore M_c, where

$$M_c^{-1} = 1 - \frac{3}{2}\left(2k_D^2\frac{l_\Gamma}{l_C}\right)^{1/3} , \tag{5.14}$$

i.e. instability occurs when M is slightly larger than unity and the melt is constitutionally supercooled.[1]

Equation (5.12) shows further that as M is increased above M_c the first modes to become unstable have wavenumber

$$\alpha = \left(\frac{k_D}{2}\frac{l_C}{l_\Gamma}\right)^{1/3} , \tag{5.15}$$

which corresponds to a (dimensional) wavelength

$$\lambda_c = 2\pi\left(\frac{2}{k_D}\right)^{1/3}\left(l_C^2 l_\Gamma\right)^{1/3} . \tag{5.16}$$

In contrast, once the stability threshold has been significantly exceeded the modes with maximum growth rate ($\partial\sigma/\partial\alpha = 0$) have a much shorter wavelength

$$\lambda_{\max} = 2\pi\left(\frac{3M}{M-1}\right)^{1/2}(l_C l_\Gamma)^{1/2} . \tag{5.17}$$

This expresses a result similar to (2.25), namely that the wavelength of maximum growth is proportional to the geometric mean of a diffusion length (here the solutal diffusion length) and a capillary length. The disparity in length scale between the marginally stable modes and the modes of maximum growth rate can be reconciled using (5.17) by recalling that M is only slightly larger than unity near marginal stability, as expressed by (5.14).

One of the most important results of this analysis is that instability is enhanced by increasing the solutal gradient, i.e. by thinning the compositional boundary layer. We shall examine the effects of various flows on the compositional boundary layer and hence deduce their influence on morphological instability. Another important result is that the characteristic length

---

[1] This situation can be reversed – the interface can be morphologically unstable while the melt is not constitutionally supercooled – in certain circumstances when the thermal conductivity of the solid is greater than that of the liquid.

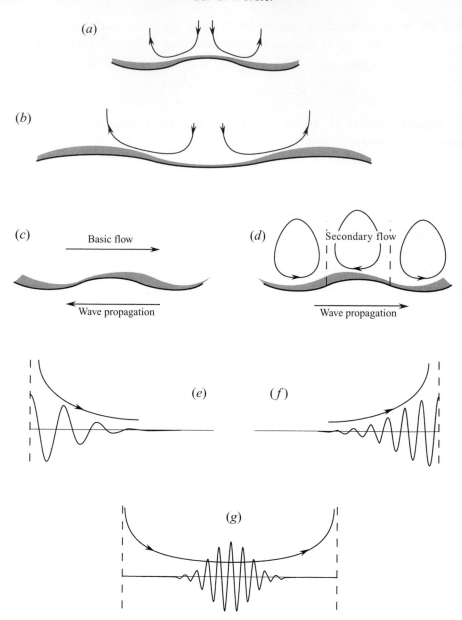

Figure 8. For caption see facing page.

scale (wavelength) of morphological instabilities is much smaller than the thickness of the compositional boundary layer (since typically $l_\Gamma \ll l_C$), which often limits the ability of flow to control morphological instability, as will be described below.

5.2 Natural compositional convection

The variation in concentration within the compositional boundary layer causes density gradients that, in a gravitational field, can drive natural buoyant convection. Consider first the case that solidification is upwards and the rejected solute is dense, for example by freezing ice from a salt solution by cooling it from below. The compositional boundary layer is statically stable to buoyancy-driven convection. However, if the solid–liquid interface becomes corrugated then the dense fluid flows down into the valleys, as shown in figure 8(*a*). A stagnation point with flow towards the solid is formed at the top of each crest, which, as we saw in § 3.1, thins the boundary layer and promotes solidification there; morphological instability is enhanced.

If the rejected solute is buoyant then the compositional boundary layer is statically unstable but may be dynamically stable if its Rayleigh number is less than critical (see Chapter 6). In this case, the flow depicted in figure 8(*a*) is reversed and morphological instability is delayed.

Once the Rayleigh number associated with the compositional boundary layer exceeds its critical value, convection occurs regardless of any deformation of the interface. The solute-conservation equation (5.2) shows that in a steadily convecting state the concentration of the melt at the interface is greatest where the solute gradient is steepest, i.e. where the boundary layer is thinnest. Consequently the equilibrium temperature is lowest at such places and the interface is depressed as shown in figure 8(*b*). This is not a morphological instability as such, because morphological changes are not

---

Figure 8. The effects of flow on morphological instabilities. In all cases solidification is upwards. (*a*) If the rejected solute is dense its buoyancy drives a flow from crests to troughs, which thins the boundary layer above crests and promotes morphological instability. (*b*) Steady compositional convection resulting from rejection of a buoyant solute on scales larger than typical morphological modes depresses the interface in regions of downflow. Morphological instabilities on smaller length scales are enhanced slightly within such depressions. (*c*) A steady shear flow compresses the compositional boundary layer on the upstream faces of interfacial corrugations causing them to propagate upstream. (*d*) The secondary flow arising from the need to satisfy the no-slip condition at the perturbed interface compresses the boundary layer on the downstream faces and causes the perturbations to propagate downstream. The first of these effects dominates at long wavelengths, the second at short wavelengths. Thus given a pure stagnation-point flow, long waves are found near diverging stagnation points (*e*) while short waves are found near converging stagnation points (*f*). In a completely confined flow, an absolute instability is found between stagnation points at stationary points of the horizontal components of the imposed flow (*g*).

intrinsic to the instability, but is rather a response of the interface to the convective flow. Whereas a morphological instability will develop into cellular and ultimately dendritic structures, the interface deformation resulting from convective motions may become less pronounced as further convective instabilities occur and turbulent motions ensue.

It is important to note that the length scale of convective instabilities is comparable to the thickness of the solutal boundary layer l_C and is therefore much longer than the scale of morphological instabilities $(l_C^2 l_\Gamma)^{1/3}$. Consequently any coupling between the two is weak, and it has been found that natural convection cannot be exploited effectively to control morphological instabilities during crystal growth. Researchers have therefore turned their attention to various forced flows in an effort to find a way of controlling instability.

5.3 Forced flows

Since the Schmidt number $Sc = \nu/D$ of most materials is very large, the compositional boundary layer at a solidification front is normally subsumed within the viscous boundary layer and the flow is locally linear. Therefore, away from confining boundaries and stagnation points, all flows can be approximated by a simple parallel shear flow on the scale of morphological instabilities.

The primary influence of a simple shear is shown in figure $8(c, d)$. The horizontal basic flow compresses the perturbed solutal boundary layer on the upstream faces of a perturbed interface, as shown in figure $8(c)$, which promotes the normal growth of those faces and results in a wave-like propagation of the perturbation upstream. Countering this tendency is the secondary flow induced by the need to satisfy the no-slip condition on the perturbed interface. The secondary flow has stagnation points between crests and troughs that compress the basic solutal boundary layer on downstream faces and thus promote a downstream propagation of the interface perturbation (figure $8d$). The first of these effects dominates at long wavelengths $\lambda > \lambda_a$ (say), while the second dominates at the shorter wavelengths $(\lambda < \lambda_a)$ characteristic of morphological instabilities. Forth & Wheeler (1989) have calculated the dispersion relation for these travelling waves and have shown that λ_a is typically of the same order of magnitude as the thickness of the compositional boundary layer l_C. In both cases the enhancement of solidification is $\pi/2$ out of phase (either negative or positive) with the interface perturbation and therefore has a neutral influence on instability. This is the dominant behaviour for large Schmidt number. At higher orders in inverse

Schmidt number, more complex interactions result in a weak suppression of the two-dimensional morphological instability. The flow has no effect on the conditions for instability of perturbations with crests aligned parallel to the direction of flow, however, so these are the preferred modes near marginal conditions.

These results on an infinite domain indicate that morphological instability in the presence of flow is *convective* (see Chapter 4) and care is needed to apply them to an understanding of what will happen in a finite domain.

In confined systems, there are stagnation points at which the tangential flow is zero. Brattkus & Davis (1988) carried out an analysis restricted to long-wavelength disturbances and found instabilities at *diverging* stagnation points, which is consistent with the fact that long waves travel upstream towards the stagnation point (figure 8e). Conversely, Bühler & Davis (1998), found that disturbances of wavelengths comparable to those of morphological instability in the absence of flow (i.e. much smaller than λ_a) are localized near *converging* stagnation points, again consistent with the fact that such waves travel downstream, towards the stagnation point (figure 8f)[1]. However, in a system confined at both ends the most *absolutely* unstable mode occurs at wavelengths such that $\partial\omega/\partial k = 0$ and at positions where $\partial\omega/\partial x = 0$ (see Chapter 4). Given the dispersion relation derived by Forth & Wheeler (1989) in which ω is directly proportional to the free-stream flow U, the latter occurs where $\partial U/\partial x = 0$, i.e. between stagnation points (figure 8g).

Overall, it seems that steady flows may do little to suppress morphological instability and can even enhance it in certain circumstances, but it can alter the characteristics of the unstable modes and may serve to confine instabilities to a small region of a growing crystal. Certain types of unsteady flows, for example orbital flows (Schulze & Davis 1995), seem more promising in controlling morphological instability.

6 Mushy layers

In most natural and metallurgical settings, solidification rates greatly exceed the critical values required for morphological instability. Planar interfaces cannot survive the build-up of solute and consequent constitutional super-cooling ahead of them, and give way to highly convoluted solid structures that form the matrix of a porous medium called a mushy layer (figure 9). The solute rejected during solidification no longer needs to be transported

---

[1] Strictly figure 8(e, f) and the associated discussion relate to the phase velocity of the perturbations. The analysis of Forth & Wheeler shows also that the group velocity of long waves is upstream and for short waves is downstream.

Figure 9. A close-up view of the interfacial region of a dendritic mushy layer. The solid phase in this case is ammonium chloride. The spacing between dendrites is about 0.3 mm while the overall depth of the mushy layer is a few centimetres.

into the bulk liquid region but can be accommodated within the interstices (pores) of the mushy layer. Local solute transport continues to play a role on the scale of individual pores, determining the size of the pores and the occurrence of side branches on the crystals forming the matrix, but the macroscopic evolution of a mushy layer is controlled predominantly by thermal balances. The structure of a mushy layer is affected by fluid motions in the melt adjacent to it and in the melt flowing through its interstices.

6.1 Relief of supercooling

We saw in §4 that a region of constitutional supercooling can develop in an alloy ahead of a solidification front. The effect of morphological instabilities is to increase the surface area of the phase boundary and thus to enhance both the liberation of latent heat and the rejection of solute. The former tends to warm the melt while the latter causes a lowering of the local freezing temperature (liquidus). These each serve to reduce the degree of constitutional supercooling. Without surface energy, morphological instabilities would occur on arbitrarily small length scales given any amount of supercooling, and it is not hard to imagine therefore that this process would continue until the degree of constitutional supercooling were reduced to zero. In practice, surface energy limits instability at some scale, as we have seen. But if the surface energy is small ($l_\Gamma \ll l_C$) then it is a good approximation to assume that the temperature and concentration of the interstitial liquid in a mushy layer lie on the liquidus

$$T = T_L(C). \tag{6.1}$$

There is some experimental support for this assumption and to date it has been almost universally adopted in mathematical models of mushy layers.

6.2 Evolution of a mushy layer without flow

The primary feature of a mushy layer is the accommodation of excess solute within its interstices and the loss thereby of solutal control on its overall extent. We can gain an understanding of this by considering a simple model in which we imagine that the mushy layer has a uniform volume fraction of solid ϕ. When the solid is pure ($C_S = 0$) and there is no fluid flow, the total amount of solute within the layer per unit horizontal area is

$$\int_0^{h(t)} (1 - \phi)C_L(T)\mathrm{d}z - hC_0 , \tag{6.2}$$

where $h(t)$ is the depth of the layer and $T(z,t)$ is the local temperature. If we make the quasi-stationary approximation then

$$T \simeq T_B + (T_L(C_0) - T_B)\frac{z}{h(t)} \tag{6.3}$$

and equation (6.2) then shows that

$$\phi = \frac{C_L(T_B) - C_0}{C_L(T_B) + C_0} \equiv \frac{1}{2\mathscr{C} - 1}, \tag{6.4}$$

assuming that the liquidus is linear, where $\mathscr{C}$ is the important concentration ratio introduced in equation (4.14).

Note that $\phi = 1$ when $C_0 = 0$ ($\mathscr{C} = 1$), corresponding to a pure melt, and that ϕ is small for large values of $\mathscr{C}$, a fact exploited in many analyses of mushy layers.

The Stefan condition at the mush–liquid interface is

$$\phi \rho_s L \dot{h} = \bar{k}(\phi)\frac{\partial T}{\partial z}\bigg|_{z=h-} = \bar{k}(\phi)\frac{T_L(C_0) - T_B}{h}, \tag{6.5}$$

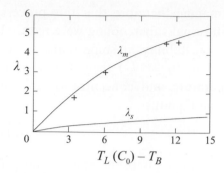
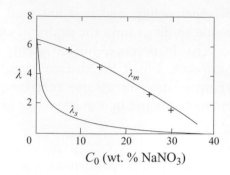

Figure 10. The rates of diffusion-controlled solidification of a binary alloy from a cooled boundary. A planar solid–liquid interface is predicted to advance as $h = 2\lambda_s\sqrt{Dt}$, where λ_s is determined from (4.14), while a mush–liquid interface is predicted to advance as $h = 2\lambda_m\sqrt{Dt}$, where λ_m is determined by an extension of the theory leading to (6.7). This theory (Huppert & Worster 1985) gives very good agreement with the data (crosses) from experiments in which mushy layers of ice crystals were formed by cooling aqueous solutions of sodium nitrate from below.

where

$$\bar{k}(\phi) = \phi k_s + (1 - \phi)k_l \tag{6.6}$$

is a mean thermal conductivity of the mushy layer, and the heat flux from the melt has been ignored. This mean conductivity is exact for a medium composed of lamellae oriented parallel to the heat flux vector (Batchelor 1974) and is a good approximation within mushy layers since the crystals forming the solid matrix have primary branches aligned with the local heat flux vector. Equation (6.5) is readily integrated to yield

$$h = \left(\frac{2}{S} \frac{\bar{k}}{k_s\phi} \kappa_s t \right)^{1/2}, \tag{6.7}$$

where the Stefan number $S = L/C_{ps}[T_L(C_0) - T_B]$, $\kappa_s = k_s/\rho_s C_{ps}$ and C_{ps} is the specific heat capacity of the solid. This result, which can be compared with (2.11) and contrasted with (4.7), highlights the fact that the growth of the mushy layer is determined by the rate of *thermal* diffusion. To make accurate predictions, care is needed in estimating the appropriate thermal conductivity of the layer. Note, for example, that the thermal conductivity of ice is about four times that of water so the mean conductivity of the mushy layer is sensitive to the solid fraction.

A slightly more detailed model along the lines above (Huppert & Worster 1985) has given results in good agreement with laboratory experiments (figure 10).

6.3 Flow in a mushy layer

The interstitial liquid within a mushy layer is free to flow in response to pressure gradients and buoyancy forces. The characteristics of its flow are fundamentally those of flows within porous media (Phillips 1991; Nield & Bejan 1999), the simplest model of which is given by Darcy's equation

$$\mu \boldsymbol{u} = \boldsymbol{\Pi}(-\nabla p + \rho \boldsymbol{g}). \tag{6.8}$$

Here $\boldsymbol{u}$ is the 'superficial' velocity or 'Darcy' velocity, which is the volume flux per unit area flowing through the medium, p is the pressure and $\boldsymbol{g}$ is the acceleration due to gravity. The permeability $\boldsymbol{\Pi} = \boldsymbol{\Pi}(\phi, \mathscr{A})$ is a second-rank tensor, reflecting the fact that the resistance to flow within the medium may be anisotropic, and is a function both of the void fraction $1 - \phi$ and the specific surface area $\mathscr{A}$ of the internal phase boundaries as well as their geometry. In many studies the permeability is assumed to be locally isotropic and the dependence on $\mathscr{A}$ is ignored for no better reason than it is a difficult quantity both to measure and to predict. We shall assume here that $\boldsymbol{\Pi} = \Pi(\phi)\boldsymbol{I}$, where $\boldsymbol{I}$ is the identity.

What distinguishes the flow in a mushy layer from that in a passive porous medium is that the solid fraction and hence its porosity can vary both in time and in space. This has two significant consequences. The first is that, as indicated above, the permeability is spatially inhomogeneous and may vary in time. We shall see later that this can lead to an interesting feedback between flow and solidification in mushy layers, typically resulting in a focusing of flows into narrow regions of low (or even zero) solid fraction. The second is that the velocity field in a mushy layer is non-solendoidal. Local mass conservation is expressed by

$$\frac{\partial \bar{\rho}}{\partial t} + \nabla \cdot (\rho_l \boldsymbol{u}) = 0, \tag{6.9}$$

where the mean density $\bar{\rho} = \phi \rho_s + (1 - \phi)\rho_l$, which can be rearranged as

$$\nabla \cdot \boldsymbol{u} = -\frac{\rho_s - \rho_l}{\rho_l} \frac{\partial \phi}{\partial t}. \tag{6.10}$$

This shows that there is a divergence of the flow as solid grows internally if $\rho_s < \rho_l$ so that there is an expansion on change of phase, as is the case for water and for silicon, for example. Conversely, many other materials contract as they solidify, which leads to a convergent flow. Clearly fluid flow is driven by this mechanism even in the absence of external forces. It can cause redistribution of solute within a casting and, in the case of contraction, can cause voids to form (see Beckermann & Wang 1995, for example). Though these are interesting effects we shall not explore them further and set $\rho_s = \rho_l$ so that $\nabla \cdot \boldsymbol{u} = 0$.

6.4 Response to an external flow

An interesting example of how the interstitial liquid can be driven by external pressure gradients is provided by a study of the morphological stability of the mush–liquid interface (the envelope of the solid matrix) in the presence of an external flow. To focus attention on the interfacial region, consider a semi-infinite mushy layer of uniform solid fraction ϕ in $z < 0$, adjacent to a liquid region ($z > 0$) flowing with uniform velocity U parallel to the interface. Consider the response to a small perturbation to the interface, $z = \zeta = \hat{\zeta} e^{i\alpha x + \sigma t}$ and, for simplicity, imagine that the flow in the liquid region is an irrotational flow $\boldsymbol{u} = \nabla \Phi$ with $\nabla^2 \Phi = 0$. The flow in the mushy region is much slower than that in the liquid, so to a leading approximation we can treat the mushy layer as impermeable to the external flow ($\boldsymbol{n} \cdot \nabla \Phi = 0$ on $z = \zeta$) and calculate that

$$\Phi = Ux - iU\zeta e^{-\alpha z} . \tag{6.11}$$

The linearized Bernoulli's equation then gives the pressure at the interface to be

$$p = -\rho\alpha U^2 \zeta . \tag{6.12}$$

There is low pressure above the crests and higher pressure above the troughs, and it is this pressure difference that drives a flow in the mushy layer (figure 11). In the absence of gravity and with ϕ constant, equation (6.8) combined with the equation of continuity, $\nabla \cdot \boldsymbol{u} = 0$, shows that the pressure in the mushy layer is harmonic, $\nabla^2 p = 0$, and equal to

$$p = -\rho\alpha U^2 \zeta e^{\alpha z} . \tag{6.13}$$

If we take the temperature field to be $T = Gz + \hat{\theta}(z)e^{i\alpha x + \sigma t}$, a perturbation from a linear gradient, the quasi-stationary thermal advection–diffusion equation gives

$$\kappa(D^2 - \alpha^2)\hat{\theta} = G\frac{\Pi}{v}\alpha^2 U^2 \hat{\zeta} e^{\alpha z} , \tag{6.14}$$

where, here, $\kappa = \bar{k}/\rho_l C_{pl}$ and C_{pl} is the specific heat of the liquid. This has solution

$$\hat{\theta} = \frac{G\Pi\alpha U^2}{2\kappa v}\hat{\zeta}z e^{\alpha z} - G\hat{\zeta}e^{\alpha z} , \tag{6.15}$$

since T is fixed at the interface by the liquidus constraint. Finally, the Stefan

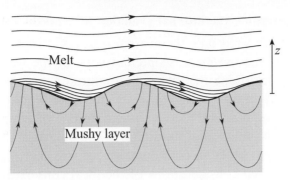

Figure 11. A schematic diagram showing the influence of an external flow on the evolution of a mushy layer. The streamlines in the melt are compressed over crests. The resulting low pressure there drives a flow in the mushy layer that interacts with the thermal field to drive a morphological instability of the interface. Note that in the liquid region away from the neighbourhood of the interface only every fourth streamline has been plotted for clarity.

condition (6.5) gives

$$\sigma = \frac{\overline{k}(\phi)}{\rho L \phi} \alpha G \left[-1 + \frac{\Pi U^2}{2\kappa \nu} \right]. \tag{6.16}$$

This simple analysis shows that there is instability at all wavelengths if the dimensionless group

$$\frac{\Pi U^2}{\kappa \nu} > 2, \tag{6.17}$$

i.e. if the external flow is sufficiently rapid. A more detailed analysis (Feltham & Worster 1999) shows that long wavelengths are stable given a mushy layer of finite depth, that the short wavelengths are stable when the external flow is a viscous shear rather than an inviscid uniform flow and that the most unstable wavelengths are comparable to the depth of the layer.

More generally, dimensional analysis shows that a high-Reynolds-number external flow has a significant influence on the evolution of the mushy layer if

$$\mathscr{U}^2 \gtrsim \sigma Da^{-1}, \tag{6.18}$$

where $\mathscr{U} = U/V$, V is the solidification rate and the Darcy number $Da = \Pi/L^2$ is the ratio of the permeability to the square of the macroscopic lengthscale $L = \kappa/V$, which is consistent with (6.17). By contrast, a low-Reynolds-number external flow can only exert a significant influence if

$$\mathscr{U} \gtrsim Da^{-1}, \tag{6.19}$$

which, since $Da \ll 1$, requires a much greater external velocity relative to the solidification rate.

6.5 Equations governing the internal evolution of a mushy layer

So far we have considered only the gross features of a mushy layer, which is sufficient to determine the macroscopic envelope of the solid matrix. However, in order to determine the internal structure of the layer, in particular how the solid fraction evolves in space and time, we require a set of differential equations that describe the local distributions of temperature $T(x,t)$, concentration of the interstitial fluid $C(x,t)$ and solid fraction $\phi(x,t)$. Such equations can be derived from considerations of free energy (Hills, Loper & Roberts 1983) or from considerations of the local conservation of heat and solute (Worster 1992a).

Conservation of heat is expressed by

$$\overline{\rho C_p}\left(\frac{\partial T}{\partial t} + \boldsymbol{u}\cdot\nabla T\right) = \nabla\cdot(\overline{k}\nabla T) + \rho_s L\frac{\partial \phi}{\partial t}, \qquad (6.20)$$

where the mean conductivity $\overline{k}$ is defined in equation (6.6),

$$\overline{\rho C_p} = \phi\rho_s C_{ps} + (1-\phi)\rho_l C_{pl} \qquad (6.21)$$

and C_{ps} and C_{pl} are the specific heat capacities of the solid and liquid respectively. Equation (6.20) is an advection–diffusion equation forced by the internal release of latent heat as solid grows within the mushy layer.

When the solid phase is pure (C_S is constant), conservation of solute is expressed by

$$(1-\phi)\frac{\partial C}{\partial t} + \boldsymbol{u}\cdot\nabla C = \nabla\cdot(\overline{D}\nabla C) + (C - C_S)\frac{\partial \phi}{\partial t}, \qquad (6.22)$$

where $\overline{D} \simeq (1-\phi)D$. This is an advection–diffusion equation for solute forced by internal release of solute into the interstices as the solid phase grows. Advection–diffusion equations in porous media are slightly modified from their counterparts in a pure phase owing to the fact that the solute is only transported through the liquid interstices (Phillips 1991).

The two equations (6.20) and (6.22) are coupled by the liquidus constraint (6.1), and this coupling is sufficient to determine the solid fraction ϕ.

6.6 The mush–liquid interface

Arguably the most difficult aspect of modelling mushy layers is to determine appropriate equations to describe the interface between a mushy region and a fully liquid region. Some approaches to the modelling of mushy regions (especially numerical modelling) utilize equations that blend smoothly from the mushy regions into the liquid regions, in which case there is no need for an explicit treatment of the interfaces (see the review by Beckermann & Wang 1995). However, there are some advantages in treating the regions separately, particularly that the equations in each region are then simpler and more amenable to analytic solution from which we can gain physical intuition. One is, however, then left with the problem of matching the solutions across the interface between the regions. The interface itself is hard to define. One can imagine it as the envelope (suitably smoothed) of the solid matrix. On the other hand, since the mushy layer is described by equations that govern properties averaged over the scale of interstitial pores, the interface is perhaps better thought of as a region of thickness comparable to the pore scale. Interfacial conditions are then expressed as jump conditions across the interfacial region. Once one appreciates that the interface physically has finite extent, it is apparent that even continuity of the dependent variables cannot be taken for granted but must either be deduced in some way or assumed and the consequences of those assumptions explored.

Two conditions follow immediately from the conservation equations (6.20) and (6.22) by integrating them across the interface and employing the divergence theorem. They are

$$\rho_s L \phi V_n = \bar{k} \boldsymbol{n} \cdot \nabla T|_m - k_l \boldsymbol{n} \cdot \nabla T|_l \tag{6.23}$$

and

$$(C - C_S)\phi V_n = (1 - \phi) D \boldsymbol{n} \cdot \nabla C|_m - D \boldsymbol{n} \cdot \nabla C|_l. \tag{6.24}$$

These express conservation of heat and solute across the interface and they reduce to the Stefan condition (4.5b) and the interfacial solute-conservation equation (4.3b) at a solid–liquid interface when $\phi = 1$. A third condition is required at mush–liquid interfaces that are solidifying (but not at those that are melting). A weak condition that suggests itself from considerations of the relief of supercooling (§6.1) is that

$$\boldsymbol{n} \cdot \nabla T|_l \geqslant \boldsymbol{n} \cdot \nabla T_L(C)|_l \tag{6.25}$$

so that the liquid is not supercooled adjacent to the interface.

On the other hand, the conservation relationships (6.23) and (6.24) can be combined to show, when $C_S = 0$, that

$$\boldsymbol{n} \cdot \nabla T_L(C)|_l - \boldsymbol{n} \cdot \nabla T|_l = \left(\frac{\rho_s L V_n}{k_l} + \frac{mC V_n}{D} - \frac{k_s}{k_l} \boldsymbol{n} \cdot \nabla T|_m \right) \phi$$
$$= \left(\frac{mC V_n}{D} - \boldsymbol{n} \cdot \nabla T|_l \right) \phi + \left(\boldsymbol{n} \cdot \nabla T|_m - \boldsymbol{n} \cdot \nabla T|_l \right)(1 - \phi). \tag{6.26}$$

The right-hand side is positive once mCV_n/D is greater than $O(\boldsymbol{n} \cdot \nabla T)$, i.e. when the morphological number is significantly greater than its critical value. Then (6.25) and (6.26) combine to show that the inequality (6.25) is in fact an equality and, further, that $\phi = 0$ at the interface.[1]

The three thermodynamic interfacial conditions that apply to a solidifying mush–liquid interface under sufficiently supercritical conditions are therefore

$$\phi = 0, \qquad \boldsymbol{n} \cdot \nabla T|_m = \boldsymbol{n} \cdot \nabla T|_l = \boldsymbol{n} \cdot \nabla T_L(C)|_l \, . \qquad (6.27a\text{--}c)$$

The fluid-mechanical conditions at a mush–liquid interface are those that apply at the boundary between a porous medium and a viscous fluid region. Only one condition is straightforward, namely that the normal mass flux is continuous, which follows directly from the continuity equation. Two other conditions are related to the transfer of stress between the two media. The difficulty is in knowing how much stress is accommodated by the solid phase of the porous medium. It is usually assumed, for example, that the solid phase absorbs all the deviatoric normal stress exerted by the fluid region, leaving the pressure field continuous between the media.

These two conditions, of normal mass flux and continuity of pressure, are sufficient to determine the flow in the mushy layer, governed by Darcy's equation. The flow in the viscous liquid region requires an additional boundary condition, determined by consideration of the tangential stresses. Where the external liquid makes contact with the solid phase of the mushy layer, its velocity is zero. Where it makes contact with the liquid phase, its velocity is continuous with the *interstitial* velocity. The adjustment from these mixed boundary conditions to a region where the external flow no longer feels the effects of individual crystals takes place on the scale of the pores of the underlying porous medium. Considerations such as these suggest that an appropriate boundary condition on the external flow is

$$\boldsymbol{t} \cdot \boldsymbol{u}_l - \boldsymbol{t} \cdot \boldsymbol{u}_m = \lambda \sqrt{\Pi} \mathscr{S} \, , \qquad (6.28)$$

where $\mathscr{S} = (\boldsymbol{n} \cdot \nabla)(\boldsymbol{t} \cdot \boldsymbol{u}_l)$ is the local shear rate, $\boldsymbol{t}$ is a unit tangent vector and λ is a constant parameter (Beavers & Joseph 1967). Note that for many external flows u_l is much larger than both u_m and $\sqrt{\Pi}\mathscr{S}$ so that the boundary condition (6.28) can often be approximated by the more familiar no-slip condition.

---

[1] Imposing $\phi = 0$ as a boundary condition *a priori* implies (6.27b, c) given the conservation relations (6.23) and (6.24) and the condition of local equilibrium in the interior of the mushy layer (6.1). However, it can lead to unphysical solutions (with $\phi < 0$ in the interior of the mushy layer) under weakly supercritical conditions when morphological instability precedes the occurrence of constitutional supercooling (Worster 1986).

$$\theta \to \theta_\infty, \qquad \Theta \to 0, \qquad u \to (\mathscr{U}, 0) \qquad (z \to \infty)$$

$$\frac{\partial \theta}{\partial t} + u \cdot \nabla\theta = \nabla^2\theta$$

$$\frac{\partial \Theta}{\partial t} + u \cdot \nabla\Theta = \epsilon\nabla^2\Theta \qquad\qquad \text{Liquid}$$

$$\frac{1}{Pr}\left(\frac{\partial u}{\partial t} + u \cdot \nabla u\right) = \nabla^2 u + R_m Da^{-1}\left(\Theta\hat{k} - R_\rho\theta\hat{k} - \nabla p\right)$$

$$[u] = \lambda\, Da\, \mathscr{S}t$$

Interface

$$[p] = [\theta] = [\Theta] = [n \cdot \nabla\theta] = [n \cdot \nabla\Theta] = 0$$

$$\phi = 0$$

Mushy layer

$$\frac{\partial\theta}{\partial t} + u \cdot \nabla\theta = \nabla^2\theta + S\frac{\partial\phi}{\partial t}$$

$$(1-\phi)\frac{\partial\Theta}{\partial t} + u \cdot \nabla\Theta = \epsilon\nabla \cdot [(1-\phi)\nabla\Theta] + (\Theta - \mathscr{C})\frac{\partial\phi}{\partial t}$$

$$u = -R_m\left(\Theta\hat{k} - R_\rho\theta\hat{k} - \nabla p\right)$$

$$\theta = \Theta$$

$$\theta = -1, \qquad\qquad n \cdot u = 0 \qquad\qquad (z = 0)$$

Figure 12. The equations and boundary conditions governing flow and solidification in a mushy layer growing from a binary melt. The boundary $z = 0$ is either the cooled boundary of a mould of fixed temperature T_B or the eutectic front with $T_B = T_E$. Lengths are scaled with κ/V, times with V^2/κ and velocities with V, where V is the rate of solidification. The dimensionless dependent variables are the velocity u, pressure p, temperature $\theta = (T - T_L(C_0))/(T_L(C_0) - T_B)$ and concentration $\Theta = (C - C_0)/(C_0 - C_B)$. The importance of each of the dimensionless parameters $S, \mathscr{C}, \theta_\infty, \mathscr{U}, \epsilon, \mathscr{A}, R_m, R_\rho$ and Da is discussed in the text.

7 Solidification and convection in mushy layers

The equations and boundary conditions discussed in the previous section are summarized in dimensionless form for a solidifying system in figure 12, with the dimensionless variables defined in the caption. The equations have been simplified by taking the thermal properties of the liquid and solid phases to be equal. There are, nevertheless, a large number of dimensionless parameters

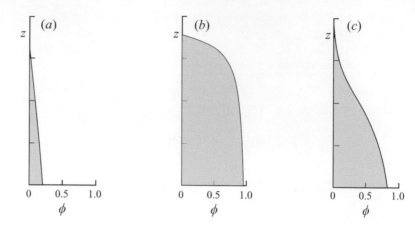

Figure 13. Solid-fraction profiles in a mushy layer when there is no convection: (a) $\mathscr{C} = 4, \theta_\infty = 1$; (b) $\mathscr{C} = 0.05, \theta_\infty = 1$; (c) $\mathscr{C} = 0.2, \theta_\infty = 0.02$. In all cases $S = 1$ and $\epsilon = 0$.

controlling the dynamical behaviour of the system. In this section we shall simply describe the importance of the different parameters and some of the phenomena that have been calculated using these equations.

Three parameters govern the structure of the mushy layer in the absence of flow. As the thickness of the mushy layer is controlled by thermal balances, it decreases as the far-field temperature θ_∞ increases and as the Stefan number S increases, since these increase the heat flux from the melt and the latent heat release respectively. The release of latent heat is further modified by the concentration ratio $\mathscr{C}$ since this, as we have seen, controls the mean solid fraction. When $\mathscr{C}$ is large, the solid fraction is inversely proportional to $\mathscr{C}$ and the total latent heat release is therefore proportional to $S/\mathscr{C}$. This can be seen from (6.4) to (6.7) which show that the thickness of a mushy layer $h \propto \sqrt{(\mathscr{C}/S)\kappa t}$ in contrast to a pure solid whose thickness $h \propto \sqrt{(1/S)\kappa t}$ (see 2.11). The internal distribution of solid fraction is controlled predominantly by $\mathscr{C}$, as illustrated in figure 13, so this is the most significant parameter determining the variation of the permeability of the layer to fluid flow.

The diffusivity ratio ϵ plays a minor role when there is no flow and can be set to zero in such cases. However, while this is a regular perturbation in the mushy layer, it is a singular perturbation of the equations in the liquid region. There is a compositional boundary layer in the liquid adjacent to the mush–liquid interface of width $\delta_c \sim \epsilon$ in the case of constant solidification rate or of width $\delta_c \sim \epsilon^{1/2}$ in the case of solidification from a cooled boundary with growth proportional to $t^{1/2}$. This boundary layer can often be resolved experimentally and can itself become unstable in a gravitational field to produce compositional convection.

We have seen how flow in the mushy layer can be generated by an imposed flow $\mathcal{U}$. However, a more common mechanism for fluid flow during solidification is natural, buoyancy-driven convection. Within the mushy layer, this is governed principally by the Rayleigh number

$$R_m = \frac{\beta \Delta C g \Pi_0 h}{\kappa \nu} \tag{7.1}$$

which is a porous-medium Rayleigh number for compositional convection, where $\beta = \rho^{-1} \partial \rho / \partial C$, $\Delta C = C_0 - C_B$ is the change in interstitial concentration across the layer, and Π_0 is a representative value of the permeability. We see from figure 12 that, since the dimensionless temperature and concentration are equal ($\theta = \Theta$) in the mushy layer, there is effectively only single-component convection there with an effective Rayleigh number $R_m(1 - R_\rho)$ where

$$R_\rho = \frac{\alpha \Delta T}{\beta \Delta C} = \frac{\alpha m}{\beta} \tag{7.2}$$

is the buoyancy ratio in which $\Delta T = T_L(C_0) - T_B$ and α is the thermal expansion coefficient. Note that, typically, $R_\rho < 1$ so that convection in the mushy layer is dictated by the solute field.

In the liquid region, the temperature and solute fields are uncoupled and convection is better described by the independent Rayleigh numbers

$$R_C \equiv R_m Da^{-1} \equiv \frac{\beta \Delta C g h^3}{\kappa \nu} \tag{7.3}$$

for solutal convection and

$$R_T \equiv -R_m Da^{-1} R_\rho \equiv -\frac{\alpha \Delta T g h^3}{\kappa \nu} \tag{7.4}$$

for thermal convection, as can be seen from the equations in figure 12. Many interesting double-diffusive effects (Turner 1979) can occur in the melt during solidification (Huppert 1990) owing to the independence of R_T and R_C.

7.1 Convective regimes

The possible states of natural convection in a solidifying alloy are summarized in figure 14 for the cases where solidification is effected by cooling a mould either from an upper or a lower horizontal boundary. Other modes of convection that occur when there is cooling through sidewalls are not discussed here (but see Huppert 1990). Note that the equations in figure 12

	Cooled from below $R_T < 0$		Cooled from above $R_T > 0$	
$C_0 < C_E$ Heavy fluid released	3.	$R_C < 0, R_m < 0$ Mushy layer No convection	5.	$R_C > 0, R_m > 0$ Mushy layer Compositional convection in liquid and mushy regions Themal convection in liquid
$C_0 = 0, C_E, 100$ Neutrally buoyant residual	1.	$R_C = 0$ No mushy layer No convection	2.	$R_C = 0$ No mushy layer Thermal convection
$C_0 > C_E$ Light fluid released	6.	$R_C > 0, R_m > 0$ Mushy layer Compositional convection in mushy region Double-diffusive convection in liquid	4.	$R_C < 0, R_m < 0$ Mushy layer Themal convection in liquid

Figure 14. The different convective regimes that can occur during solidification of a binary alloy at a cooled, horizontal boundary. The numbered cases 1–6 are discussed in turn in the text.

and the definitions of Rayleigh numbers therein are written for the case where gravity acts towards the cooled boundary, i.e. for when cooling is through the lower boundary. Cases in which the upper boundary of the mould is cooled are described by changing the sign of the Rayleigh number. The various boxes in figure 14 are now described in turn.

Case 1. If the melt is pure ($C_0 = 0$ or $C_0 = 100$) or if it has exactly the eutectic concentration ($C_0 = C_E$) then the solid formed has the same composition as the melt, there is no rejected solute and the solute field plays no role in the solidification or convection. If, in addition, the mould is cooled from below then the melt is statically stable to thermal convection.[1] There is therefore no convection, and solidification proceeds at a planar interface as described in § 2.2.

Case 2. In this case there is again no solute rejection, no mushy layer can form (provided the melt is not initially supercooled) and there is no compositional convection. Since the mould is cooled from above, the melt is unstable to thermal convection and its solidification is described in § 3.2.

Case 3. If the melt is neither pure nor of eutectic composition then solute is rejected during solidification and, in the general case, a mushy layer will form. When the melt has an initial composition less than eutectic (where C measures the component that causes the density of the melt to increase) then the residual melt is denser than the initial melt. The liquid is therefore stable

---

[1] Note that in the special case of pure water there is density maximum at 4°C so there is thermal convection when ice is formed by freezing pure water from below.

to compositional convection when the mould is cooled from below. In this case the thermal field is also stable so there is no convection and the mushy layer grows as described in §6.2.

Case 4. The solute field is also stable in the case that the melt has an initial composition that is greater than eutectic (so that the residual melt is buoyant) and the mould is cooled from above. The cooling drives convection in the liquid region only (the mushy layer remains stagnant) and solidification proceeds similarly to that in case 2 but with the latent heat release now distributed throughout the mushy layer (Kerr *et al.* 1990*a*). There are some very interesting additional effects in this case associated with supercooling at the mush–liquid interface, which will be described in the context of lava lakes in the next section.

Case 5. When the mould is cooled from above and the residual melt is denser than the initial melt then thermal and compositional convection act in concert. Thermal convection in the liquid region is augmented slightly by the solutal buoyancy from the compositional boundary layer near the mush–liquid interface but otherwise proceeds similarly to that in cases 2 and 4. The new effect here is that the interstitial liquid in the mushy layer is now also statically unstable. Convection can be driven from the interior of the mushy layer leading to interesting modifications of its microstructure, as described below in general and also in the context of the formation of sea ice.

Case 6. Directional casting of high-performance turbine blades (for example) is executed by withdrawing a mould vertically downwards from a hot furnace, so that the mould is cooled from below and the thermal field is convectively stable. If the residual melt is less dense, however, then compositional convection can occur in both the liquid and mushy regions. The convection in the mushy layer is similar to that in case 5 with the effective Rayleigh number simply being $R_m(1-R_\rho)$ rather than $R_m(1+R_\rho)$. The stable thermal field in the liquid region can, however, modify the compositional convection there and cause the formation of double-diffusive 'fingers' (Turner 1979; Chen & Chen 1991; and see Chapter 6).

7.2 Convection within a mushy layer

The last two cases described above both involve convection of the interstitial liquid within the mushy layer (reviewed by Worster 1997). To analyse such convection in detail it is necessary to solve the full set of equations displayed in figure 12. However, we can make a preliminary estimate of the conditions under which internal convection will occur by analysing the following much-reduced model.

A major simplification is obtained by arbitrarily ignoring the fact that the mush–liquid interface is a free boundary and fixing it at the dimensionless position $z = 1$. The remaining simplifications follow from asymptotic limits of the governing equations. In particular we take $\epsilon \to 0$ and $Da \to 0$.

If $\mathscr{C} \gg 1$ then $\phi \ll 1$ and the solute-conservation equation is approximately

$$\frac{\partial \theta}{\partial t} + \boldsymbol{u} \cdot \nabla \theta = -\mathscr{C} \frac{\partial \phi}{\partial t} \tag{7.5}$$

whence the heat-conservation equation becomes

$$\Omega \left(\frac{\partial \theta}{\partial t} + \boldsymbol{u} \cdot \nabla \theta \right) = \nabla^2 \theta, \tag{7.6}$$

where $\Omega = 1 + S/\mathscr{C}$ and we have taken $S \gg 1$ with $S/\mathscr{C} = O(1)$. The heat equation is thus decoupled from the solute equation, as derived by Emms & Fowler (1994). Since $\phi \ll 1$ throughout the mushy layer, the permeability is approximately constant. If we also assume that $R_\rho \ll 1$ then Darcy's equation becomes

$$\boldsymbol{u} = -R_m(\nabla p + \theta \boldsymbol{k}), \tag{7.7}$$

where $\boldsymbol{k}$ is a unit vector in the z-direction. The relevant boundary conditions given the limits stated above are

$$\theta = -1, \quad \boldsymbol{n} \cdot \boldsymbol{u} = 0 \quad (z = 0), \qquad \theta = p = 0 \quad (z = 1). \tag{7.8}$$

These equations are mathematically identical to the equations for convection in a passive porous medium. By writing $\Omega \boldsymbol{u} = \boldsymbol{v}$ and $t = \Omega \tau$, so that

$$\frac{\partial \theta}{\partial \tau} + \boldsymbol{u} \cdot \nabla \theta = \nabla^2 \theta \quad \text{and} \quad \boldsymbol{v} = -(R_M \Omega)(\nabla p + \theta \boldsymbol{k}), \tag{7.9}$$

we see that the Rayleigh number is simply modified by the factor Ω. The stability of this system was analysed by Lapwood (1948) who showed that convection begins once

$$\Omega R_m > \Omega R_{\text{crit}} \simeq 27.1. \tag{7.10}$$

More generally, the critical Rayleigh number for linear convective instability $R_{\text{crit}} = R_{\text{crit}}(S, \mathscr{C}, \theta_\infty, \epsilon, R_\rho, Da)$. Some of this parameter space has been explored (Worster 1992b; Chen, Lu & Yang 1994; Emms & Fowler 1994) and it has been shown that, over a large range of parameters, provided one rescales the Rayleigh number in terms of the mean permeability of the layer and its undisturbed depth, the critical Rayleigh number R_{crit} is, within a factor of 2, equal to about 10. This much-reduced model embodies the interesting feature that systems with large Stefan numbers (or small $\mathscr{C}$) are

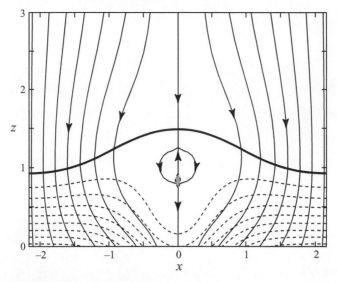

Figure 15. The streamlines (thin solid lines with arrows) and contours of solid fraction (dashed lines) in a steadily solidifying and convecting mushy layer below a liquid melt. Streamlines are shown in a frame of reference moving with the phase boundaries. The mush–liquid interface (thick solid line) is deformed upwards where the flow towards the mushy layer (relative to the interface) is weaker. Below the upwardly deflected interface the solid fraction is reduced and has become negative in the small shaded region near the bottom of the recirculating region interior to the mushy layer. It is anticipated that the shaded region, which must be liquid, develops into a chimney (figure 16) as the amplitude of convection increases.

more prone to convection. The reason for this highlights the fact that the mushy layer is a *reactive* porous medium. Although the driving buoyancy results from the solute field, the dissipation of that buoyancy is mediated by the thermal field as follows. When a parcel of interstitial fluid rises, it approaches thermal equilibrium with its surroundings by diffusion of heat but does not similarly exchange solute with its new surroundings. Rather, since an approach to thermal equilibrium without any phase change would leave the interstitial liquid undersaturated (above the liquidus), equilibrium is restored by dissolution of the solid matrix. This causes the interstitial liquid to become more dense (in case 6), and thus dissipates the buoyancy. Systems with large Stefan number can dissolve less of the solid matrix for a given thermal perturbation and are therefore more unstable.

The dissolution just described is manifest in equation (7.5), which shows that the rate of solidification ($\partial \phi / \partial t$) is reduced where the flow has a component parallel to the local temperature gradient, i.e. where the interstitial liquid flows from cooler to warmer regions of the mushy layer. This corresponds to downflow in case 5 and upflow in case 6 above.

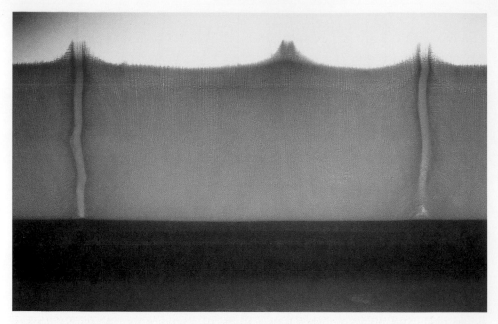

Figure 16. Photograph of a mushy layer of ammonium-chloride crystals grown from aqueous solution showing sections through two complete chimneys and the vent of a third. The black region at the bottom of the photograph is eutectic solid and shows that the interface of this region with the mushy layer is quite planar. The mush–liquid interface, by contrast, is deflected upwards at the sites of chimneys to form a volcano-like vent. A plume of depleted fluid issues from each chimney into the overlying melt. This picture turned upside down is similar to what is seen in sea ice, in which case dense brine drains from the chimneys, which are known as brine channels.

Such dissolution (or reduced solidification) increases with the amplitude of the convective motion until the solid fraction becomes equal to zero at some point in the mushy layer. Recent numerical calculations (Schulze & Worster 1999) have shown that this can occur in the interior of the layer, as shown in figure 15. At even larger convective amplitudes the region of zero solid fraction can form a narrow cylindrical conduit, or *chimney* through which most of the outflow from the mushy layer is channelled (see figure 16). Convection through fully developed chimneys, particularly the theoretical determination of the location of the chimney wall, remains a topic of current research.

7.3 The early evolution of sea ice

Many of the fundamental ideas presented in this chapter are illustrated in two case studies with which I shall conclude. The first is a study of the growth of sea ice on the surface of the polar oceans. The Arctic ocean is covered with a layer of ice with a mean thickness of about 3 m. Owing to wind stresses, the ice is in a very dynamic state, being continually fractured to expose sea water to the atmosphere. The exposed water in these cracks, or *leads*, quickly freezes over and the ice begins to grow downwards, reaching a thickness of about 1 m in the first winter season if left undeformed. It is in the first 24 hours or so of growth, while the ice is less than about 10 cm thick, that the most significant heat transfer takes place between the ocean and the atmosphere, and these episodes of lead formation and early ice growth account for about half the total heat budget. It is therefore very important to understand the processes controlling the dynamics of this early growth.

Since sea water is a mixture of water and dissolved salts (an alloy in the terminology adopted here), sea ice is a mushy layer comprising a matrix of ice crystals and interstitial brine. Because brine is denser than fresh water, growing sea ice is an example of case 5 from § 7.1. We can therefore anticipate that, under appropriate conditions, the interstitial brine will convect out of the layer of sea ice and contribute to the convective state of the underlying ocean.

A laboratory study of sea ice (Wettlaufer, Worster & Huppert 1997, illustrated and described in figure 17) has shown that there is no measurable brine flux from the mushy layer until it has exceeded a critical thickness h_c that depends upon the applied surface temperature (atmospheric temperature) and initial concentration of the salt solution. Before the critical thickness has been reached, the mushy layer is stagnant and its growth can be described approximately by the model presented in § 6.2, modified slightly by the heat flux from the ocean, as described in § 3.2. An important parameter influencing the remote sensing of sea ice by satellite, using radar, is its solid fraction, which can be readily calculated from the equations presented in section § 6.5, as shown in figure 13.

The critical thickness h_c can in principal be determined from the stability analysis described in the previous section. However, a problem with this is that the permeability of the mushy layer is difficult to measure directly. An alternative approach is to notice that equation (7.10) implies that

$$[(C_0 - C_B)h_c]^{-1} \propto \Pi(\phi) \qquad (7.11)$$

or that $h_c \Delta C$ is a function only of the solid fraction and material parameters. A plot of the data from many experiments starting with different initial concentrations and having different surface temperatures (figure 18) shows

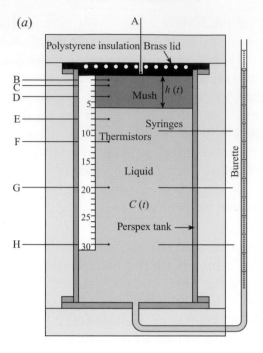

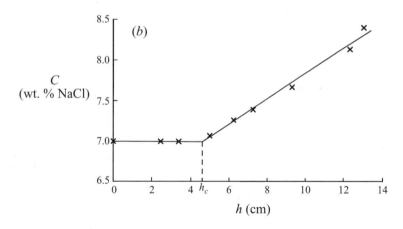

Figure 17. (*a*) Experimental apparatus used by Wettlaufer *et al.* (1997) to study the early growth of sea ice. A mushy layer of ice crystals was grown from aqueous solutions of NaCl by cooling the upper boundary of the apparatus. Temperatures were recorded by thermistors A–H and samples of liquid were withdrawn periodically using syringes in order to measure the evolving concentration $C(t)$ of the liquid region. These data are plotted against the thickness of the mushy layer $h(t)$ in (*b*). Initially all the brine rejected by the growing ice crystals remains within the interstices of the mushy layer but it subsequently drains once the thickness of the layer has exceeded a critical value h_c.

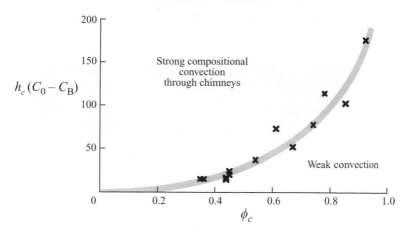

Figure 18. The critical conditions for brine drainage from sea ice.

that equation (7.11) gives a good collapse of all the data onto a single curve. This curve gives reasonable values for the permeability as a function of the solid fraction (increasing rapidly as the solid fraction approaches zero). In particular, if R_{crit} is taken to be equal to 10 then the permeability of typical young sea ice (having a solid fraction between 0.6 and 0.8) is predicted to be about 10^{-10} m^2, which is reasonable given the scale of observed microstructures (platelets approximately 1 mm thick separated by about 0.1 mm).

Once the critical thickness is exceeded, brine begins to drain from the sea ice and does so via narrow *brine channels*, which are the chimneys described in the previous section. Brine channels are home to a host of organisms which feed on nutrients from the sea water that flushes through them. Theoretical and experimental studies of convection in mushy layers suggest that the flow permeates the sea ice surrounding brine channels, originating from the underlying ocean and draining through the channels. During its passage through the porous sea ice, continued solidification of ice enriches the remaining liquid both in salt and in nutrients. Brine channels are therefore rich harvesting grounds in addition, no doubt, to being relatively safe havens from predators.

The brine draining from sea ice raises the density of the oceanic mixed layer beneath it and therefore contributes to its deepening during the winter months. The seasonal dynamics of the mixed layer is a significant factor affecting climate, so the small-scale processes occurring within the mushy layer that is sea ice may have important global consequences.

Figure 19. Photograph of a lava lake in Hawaii.

7.4 Mineral segregation in lava lakes

Nature provides many examples of large-scale casting processes. The largest is the growth of the inner core of the Earth (see Chapter 9), while numerous smaller examples are given by the solidification of magmas and lavas to form igneous rocks. One relatively uncomplicated example is the solidification of lava that has ponded within the crater of a volcano to form a *lava lake* (figure 19). The lake is cooled predominantly by contact with the atmosphere above it. We can ignore the relatively small heat transfer to the rock beneath and to the sides of it, which complicate considerations of solidification in intrusive magmas (dykes and sills). The lava lake therefore fits well the ideal case 4 of §7.1, the first minerals to solidify typically being the densest.

Examination of solidified lava lakes shows that, in addition to crystals growing adjacent to the cooled top to form a crust, a significant proportion of the crystallization occurs in its interior. Similar behaviour occurs in metallic castings, in which there is also a textural division between *columnar* crystals formed adjacent to the boundaries of a mould, being elongated and aligned normal to the boundaries, and *equiaxed* crystals that grow in the interior. The marginal crystals in a lava lake or magma chamber and the

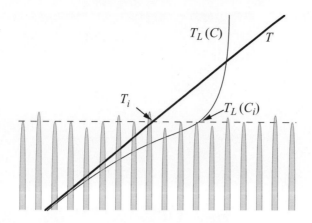

Figure 20. Schematic diagram of the interfacial region of a mushy layer showing the mean temperature T and mean liquidus temperature $T_L(C)$. The horizontal dashed line shows the position of maximum undercooling $T_L(C) - T$. Dendrites further into the liquid than this grow slower than the mean and are caught up by the others. Those that recede from this position grow slower and cease to contribute to the interfacial conditions. This reasoning suggests that the mush–liquid interface is located where $n \cdot \nabla[T_l(C) - T] = 0$, which is mathematically identical to the condition of marginal equilibrium. In systems with slow kinetics the interfacial undercooling is significant and, coupled with convection of the melt, can have large-scale effects in a casting.

columnar crystals in a metallic casting form the matrices of mushy layers during solidification.

There is a problem in describing how the interior crystals can grow given the theory of solidification presented so far. The assumptions of internal equilibrium (6.1) and, more importantly, of marginal equilibrium (6.25) ensure that the liquid region can never be cooled below its liquidus. Crystals cannot grow in suspension in such a liquid and solidification only occurs within the mushy layer.

However, as discussed in §2.3, the surface of a growing crystal is at a temperature below the freezing temperature of the melt. If the local normal growth of each crystal within a mushy layer obeys an equation of the form of (2.15) then the solid fraction must evolve according to an equation of the form

$$\frac{\partial \phi}{\partial t} = \mathscr{A}\mathscr{G}[T_L(C) - T], \qquad (7.12)$$

where $\mathscr{A}$ is the specific surface area of the internal phase boundaries. Morphological instabilities within the mushy layer serve to increase $\mathscr{A}$, and it is clear from (7.12) that if $\mathscr{A}\mathscr{G}$ is large then $T \simeq T_L(C)$, which is

Figure 21. Photograph of an experiment (Kerr *et al.* 1990*b*) in which an aqueous solution of sodium sulphate was solidified by cooling it at the top boundary of a rectangular mould. At the top of the mould is a white region of eutectic solid above a mushy layer of hydrated sodium-sulphate crystals $Na_2SO_4.10H_2O$. Supercooling of the solution, as explained in the text, allows further crystallization of equiaxed crystals at the floor of the mould.

the condition of local, internal equilibrium (6.1). However, near the interface with the liquid region the morphological instabilities on the primary dendrites leading to side branches are not fully developed and one can anticipate that higher degrees of supercooling prevail in the interfacial region, as sketched in figure 20. It is clear, by differentiating the right-hand side of (7.12) that the maximum rate at which the mush–liquid interface can advance occurs when the tips of the primary dendrites sit where the undercooling is largest,

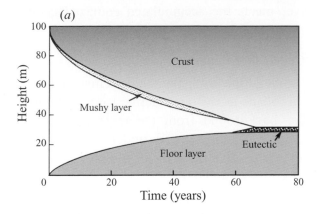
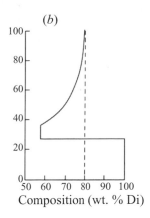

Figure 22. (*a*) The predicted evolution of a hypothetical lava lake composed of Diopside (Di) and Anorthite (An) (Worster *et al.* 1993). A eutectic crust and underlying mushy layer grow downwards from the upper surface of the lake, while further crystals grow in the interior and settle to the floor. The composition of the melt evolves as it is depleted of the minerals forming the interior crystals. (*b*) The predicted distribution of minerals in the lava lake once it is completely solidified.

i.e. where

$$\boldsymbol{n} \cdot \nabla\left[T_L(C) - T\right] = 0. \qquad (7.13)$$

This provides further justification of the condition (6.27*c*) derived previously from considerations of marginal equilibrium and applies regardless of the degree of interfacial supercooling. What is apparent in figure 20 is that $T_i \neq T_L(C_i)$. In consequence, the solid fraction is non-zero at the interface.

All this has negligible influence on the growth of the mushy layer. However, if the liquid region is flowing then supercooled liquid can be swept from the neighbourhood of the interface into the bulk of the region. The whole liquid region can thereby become supercooled, allowing additional solidification of crystals from any suitable nucleation sites within that region (figure 21, Kerr *et al.* 1990*b*). In the case of lava lakes, there are numerous small crystals (phenocrysts) that can act as nucleation sites for such secondary crystallization. In metal castings it is common to add small particles to the melt to promote such growth in cases where this is desired.

There is a further twist to the tale in that as the suspended crystals grow and settle they leave the melt depleted of the high-melting point minerals that form them. The composition of the melt thus evolves and its time history is frozen into the mushy layer as it grows downwards from the surface of the lake. The bulk composition of the frozen lake, though the lava was initially uniform, has a continuous stratification in its upper regions and changes abruptly where the mushy layer meets the accumulated equiaxed crystals (figure 22).

8 Concluding remarks

In this chapter we have visited some of the fundamental ideas involved in the solidification of fluids. Many problems in this area derive from metallurgy, and students can gain a fuller background from the texts by Flemings (1974) and Kurz & Fisher (1986). To gain a better understanding of the mathematical techniques, useful references are Alexiades & Solomon (1993), Carslaw & Jaeger (1959), Crank (1984) and Hill (1987). There is a broad range of excellent reviews covering many of the modern developments in the subject in the *Handbook of Crystal Growth* edited by Hurle (1993).

An understanding of the solidification of fluid melts is becoming increasingly important as we try to develop better ways to process more exotic materials. It is similarly important in many geophysical contexts. There are many unsolved problems related to solidification from atomic to planetary scales and the fluid dynamist can make significant contributions in predicting behaviour and developing controlling strategies for the future.

Acknowledgements

I am grateful to Herbert Huppert and Stephen Davis, who have inspired different aspects of my interest in the solidification of fluids and to George Batchelor for encouraging me in its pursuit. My research in this area has been additionally stimulated by a number of co-workers including Dan Anderson, Danny Feltham, Ross Kerr, Tim Schulze and John Wettlaufer. I am especially grateful to them, and to Javier Jimenez, Keith Moffatt, Alan Rempel and Bill Schultz for their critical reading of earlier versions of this chapter.

References

ALEXIADES, V. & SOLOMON, A. D. 1993 *Mathematical Modeling of Melting and Freezing Processes.* Hemisphere.

BATCHELOR, G. K. 1967 *An Introduction to Fluid Dynamics.* Cambridge University Press.

BATCHELOR, G. K. 1974 Transport properties of two-phase materials with random structure. *Ann. Rev. Fluid Mech.* **6**, 227–255.

BEAVERS, G. S. & JOSEPH, D. D. 1967 Boundary conditions at a naturally permeable wall. *J. Fluid Mech.* **30**, 197–207.

BECKERMANN, C. & WANG, C. Y. 1995 Transport phenomena in alloy solidification. In *Annual Review of Heat Transfer VI* (ed. C. L. Tien), pp. 115–198, Begell House.

BRATTKUS, K. & DAVIS, S. H. 1988 Flow-induced morphological instabilities: stagnation-point flows. *J. Cryst. Growth* **89**, 423–427.

BÜHLER, L. & DAVIS, S. H. 1998 Flow induced changes of the morphological stability in directional solidification: localized morphologies. *J. Cryst. Growth* **186**, 629–647.

CARSLAW, H. S. & JAEGER, J. C. 1959 *Conduction of Heat in Solids*. Cambridge University Press.

CHEN, F. & CHEN, C. F. 1991 Experimental study of directional solidification of aqueous ammonium chloride solution. *J. Fluid Mech.* **227**, 567–586.

CHEN, F., LU, J. W. & YANG, T. L. 1994 Convective instability in ammonium chloride solution directionally solidified from below. *J. Fluid Mech.* **276**, 163–187.

CRANK, J. 1984 *Free- and Moving-Boundary Problems*. Clarendon Press.

DAVIS, S. H. 1990 Hydrodynamic interactions in directional solidification. *J. Fluid Mech.* **212**, 241–262.

EMMS, P. W. & FOWLER, A. C. 1994 Compositional convection in the solidification of binary alloys. *J. Fluid Mech.* **262**, 111–139.

FELTHAM, D. L. & WORSTER, M. G. 1999 Flow induced morphological instability in a mushy layer. *J. Fluid Mech.* **391**, 337–357.

FLEMINGS, M. C. 1974 *Solidification Processing*. McGraw-Hill.

FORTH, S. A. & WHEELER, A. A. 1989 Hydrodynamic and morphological stability of the unidirectional solidification of a freezing binary alloy: a simple model. *J. Fluid Mech.* **202**, 339–366.

GLICKSMAN, M. E., CORIELL, S. R. & McFADDEN, G. B. 1986 Interaction of flows with the crystal–melt interface. *Ann. Rev. Fluid Mech.* **18**, 307–335.

GLICKSMAN, M. E. & MARSH, S. P. 1993 The Dendrite. In *Handbook of Crystal Growth 1, Part B: Transport and Stability* (ed. D. T. J. Hurle). North Holland.

HILL, J. M. 1987 *One-dimensional Stefan Problems: An Introduction*. Longman.

HILLS, R. N., LOPER, D. E. & ROBERTS, P. H. 1983 A thermodynamically consistent model of a mushy zone. *Q. J. Mech. Appl. Math.* **36**, 505–539.

HUPPERT, H. E. 1990 The fluid mechanics of solidification. *J. Fluid Mech.* **212**, 209–240.

HUPPERT, H. E. & WORSTER, M. G. 1985 Dynamic solidification of a binary melt. *Nature* **314**, 703–707.

HUPPERT, H. E. & WORSTER, M. G. 1991 Vigorous motions in magma chambers and lava lakes. In *Chaotic Processes in the Geological Sciences* (ed. D. A. Yuen). Springer.

HURLE, D. T. J. (Ed.) 1993 *Handbook of Crystal Growth. Volume 1: Fundamentals, A: Thermodynamics and Kinetics, B: Transport and Stability*. North Holland.

KERR, R. C., WOODS, A. W., WORSTER, M. G. & HUPPERT, H. E. 1990a Solidification of an alloy cooled from above. Part 1. Equilibrium growth. *J. Fluid Mech.* **216**, 323–342.

KERR, R. C., WOODS, A. W., WORSTER, M. G. & HUPPERT, H. E. 1990b Solidification of an alloy cooled from above. Part 3. Compositional stratification within the solid. *J. Fluid Mech.* **218**, 337–354.

KIRKPATRICK, R. J. 1975 Crystal growth from the melt: a review. *Am. Mineral.* **60**, 798–814.

KURZ, W. & FISHER, D. J. 1986 *Fundamentals of Solidification*. Trans. Tech. Publications.

LANGER, J. S. 1980 Instabilities and pattern formation in crystal growth. *Rev. Mod. Phys.* **52**, 1–28.

LAPWOOD, E. R. 1948 Convection of a fluid in a porous medium. *Proc. Camb. Phil. Soc.* **44**, 508–521.

MULLINS, W. W. & SEKERKA, R. F. 1964 Stability of a planar interface during solidification of a dilute binary alloy. *J. Appl. Phys.* **35**, 444–451.

NIELD, D. A. & BEJAN, A. 1999 *Convection in Porous Media*. Springer.

PHILLIPS, O. M. 1991 *Flow and Reactions in Permeable Rocks*. Cambridge University Press.

SCHULZE, T. P. & DAVIS, S. H. 1995 Shear stabilization of morphological instability during directional solidification. *J. Cryst. Growth* **149**, 253–265.

SCHULZE, T. P. & WORSTER, M. G. 1999 Weak convection, liquid inclusions and the formation of chimneys in mushy layers. *J. Fluid Mech.* **388**, 197–215.

TURNER, J. S. 1979 *Buoyancy Effects in Fluids*. Cambridge University Press.

UMANTSEV, A. R. 1985 Motion of a plane front during crystallization. *Sov. Phys. Crystallogr.* **30**(1), 87–91.

WETTLAUFER, J. S., WORSTER, M. G. & HUPPERT, H. E. 1997 Natural convection during solidification of an alloy from above with application to the evolution of sea ice. *J. Fluid Mech.* **344**, 291–316.

WORSTER, M. G. 1986 Solidification of an alloy from a cooled boundary. *J. Fluid Mech.* **167**, 481–501.

WORSTER, M. G. 1992*a* The dynamics of mushy layers. In *Interactive Dynamics of Convection of Solidification*, (ed. S. H. Davis, H. E. Huppert, W. Müller & M. G. Worster), NATO ASI ser. E219, pp. 113–138. Kluwer.

WORSTER, M. G. 1992*b* Instabilities of the liquid and mushy regions during solidification of alloys. *J. Fluid Mech.* **237**, 649–669.

WORSTER, M. G. 1997 Convection in mushy layers. *Ann. Rev. Fluid Mech.* **29**, 91–122.

WORSTER, M. G., HUPPERT, H. E. & SPARKS, R. S. J. 1993 The crystallization of lava lakes. *J. Geophys. Res.* **98** (B9), 15891–15901.

XU, J.-J. 1994 Dendritic growth from a melt in an external flow: uniformly valid asymptotic solution for the steady state. *J. Fluid Mech.* **263**, 227–243.

Institute of Theoretical Geophysics, Department of Applied Mathematics and Theoretical Physics, Silver Street, Cambridge CB3 9EW, UK

9

Geological Fluid Mechanics

HERBERT E. HUPPERT

1 The Earth

Since antiquity there has been investigation of and speculation about the Earth on which we live. How did it form; how does it evolve; and how can its riches be exploited? It was relatively easy to understand some of the fundamentals of the enveloping atmosphere, of order 10 km thick, because of its optical transparency. Some of these fundamentals and the ensuing consequences are described in Chapters 6 and 11. Satisfactory investigations of the oceans, which cover 70% of the globe to a mean depth of 4 km, were more difficult. Some of these are described in Chapter 10. The 'solid' Earth, of mean radius 6371 km, whose volume and mass greatly exceed those of either the atmosphere or the oceans (see table 1), has been the most difficult to examine. In large part this is because almost all of the globe is inaccessible to direct observation. Inferences have to be drawn from observations at (or near) the Earth's surface, appropriately combined with theoretical reasoning. Fluid mechanics plays a considerable and ever increasing role in this investigation.

The relatively new subject of geological fluid mechanics is concerned with applying fundamental fluid-mechanical concepts to following the motion of the fluid material that upon either solidification or sedimentation become the rocks that make up the Earth. A full understanding of the subject comes from a combination of theoretical analysis, data from laboratory experiments and field observations; and this breadth of essential input is partly what has made development of the area so exciting in the last decades of the previous millennium.

Much of the subject is motivated by the motions of: magma – the geologist's term for fluid rock when within the Earth; lava – the name for magma erupted from volcanoes at the surface of the Earth; and particulate suspen-

	Inner Core	Outer Core	Mantle	Total 'Solid' Earth	Oceans	Atmosphere
Mass (kg)	9.7×10^{22}	1.9×10^{24}	4.0×10^{24}	6.0×10^{24}	1.4×10^{21}	5.1×10^{18}
Volume (km^3)	7.6×10^{9}	1.7×10^{11}	9.1×10^{11}	1.1×10^{12}	1.4×10^{9}	

Table 1. The mass and volume of various regions of the Earth. Note the relatively small values for the oceans and atmosphere.

sions in either the oceans or atmosphere due to a variety of natural processes including volcanic eruptions. Both chemically and physically, magma (and lava) can vary enormously from one location within (or on) the Earth to another. The major constituent, silica, can be as low as 45% or as high as 75%, which might be compared with the very much smaller range of the major constituents of the atmosphere, nitrogen and oxygen, or those of the oceans, water and salt. The viscosity varies over many orders of magnitude with both composition and temperature. A silica-poor, high-temperature magma may have a viscosity as low as 5×10^{-6} m^2 s^{-1}, while a silica-rich, low-temperature magma may have a viscosity of 10^8 m^2 s^{-1} or even more as it finally solidifies. In detail, many magmas are no doubt non-Newtonian, and the values above are but constructs. However, much can be learnt from purely Newtonian models of magma and lava flows and much, but definitely not all, of the material presented in this chapter is developed on the basis of Newtonian fluid mechanics, which is a good approximation in many cases.

An extensive terminology for magmas and lavas has developed, based mainly on their chemical composition. Some understanding of this (not always uniquely defined) terminology is necessary to read the geological literature. For the purposes of this chapter, however, it suffices to define two of the most common types. *Basalts* are magmas derived from the Earth's mantle and are composed of approximately 50% silica, have temperatures between 1100°C and 1250°C, densities between 2.60 and 2.75 g cm^{-3} and viscosities between 0.003 and 0.1 m^2 s^{-1}. *Rhyolites* are magmas that can become granite on solidification within the Earth and originate from melting of the continental crust or by chemical evolution of basalt. Rhyolites have a silica content of between 70% and 75%, have temperatures between 700°C and 1000°C, densities of approximately 2.3 g cm^{-3} and viscosities between 1 and 10^8 m^2 s^{-1}.

The formation of the Earth (and the other terrestrial planets) was a by-product of the formation of the Sun. The Sun, along with a number of stars, was formed by gravitational instability of the dense rotating interstellar molecular medium. It is commonly believed that the terrestrial planets then grew by agglomeration of so-called planetesimals, which formed in the solar nebula and ranged in size from a few metres to a few thousands of kilometres. Calculations indicate that by this process a body of 10^{23} kg can result after about 10^5 years. Of the order of 100 such planetary embryos then merged in the next 10^8 years to form the early Earth. During this period giant impacts occurred between bodies of similar mass, which caused melting and reconfiguration of the Earth to great depths. One of these giant impacts is believed to have resulted in the orbital injection of material from which the Moon formed. These processes ceased about 4.5×10^9 years ago, the date recognized as the 'birth' of the Earth as we know it today.

It is generally believed that after each giant impact the Earth was left molten. On cooling, by radiation into Space, an outer crust formed first, though no geological record remains of this early crust. The current continental crust formed later, beginning around 4×10^9 years ago, and is now typically between 25 to 40 km thick. Oceanic crust, which is much younger and has been frequently recycled, is only about 5 km thick. At the base of the crust lies the Mohorovičić discontinuity (known after its discoverer) or Moho for short. Its existence was identified by its ability to reflect and scatter seismic waves, an important tool in the investigation of the physical and chemical properties inside the Earth. The region between the Moho and a depth of 670 km is known as the upper mantle. At this depth a phase transition occurs and below this, to a depth of 2920 km, extends the lower mantle. The mantle is rich in silicates with the most common upper mantle mineral being olivine, which at its simplest is a solid solution of magnesium and iron silicate. Below the mantle is the iron-rich liquid outer core, as first announced by Jeffreys in 1926 (see figure 1). At the centre of the Earth there is a solid inner core which is closer to pure iron than even the outer core and currently extends to a radius of 1221 km. The existence of this solid inner core was first suggested in 1936, from seismic evidence, by Inge Lehmann, who died recently in her 105th year. This differentiation into a core dominated by iron and a mantle dominated by silica occurred when the Earth was molten and the heavy iron 'sank' to the centre.

Of course virtually no part of the Earth is static – from the inner core to the outer stratosphere it is almost all in motion, in some parts quite vigorous. The temperature of the centre of the Earth is currently somewhere between about 5000°C and 6500°C (which represents a fascinatingly large range of uncertainty about a particular and important geophysical quantity). The temperature at the boundary between the inner and outer cores is thought to be about 250°C lower. It has been suggested that this temperature difference

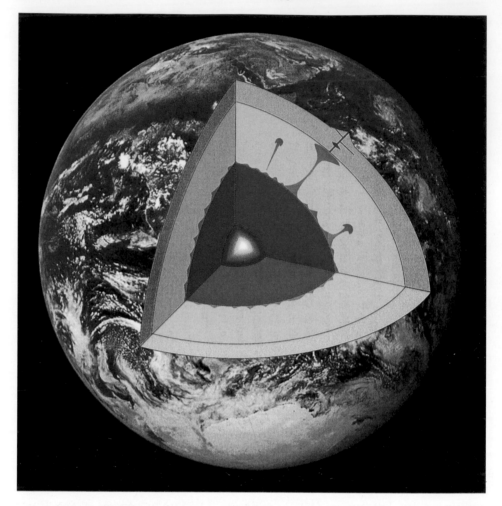

Figure 1. A diagrammatic Earth showing the solid inner core, the liquid outer core, the mantle, the lithosphere and the covering clouds. Also depicted are deep mantle plumes and erupting volcanoes.

drives convection in the inner core, though there is as yet no observational evidence for this. As a consequence of the slow cooling of the Earth, the iron-rich liquid outer core cools and part of it solidifies to form almost pure iron at the inner/outer core boundary to release fluid depleted in iron and hence less dense than its surroundings.[1] This is an archetypical situation for com-

---

[1] Solidification takes place at the hotter surface of the boundary of the outer core, rather than at the approximately 1500°C cooler boundary with the mantle because of effects due to pressure. The amount of iron produced at the inner/outer boundary exceeds by about four orders of magnitude the total iron and steel production of mankind on the surface of the Earth; nevertheless the radius of the inner core increases at a rate estimated to be only about 1 mm/year.

positional convection (Huppert & Sparks 1984) as described in Chapter 8. The rising plumes, which may break up into blobs, drive strong convective motions in the rotating outer core and maintain the Earth's magnetic field, by a dynamo mechanism described in Chapter 7 (and further in §6). Driven partly by heat losses from the core and partly by heat generation from radioactive decay in the interior of the mantle, there is a complicated, unsteady bulk motion in both the lower and upper mantle with typical velocities of a few centimetres a year – about as fast as one's fingernails grow.

These motions lead to an important concept in understanding the Earth, the theory of plate tectonics, which was developed in its full form in the mid 1960s. Nineteenth century scientists, notably Alfred Wegener, had developed the rudimentary notions of continental drift, but the suggestion that the solid rocks of the Earth behave like a fluid seemed so implausible that the notion of continental drift was not initially accepted. However, the idea that the hot crystalline rocks of the mantle could flow in convective currents slowly began to be appreciated, partly due to the influence of the great British geologist Arthur Holmes. The major breakthrough came in the mid 1960s with proof, from the magnetic signatures of volcanic rocks on the sea floor, that continents could spread apart at the mid-ocean ridges by sea-floor spreading. The concept of plate tectonics rapidly evolved as it was recognized that only the outermost 100 km or so of the Earth (the lithosphere) was cold and rigid, whereas the bulk of the Earth's mantle was sufficiently hot that, despite its crystalline nature, it could convect and flow. The Earth's surface is broken into lithospheric plates which are constantly moving apart and (elsewhere) colliding as a consequence of motion driven both by their own high density compared to the deeper, hotter Earth and by convection in the Earth's interior. Volcanoes and earthquakes are found to be mostly located in great zones marking the plate boundaries. This chapter is not the place to detail the excitement in the mid 1960s as the theory evolved, nor to describe exactly how plate tectonics works. An excellent account, however, can be found in Gubbins (1990). The additional point to note here is that there also exist in the mantle and the crust much more rapid, smaller-scale fluid motions of enormous importance. The investigation of these motions are central to this chapter.

Much of the above introduction is treated at length in standard books on Earth sciences. Interested readers might like to dip into Anderson (1989), Brown, Hawkesworth & Wilson (1992) or Press & Siever (1986). A short description of how some of the current physical facts about the Earth were determined, with an unashamed bias towards contributions made by Cambridge geophysicists, is contained in Huppert (1998*a*).

One important way by which an Earth scientist learns about the physics and chemistry of the interior of the Earth is from volcanic eruptions. The episodic eruption of magma at the Earth's surface immediately indicates that some parts of the 'solid' interior must be molten. Investigations have shown that there are large storage reservoirs of (at least partially) liquid rock known as magma chambers at depths of between a few to perhaps tens of kilometres beneath all volcanoes, both those that erupt on the continents and the many more numerous volcanoes on the floor of the oceans. The chambers may range in horizontal dimensions from hundreds of metres to more than a hundred kilometres and in vertical dimensions from tens of metres to many kilometres. Each magma chamber, which may contain as much as 10^5 km^3 of molten rock, possibly with a considerable crystal content, all at temperatures from 800°C to over 1200° C, acts as the energy source for the overlying volcano. Many interesting and important physical and chemical processes occur in a chamber, some of them leading to an eruption. The purpose of this chapter is to present a description of some of the fundamental processes that occur in these situations. An earlier review, from a different perspective, of the role played by fluid mechanics in geology is contained in Huppert (1986).

2 Fluid processes in magma chambers

The central problem in magma chamber dynamics is the determination of the physical (and chemical) evolution of a large body of cooling and crystallizing, multi-component liquid. This section aims to build a physical understanding of the fundamentals of magma-chamber dynamics by describing sequentially a range of fluid processes.

2.1 Conduction

The cooling of magma against surrounding country rock is one important aspect of many processes. As an introductory, very much simplified, problem, consider the half-space $z > 0$ at uniform temperature T_+ suddenly brought at time $t = 0$ into contact with the half-space $z < 0$ at a lower uniform temperature T_-. Each half-space could be either solid or liquid – but the effects of any motion are neglected if it is liquid. The temperature $T(z, t)$ must then adjust purely by the conductive transfer of heat. Assume, for simplicity, that the thermal diffusivity of the material of each half-space is identical and denoted by κ. The situation is then described by the one-dimensional

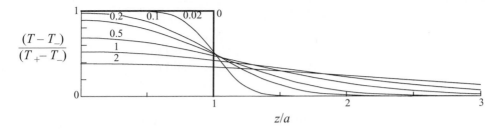

Figure 2. The non-dimensional temperature as a function of position for various values of $\kappa t/a^2$ as a result of the temperature of a layer of thickness $2a$ being initially raised to a temperature excess $T_+ - T_1$ above the ambient (taken from Carslaw and Jaeger, 1980).

thermal conduction equation

$$T_t = \kappa T_{zz} \tag{2.1}$$

with initial and boundary conditions

$$T = T_- \quad (z < 0) \quad T = T_+ \quad (z > 0) \quad (t = 0), \tag{2.2a, b}$$

$$T \to T_- \quad (z \to -\infty) \quad T \to T_+ \quad (z \to \infty) \quad (t > 0), \tag{2.3a, b}$$

$$T, \ T_z \ \text{continuous at } z = 0 \quad (t > 0). \tag{2.4a, b}$$

Because there is no specified length scale in the problem, the solution of the partial differential system (2.1)–(2.4) cannot be a function of the two independent variables z and t separately but there must be a *similarity* solution in terms of one variable which is a suitable combination of z and t. In this case a suitable similarity variable is $\eta = \frac{1}{2}z(\kappa t)^{-1/2}$, which transforms (2.1)–(2.4) into the ordinary differential system

$$\frac{d^2 T}{d\eta^2} - \eta \frac{dT}{d\eta} = 0, \tag{2.5}$$

$$T \to T_- \quad (\eta \to -\infty), \quad T \to T_+ \quad (\eta \to \infty), \tag{2.6a, b}$$

$$T, \ T_\eta \ \text{continuous at } \eta = 0. \tag{2.7a, b}$$

The solution of (2.5)–(2.7) (in terms of η) is

$$T(\eta) = \tfrac{1}{2}(T_+ + T_-) - \tfrac{1}{2}(T_+ - T_-)\mathrm{erf}\,(\eta), \tag{2.8}$$

where $\mathrm{erf}(x) = 2\pi^{-1/2} \int_0^x e^{-u^2} du$ is the error function. The relationship (2.8) indicates that the temperature of the boundary, $T(0, t)$, immediately adjusts to the mean temperature of the two half-spaces and remains at that value. This quantitative relationship is one of the best known to geologists.

a	1 mm	1 cm	10 cm	1 m	10 m	100 m	1 km
$t_c = a^2/\kappa$	1 s	2 min	3 h	10 days	3 yr	300 yr	3×10^4 yr
t_v (lab)		1 min*	10 min	1 hr	1 day		
t_v (magma 1)		5 min*	1 hr*	5 hr	3 days	25 days	1 yr
t_v (magma 2)					2 yr	20 yr	200 yr

Table 2. Values for the conductive time scale $t_c = a^2/\kappa$ for $\kappa = 10^{-6}$ m^2 s^{-1} and the convective time scale $t_v = 25 t_c Ra^{-1/3}$, where Ra is the Rayleigh number (with $g = 10$ m s^{-2}), for typical laboratory values ($\Delta T = 20$ K, $v = 10^{-6}$ m^2 s^{-1} and $\alpha = 2 \times 10^{-4}$ K^{-1}) and magmatic values. Magma 1, for which $v = 10^{-3}$ m^2 s^{-1} is a rather inviscid magma, while magma 2, for which $v = 10^3$ m^2 s^{-1}, is quite viscous. For both magmas $\Delta T = 10^{-3}$ K and $\alpha = 5 \times 10^{-5}$ K^{-1}. A star against a value indicates that $Ra < 10^6$ and the value for t_v is either unreliable or meaningless.

Consider now the slightly more complicated problem where the half-spaces $z > a$ and $z < -a$ are initially at T_-, while the material in $-a < z < a$ is initially at T_+ ($> T_-$). Because of the introduction of the length scale a, the solution is not expressible in terms of one similarity variable, but nevertheless can easily be determined to be (Carslaw & Jaeger 1980, p. 54)

$$T(z,t) = T_- + \tfrac{1}{2}(T_+ - T_-)\left\{ \operatorname{erf} \frac{a-z}{2(\kappa t)^{1/2}} + \operatorname{erf} \frac{a+z}{2(\kappa t)^{1/2}} \right\}, \qquad (2.9)$$

which is graphed as a function of z for various values of $\kappa t/a^2$ in figure 2. The time scale for the temperature evolution by conduction is given by $t_c = a^2/\kappa$, values for which are given in table 2 for various values of a with $\kappa = 0.01$ cm^2 s^{-1}, a representative value for magmas and surrounding country rocks. The variation is seen to be considerable. This simple model is of some applicability to the cooling of either lavas or relatively long, thin intrusive sheets of magma in what are called dykes or sills by geologists.[1] But consideration of further physical processes is needed before it is applicable to (thicker) magma chambers. Two of the main ones are the effects of convection and crystallization in the magma, both of which are dominant processes in real magma chambers.

2.2 Thermal convection

In order to analyse some of the important effects of purely thermal convection in magma chambers, Jaupart & Brandeis (1986) carried out a series of

---

[1] Solutions to a whole series of conduction problems are presented in Carslaw & Jaeger (1980) and reviewed in the context of the Earth sciences by Jaeger (1968).

experiments in which a layer of initially hot silicone oil was confined between two rigid horizontal boundaries whose temperature at $t = 0$ was suddenly decreased and maintained constant, at typically some 20 K less than the initial temperature of the isothermal oil. Based on an initial temperature difference $\Delta T_i \simeq 20$ K, the initial Rayleigh number $Ra_i = \alpha g \Delta T_i d^3 / \kappa v$ was of the order of 10^8, where g is the acceleration due to gravity, α and v are the coefficients of thermal expansion and kinematic viscosity respectively and $d = 10$ cm was the distance between the boundary plates. At such high Rayleigh numbers the resulting convective motions are turbulent (see Chapter 6) and the heat transfer is dominated by the convective component. Convective plumes penetrated the fluid from a layer near the upper surface, leading to an almost isothermal interior, while a thin stagnant thermally stable boundary layer evolved at the base.

With the employment of the famous four-thirds flux law for turbulent convection (see Chapter 6) and the equating of this heat loss to the rate of decrease of heat in the interior, the thermal balance, in terms of the uniform temperature $T(t)$ of the interior, becomes

$$\rho C_p d \frac{\mathrm{d}T}{\mathrm{d}t} = -0.12k \left(\frac{\alpha g}{\kappa v} \right)^{1/3} (T - T_B)^{4/3}, \tag{2.10}$$

where T_B is the (maintained constant) temperature of the boundaries, C_p is the specific heat per unit mass, ρ is the density and $k = \rho C_p \kappa$ is the thermal conductivity. Non-dimensionalizing temperature with respect to the initial temperature difference and time with respect to the conduction time based on the thickness of the layer to introduce the variables

$$\theta(\tau) = [T(t) - T_B]/\Delta T_i \quad \text{and} \quad \tau = \kappa t / d^2 \equiv t/t_c, \tag{2.11a–c}$$

we can write the solution of (2.10) as

$$\theta(\tau) = (1 + 0.04 Ra_i^{1/3} \tau)^{-1/3}, \tag{2.12}$$

a relationship which is in good agreement with the experimental data. The large constant before τ shows that the conduction time scale is much too long for this turbulently convecting system and the appropriate time scale is not t_c but $t_v = 25(d^2/\kappa)R_i^{-1/3}$ (see table 2).

The development of the stagnant boundary layer at the base can be determined by solution of the conduction equation with appropriate boundary conditions. An approximate solution is given in Jaupart & Brandeis (1986). For our purposes it suffices to state that because it results from the solution of the diffusion equation, the thickness of the boundary layer δ scales with

$(\kappa t_v)^{1/2}$, i.e. $\delta/d \sim R_i^{-1/6}$. Thus for a typical magma chamber a stagnant lower boundary layer may occupy of order 1% of the depth.

2.3 Crystallization and compositional convection

As a magma cools in a chamber it begins to solidify, preferentially at the colder boundaries, but also in the warmer interior. Magma is composed of many chemical components, with those taken into the solid being almost always different from those in the neighbouring fluid. This generally means that the density of the depleted fluid close to the solidifying interface is different (either greater or less) than that of fluid nearby. This buoyancy difference (due to crystallization) leads to what is called *compositional convection*, as described in greater detail in Chapter 8. Because the density difference due to the compositional differences is generally very much larger than that due to the associated thermal differences, compositional convection is generally much more vigorous than, and dominates, any thermal convection present. The convection can mix the magma, carry small crystals along with it, and introduce strong chemical stratification, with almost non-interacting regions of magma, by the processes of double-diffusive convection (Huppert & Turner 1981a; Chapter 6).

Exactly how the magma evolves depends on the geometry of the chamber and on the sign of the density difference of the released fluid. Even the simplest case of cooling an initially uniform layer at a horizontal boundary leads to a 2×3 matrix of possibilities (Huppert & Worster 1985). The cooled boundary can be at the top of the layer, leading to an unstable thermal field, which is liable to thermal convection, or at the base of the layer, which results in a stable thermal field. In addition, the released fluid can be either positively or negatively buoyant or, as a somewhat special case, of equal density to the neighbouring fluid. Each of these six cases needs, and has been subject to, special investigation. The analyses include incorporation of the effects of mushy layers (Chapter 8) and thermodynamic non-equilibrium. In each case the quantitative predictions of the mathematical models have been in good agreement with the results of specially designed laboratory experiments using aqueous systems such as solutions of KNO_3, Na_2CO_3 or NH_4Cl (Huppert 1990, which contains numerous colour photographs of such experiments). Possibly the most surprising result, which has direct application to the interpretation of rock layers found at the base of the frozen remains of some magma chambers, is that the cooling from above of a layer of fluid which releases less-dense fluid on crystallization can lead to the evolution of a crystal layer on the floor of the container (Kerr *et al.* 1989). This is because undercooling in the interior of the fluid due to non-equilibrium effects at the crystallizing interface near the top of the layer (which must occur, at least to some extent, for crystallization to proceed) drives further crystallization at the base, as depicted in figure 21 of Chapter 8.

If the cooling and crystallization take place at a vertical sidewall, the released fluid flows either up or down through the crystal mush, depending on the sign of its buoyancy, to form a separate layer at the top or bottom. With time, a strong vertical stratification can result, as described by Turner & Campbell (1986). The horizontal temperature gradient then couples with the vertical compositional gradient to lead to the inevitable double-diffusive layering (see Chapter 6), with vigorously convecting almost uniform layers separated by thin interfaces across which there are (relatively) large changes in temperature and composition.

An interesting laboratory experiment, with results directly applicable to effects due to the sloping retaining walls of magma chambers, illustrates many of the above features. Huppert *et al.* (1986*b*) cooled an initially aqueous solution of Na_2CO_3 at an inclined plane, which was inserted into the container to divide the fluid into two geometrically identical halves. Upon crystallization on the upper surface of the inclined plane, the released less-dense fluid rose, mixed with fluid of the upper region and induced a vertical stratification, which with time broke up into a series of double-diffusive layers. Crystallization on the lower surface of the plane also resulted in the release of less-dense fluid, which could not rise because of the constraint of the impermeable plane above it. This fluid wound its way through the porous mushy layer to be deposited as a separate, ever growing layer at the top. As time proceeded the density of the released fluid decreased and hence displaced downwards fluid at the top of this layer by the 'filling box' mechanism (Chapter 6). The resultant crystal shapes on the two sides of the plane were quite different. The geometry of the two regions was identical, but the convective effects caused the solidification processes above and below the plane to be very different.

These general ideas, on generating a stable stratification in magma chambers, can be applied to understanding the interpretation of compositionally zoned volcanic products. In large-magnitude explosive eruptions, tens to hundreds of km^3 of magma can be erupted in a few hours or days, often emptying a considerable proportion of the magma chamber. Such volcanic products are typically zoned in a systematic way in both chemical composition and inferred magma temperature. This indicates that the top of the chamber contained cool evolved magma, typically enriched in silica and volatiles, and the chamber was then stratified, with hotter and relatively silica-poor magma occurring at depth (Sparks, Huppert & Turner 1984). Such zoning is very common and can be satisfactorily explained by sidewall crystallization and compositional convection.

2.4 Replenished chambers

Replenishment of a magma chamber by new hotter magma from below can revitalize the motions and initiate new processes. New magma can enter in batches in the form of solitary waves due to *compaction* deeper in the Earth, as described in § 6 and Spiegelman (1993), but continuous 'seepage' of new magma may also be possible.

A particular situation, which illustrates how a new, and possibly counter-intuitive, process can arise in a replenished magma chamber and indicates the dominant role of fluid mechanics, was considered theoretically by Huppert & Sparks (1980) and tested experimentally by Huppert & Turner (1981*b*). In the 1970s geologists had accumulated reliable evidence that the erupted output from many basaltic magma chambers was typically composed of approximately 50% silica and about 6% to 9% magnesium oxide (MgO). Different lines of argument suggested that the input magma had slightly less silica (c. 48%) but double or even triple the MgO content (about 12% to 18%). What is the reason for this large difference in MgO content?

Huppert & Sparks (1980) considered the illustrative case of a chamber filled with magma containing around 9% MgO at a temperature around 1250°C. New magma from deeper in the mantle, composed of 18% MgO at a temperature around 1400°C, is episodically intruded into the base of the chamber. This new magma, although hotter than the resident magma, is more dense, owing to compositional differences, and hence forms a separate layer at the base. The bottom layer gradually cools because of its contact with the colder upper layer, while remaining distinct because of the com-positional difference. As it cools, the lower layer crystallizes to form small olivine crystals, which preferentially extract MgO from the melt, and thereby becomes less dense. The olivine is (mainly) kept in suspension in the melt by the vigorous convection in the lower layer. As the temperature difference between the layers approaches zero, the vigour of the convection eases, the relatively heavy olivine crystals fall to the base, the density difference across the double-diffusive interface vanishes and the MgO-depleted liquid rises into the main interior of the chamber, leaving behind a layer of compacted olivine crystals.

Huppert & Turner (1981*b*) modelled the essentials of this process in the laboratory, by feeding into a large reservoir a hot heavy layer of aqueous

KNO_3 beneath a colder less-dense layer of $NaNO_3$. As the lower layer cooled beyond the solidification point, crystals of KNO_3 grew in the presence of strong compositional convection in the lower layer. The density of the saturated aqueous KNO_3 decreased (at a rate in good agreement with theoretical calculations) until it reached that of the upper layer, at which point the liquid of the lower layer rose to mix with that of the upper – leaving behind the KNO_3 crystals.

These processes were consistent with evidence from old solidified magma chambers, such as on the island of Rum, off northwest Scotland, which displayed distinctive alternating layers of crystals representing solidification of magmas at alternately high and low temperatures (and hence high and low MgO contents). There are numerous other geological examples of this process and the concept of a *density trap*, as it is sometimes referred to by Earth scientists, is frequently invoked in describing observations in the field.

The details of the injection of relatively heavy magma into a less-dense ambient from either a point or line source to form a turbulent fountain can be investigated using well-known concepts of turbulent plume theory (see Chapter 6 and Turner 1979). When the Reynolds number of the input, Re_i, is very large and viscous effects can be neglected in both input and ambient fluid, of density ρ_i and ρ_a respectively, the height of rise of the fountain, h_f, is determined from the specific momentum and buoyancy fluxes at the source,

$$M_s = w_i^2 r_s^2 \qquad \text{and} \qquad B_s = w_i g_i' r_s^2, \qquad (2.13a, b)$$

as

$$h_f = 2.5 M_s^{3/4} B_s^{-1/2} = 2.5 w_i (r_s / g_i')^{1/2}, \qquad (2.14a, b)$$

where w_i is the input vertical velocity at the source of radius r_s of fluid with initial reduced gravity $g_i' = (\rho_i - \rho_a) g / \rho_a$. Campbell & Turner (1986) conducted a series of experiments, backed up by dimensional arguments, to investigate the influence of the viscosity of the input and ambient fluid, denoted by v_i and v_a respectively. They found that when $Re_i \geqslant 400$ mixing between input and ambient fluid is controlled exclusively by the Reynolds number of the ambient $Re_a = w_i r_s / v_a$ (and the value of Re_i has no significant influence). Mixing between the fluids, and the height of the fountain, both relative to the values at infinite ambient Reynolds number, is reduced by 10% (50%) as Re_a decreases below 70 (30).

For a magma chamber fed by a simple vertical conduit of radius r_m, the steady efflux can be calculated by balancing two quantities: the potential energy available to the magma of density ρ_m to rise because of its positive buoyancy with respect to the surrounding country rocks of density ρ_{cr}; and the energy lost due to dissipative effects as the magma flows with mean velocity w along the rough sidewalls of the conduit. On the assumption that the flow in the conduit is turbulent this leads to an efflux rate

$$Q = \pi w r_m^2 = 2\pi (g'_{cr}/f)^{1/2} r_m^{5/2}, \qquad (2.15a,b)$$

where g'_{cr} is the reduced gravity of the surrounding country rock relative to the magma and f is a friction coefficient, which typically lies between 0.01 and 0.08. Combining (2.14b) and (2.15b), we obtain (within the Boussinesq approximation)

$$h_f = 5\left(\frac{\rho_{cr} - \rho_m}{\rho_m - \rho_a}\right)^{1/2} \frac{r_m}{f^{1/2}}, \qquad (2.16)$$

where $\rho_a < \rho_m < \rho_{cr}$. With the insertion of typical values in (2.16), $h_f \sim 25 r_m$, which might vary between 25 m and 250 m. Thus for small conduits (and hence relatively small effluxes) effects due to the input fountain are also small. Only when the conduit radius (and associated Reynolds number) is quite large do fountains, and the associated mixing between input and ambient magmas, play an important role.

Input of magma less dense than the ambient occurs in some situations and for these cases the form of the flow of input magma has been investigated as a function of the Reynolds numbers of the input Re_i and ambient Re_a (Huppert *et al.* 1986a). For sufficiently low values of Re_i and Re_a (each less than about 10), the input rises as a laminar conduit. With an increase in either Reynolds number the conduit becomes unstable and entrains ambient magma. For sufficiently large Reynolds numbers the input becomes turbulent and considerable entrainment can occur.

These ideas have been applied to understanding the observed mixing of basaltic magmas, the origin of some economic ore deposits such as platinum sulphides (Campbell, Naldrett & Barnes, 1983) and the difficulty of erupting pure unevolved magmas.

2.5 Melting

Because temperature increases with depth within the Earth and relatively little heat is lost from a magma as it ascends, molten magma may enter a chamber at a temperature above the melting temperature of some of the components of the surrounding rock. Melting of the surrounding rock may then take place at the roof, base or sides of the chamber and the melting process will be strongly influenced by whether the melt is more or less dense

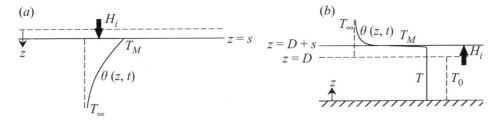

Figure 3. (a) The resulting temperature profile $\theta(z,t)$ when a constant heat flux H_i is incident on the boundary of a solid with melting temperature T_M. (b) The melting of a solid roof due to a thermally convecting fluid beneath it. The released melt is of greater density and is miscible with the fluid.

than the ambient magma, or of identical density (which is rather unlikely). An important example occurs following the input of basaltic magma at around 1300°C into the granitic continental crust, which leads to a (granitic) melt that is about 0.4 g cm^{-3} less dense than the basaltic ambient.

As a fundamental example of melting, consider a heat flux H_i incident on a planar surface of a solid with melting temperature T_M (as sketched in figure 3a). Denoting the position of the surface at time t as $z = s$ and the temperature of the solid as $z \to \infty$ as T_∞, the temperature profile $\theta(z,t)$ in the solid and the rate of melting are determined from

$$\theta_t = \kappa \theta_{zz} \qquad (z > s), \tag{2.17}$$

$$\theta(z,t) = T_M \quad (z = s), \quad \theta \to T_\infty \quad (z \to \infty) \tag{2.18a,b}$$

(on the assumption that the surface remains planar). A solution of (2.17) and (2.18), known sometimes as the ablation solution, valid away from $t = 0$, is given by

$$\theta(z,t) = T_\infty + (T_M - T_\infty) \exp[-V(z - s)/\kappa], \tag{2.19}$$

where $V = ds/dt$ is the melting rate, assumed constant. Equating the difference in heat fluxes across the interface to the rate of latent heat release (see Chapter 8),

$$H_i - k\theta_z|_{z=s} = \rho L V, \tag{2.20}$$

where L is the latent heat per unit mass of solid, and substituting (2.19) into (2.20), we determine that the melt rate

$$V = \frac{H_i}{\rho[L + C_p(T_M - T_\infty)]} \equiv \frac{H_i}{\mathcal{L}_*}. \tag{2.21a,b}$$

This relationship along with (2.19) represents a long-time solution; solutions satisfying particular initial conditions are derived in Huppert (1989).

Consider now a semi-infinite flat solid roof, of melting temperature T_M, overlying a hot fluid, initially of depth D and at temperature T_0 ($> T_M$), the bottom of which, at $z = 0$, is thermally insulated (as sketched in figure 3b). If the density of the melt is greater than that of the original fluid and is miscible with it, the melt will mix with the fluid and the melting is determined entirely by thermal processes. The heat flux incident on the base of the melting roof is given by (cf. (2.10))

$$H_i = 0.12k \left(\frac{\alpha g}{\kappa \nu}\right)^{1/3} (T - T_M)^{4/3} \sim \Gamma \Delta T^{4/3}, \qquad (2.22a, b)$$

where T is the temperature of the hot underlying fluid, $\Gamma \equiv 0.12k(\alpha g/\kappa \nu)^{1/3}$, $\Delta T = T_0 - T_M$, and (2.22b) is correct only initially, but gives a good indication of the scale of H_i for considerably longer. Combining (2.21) and (2.22), we obtain

$$\dot{s} = \Gamma \Delta T^{4/3}/\mathscr{L}_*, \qquad (2.23)$$

which indicates that, at least initially,

$$s = (\Gamma \Delta T^{4/3}/\mathscr{L}_*)t. \qquad (2.24)$$

The heat gained by the melt is taken from the fluid into which it mixes, which indicates that $\Delta T \delta s \sim -D\delta T$ and thus the non-dimensional temperature of the lower layer

$$\Theta \equiv (T - T_0)/\Delta T \sim -s/D \propto -t, \qquad (2.25a, b)$$

i.e. Θ increases linearly with time. A more accurate quantification of this process is presented by Huppert & Sparks (1988a), who also describe laboratory experiments in which a wax roof was melted by a hot aqueous salt solution to yield data on melt rates and fluid temperature which are in very good agreement with their theoretical predictions.

If the released melt is less dense than the hot fluid layer, it forms a separate layer beneath the roof, as sketched in figure 4. Initially heat is transferred by conduction across the melt layer as depicted in figure 4(a). As the Rayleigh number of the layer increases, however, convection sets in. If the Rayleigh number becomes sufficiently large, as it does in natural situations, the convection becomes turbulent and the heat transfer can be analysed using the four-thirds formulation. For illustrative purposes, we consider only the vigorously convecting state, as depicted in figure 4(b). (The conductive state is analysed fully by Huppert & Sparks 1988a.) We seek to determine

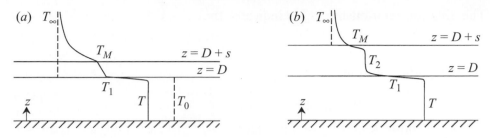

Figure 4. The melting of a solid roof due to a thermally convecting fluid beneath it. The released melt is of smaller density than the fluid beneath it. (*a*) The Rayleigh number of the melt layer Ra is sufficiently small that heat is transferred through the layer by conduction. (*b*) Ra is so large that the heat is transferred by vigorous convection.

the temperatures T_2 and T of the melt and fluid layer, respectively, the interfacial temperature T_1 and the thickness of the melt layer s as functions of time. For simplicity we shall assume here that the various phases all have the same values of physical parameters such as specific heat, thermal diffusivity, etc.

Because the upward heat flux from the hot fluid layer must equal that from the base of the melt layer

$$T_1 = \tfrac{1}{2}(T + T_2). \tag{2.26}$$

The heat flows primarily into the base of the roof and, under most circumstances, a very small portion of it is used to raise the temperature of the melt layer through which it passes. With the neglect of this effect

$$T_2 = \tfrac{1}{2}(T_1 + T_M) = \tfrac{2}{3}T_M + \tfrac{1}{3}T. \tag{2.27a,b}$$

Conservation of heat in the hot fluid layer requires

$$\rho C_p D \frac{dT}{dt} = -\Gamma (T - T_1)^{4/3} = -\left(\tfrac{1}{3}\right)^{4/3}\Gamma (T - T_M)^{4/3}, \tag{2.28}$$

which, subject to

$$T = T_0 \qquad (t = 0), \tag{2.29}$$

(where T_0 is the temperature of the layer when vigorous convection sets in at $t = 0$), has the solution

$$T = T_M + (\beta_1 + \beta_2 t)^{-3}, \tag{2.30}$$

in terms of the two constants

$$\beta_1 = (T_0 - T_M)^{-1/3} \quad \text{and} \quad \beta_2 = \left(\tfrac{1}{3}\right)^{7/3}\Gamma/(\rho C_p D). \tag{2.31a,b}$$

The ablation relationship (2.21) indicates that

$$\dot{s} = \Gamma (T_2 - T_M)^{4/3}/\mathcal{L}_*, \qquad (2.32)$$

which has the solution

$$s(t)/D = \rho C_p[\beta_1^{-3} - (\beta_1 + \beta_2 t)^{-3}]/\mathcal{L}_*. \qquad (2.33)$$

As $t \to \infty$, both T and $T_2 \to T_M$ while $s \to \rho C_p D(T_0 - T_M)/\mathcal{L}_*$. Expressed alternatively, after sufficient time has elapsed all the available heat in the hot fluid layer has been used to melt a specific finite thickness of the roof (and a negligibly small amount has been conducted away).

To make the model directly applicable to natural situations, Huppert & Sparks (1988a) evaluated the effect of allowing the lower fluid layer to crystallize and its viscosity to increase as the temperature falls. They used these results in two companion papers (Huppert & Sparks 1988b, c) which discuss the generation of granite by the melting of the continental crust due to the input of hot basaltic magma. The melting can be quite rapid, particularly if the crustal rocks are close to their melting temperature. The generation of large volumes of silicic magma, ready for eruption, can thus be accomplished in periods of decades when relatively hot crust is invaded by basaltic magma (rather than the tens of thousands of years previously conjectured by some geologists). The layer of rhyolite magma both grows and cools with time. The rhyolite magma must continually crystallize as it forms, with melting confined to the boundary of the system. The fundamental fluid dynamical concepts have also been applied quantitatively to the melting of ice sheets by basaltic lava eruptions (Hoskuldsson & Sparks 1997).

Melting for the floor of a container has been considered by both Huppert & Sparks (1988c) and Kerr (1994). Further research is needed on this topic to analyse the effects of melting a multi-component solid.

2.6 Volatiles

Dissolved gases or 'volatiles', particularly water and carbon dioxide, can play an important role in magmatic systems, both in magma chambers and, more centrally, in conduits feeding a volcanic eruption (as described in the next chapter). This is a consequence of the fact that smallish (but definitely non-zero) amounts of H_2O, CO_2, SO_2 and the like can be dissolved in magma. Due to a variety of processes, which include the most important ones of pressure release and crystallization, the magma can first become saturated and then release the dissolved volatiles as a gaseous phase (cf. opening, i.e. releasing the pressure in, a bottle of champagne). Because the density of gas is so much less than that of the liquid from which it originated (by approximately three orders of magnitude), the bulk density of gas plus liquid drops dramatically, or the pressure on the surrounding walls increases considerably, or, if the container is open, the volume increases rapidly (as in the champagne situation). All three of these can occur in various combinations.

Central to the argument is the solubility relationship, which can be expressed as

$$n_s = k_s \, p^{1/2}, \qquad (2.34)$$

where n_s is the mass fraction of solute, water for example, in the magma at saturation, p the pressure and k_s a solubility constant, of the order 0.0014 $\text{bar}^{-1/2}$. For a system which crystallizes entirely anhydrous crystals, the dissolved water becomes driven into an ever-decreasing liquid mass and the saturation level will decrease, as given quantitatively by $n_s = k_s(1 - X) \, p^{1/2}$, where X is the mass fraction of the crystalline phases. If the total mass fraction of water, N, exceeds n_s the difference will exsolve as gas, initially in the form of very small bubbles at nucleation sites. With the use of the perfect gas laws and the assumption of thermodynamic equilibrium, the bulk density ρ_b of gas plus liquid plus crystals can be expressed as

$$\rho_b^{-1} = RT(N - n_s)/p + (1 - N + n_s)/\rho, \qquad (2.35)$$

where $\rho(p, T, X)$ is the density of the crystals plus melt, T is the temperature and R the universal gas constant. Values of ρ_b as a function of X for various values of N for $p = 1.5$ kbar (~ 500 m depth) are graphed in figure 5. For this value of p a magma with less than approximately 2.3% by weight of water crystallizes (to up to 50% by weight) without becoming saturated. For a range of higher water contents, at a specific crystallization level X_c (related directly to temperature), the magma becomes saturated. For $X > X_c$ water exsolves and the bulk density drops precipitously. Even for temperatures sufficiently high that no crystallization has yet taken place, it follows by extrapolation from figure 5 that a maximum of approximately 5.4% of water can be dissolved (at that pressure).

The replenishment of a magma chamber by relatively heavy, *wet* undersaturated magma can be investigated using these concepts (Huppert, Sparks & Turner 1982). Turbulent transfer of heat from the new lower layer to the upper layer leads to crystallization and exsolution of volatiles in the lower layer. As figure 5 shows, initial water contents of only a few per cent are sufficient for the bulk density of the lower layer to fall significantly, possibly to become equal to that of the upper layer and thereby lead to overturning

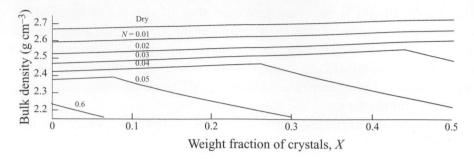

Figure 5. The bulk density of a melt as a function of crystal content X for various values of the total fractional water content by weight N at a pressure of 1.5 kbar, which corresponds to a depth within the Earth of about 500 m (taken from Huppert *et al.* 1982).

and intimate mixing between the two different magmas. There are numerous petrographic observations of hybrid rocks associated with volcanic eruptions worldwide which can be explained by such a mechanism of mixing. For example, one of the most common volcanic rock types on Earth is formed at island arcs where plates collide. The magma is called *andesite* and is intermediate in composition between basalt and rhyolite. Commonly, andesites are mixtures of basalt and rhyolite. The mechanism just described provides a possible explanation for this mixing.

The fundamental ideas and the subsequent eruption which such mixing can trigger was demonstrated experimentally by Turner, Huppert & Sparks (1983), who repeated the experiment performed by Huppert & Turner (1981*b*) and described in § 2.4, but added a small amount of HNO_3 to the lower layer and some Na_2CO_3 to the upper layer. These chemicals were held separate by double-diffusive effects operating in the two layers until the interface between them broke down. Subsequently, the acid and base produced CO_2 which bubbled vigorously through the system and frothed out of the simple volcanic crater made of Perspex which Turner *et al.* had placed over the top of the container.

3 The propagation of magma through the crust

An increase in pressure in a magma chamber beneath a volcano, due either to local processes, such as discussed at the end of the last section, or to large-scale tectonic movement, can cause the magma to rise through the Earth's crust and erupt at the surface. The magma flows in conduits or dykes in a manner controlled by elastic, fluid dynamic and thermodynamic effects. Basaltic magmas, such as those in Hawaii, in Iceland and along the mid-ocean ridges, are relatively low in both volatile content and viscosity and produce non-explosive eruptions of comparatively dry magma. The dry magma often comes to the surface through long (between 1 and 10 km), relatively straight fissures, which are of order one metre in width and may extend from between 10 to 100 km in the other horizontal direction (figure 6). Silicic magmas, on the other hand, such as those erupted on the continents and from descending lithospheric plates, tend to have relatively high volatile content and viscosity. They hence tend to produce more explosive eruptions, such as occurred at the Soufriere Hills volcano on Montserrat, Mt Pinatubo in the Phillipines, Redoubt in Alaska and Unzen in Japan. The magma conduits from chamber to surface tend to be of roughly circular cross-section and the erupted material is a complicated mixture of solid ash particles, liquid magma and gas.

We commence our description of the flow of magma through the crust by considering effects induced in dry magmas and then discuss the extra effects accompanying the transport of wet magmas, which can exhibit considerable pressure exsolution as they rise.

Any fluid flowing in a laminar manner due to a (local negative) pressure gradient ∇p in either a channel of slowly varying width or a pipe of local radius a has velocity profile

$$\boldsymbol{u} = -\gamma_u (\nabla p/\mu)(a^2 - \zeta^2), \tag{3.1}$$

where μ is the coefficient of dynamic viscosity and ζ is either the coordinate across the channel (with the walls at $\zeta = \pm a$) in which case $\gamma_u = \frac{1}{2}$, or the radial coordinate, in which case $\gamma_u = \frac{1}{4}$. The associated volume flux $\boldsymbol{Q}$, is given by $\boldsymbol{Q} = -\gamma_Q \nabla p/\mu$, where $\gamma_Q = a^3/3$ for a two-dimensional channel and $\gamma_Q = \pi a^4/8$ for circular geometry.

If, on the other hand, the Reynolds number of the flow based on the radius a exceeds a value of order 1000, the flow is turbulent, the mean velocity profile is effectively uniform and the volume flux is given in terms of a friction coefficient f by

$$\boldsymbol{Q} = -2\pi \nabla p/(\rho f |\nabla p| a^5)^{1/2}. \tag{3.2}$$

3.1 Two-dimensional fissure eruptions: thermodynamics and fluid mechanics

This subsection concentrates on two ingredients of magma flow in dykes: the fluid mechanics of flow along an already open dyke; and the heat transfer from the dyke to the surrounding country rock, which can result in either

(a)

(b)

Figure 6. (*a*) A fissure eruption in Hawaii. (*b*) The remains of a long fissure in the Four Corners region of the United States.

solidification of the magma or melting of the country rock. Elastic effects in the surrounding rock, in particular the mechanisms by which a fluid dyke is initiated in solid rock, will be discussed in the next subsection.

The starting point of the model is a long two-dimensional dyke of initially uniform width W_i through which relatively hot Newtonian fluid of uniform viscosity μ (~ 100 Pa s) is driven by a pressure drop, ΔP. Heat is transferred by both advection and conduction. Because the width of a dyke W (~ 1 m) is so much less than its length L (~ 1 km), conduction along the dyke can be neglected (with respect to conduction across the dyke). In the surrounding rock, heat can only be transferred by conduction, as determined by the two fixed temperatures of T_∞ (typically between 0 and 1000°C) in the far field and T_W (~ 1150°C) at the dyke wall. With flow rates $\bar{u}$ of order 1 m s^{-1}, the Reynolds number (of order 10) is sufficiently low for the flow to be laminar.

The time for the magma in the main dyke flow to traverse the length L is $L/\bar{u}$ (~ 20 min), in which time a thermal boundary layer of width $\delta = (\kappa L/\bar{u})^{1/2} \sim 3$ cm $\ll W_i$ is formed, where the thermal diffusivity in the magma, $\kappa \sim 10^{-2}$ cm^2 s^{-1}. Within the boundary layer the main Poiseuille flow appears as a uniform shear flow, of strength γ_s (~ 1 s^{-1}). Each of W, δ and γ will vary gradually along the dyke and slowly with time.

Consider a locally Cartesian coordinate system, moving horizontally with the dyke wall, which employs a vertical z-axis, with $z = 0$ at the base of the dyke, and a horizontal y-axis, with $y = 0$ at the dyke wall, so that $y > 0$ in the magma and $y < 0$ in the surrounding solid. The wall will migrate with velocity v due either to solidification of magma against the wall ($v > 0$) or to melting of the solid by the flow ($v < 0$). With identical values of the thermal diffusivities in magma and solid, the initial-value problem for the temperature in both fluid and solid $T(y, z, t)$ can then be stated, using the usual boundary-layer assumptions, as

$$T_t - vT_y + \gamma_s y T_z = \kappa T_{yy} \qquad (y, t > 0), \qquad (3.3)$$

$$T_t - vT_y = \kappa T_{yy} \qquad (y < 0, t > 0), \qquad (3.4)$$

along with boundary and initial conditions

$$T = T_W \ (y = 0), \quad T \to T_0 \ (\text{either } z \text{ or } t = 0, \ y > 0, \text{ and } y \to \infty), \quad (3.5a, b)$$
$$T \to T_\infty \ (t = 0, \ y < 0 \text{ and } y \to -\infty), \qquad (3.5c)$$

where T_0 (~ 1200°C) is the temperature of the magma at the base of the dyke.

The velocity of migration of the interface between liquid and solid is proportional to the difference in the conductive fluxes across the wall, as expressed by (see Chapter 8)

$$(L/C_p)v = \kappa[T_y(0-,z,t) - T_y(0+,z,t)], \tag{3.6}$$

where L is the latent heat per unit mass of either melting or solidification and C_p the specific heat per unit mass. With the use of standard relationships for Poiseuille flow, the volume flow rate $Q(t)$ and local dyke width $W(z,t)$ are related to ΔP and γ_s through

$$\Delta P = 12\mu Q(t) \int_0^L W^{-3}(z,t)\,\mathrm{d}z \quad \text{and} \quad \gamma_s(z,t) = 6Q(t)W^{-2}(z,t). \tag{3.7a,b}$$

Owing to either solidification or melting at the walls, the width of the dyke gradually changes according to

$$W(z,t) = W_i - 2\int_0^t v(z,t')\,\mathrm{d}t'. \tag{3.8}$$

The solutions of (3.3)–(3.8) depend on the Stefan number of the magma $S_0 = L/[C_p(T_0 - T_W)] \sim 20$ and that of the solid $S_\infty = L/[C_p(T_W - T_\infty]$, which typically takes values between 1 and 10 dependent on the far-field temperature of the country rock due to previous dyking episodes. At first both fluid and solid temperatures adjust to T_W, during which time fluid *must* solidify at the wall because of the very large (initially infinite) conductive heat flux into the wall (cf. the initial-value response of hot fluid running over a cold floor as discussed by Huppert 1989). Subsequently the continual supply of hot fluid transfers heat into the wall, which may melt it, while the solid continues to adjust to T_W.

Considerable analysis of the system (3.3)–(3.8) has been carried out (Bruce & Huppert 1989; Lister & Dellar 1996), which when complemented by numerical integration leads to the following results. For dykes less than a critical width (dependent on the input parameters) the initial solidification at the walls continues until the dyke becomes blocked, which first occurs at the surface, and the eruption ceases. For dykes originally broader than this critical width, before solidification can close the conduit, the advected heat flux melts back the newly solidified magma and melting occurs along the entire length of the dyke. The width then gradually increases (as long as the driving pressure is maintained). These conclusions are summarized quantitatively in figure 7.

Such analyses of the fluid dynamics and heat transfer in dykes can help explain many aspects of dyke geology and basaltic lava eruption. Typically, dykes are observed from seismic data to propagate at speeds of 0.1 to 1 m s^{-1}.

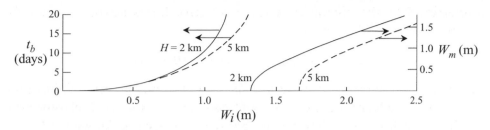

Figure 7. The time t_b for a two-dimensional fissure to solidify, or the minimum width W_m it attains, as a function of its initial width W_i for a dyke of length H intruding magma at initial temperature 1200°C into country rock at 100°C with a melting temperature of 1150°C (taken from Bruce & Huppert 1989).

Such speeds are broadly consistent with the flow conditions expected along dykes of width 0.5 to 2 m. The strong sensitivity of flow and heat transfer to dyke width can also help explain the localization of flow in basaltic fissure eruptions. As solidification constricts the narrower part of the dyke and chokes off flow, melting can widen the widest parts, eventually focusing flow locally. In long-lived eruptions, melting in focused regions can lead to formation of cylindrical conduits.

3.2 Fluid mechanics and elasticity

In reality the magma needs to force its way through the solid surrounding rocks: dykes are not supplied, but need to be made. This brings into consideration elastic responses in the rock. A series of solutions incorporating effects of both fluid mechanics and elasticity are now developed. The conclusions are that with values of natural physical parameters the flow is mainly characterized by a local balance between buoyancy and viscous forces; elastic forces play a secondary role except near the tip of the dyke.

Consider the laminar flow of fluid through a channel of width $W(x, z, t)$ with respect to horizontal and vertical coordinates x and z. Local continuity requires that

$$\frac{\partial W}{\partial t} = -\nabla \cdot \boldsymbol{Q} = \frac{1}{12\mu}\nabla \cdot (W^3 \nabla p). \qquad (3.9a, b)$$

If the dyke is two-dimensional – its width independent of x – the theory of linear elasticity indicates that the pressure p_e on the fluid due to elastic deformation of the surrounding solid (assumed to be of infinite extent) is given by

$$
p_e = -m\mathcal{H}\left\{\frac{\partial W}{\partial z}\right\} \equiv -\frac{m}{\pi}\int_{-\infty}^{\infty}\frac{\partial W(s,t)}{\partial s}\frac{ds}{s-z}, \qquad (3.10a,b)
$$

where $\mathcal{H}\{f\}$ is the Hilbert transform (Miles 1971) of $f(z,t)$ with respect to z, and m is a material constant of the rock $\sim$ 30–50 GPa for basalts and 10–20 GPa for granites. The magma forces its way through the solid by the residual pressure due to buoyancy, $p_b = -\Delta\rho gz$, which arises because the magma density ρ_m is less than that of the solid ρ_s by a (positive) amount $\Delta\rho \sim 300$ kg m^{-3} for granitic melt flowing through basalt. The pressures due to buoyancy and elasticity add linearly to yield the total pressure p driving the flow. Substituting this result into (3.9), Spence, Sharp & Turcotte (1987) obtained the governing integro-differential equation

$$
\frac{\partial W}{\partial t} + \frac{1}{12}\Delta\rho\frac{g}{\mu}\frac{\partial W^3}{\partial z} = -\frac{m}{12\mu}\frac{\partial}{\partial z}\left[W^3\frac{\partial}{\partial z}\mathcal{H}\left\{\frac{\partial W}{\partial z}\right\}\right]. \qquad (3.11)
$$

The two terms on the left represent fluid-mechanical effects; the one on the right elastic effects.

Suppose that the dyke is fed by a line source which releases a constant flux q per unit length. The dyke will then rise at a constant speed, say c, which can be found by determining a travelling wave solution of the form $W(z,t) = W(\eta \equiv ct - z)$. Far from the front of the dyke W will tend to a constant value, say W_∞, where the elastic term on the right of (3.11) is dominated by the fluid-mechanical terms on the left because of the smaller number of derivatives involved. In this limit $\partial p/\partial z = -\Delta\rho g$ and from (3.9b) $W_\infty = (12\mu q/g\Delta\rho)^{1/3}$ (independent of the value of the elastic constant m). Substituting the postulated wave form of W into the left-hand side of (3.11), integrating the result and evaluating the resulting constant by realizing that ahead of the dyke its width is zero, we determine that $c = q/W_\infty$ (as it must from continuity) and $dp/d\eta = 12\mu c/W^2$.

For small distances ζ in the vicinity of the crack tip there must be a singularity of the form $p \sim -\Lambda\zeta^{-1/2}$ where Λ is a material constant known as the stress-intensity factor $\sim 6 \times 10^6$ Pa m$^{1/2}$. A weaker singularity at the front does not allow the crack to propagate, while a stronger singularity would suggest it propagates far too fast.

With the inversion of the Hilbert transform (3.10), use of $dp/d\eta = 12\mu c/W^2$ and some algebraic manipulations, the problem can be expressed as a nonlinear integral equation for $W(\eta)$ subject to a side condition incorporating a non-dimensional stress intensity factor $\Lambda^* = \Lambda/\left(\frac{3}{8}m^3\mu q\right)^{1/4} \sim 10^{-2}$ for natural values of the parameters. This small value suggests that the

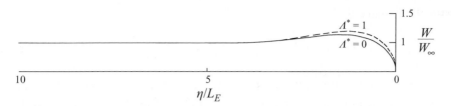

Figure 8. The (half-)profile of a crack propagating *vertically* under the influence of buoyancy and elasticity for $\Lambda^* = 0$ and 1 (taken from Lister & Kerr 1991).

singularity at the front plays a negligible role in the shape (or propagation) of the dyke. The solution to this problem must be obtained numerically (Lister 1990a); and the resulting dyke profile for $\Lambda^* = 0$ (and 1 for comparison) is presented in figure 8. The front of the propagating crack is bulbous with a maximum width of $1.3W_\infty$. Elastic effects play a negligible role behind the bulbous nose whose length, proportional to an elastic length $L_E = (mW_\infty/g\Delta\rho)^{1/2}$, is of order 1 km.

The release of a flux Q from a *point* (in contrast to a line) source can be analysed using similar concepts. With respect to perpendicular horizontal axes x and y and a vertical axis z with the origin of coordinates at the source, consider a planar steady-state dyke described by $|y| < \frac{1}{2}W(x,y)$ for $|x| < B(z,t)$. After the dyke has propagated some distance, $W \ll B \ll h$, where h is a representative height, and a similarity solution can be found (Lister 1990b) of the form

$$B(z) = \beta\, z^{3/10} \text{ and } W(x,z) = \gamma\, z^{-1/10}(1-\zeta^2)^{3/2}, \quad \text{where} \quad \zeta = \beta^{-1}xz^{3/10},$$
(3.12)

$\beta = 2.6(Q\mu\mathrm{m}^3)^{1/10}/(g\Delta\rho)^{2/5}$ and $\gamma = 0.18[Q^3\mu^3/m(g\Delta\rho)^2]^{1/10}$. For a range of Q between 1 and 10^6 m³ s⁻¹ and with representative values of the physical parameters, W typically ranges between a few centimetres and a few metres, while B ranges between a few kilometres and several tens of kilometres. These values are consistent with observations of dykes, whose widths range from 0.1 to 10 m, with the most common widths being between 0.5 and 2 m. It should be noted, however, that over a considerable portion at the lower ends of these ranges, which reflect the lower values of Q, solidification at the edges and thermal erosion near the centre would play an essential, and as yet unincorporated, role.

Throughout this subsection the densities of the magma and the surrounding solid have been considered constant. In reality the density of solid rock tends to increase with depth. One can thus envisage magma propagating upwards through the lithosphere as a result of its (decreasing) excess buoyancy until the density of the magma equals that of the surroundings, whereupon the magma will intrude laterally at what has been called the level of neutral buoyancy (LNB). This is similar to the lateral intrusion of fluid at the top of a plume in a stratified environment, as described in Chapter 6 and in §4.3. Elegant similarity solutions for a fluid-filled crack propagating at its LNB, either at an interface between two semi-infinite solid layers of different densities or into a density-stratified solid medium are reviewed by Lister & Kerr (1991). Field observations indicate, however, that the density of some magmas that are considered to have intruded laterally into the surrounding rock is quite different from that of the surrounding solid medium. This reflects a typical little cameo of geophysics: an attractive theory suggests that observationalists re-examine their findings, which leads to the theory being found deficient, at least in some of its applications, until a more complete (and maybe less elegant) series of processes is incorporated.

3.3 The ascent of wet magmas

As discussed in §2.6 and at the beginning of this section, the release of volatiles can play a large role in magma dynamics, especially in relatively viscous silicic magmas. Owing to exsolution, such magmatic systems generate the most explosive eruptions. The bubbles that form range in size from 10^{-4} to 1 cm, which, due to their buoyancy, would rise relative to the surrounding viscous magma at speeds between 10^{-15} and 10^{-6} m s^{-1}. This is so slow that it is generally assumed that the bubbles travel with the magma and form a homogeneous mixture. For every 1% by weight of volatiles exsolved, the viscosity of the mixture increases by an order of magnitude until, at volatile contents of about 72% by volume, the mixture behaves more like a foam than a simple Newtonian liquid incorporating numerous bubbles, as reviewed by Woods (1995).

The steady eruption at mean speed u of magma of density ρ from a chamber along a vertical conduit of prescribed shape can be described by the one-dimensional equations

$$\rho u A = Q, \tag{3.13}$$

which represents the conservation of mass flux Q in a conduit of cross-sectional area A, and

$$\rho u \frac{\mathrm{d}u}{\mathrm{d}z} = -\frac{\mathrm{d}p}{\mathrm{d}z} - \rho g - f, \tag{3.14}$$

which represents conservation of momentum, where z is a vertical coordinate and f denotes the frictional dissipation. The motion of bubbly liquids is still not sufficiently well understood to determine f completely. A simple approximation, while the gas content is reasonably small and the flow laminar, would be to set $f = 8\pi\mu u/A$, as appropriate for a Newtonian viscous liquid. This relationship has been augmented by resort to empirical functions which incorporate effects of the total weight fraction of water N and the fraction by volume of gas, or void fraction, ϕ. As ϕ increases the magma becomes increasingly foamy until, at a void fraction of approximately 75%, the gaseous magma fragments and changes from a bubbly liquid to an ash- and liquid-laden gas, with greatly reduced viscosity. Thereafter, frictional effects are sufficiently small that f can be ignored.

A relationship among pressure, density and volatile content can be formulated as explained in §2.6 to lead to

$$\rho^{-1} = RT(N - n_s)/p + (1 - N + n_s)/\rho_l, \tag{3.15}$$

where ρ_l is the density of the liquid. This relationship can be rearranged to indicate that

$$\phi^{-1} = 1 + (1 - N + n_s)p_s/[(N - n_s)RT\rho_l]. \tag{3.16}$$

Differentiating (3.13) with respect to z after taking the logarithm of both sides and (3.15) with respect to ρ on the assumption that the flow in the conduit is sufficiently rapid that T remains effectively constant, we can write (3.14) as

$$\frac{dp}{dz} = \left[\left(\frac{\rho u^2}{A}\right)\frac{dA}{dz} - \rho g - f\right] \bigg/ \left(1 - \frac{u^2}{a_s^2}\right), \tag{3.17}$$

where $a_s^2 = dp/d\rho$, the square of the speed of sound in the bubbly magma, is given by

$$a_s^2 = (p/\rho)^2 \left[\left(N - \tfrac{1}{2}n_s\right)RT - (p/\rho_l)\right]^{-1}. \tag{3.18}$$

The speed of sound in this multi-phase mixture is generally considerably less than that in a pure gas because the mixture displays the compressibility of a gas with the much higher density, and hence inertia, of a solid or liquid. Equation (3.17), along with the subsidiary equations (2.34), (3.13) and (3.18), can be integrated numerically starting from a given pressure excess (typically between 0 and 100 bar) at the base of the conduit and an assumption of an empirical relationship for the viscosity of the system as a function of ϕ and n_s. So far this has only been done for a straight-walled chamber, $dA/dz \equiv 0$, along with the assumption that the flow does not become supersonic within the conduit. It is well known from the field of gas dynamics that flows become supersonic at either a flow constriction ($dA/dz < 0$) or at an open

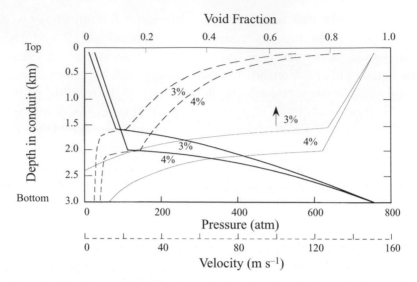

Figure 9. The velocity (dashed lines), pressure (thick lines) and void fraction (thin lines) as a function of depth due to an eruption of magma at 1000 K with two different water contents in a cylindrical conduit of radius 20 m and length 3 km. Note the fragmentation and dramatic change of behaviour at a void fraction of 75% (courtesy of A. W. Woods).

end. Real magmatic conduits no doubt have numerous constrictions and so many interesting results are still to be found. Typical solutions, assuming that the flow is sonic at the vent, 3 km above the base, are graphed in figure 9.

As the magma rises and decompresses, ϕ and u increase and, up to the fragmentation level, the viscosity of the magma increases. Beyond the fragmentation level the flow is effectively inviscid and so the resulting pressure gradient is very much less and virtually constant. The flow exits at a pressure considerably in excess of the atmospheric pressure (typically between 10 and 50 times greater) with $u = a_s$. Numerical evaluation of (3.18) indicates that for $0.03 < N < 0.07$ and $1 < p < 100$ bar, a_s does not depart by more than 5% from $0.93(NRT)^{1/2}$, which gives a way of estimating the exit speeds without detailed solution of the governing equations. For example, for $T = 10^3$ K, $a_s = 115(200)$ m s^{-1} for $N = 0.03(0.06)$. This small variation in exit speed, coupled with the small change in magma density, indicates (through (3.13)) that the mass eruption rate is primarily related to the cross-sectional area of the vent, A. Because the exit flow is heavily overpressured a significant decompression phase must be experienced close to the vent. This is really part of the eruption column dynamics, which is treated in the next section.

Finally, extra effects, not yet fully understood, can arise due to crys-

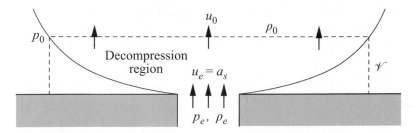

Figure 10. The decompression phase wherein a flow of pressure p_e and density ρ_e exits a conduit at the local sonic speed a_s and decompresses to the local atmospheric pressure p_0, to obtain density ρ_0 and velocity u_0. The temperature during the decompression phase remains essentially constant.

tallization, supersaturation or the delayed nucleation of exsolved bubbles: typical kinetic and disequilibrium effects. Preliminary experiments suggest that supersaturations corresponding to over 100 MPa may be feasible. This would constrain the presence of bubbles until just below the fragmentation level and alter the details of the numerical calculations considerably. Some discussion of these effects is presented in Chapter 3 of Sparks *et al.* (1997).

4 Fluid mechanics and thermodynamics of volcanic eruption columns

Once out of the volcanic conduit, the explosive mixture of gas and ash intrudes into the atmosphere and reacts with it. Recent eruption columns have risen as much as 45 km before invading laterally into the atmosphere to cause local, and even global, changes in the weather and climate. This section presents the main fluid-mechanical concepts used to describe eruption columns, explains why the initially heavier-than-air explosive mixture can penetrate so far into the atmosphere and obtains quantitative relationships for the properties of volcanic eruption columns.

4.1 The decompression phase

The turbulent multi-phase eruption jet is modelled to exit from the top of the vent at sonic speed and exit pressure p_e considerably higher than the atmospheric pressure p_a. A decompression phase must then follow. Consider the vent to open directly into the atmosphere, rather than through a large expanding crater, a situation which will be considered explicitly later. The decompression takes place, as sketched in figure 10, over a relatively short distance (less than a few hundred metres) during which the flow can be considered steady and effects of gravity, friction and fluid entrainment neglected. (This is *not* appropriate higher up in the eruption column.)

The governing equations (Woods 1995) are then mass continuity, expressed by (3.13), momentum continuity

$$\rho \boldsymbol{u} \cdot \nabla \boldsymbol{u} = -\nabla p \qquad (4.1)$$

and conservation of enthalpy

$$\rho \boldsymbol{u} \cdot \nabla [C_v T + (p/\rho) + u^2/2] = 0, \qquad (4.2)$$

where C_v is the specific heat at constant volume. Integrating (4.1) over the control volume $\mathscr{V}$ with the use of (3.13), we determine the velocity at the end of the decompression phase u_0, when the pressure has fallen to p_0, as

$$u_0 = u_e + A_e(p_e - p_0)/Q \approx u_e + A_e p_e/Q = u_e + p_e/(\rho_e u_e), \quad (4.3a\text{–}c)$$

where the subscript e denotes values at the exit point at the top of the vent (and $u_e = a_s$).

Numerical examination of the terms within the square brackets of (4.2) indicates that the specific heat $C_v T$ is of order 10^6 J kg^{-1}, while each of the other two terms is very much less and of order 10^4 J kg^{-1}. Thus the temperature is virtually constant across the decompression region and can be equated to that at the vent, which in turn is virtually identical to that of the original magma in the chamber $T \sim 1000°$C.

At the relatively low pressure of the atmosphere, (2.34) indicates that most of the originally dissolved volatiles have been exsolved by the end of the decompression phase ($n_s \ll N$) and the volume of the gas phase greatly exceeds that of the other phases. Thus the second term on the right of (3.15) is negligible compared to the first term and the density of the flow at the end of the decompression phase $\rho_0 \simeq p_0/(NRT)$. Using the form of this relationship to describe *very approximately* the exit conditions, we substitute $p_e \approx \rho_e/(NRT)$ into (4.3) to determine that $u_0 \simeq 1.9(NRT)^{1/2}$. An accurate numerical evaluation shows that for $0.02 < N < 0.07$ and $p_0 < p_e/10$, $1.7 < u_0/(NRT)^{1/2} < 1.9$, with velocities falling below the lower value for relatively low exit pressures, which correspond to low values of eruption rates Q or vent radii. The area of the jet at the end of the decompression stage, that is at the effective base of the eruption column, A_0, is hence given by $A_0 = Q/(u_e \rho_0) \simeq 0.5 Q p_0 (NRT)^{1/2}$.

If the eruption conduit opens into a large crater, additional phenomena are possible. The sonic flow at the vent will become supersonic within the diverging walls of the crater and the term in dA/dz dominates (3.17). The pressure decreases and, analogously to classical gasdynamic flows from nozzles (Liepmann & Roshko 1957), three different types of flow can issue from the top of the crater. First, the pressure may remain above atmospheric and an overpressured supersonic flow results. At lower exit pressures the flow can decompress relatively further in the crater and an underpressured supersonic flow results. In both these situations a short adjustment region above the crater is required to bring the flow to atmospheric pressure. At even lower exit pressures, a shock develops within the conduit and the material leaves the crater subsonically at atmospheric pressure.

4.2 Eruption columns

Vertical exit velocities of a few hundred metres per second for a heavily ash-laden and gas-dominated flow are impressive, but are nowhere near sufficiently large to take the eruption column many tens of kilometres into the atmosphere, as is observed, merely by converting the available kinetic energy into potential energy. (At the upper range, $u_e = 400$ m s^{-1}, $u_e^2/2g \simeq 8$ km, almost six times too small.) Eruption columns penetrate deep into the stratosphere by converting *thermal* energy into potential energy in a way that will now be described.

The mass fraction of small solids – the volcanic ash – is typically between 0.95 and 0.98, although the volume fraction may be only of order 10^{-4}. The large mass fraction makes the density of the eruption column at its base typically between 20 and 50 times that of the surrounding atmosphere. As the flow in the eruption column develops, the plume entrains the relatively cold surrounding air (as described in Chapter 6). The small hot ash particles readily transfer their heat to the engulfed air and so the bulk density of the plume decreases with height (because the heated air is less dense), while at the same time its upward velocity decreases, partially due to gravitational effects operating on the relatively heavy material and partially due to the necessity of imparting upward momentum to the initially stationary entrained air. The competition between the decreasing density and velocity lead to the two fundamental styles of volcanic eruption. Either the decreasing density dominates and the material of the column becomes relatively buoyant, to lead to what is known as a Plinian eruption column, examples of which include the eruptions of Vesuvius in AD79 documented by Pliny, St. Helens in 1980 and El Cichon in 1982; or the upward velocity ceases before the material becomes buoyant and a collapsed fountain develops which results in a ground-hugging, ash-laden surge known as a pyroclastic flow, such as occurred at Taupo in AD186, Ngauruhoe in 1973 and Pinatubo in 1991. A full description of these flows will be delayed until the next section.

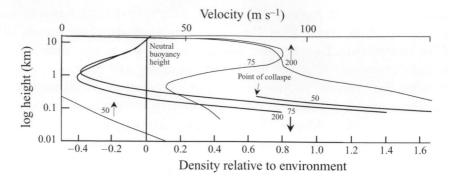

Figure 11. The velocity and density deficiency at the base of the column in an eruption column with initial velocities $u_0 = 50$, 75 and 200 m s^{-1}. Initial values of the mass flux, temperature and water content are 10^{-1} kg s^{-1}, 1000 K and 3% respectively (taken from Woods, 1995).

The equations of motion in the entraining plume, which rises through an atmosphere of spatially-varying properties, are based on the pioneering entrainment assumption of Morton, Taylor & Turner (1956) as described further in Chapter 6 and by Turner (1979, 1986). Allowing for the considerable changes in density involved, i.e. *not* making the Boussinesq approximation, we write the conservation of mass and momentum equations with respect to the vertical z-axis in standard form as

$$\frac{d}{dz}(\rho b^2 u) = 2\alpha \rho_a(z) b u \qquad \text{and} \qquad \frac{d}{dz}(\rho b^2 u^2) = \rho_a b^2 g', \qquad (4.4a,b)$$

where $\rho(z)$ is the mean density in the column of radius $b(z)$, in which the fluid is propagating with mean vertical velocity $u(z)$ through an atmosphere specified by its density $\rho_a(z)$, α is the entrainment constant $\simeq 0.1$ and $g'(z) = (\rho_a - \rho)g/\rho_a$. To (4.4) must be added the energy equation (Woods 1995)

$$\frac{d}{dz}\left[\rho b^2 \left(C_p T + \tfrac{1}{2}u^2 + gz\right) u\right] = 2\alpha \rho_a(z) b u (C_p T_a + gz), \qquad (4.5)$$

where C_p is the specific heat at constant pressure, $T(z)$ and $T_a(z)$ are the temperatures within the plume and surrounding atmosphere respectively, with $T_a(z)$, $\rho_a(z)$ and $p_a(z)$ linked by

$$p_a(z) = \rho_a(z) R T_a(z) = p_0 - g \int_0^z \rho_a(z') dz'. \qquad (4.6a,b)$$

The ordinary differential system (4.4)–(4.6) needs to be integrated numerically from $z = 0$ given initial values of the mass flux Q, momentum flux Qu_0, energy flux $\mathscr{E}$ and a functional form for one of $\rho_a(z), p_a(z)$ or $T_a(z)$. The 'standard' calculation assumes an initially decreasing temperature profile in

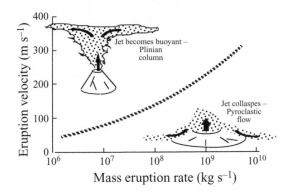

Figure 12. The separation in the plane of eruption velocity against mass eruption rate between a Plinian column and a collapsed fountain.

the atmosphere with constant lapse rate 6.5 K km^{-1} up to a tropopause of between 8 and 17 km, followed by a constant-temperature regime up to the base of the stratosphere at 21 km, with a temperature profile increasing at 2 K km^{-1} beyond that.

A typical set of results is shown in figure 11, which presents the velocity and density deficiency in the jet for three different values of u_0. For the largest of these (200 m s^{-1}) the eruption column density falls off rapidly due to entrainment and the material in the column becomes buoyant at a height of about 200 m. For the smallest u_0 (50 m s^{-1}) the column runs out of upward momentum (coincidentally at about the same height) and a collapsing fountain occurs. Figure 12 presents the broad curve that separates Plinian behaviour from the occurrence of a collapsed fountain. A Plinian column is favoured by lower mass eruption rates and higher eruption velocities.

The original theory of Morton *et al.* (1956) calculated the height of rise H of a buoyant plume in a stratified environment to be given by

$$H = 5F_0^{1/4}N^{-3/4}, \tag{4.7}$$

where F_0 is the initial specific buoyancy flux, here given by $F_0 = Q(T - T_0)g/(\rho_0 T_0)$, and N is the (assumed constant) buoyancy frequency of the atmosphere. Figure 13 presents the theoretical relationship (4.7) using standard atmospheric values, which leads to $H = 260Q^{1/4}$ m if Q is expressed in kg s^{-1}. For comparison, the figure plots the maximum penetration heights of 12 volcanic eruptions that occurred this century. The agreement is remarkably good; and for the 45 km high Bezymianny eruption represents an extrapolation (for the use of the value of entrainment coefficient $\alpha \simeq 0.1$) over more than five orders of magnitude from the 20 cm measurements in the laboratory.

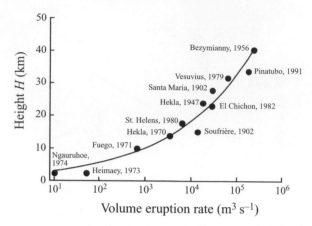

Figure 13. The curve is the theoretical relationship for the height of an eruption column H as a function of the volume eruption rate for standard atmospheric parameter values. The data are from observations of 12 volcanic eruptions in the twentieth century.

Other physical effects play an additional, albeit secondary, role in real eruption columns. The main ones are moisture in the atmosphere and particle fall-out in the plume. The former requires incorporation of standard *wet* thermodynamics in the atmospheric modelling, as laid out in Sparks *et al.* (1997). A discussion of the latter is deferred until the next section. Note that (horizontal) winds in the atmosphere tend to play a rather small role on the column itself, because the (vertical) velocities and turbulent intensities in the plume are so large. Generally, the scale of the vertical velocity w_s takes values between 40 and 400 m s^{-1}. If the typical *horizontal* wind velocity U is very much less than w_s it has very little influence on the plume. If $w_s \ll U$ a weak, bent-over plume develops and the eruption penetrates far less into the atmosphere. Wind may also be important, however, in the dispersion of the final intrusion, known as an umbrella cloud, as will now be described.

4.3 The development of an umbrella cloud

The mushroom-shaped top of an eruption column, which slowly intrudes laterally at its own density level into the stratified atmosphere, is one of the more awesome sights of a volcanic eruption (figure 14). The thickness of the umbrella cloud, h, determined from observations on volcanic eruptions appears to be roughly $\frac{1}{4}H$, with approximately half the cloud above H and half below that height.

If there is no wind, the cloud intrudes radially, driven by the horizontal pressure gradient which arises because of the different vertical hydrostatic

(*a*)

(*b*)

Figure 14. (*a*) The umbrella cloud resulting from the eruption of Mt. Redoubt, Alaska in 1991. (*b*) A simulatory laboratory experiment in which a plume of fresh water laden with particles is released into a salinity gradient.

pressure gradients in the cloud of mean density $\bar{\rho}$ and the atmosphere. Modelling the cloud as an expanding disc or cylinder of thickness $h(t)$ and radius $R(t)$ and neglecting both entrainment of the ambient and particle fall-out, effects which will be discussed in the next section, we can write the mass conservation equation as

$$\pi\bar{\rho}\frac{\mathrm{d}}{\mathrm{d}t}(R^2h) = Q_H, \qquad (4.8)$$

where Q_H is the mass flux of the intruding cloud (which is very much larger than that at the base of the column because of entrainment). Integrating (4.8) on the assumption that the umbrella cloud already has a radius R_0 at the start of the lateral intrusion at $t = 0$, we find that the volume of the cloud, $\pi R^2 h$, increases linearly with time as

$$\pi R^2 h = (Q_H/\bar{\rho})t + V_0, \qquad (4.9)$$

where $V_0 = \pi R_0^2 h_0$ is the initial volume of the cloud.

The cloud is an example of a gravity current, as discussed in much more detail in the next section, which, on the large scale considered here, propagates under a balance between inertial and buoyancy forces (and frictional effects are negligibly small). By dimensional analysis, or by use of Bernouilli's theorem, it can be shown that the horizontal momentum equation can be replaced by the Froude condition

$$\frac{\mathrm{d}R}{\mathrm{d}t} = Fr(g'h)^{1/2}, \qquad (4.10)$$

where the reduced gravity $g' = (\bar{\rho} - \rho_H)g/\rho_H$, ρ_H is the density of the atmosphere at height H and the Froude number Fr is constant $\simeq 1.19$ (§ 5.2). In the vicinity of H the stratification in the atmosphere is effectively linear and so $g' \simeq N_H^2 h$, where N_H is the buoyancy frequency of the atmosphere at a height H. Substituting this relationship into (4.10), we find that $h = \mathrm{d}R/\mathrm{d}t/(Fr N_H)$, which upon substitution into (4.9) and integration yields

$$R^3(t) = \frac{3Fr}{2\pi}N_H\left(\frac{Q_H}{\bar{\rho}}\right)t^2 + \frac{3Fr}{\pi}N_H V_0 t + R_0^3. \qquad (4.11)$$

There is very good agreement between this relationship and data taken from the eruption clouds of St. Helens in 1980, Redoubt in 1990 and Pinatubo in 1991 (Sparks *et al.* 1997).

5 Gravity currents: pyroclastic flows, turbidity currents, lava domes

Gravity currents occur whenever fluid of one density flows primarily horizontally into fluid of a different density. They are driven by horizontal

Figure 15. A hot dense pyroclastic flow from the eruption of Mt. Unzen, Japan in 1991.

pressure gradients which result from the buoyancy (either relatively positive or negative) of the current. (Primarily vertical propagation, driven by vertical pressure gradients due to horizontal buoyancy differences is studied mainly as plumes.) Gravity currents occur in a wide range of natural (and industrial) situations, including: the spread of oil on the surface of water; the motion of a dense, ash-laden pyroclastic flow along the ground (figure 15); the flow of sand- and silt-laden water from the continental shelves across the ocean floor in what is termed a turbidity current; the propagation of relatively dense sea breezes in the atmosphere; and the slow motion of thick sticky lava in a volcanic crater. A nice review of the field is given by Simpson (1997).

Buoyancy forces are always important in the motion of gravity currents, which can flow either at large Reynolds numbers, when the buoyancy forces are balanced by inertia forces, or at low Reynolds numbers, when the buoyancy forces are balanced by viscous forces (and inertial forces are negligible).[1] Each current can be essentially two-dimensional, axisymmetric or influenced by topography. Some gravity currents are the result of a rather rapid release of a given volume of fluid; some are due to a continual flux of fluid; and other possibilities have also been investigated. This section develops

---

[1] Flows at intermediate values of the Reynolds number, where all three forces are roughly balanced, are rarely encountered.

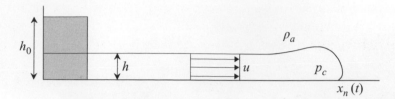

Figure 16. A sketch of a gravity current propagating at high Reynolds number initiated by the instantaneous release of a fixed volume of fluid behind a lock gate.

fundamental concepts used to describe gravity currents and discusses a few geological applications of recent interest.

5.1 Compositional, large-Reynolds-number gravity currents

The simplest gravity currents are driven by differences in composition, such as salt. The fact that the currents tend to be very much longer than they are thick immediately suggests the use of the shallow-water equations (Whitham 1974) wherein the vertical pressure gradient is hydrostatic. Under this assumption, the equations for conservation of mass and momentum (neglecting any entrainment of the ambient which might occur) for a two-dimensional current of density ρ_c and height $h(x,t)$ propagating with horizontal velocity $u(x,t)$ below a layer of fluid of density $\rho_0 < \rho_c$ (as sketched in figure 16) are

$$\frac{\partial h}{\partial t} + \frac{\partial}{\partial x}(uh) = 0 \quad \text{and} \quad \frac{\partial}{\partial t}(uh) + \frac{\partial}{\partial x}(u^2 h) + g'h\frac{\partial h}{\partial x} = 0, \quad (5.1a,b)$$

where the current propagates under horizontal pressure gradient $\partial p/\partial x = -\rho_0 g' \partial h/\partial x$ and reduced gravity $g' = (\rho_c - \rho_0)g/\rho_0$. In order to solve equation (5.1) initial conditions are required – generally describing how the current was initiated – and two boundary conditions – one at each end of the current. The condition at $x = 0$, where in the laboratory there is a vertical wall, is generally $u(0,t) = 0$. The use of Bernoulli's theorem indicates that the velocity at the nose of the current u_N and the depth at the head h_N are related by

$$u_N = Fr(g'h_N)^{1/2}, \quad (5.2)$$

where the Froude number Fr is a constant, determined by perfect-fluid theory to be $\sqrt{2}$ for an infinitely deep upper layer or by experiments on real fluids to be 1.19.[1] The difference between these two values represents the

---

[1] Under quite general conditions in an infinitely deep ambient, u_N can only be a function of g' and h_N. Dimensional arguments then indicate that the only non-dimensional quantity, $u_N/(g'h_N)^{1/2}$, must be a constant (because there is nothing else for it to depend upon).

effects of turbulent Reynolds stresses and viscous drag in the vicinity of the head in a real fluid, which bring about additional momentum transfer at the head and hence retard the flow. (If the undisturbed ambient layer is of finite height $H_a < h_N/0.075$, $Fr = 0.5(h_N/H_a)^{1/3}$.)

There exists a useful family of similarity solutions to (5.1) and (5.2). For a current of fixed volume A per unit width intruding into a layer of very large depth, the solution can be found by substituting into (5.1) and (5.2) the forms

$$g'h(x, t) = \dot{x}_N^2(t)\mathcal{H}(\sigma) \quad \text{and} \quad u(x, t) = \dot{x}_N(t)\mathcal{U}(\sigma), \qquad (5.3a, b)$$

where the similarity variable $\sigma = x/x_N(t)$, to determine that

$$\mathcal{H}(\sigma) = \tfrac{1}{4}(\sigma^2 - 1) + Fr^{-2}, \quad \mathcal{U}(\sigma) = s \quad \text{and} \quad x_N(t) = \mathcal{C}(g'A)^{1/3}t^{2/3}, \qquad (5.4a\text{--}c)$$

where $\mathcal{C} = [27Fr^2/(12 - 2Fr^2)]^{1/3}$ and $x_N(t)$ is the length of the current. This solution is valid some time after the release of the current. If the release takes place by instantaneously lifting a lock gate in front of a rectangular region of heavy fluid, as is frequently the case in laboratory experiments, the current starts in a 'slumping phase', wherein the initial column collapses, a nose is driven forward at virtually constant speed and a return bore in the upper fluid propagates backwards to conserve volume. Once the bore has reflected off the back wall and caught up with the front of the current the similarity form of solution (5.4) becomes valid.

Mathematically different, but qualitatively similar, similarity solutions exist if either the volume increases as a power law in time due to an input flux at the origin or there is flow in an axisymmetric geometry, or both (Bonnecaze, Huppert & Lister 1993; Bonnecaze *et al.* 1995).

A less rigorous, but extremely useful, approach is to consider a simple 'box' model of the flow, which represents the current as a series of rectangles of equal volume with no horizontal variation of properties within the flow. In two dimensions this requires the solution of

$$x_N h_N = A \quad \text{and} \quad \dot{x}_N(t) = Fr(g'h_N)^{1/2}, \qquad (5.5a, b)$$

which, with initial condition $x_N(0) = 0$, is identical to (5.4c) except that $\mathcal{C} = (3Fr/2)^{2/3}$. The difference between these two values of $\mathcal{C}$ is no more than 10% for $1.2 < Fr < \sqrt{2}$. In axisymmetric geometry, the radial extent $r_N(t)$ of an instantaneously released fixed volume of fluid V is given by $r_N(t) = \mathcal{C}_r(g'V)^{1/4}t^{1/2}$, where the box model and similarity solution values of $\mathcal{C}_r$ are $(4Fr^2/\pi)^{1/4}$ and $(4Fr^2/\pi)^{1/4}[4/(4 - Fr^2)]^{1/4}$. A good qualitative feel for the solutions can also be obtained from evaluating and then equating the total buoyancy and inertial forces in the current. This approach is described in the Appendix of Huppert (1982).

Entrainment of ambient fluid into the flow has been investigated by experimentally following the intrusion of an alkaline current into an acidic ambient. Entrainment was seen to take place almost entirely at the head of the current owing to shear instabilities on the interface between the current and the ambient and by the over-riding of the (relatively less dense) ambient fluid by the head. An entrainment or dilution ratio E, defined as the ratio of the volumes of ambient and original fluid in the head, which hence must be non-negative, can be shown by dimensional analysis, and confirmed by experiment, to be independent of g', and to be given in two dimensions by

$$E = [1 - c_1 y_N / A_S^{1/2}]^{-c_2} - 1, \qquad (5.6)$$

where A_S is the volume per unit width of fluid in the head at the end of the slumping phase (which occurs when the current has propagated about ten lock lengths), y_N is the position of the head beyond the slumping point, and $c_1 \approx 0.05$ and $c_2 \approx 1.5$ are empirical constants determined by the roughness of the floor. Note that (5.6) is consistent with the rather surprising result that the entrainment is essentially zero in the slumping phase (when $y_N < 0$).

5.2 Particle-driven gravity currents

When heavy (or possibly relatively less dense) particles drive the flow the major new addition is that the particles fall (or rise) out of the flow and the driving buoyancy continually decreases. The approach most frequently taken to analyse the sedimentation if the concentration is not too large is to assume that the (high-Reynolds-number) flow is sufficiently turbulent to maintain a vertically uniform particle concentration in the main body of the current. However, at the base of the flow, where the fluid velocities diminish appreciably, the settling of particles occurs at the (low-Reynolds-number) Stokes velocity V_s in otherwise quiescent fluid. Quantitatively, this indicates that, neglecting particle *advection* for the moment and assuming that the particles are all of one size, if N_p (which is possibly a function of time and position) denotes the total number of particles per unit horizontal area in a layer of depth h, the change of N_p in time δt, δN_p, due only to the sedimentation is given by $\delta N_p = -V_s C_0 \delta t$, where C_0 is the (number) concentration (per unit volume) just above the base of the flow. Vigorous turbulent mixing implies that $C_0 = N_p/h$, which (on taking the appropriate infinitesimal limits) indicates that $\dot{N}_p = -V_s N_p/h$, a relationship which has been carefully verified by experiments (Martin & Nokes 1988). The incorporation of advection of the particles by the mean flow then results in

$$\frac{D\Phi}{Dt} \equiv \frac{\partial\Phi}{\partial t} + \boldsymbol{u}\cdot\nabla\Phi = -V_s\Phi/h, \qquad (5.7a,b)$$

where Φ is the volume concentration of particles.

Shallow-water equations, akin to (5.1), and incorporating (5.7), can be derived (Bonnecaze *et al.* 1993, 1995). There are no similarity solutions and recourse, in general, has to be made to numerical solution, although it is also possible to develop asymptotic, analytic solutions, based on the smallness of $\beta_s = V_s/(g_0'h_0)^{1/2}$, where g_0' is the *initial* reduced gravity of the system (Harris, Hogg & Huppert 2000).

Insightful box-model solutions are relatively straightforward to obtain. In two dimensions this requires, for the instantaneous release of a fixed volume A per unit width of particle-laden fluid, the solution of

$$\dot{x}_N = Fr(g_p'\Phi A/x_N)^{1/2} \quad \text{and} \quad \dot{\Phi} = -V_s x_N\Phi/A, \qquad (5.8a,b)$$

where $g_p' = (\rho_p - \rho_a)g/\rho_a$, ρ_p is the density of the particles, and the density of the interstitial fluid in the current has been assumed to be identical to that of the ambient. Appropriate initial conditions are

$$x_N = 0 \quad \text{and} \quad \Phi = \Phi_0 \quad (t=0). \qquad (5.9a,b)$$

Dividing (5.8b) by (5.8a) and integrating the resulting ordinary differential equation subject to (5.9), we obtain

$$\Phi(t) = \left(\Phi_0^{1/2} - \lambda_p x_N^{5/2}\right)^2, \qquad (5.10)$$

where $\lambda_p = 0.2V_s/(Fr^2 g_p'A^3)^{1/2}$, from which we can deduce immediately that the current ceases to flow ($\Phi = 0$) at $x_N = l_\infty \equiv (\Phi_0^{1/2}/\lambda_p)^{2/5}$. Introducing non-dimensional variables $\Phi = \phi/\phi_0$ and $\xi = x_N/l_\infty$, substituting (5.10) into (5.8a) and using (5.9a), we determine that

$$\tau = \int_0^\xi s^{1/2}(1 - s^{5/2})^{-1}ds \equiv \mathscr{F}(\xi) \qquad (5.11)$$

in terms of a dimensionless time τ given by $\tau = Fr(g_p'A\Phi_0)^{1/2}(x_N/l_\infty)^{-3/2}t$.

In order to evaluate the resulting deposit distribution, we argue that in time δt, a mass per unit width $\delta M = -\rho_p A\delta\Phi$ is deposited uniformly over a length x_N to lead to a deposit density $\delta\eta = -\rho_p A\delta\Phi/x_N$. Thus the total deposit density (of dimensions ML^{-2}) after the flow has ceased is given by

$$\eta = -\rho_p A \int_{x_N}^{l_\infty} z^{-1}\frac{d\Phi}{dz}dz = \frac{25\phi_0\rho_p A}{12xl_\infty}\left(1 - \tfrac{8}{5}\xi^{3/2} + \tfrac{3}{5}\xi^4\right). \qquad (5.12a,b)$$

Similar results can be obtained for axisymmetric particle-driven gravity currents. Evaluation of the details are left to the reader as an exercise, with the answers given in Huppert & Dade (1998).

The erosion of a sedimentary bed due to the pick-up of particles can play an important role in particle-driven flows. The erosion of a bed by a shear flow, in such a way as to increase the buoyancy, and hence the shear, to lead to a self-accelerating current, a process often called autosuspension, was first considered, independently, by Pantin (1979) and Parker (1982). They derived a fifth-order nonlinear ordinary differential system to describe the evolution of the flow and analysed the behaviour of the resulting solutions using phase-plane techniques. Unfortunately, the values of some of the parameters that appear in the theory are not known and the predicted criteria for particle erosion have not yet been subjected to careful experimental investigation.

An expanded version of this subsection, which brings out additional concepts and examples can be found in Huppert (1998b).

5.3 Some geological applications

There are many geological situations in which particle-driven gravity currents play a fundamental role. One example concerns the motion of pyroclastic flows, resulting from the collapse of volcanic eruption columns. The largest flows have been modelled as isothermal relatively low-concentration entities which spread radially along the ground. A particular problem posed, and answered, by Dade & Huppert (1996) was the determination of the initial conditions of the flow, and especially the initial particulate concentration, given the observed radial distribution of the final deposit. Having set up a general framework, Dade & Huppert applied it specifically to analysing one of the largest eruptions in the last 10 000 years, the eruption of Taupo on the North Island of New Zealand in AD 186. Approximately 30 km^3 of solid material was distributed in a roughly axisymmetric fashion around a vent up to a radius of 80 km as a result of the eruption. The total volumetric flux, after column collapse and the associated entrainment of air, was found to be of order 40 km^3 s^{-1}, over a period of approximately 15 minutes. The initial solids concentration in the pyroclastic flow was around 0.3% by volume – a result consistent with the initial assumption of low particle concentrations.

This low concentration, which was greeted with surprise by some Earth scientists, is consistent with the idea, well known to geologists, that some pyroclastic flows can, quite suddenly, lift up into the atmosphere to form what are called co-ignimbrite plumes. These occur because the small hot ash particles can heat the air in the flow sufficiently that the bulk density of hot air plus particles exceeds that of the relatively cold particle-free air of the surrounding atmosphere. The general analysis of such particle-driven flows with less-dense interstitial fluid can be analysed using the concepts (box models, shallow-water theory, etc.) of the last subsection (Sparks *et al.* 1993; Hogg, Huppert & Hallworth 1999). A simple but instructive calculation equates the density difference purely due to thermal differences $\Delta \rho_T$ to the density difference due to the particulate concentration $\Delta \rho_c = C \rho_p$, where C is the fractional volume concentration of particles of density ρ_p. The equality of these two contributions to the density would strictly be appropriate just at lift-off, but it allows an approximate estimate of the particle concentration to be obtained. Thus with the densities of air at $1000°C$ and $10°C$ being 0.28 and 1.3 kg m^{-3} respectively and $\rho_p \sim 2500$ kg m^{-3}, $C \sim 0.0004$, which indicates that the calculated concentration of particles in the Taupo eruption, although considered low by some geologists, was already considerably larger than that in many other such flows.

Particle-driven flows with less-dense interstitial fluid can also occur when sand- and silt-laden fresh water rivers discharge into a (salty) ocean. The particulate concentration can be sufficiently large that the discharge flows along the bottom of the ocean for many kilometres until, when sufficient particles have fallen out of the flow, the interstitial fluid rises and thereby mixes fresh water into the ocean a considerable distance offshore. Such a situation is believed to be permanently operating in some ten of the world's largest rivers, many of them in China, with the most famous example being the Yellow River.

Large suspension-driven flows at the bottom of the oceans, known as turbidity currents, have been well documented by geologists and are the main mechanism by which sediment from land is transported into the deep sea. Volumes as large as 500 km^3 of sand and silt can propagate many hundreds of kilometres across the ocean floor, as was first effectively realized in 1929 when, owing to an earthquake on the Grand Banks off the eastern coast of the USA, submarine cables were broken sequentially by the resulting turbidity current as it propagated across the Atlantic. Sequential turbidite flows can lay down a series of beds which can act as reservoirs for oil. One of the larger turbidite flows to have been continuously traced on the ocean floor is found on the Hatteras Plain off the eastern coast of North America and is known as the 'Black-Shell' turbidite because of the many small black shells that litter the deposit. The turbidite lies in water 5.5 km deep, covers an area of at least 44×10^3 km^2 and extends for more than 500 km along a fairly straight two-dimensional channel flanked by abyssal hills. Using the observed thickness of the deposit as a function of the distance along the Hatteras Abyssal Channel, a mean fall speed V_s of 0.08 cm s^{-1} (corresponding to silt-sized particles with an effective diameter of 32 μm) and the box-model results developed in §5.2, Dade & Huppert (1994) showed that the deposit resulted from an initial surge 30 km long, 300 m high and approximately 200 km wide (≈ 2000 km^3) containing particulate matter which made up 5% by volume, and 13% by weight, of the surge.

As described at the end of §4, a volcanic eruption plume which intrudes laterally into the atmosphere at its level of neutral buoyancy, contains many small (heavy) ash particles. Consider the penetration as a steady axisymmetric flow with radial velocity u_r of a layer of turbulent fluid of thickness h. The rate of change of the mass M of suspended particles in the current with radial distance r will be given, following (5.7b), by $\mathrm{d}M/\mathrm{d}r = -V_s M/(u_r h)$. Because the flow is steady, the flux $Q = 2\pi r h u_r$ is constant and so

$$M = M_0 \exp\left[-\pi V_s(r^2 - r_0^2)/Q\right], \tag{5.13}$$

where M_0 is the mass of particles in the current at radius r_0. The agreement between the prediction of (5.13) and data from both laboratory experiments and from the measured sedimentation density of, for example, the deposit on the island of San Miguel in the Azores, from the umbrella cloud associated with the eruption of the Agua de Pau Volcano about 3000 BCE, is surprisingly good (Sparks *et al.* 1997).

5.4 Low-Reynolds-number gravity currents

When viscous forces dominate inertial forces, the well-known concepts of lubrication theory (see Chapter 1) can be used to determine the flow. The formulation is then similar to that employed in analysing parallel flow, as described at the beginning of §3. The paradigm situation (Huppert 1982) considers the spreading in either a two-dimensional or axisymmetric geometry of a Newtonian fluid above a horizontal base and seeks the shape and resultant rate of spreading of the current under the assumption that its horizontal extent greatly exceeds its thickness. The pressure p within the current is hydrostatic and thus the pressure gradient driving the flow is proportional to the slope of the height of the (unknown) free surface h. Hence for viscous fluid of density ρ and kinematic viscosity ν extruded from a line source in a two-dimensional fashion, as sketched in figure 17, the parabolic horizontal velocity profile is directly proportional to $\partial h/\partial x$, where x is the horizontal coordinate perpendicular to the source (with $x = 0$ at the source). Use of the depth-integrated continuity equation (with details not given here because they are supplied in Huppert 1982, 1986) then indicates that $h(x, t)$ satisfies

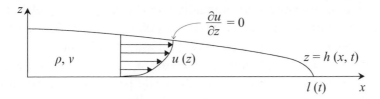

Figure 17. A sketch of a viscous (low-Reynolds-number) gravity current propagating over a horizontal surface.

the singular nonlinear diffusion equation

$$\frac{\partial h}{\partial t} - \frac{g}{3v}\frac{\partial}{\partial x}\left(h^3\frac{\partial h}{\partial x}\right) = 0, \tag{5.14}$$

with the one boundary condition $h[l(t), t] = 0$, where $l(t)$ is the length of the current (cf. equation (3.17) of Chapter 1). If the extrusion of fluid takes place so that the cross-sectional volume per unit width of the current increases with time as qt^α where α is a (non-negative) constant, (5.14) must be solved subject to the global continuity equation

$$\int_0^{l(t)} h(x, t)\,\mathrm{d}x = qt^\alpha. \tag{5.15}$$

With difficulty, numerical solutions of (5.14) and (5.15) can be obtained for any given initial conditions. Alternatively, similarity solutions, to which all solutions with sufficiently regular initial conditions will tend, except possibly at $x = 0$, are easy to obtain. Indeed, determination of the independent similarity variable is sufficient to show that

$$l(t) = \eta_N(\alpha)(gq^3/3v)^{1/5}t^{(3\alpha+1)/5}, \tag{5.16}$$

where the constant $\eta_N(\alpha)$ can be calculated to be a monotonically decreasing function of α with $1.41 > \eta_N > 0.85$ for $0 < \alpha < 2.5$. (Equation (5.16) can also be obtained by globally equating viscous forces to buoyancy forces.)

For an axisymmetric current of radius $r_N(t)$, extruded from the origin so that the volume of the current is Qt^α, the argument outlined above leads to the result

$$r_N(t) = \xi_N(\alpha)(gQ^3/3v)^{1/5}t^{(3\alpha+1)/8}, \tag{5.17}$$

where $\xi_N(\alpha)$ is also a monotonically decreasing function with $0.90 > \xi_N > 0.65$ for $0 < \alpha < 2.5$. In particular, for a viscous current of constant volume, $r_N(t) \propto t^{1/8}$. Numerous laboratory experiments confirm the accuracy of the theoretical approach and predictions.

One of the motivations of the original analysis was to apply the results to examine and interpret the spreading of lava domes, which are often extruded in volcanic craters. For example, commencing in late April 1979, a lava dome spread across the relatively flat base of the 1200 m high Soufrière of St. Vincent in the West Indies, following a series of explosive eruptions starting on 13 April. The lava dome slowly increased in volume, and spread horizontally for the next five months, by which time the pancake-shaped dome had a height of 130 m, a mean diameter of 870 m and a volume of 50×10^6 m^3. Careful measurements during this time indicated that the volume $V(t)$ and radius $r(t)$ as functions of time were represented well by $V(t) = 930t^{0.66}$ and $r(t) = 90t^{0.39}$ (in SI units).

The theoretical result (5.17) can then be used in two different ways. First, the power-law representation of the measured volume indicates that for this dome $\alpha = 0.66$, which should lead to a radial time dependence of $(3\alpha + 1)/8 = 0.37$, in good agreement with the measured value of 0.39. Secondly, the observed coefficient of 90 in the radial spread can be substituted into (5.17) to determine a value of 6×10^7 m^2 s^{-1} for ν.

An isothermal Newtonian fluid captures some of the features of some spreading lava domes, and in so doing acts as the foundation for more developed models which incorporate extra effects. Amongst the effects that need to be included are the cooling of the lava as it flows (possibly in a non-Newtonian way) to form a resistive crust on the surface. This area of research has advanced mainly due to extensive laboratory experiments, with the results backed up by scaling arguments – detailed solutions to governing equations are still awaited. One aim of the research has been to describe the different flow morphologies observed in recently active domes, which include, aside from the Soufrière of St. Vincent, Unzen, Merapi and others.

Fink & Griffiths (1990) conducted a series of experiments in which liquid polyethylene glycol (PEG) was extruded at a constant rate from a point source to flow along a horizontal boundary below a layer of effectively cold water. The water temperature was maintained below the freezing temperature of the PEG. For those experiments with the warmest layer temperatures, the PEG did not solidify but spread as calculated above for an isothermal constant-viscosity (Newtonian) fluid. In experiments with cooler layers, a solid PEG crust formed and one of four different morphologies developed. These were categorized by Fink & Griffiths as: levées; folds; rifts; and pillows. The transition between the different morphologies (and from the formation of no crust at all) was parameterized in terms of a single variable Ψ, which is the ratio of the time for the initially warm PEG to solidify to an advection time scale.

Fink & Griffiths (1998) extended these experiments to mimic a (plastic) yield strength in magma by adding kaolin powder to the liquid PEG. The existence of the resulting internal yield stress had a strong influence on the behaviour of the flows, which displayed morphologies quite different from the previous flows. They argued that the latter experiments are better analogues of the more-viscous magma domes and, in particular, the numerous lava domes recently observed on Venus.

6 Extra topics

There remain numerous areas of geological fluid mechanics which have not been described here. Many (but not all) of them have as a common feature the flow of fluid through a porous medium. The Navier–Stokes momentum equation is then replaced by Darcy's law (or some extension of it), whereby the fluid transport velocity is directly proportional to the driving pressure gradient, with the constant of proportionality being the permeability of the medium divided by the dynamic viscosity of the fluid (Phillips 1991). A large number of interestingly different flows can result.

Deep-sea vents and the hydrothermal circulation. One of the most exciting recent events in the Earth sciences has been the discovery by small manned submersibles of the existence of a chain of vents littering the bottom of the sea, centred on the mid-ocean ridges. As depicted in figure 18, hot water gushes through holes ~ 70 cm in diameter, at velocities of ~ 5 m s^{-1}, into an ocean whose ambient temperature (at that depth) is quite steady at approximately 2°C. These turbulent, entraining plumes, which are part of the hydrothermal circulation, are driven by the heat lost from the hot magma chambers several kilometres beneath the ridge axes. Relatively cold sea water penetrates the oceanic crust over a broad area several tens of kilometres on either side of the ridge axis. As water flows down through the crust it is heated to temperatures as high as 400°C without vaporizing, because of the high pressures. At such temperatures the density of the water is ~ 0.7 times that of cold sea water and hence extremely buoyant. The venting at individual sources represents the return flow. Calculations of thermal flow in a porous medium driven by a heat source, in some ways similar to the high-Rayleigh-number calculations in Chapter 6 (but see also Phillips 1991, §7.6) reflect many of the observed features of the flow. There are indications, however, that the crust is heavily fractured and that the flow is strongly focused along numerous cracks, which will require augmentation of the governing equations.

The turbulent discharge of this less-dense hot water at the bottom of the ocean is another example (following §4.2) of a buoyant plume as described

Figure 18. A typical black smoker vent.

in Chapter 6. Chemical analyses indicate that the mixing of the entrained sea water into the hydrothermal fluid of the plume forms a black precipitate, which has led to the plumes being called 'black smokers'. (A very small number of plumes precipitate white particles and are hence known as white smokers.) Some of the heavy particles fall out of the plume and calculations indicate that more than a half are then re-entrained back into the plume.

After a rise of a few hundred metres, the plumes reach their level of neutral buoyancy, as quantitatively predicted by (4.7), and flow out horizontally into the ocean.

Aside from the chemical and physical importance (and their great fluid-mechanical interest), black smokers have introduced new concepts into biology. At that depth in the ocean, sunlight is totally absent and chemosynthesis replaces photosynthesis as the mechanism for life. Completely new biological organisms have been found around the vents where they thrive on the hot chemically rich environment. It has been seriously suggested that they will represent a significant food source in the future.

Flow and reactions in porous sedimentary rock. A foundation for those aspects of flow in a porous medium that are relevant to sedimentary geology is clearly described by Phillips (1991) with a stimulating summary given in Phillips (1990). One of the central themes of the book and paper is the analysis of slow flows that react thermally and chemically with the solid matrix through which they percolate (cf. the flow through a mushy layer described in Chapter 8). The controlling processes in such flows are the fluid mechanical ones of advection and diffusion, which specify the rate at which reactants can be supplied to the reaction sites, and the chemical ones which specify the rates of reaction. Phillips categorizes three different types of flow. First, isothermal reaction fronts, which propagate like broad shock waves, altering the chemical composition (of both the fluid and the enveloping matrix) from a constant, specified value ahead of the travelling front to a different constant value behind it. These can occur, for example, as a result of the reaction which turns limestone or calcite into dolomite, or, maybe more generally, when an acidic aqueous solution dissolves material in the solid rock matrix.

With $c(\mathbf{r}, t)$ and $\mathbf{u}(\mathbf{r}, t)$ denoting the concentration of the reactant (e.g. magnesium ions, acid, ...) and transport velocity in the fluid as functions of space and time, the governing conservation equation becomes

$$\phi \frac{\partial c}{\partial t} + \nabla \cdot (\mathbf{u}c) = \phi D_e \nabla^2 c + \phi Q_c, \qquad (6.1)$$

where ϕ is the porosity (fluid volume fraction), D_e the effective diffusivity of concentration, and the source term Q_c may be a function of the concentration of each of the reactants, temperature, pressure, space, time and possibly other variables. Note that ϕ (typically $\ll 1$) appears as a pre-multiplicative factor in front of those terms that reflect that c (the acid concentration, say) is confined to the fluid. As a consequence, as seen on dividing (6.1) by ϕ, variations in the acid concentration are advected with fluid velocity $\mathbf{q}/\phi$, which is much larger than the transport velocity q. With $m(r, t)$ denoting the concentration of the other reactant in the solid (e.g. $CaCO_3$; let us call it mineral), the governing equation becomes

$$(1 - \phi)\frac{\partial m}{\partial t} = -\gamma_c(1 - \phi)Q_c, \tag{6.2}$$

where γ_c is a (stoichiometric) constant dependent on the particular chemistry of the reaction under investigation.

With the simplest assumption, that $Q_c \propto mc$, Phillips (1991), Hinch & Bhatt (1990) (and others) treat ϕ as a constant and develop analytical one-dimensional travelling wave solutions of the form

$$c = F(\zeta = x - Vt), \qquad m = G(\zeta), \tag{6.3a, b}$$

with $F \to 0$ (all the acid is used), $G \to m_0$ (the initial mineral concentration) as $\zeta \to \infty$ and $F \to c_0$ (the initial acid concentration), $G \to 0$ (mineralization is complete) as $\zeta \to -\infty$. The resultant velocity with which the front moves, V, is typically a small fraction of the velocity of the acid.

Hinch & Bhatt go on to investigate the linear stability of the front by allowing the permeability (but not the porosity) to be a function of m. This is a general Saffman–Taylor situation (see Chapter 2) and hence if the permeability decreases with m (i.e. in the direction of motion) the front is unstable; otherwise it is stable.

In other situations there are initial spatial gradients of m. A simple travelling wave representation is then not possible because the reaction can continue at different rates within the matrix. Phillips terms this a gradient reaction in which the rate of reaction or deposition in the pores is proportional to the gradient of temperature and to the interstitial fluid velocity in the direction of the gradient. In a faulted or fractured medium, the fluid velocities in the fracture network are typically very much larger than those in the matrix, and the rates of mineral deposition are correspondingly larger, leading to the characteristically vinuous structure of mineral deposits. Finally, Phillips defines mixing zone reactions which occur as a result of intimate mixing between different fluids as they seep through a porous matrix, such as can occur when sea water infiltrates a coastal aquifer initially saturated with fresh water. The ability of areas of relatively high permeability to attract and focus flow from low-permeability regions draws streamlines closer together and hence promotes mixing.

Geothermal reservoirs and their replenishment. Another exciting area of flow in porous media concerns advection and convection in geothermal reservoirs made up of fractured rock at temperatures of up to 400°C which are located in the upper 10 km of the Earth's crust. A number of these on land have been used over the last thirty years or so as thermal energy sources on a commercial scale, although the use of naturally heated water has been common in spas and even in some forms of central heating for a few millennia. One fascinating problem associated with such areas is the formation of geysers, such as at Yellowstone National Park in the United States, which can erupt with remarkable regularity although the essential mechanisms are still not understood. Another problem concerns the input of cold water to replenish warm water extracted by commercial hydrothermal heating plants.

A simple situation, which nevertheless illustrates many of the essential features, occurs when cold water at temperature T_i is input from a small source with flux Q (of dimensions $L^2 T^{-1}$) in a saturated hot two-dimensional porous medium at temperature $T_\infty (> T_i)$. The thermal balance for the temperature T, which is identical in the solid and adjoining fluid, can be written as

$$\overline{\rho C_p}\frac{\partial T}{\partial t} + (\rho C_p)_l(Q/2\pi r)\frac{\partial T}{\partial r} = \overline{\rho C_p \kappa} r^{-1}\frac{\partial}{\partial r}\left(r\frac{\partial T}{\partial r}\right), \qquad (6.4)$$

Woods (1999), where r is the radial coordinate and an overbar indicates a mean value for the system weighted with respect to the volume occupied, i.e. for any variable σ say, $\overline{\sigma} = \phi\sigma_l + (1-\phi)\sigma_s$, where σ_l and σ_s are the values of σ in the liquid and solid (matrix) respectively (see Chapter 8). The different coefficients in (6.4) reflect the fact that heat diffuses throughout the matrix but is advected only by the fluid motion through the pores.

Because there is no externally imposed length scale there is a similarity solution to (6.4) in terms of $\eta = r/2(\kappa_m t)^{1/2}$, where $\kappa_m = \overline{\rho C_p \kappa}/\overline{\rho C_p}$, and the one non-dimensional parameter $\lambda = [(\rho C_p)_l/\overline{\rho C_p}](Q/\kappa_m) \sim (Q/\kappa_m)$, given by

$$T(\eta) = T_i + (T_\infty - T_i)P\left(\tfrac{1}{2}\lambda, \eta^2\right), \qquad (6.5)$$

where $P(a, x)$ is the incomplete gamma function. For large values of λ, which correspond to input fluxes much larger than thermal diffusion, there is negligible conduction of heat from the far field. Instead the input liquid is heated to T_∞ by extraction of heat from the rock matrix near the source. The temperature jumps quite rapidly from T_i to T_∞ over a thin region at a radius (given by $\eta = (\lambda/2)^{1/2} \equiv \eta_T$) considerably smaller than that to which the input liquid has penetrated (given by $\eta = \eta_T/(\pi\phi)^{1/2}$). For $\lambda \ll 1$ the liquid front advances more slowly than the rate at which heat is diffused and the problem approaches the purely thermal diffusion situation discussed by Carslaw & Jaeger (1980). For almost all real geothermal systems $\lambda \gg 1$.

If the porous rock is at a temperature above the boiling point of the input liquid, vapour may be produced ahead of the advancing front. The motion may then be determined by combining Darcy's law for the flow of vapour in the pore spaces, the perfect gas law, and the mass conservation equation, in the form

$$u = (-k/\mu_v)\nabla P, \qquad p = \rho_v RT; \qquad \phi\frac{\partial \rho_v}{\partial t} + \nabla \cdot (\rho_v u) = 0, \qquad (6.6a\text{--}c)$$

where subscript v denotes the vapour phase. This leads to the nonlinear diffusion equation

$$\frac{\partial}{\partial t}(p/T) = \Gamma \nabla \cdot (p\nabla p/T), \qquad (6.7)$$

where $\Gamma = k/(\phi\mu_v)$ and k is the permeability of the porous medium. Again, on the assumption that the flow is radially symmetric, a similarity solution is obtainable, which has a different structure for small, intermediate and large values of Γ. In practice such boiling fronts can become unstable to yet another manifestation of a Saffman–Taylor instability.

The evolution of the inner core. The largest fluid region of the Earth is the liquid outer core, in which there are vigorous convectively-driven motions which maintain the all-important magnetic field of the Earth as described in Chapter 7. The motions are driven by the slow cooling of the Earth, which causes the iron-rich liquid, with minor constituents of between 5% and 15% of sulphur, oxygen, nitrogen and nickel, to solidify almost pure iron on the boundary of the inner core, as mentioned in §1. The convection is partially thermal, driven by the heat transfer associated with the rise of warm fluid (and sinking of cold fluid) within the outer core, and partially compositional, driven by the release of relatively less-dense fluid due to the incorporation of the heavy iron component into the inner core.

A number of (rather complicated) numerical models have been introduced to study this evolution and to evaluate the radius of the inner core $r_c(t)$ as a function of time. A simplifying step was taken by Buffett *et al.* (1996) who developed a new theoretical model for which the results could be obtained analytically, thus allowing a general understanding to be developed of the role of the various parameters, whose explicit values are not very well known. Buffett *et al.* based their model on global heat conservation and the realization that the cooling of the core is regulated by the heat flux $f_m(t)$ that is taken away from the core–mantle boundary by the motions in the sluggish, more massive, mantle. They considered the convection in the outer core to be sufficiently vigorous that the fluid is well mixed there, with a uniform potential temperature (the temperature after subtracting the adiabatic variation) of $T(t)$, which decreases slowly with time. This temperature is equal to the solidification temperature $T_s[r_c(t)]$ of the inner core, which is a strong function of the pressure at $r = r_c$. The thermal evolution of the (radially symmetric) solid inner core can then be determined by solving a thermal diffusion equation in terms of the imposed temperature T_s at the boundary to yield the resultant heat flux at the boundary between the inner and outer core. From this calculation they determined a (fifth-order) polynomial for $r_c(t)$, with an approximate solution

$$r_c(t) = r_b \left\{ \int_{t_0}^{t} f_m(t') \, dt' / \mathcal{M} \right\}^{1/2}, \tag{6.8}$$

where r_b ($= 3454$ km) is the radius of the outer core, t_0 is the initiation time of the growth of the inner core ($\sim 2 \times 10^9$ years ago), and all other variables are grouped in $\mathcal{M} = (2\pi/9) r_b^3 C_p G \rho^2 (\partial T_s / \partial \rho)$ where G is the gravitational constant. Buffett *et al.* (1996) verified this result by a full numerical computation and investigated the relative importance of thermally and compositionally driven convection. They concluded that in the early Earth, when the inner core was much smaller than it is today, thermal convection dominated. Their calculations, extended by Lister & Buffett (1995), indicate that at present the contribution made by compositional convection to the ohmic dissipation, and hence the relative amount of energy available to drive the geodynamo (see Chapter 7), represents approximately three-quarters of the total.

Compaction. The initial production of melt from a solid, in infinitesimal amounts, occurs at the boundaries between individual grains. When this happens in the Earth, some of the melt migrates away from the matrix from which it was formed and eventually finds itself, as part of a much greater volume, flowing into magma chambers, up dykes and partaking in a volcanic eruption. The investigation of the mechanisms by which small amounts of melt percolate through a solid matrix has generated enormous interest over the last two decades amongst Earth scientists, applied mathematicians and fluid dynamicists (as well as being of considerable relevance to areas of metallurgy, petroleum engineering and soil science).

In order for the melt to migrate, the matrix must deform, so as to conserve volume. The resulting motion can be analysed using the concept that the matrix be described as a high-viscosity compressible fluid and the dynamics of melt and matrix be considered separately. Each is assumed to obey the standard conservation laws, with the interaction between the two fluids coupling their motion. This approach was pioneered by McKenzie (1984), who derives carefully the following *compaction equations*, which have now become standard.

In terms of the densities of fluid and solid, ρ_f and ρ_s, and porosity $\phi(\mathbf{r}, t)$, the mass conservation equations for the two phases are

$$\frac{\partial}{\partial t}(\rho_f \phi) + \nabla \cdot (\rho_f \phi \mathbf{v}) = \mathcal{Q} = -\frac{\partial}{\partial t}[\rho_s(1 - \phi)] - \nabla \cdot [\rho_s(1 - \phi)\mathbf{V}], \quad (6.9a, b)$$

where $\mathbf{v}$ and $\mathbf{V}$ are the velocities of melt and matrix, and $\mathcal{Q}$ the mass transfer rate, or melting rate, from matrix to melt. The melt, being much less viscous than the matrix, is transported by the matrix deformation as well as flowing relative to it, as described by Darcy's law in the form

$$\phi(\mathbf{v} - \mathbf{V}) = -[a^2 K(\phi)/\mu_f]\nabla(p - \rho_f g z), \quad (6.10)$$

where a is an average grain size and the dimensionless permeability $K(\phi)$ depends on the geometry of the porous network. The corresponding low-Reynolds-number momentum conservation equation for the (fluid) matrix is written as

$$0 = \nabla(p - \rho_f g z) - (1 - \phi)(\rho_s - \rho_f)\mathbf{g} + \nabla \cdot \left\{ \eta \left[\nabla \mathbf{V} + (\nabla \mathbf{V})^T \right] + (\zeta - \tfrac{2}{3}\eta)(\nabla \cdot \mathbf{V}) \right\}, \quad (6.11)$$

where ζ and η, the bulk and shear viscosities of the matrix, may also be functions of ϕ.

Equations (6.9)–(6.11) display a rich variety of solutions. An important length scale which arises from the equations, as first determined by McKenzie on analysing the simplest problem of the behaviour of a layer with constant porosity above an impermeable boundary, is the *compaction length* $\delta_c = [(\zeta + \tfrac{4}{3}\eta)K/\mu_f]^{1/2}a$ (typically between 0.1 and 1 km for geological situations), which is independent of the all-important density contrast, $\rho_f - \rho_s$ between melt and matrix, although this becomes a controlling factor in the relative velocity and (hence time scale) of the resultant motions. If the porosity decreases with height, finite-amplitude solitary waves can be initiated which propagate upwards at velocities dependent on their amplitude. This behaviour has been reproduced in laboratory experiments by Scott, Stevenson & Whitehead (1986) in which buoyant relatively inviscid water was released at the base of a layer of much more viscous and deformable glycerine. In general, the waves can, dependent on the parameters, have phase velocities greater or less than the fluid velocities and group velocities that are directed either parallel or anti-parallel to gravity. The compaction equations have been used not only to examine the physics of melt extraction from the mantle, but also to study the associated chemical signals. Some of this work is reviewed in the papers which appear in Cann, Elderfield & Laughton (1997).

There are many more topics of immediate concern in understanding the

Earth where fluid mechanics plays a central role. These include such topics as the propagation of plumes through the mantle (Jackson 1998), mountain building and deformation of continents (England & Jackson 1989), deposition and evolution of ore deposits (Phillips 1991) and the role of convection in maintaining the motion of the continental plates (Peltier 1989), plus many others. At the change of the millennium we are just beginning to discover the main processes involved in these fluid motions. A combination of penetrating physical reasoning, powerful applied mathematics and imaginative laboratory experiments will be required to reveal the full range of the further fundamental geophysical mechanics that control the evolution and behaviour of our planet.

Acknowledgements

I am grateful to the late George Batchelor, who over thirty years gave generously of his time to show me how to carry out creative research in real fluid mechanics. Steve Sparks introduced me to geology and has been an ever insightful friend and colleague, as has Stewart Turner. Mark Hallworth has helped immeasurably in much of my research. I owe a particular debt to many students and post-docs with whom I have shared many stimulating intellectual discussions. These include Roger Bonnecaze, Paul Bruce, Brian Dade, Andrew Hogg, Ross Kerr, John Lister, Andrew Woods and Grae Worster. Very helpful reviews of an earlier version of this chapter were sent to me by Ian Campbell, Yves Couder, Brian Dade, Ross Griffiths, Ross Kerr, John Lister, Dan McKenzie, John Miles, Keith Moffatt, Owen Phillips, Steve Sparks, Dave Stevenson, Stewart Turner, Andrew Woods and Grae Worster.

Bibliography

ANDERSON, D. A. 1989 *Theory of the Earth.* Blackwell Scientific, Oxford.

BONNECAZE, R. T., HALLWORTH, M. A., HUPPERT, H. E. & LISTER, J. R. 1995 Axisymmetric particle-driven gravity currents. *J. Fluid Mech.* **294**, 93–121.

BONNECAZE, R. T., HUPPERT, H. E. & LISTER, J. R. 1993 Particle-driven gravity currents. *J. Fluid Mech.* **250**, 339–369.

BROWN, G. C., HAWKESWORTH, C. J. & WILSON, R. C. L. 1992 *Understanding the Earth.* Cambridge University Press.

BRUCE, P. M. & HUPPERT, H. E. 1989 Thermal control of basaltic fissure eruptions. *Nature* **342**, 665–667.

BUFFETT, B. A., HUPPERT, H. E., LISTER, J. R. & WOODS, A. W. 1996 On the thermal evolution of the Earth's core. *J. Geophys. Res.* **101**, 7989–8006.

CAMPBELL, I. H., NALDRETT, A. J. & BARNES, S. J. 1983 A model for the origin of platinum-rich sulphide horizons in the Bushveld and Stillwater complexes. *J. Petrol.* **24**, 133–165.

CAMPBELL, I. H. & TURNER, J. S. 1986 The influence of viscosity on fountains in magma chambers. *J. Petrol.* **27**, 1–30.

CANN, J. R., ELDERFIELD, H. & LAUGHTON, A. 1997 Mid-ocean ridges: dynamics of processes associated with creation of new ocean crust. *Phil. Trans. R. Lond. Soc.* **355**, 213–486.

CARSLAW, H. S. & JAEGER, J. C. 1980 *Conduction of Heat in Solids.* Oxford University Press.

DADE, W. B. & HUPPERT, H. E. 1994 Predicting the geometry of channelised deep-sea turbidites. *Geology* **22**, 645–648.

DADE, W. B. & HUPPERT, H. E. 1996 Emplacement of the Taupo ignimbrite by a dilute turbulent flow. *Nature* **381**, 509–512.

ENGLAND, P. & JACKSON, J. 1989 Active deformation of the continents. *Ann. Rev. Earth and Planet. Sci.* **17**, 197–226.

FINK, J. H. & GRIFFITHS, R. W. 1990 Radial spreading of viscous gravity currents with solidifying crust. *J. Fluid Mech.* **221**, 485–510.

FINK, J. H. & GRIFFITHS, R. W. 1998 Morphology, eruption rates, and rheology of lava domes: insights from laboratory models. *J. Geophys. Res.* **103**, 527–545.

GUBBINS, D 1990 *Seismology and Plate Tectonics.* Cambridge University Press.

HARRIS, T. C., HOGG, A. J. & HUPPERT, H. E. 2000 A mathematical framework for the analysis of particle-driven gravity currents. *Proc. R. Soc. Lond.* A (sub judice)

HINCH, E. J. & BHATT, B. S. 1990 Stability of an acid front moving through porous rock. *J. Fluid Mech.* **212**, 279–288.

HOGG, A. J., HUPPERT, H. E. & HALLWORTH, M. A. 1999 Reversing buoyancy of particle-driven gravity currents. *Physics of Fluids* **11**, 2891–2900.

HOSKULDSSON, A. & SPARKS, R. S. J. 1997 Thermodynamics and fluid dynamics of effusive subglacial eruptions. *Bull. Volc.* **59**, 219–230.

HUPPERT, H. E. 1982 The propagation of two-dimensional and axisymmetric viscous gravity currents over a rigid horizontal surface. *J. Fluid Mech.* **121**, 43–58.

HUPPERT, H. E. 1986 The intrusion of fluid mechanics into geology. *J. Fluid Mech.* **173**, 557–594.

HUPPERT, H.E. 1989 Phase changes following the initiation of a hot, turbulent flow over a cold, solid surface. *J. Fluid Mech.* **198**, 293–319.

HUPPERT, H. E. 1990 The fluid mechanics of solidification. *J. Fluid Mech.* **212**, 209–240.

HUPPERT, H. E. 1993 Bulk models of solidification. In *The Handbook of Crystal Growth* (ed. by D. T. J Hurle), North Holland.

HUPPERT, H. E. 1998a Cambridge geophysics: past and active volcanoes. In *Cambridge Contributions* (ed. S. Ormrod). Cambridge University Press.

HUPPERT, H. E. 1998b Quantitative modelling of granular suspension flows. *Phil. Trans. R. Soc.* **356**, 2471–2496.

HUPPERT, H. E. & DADE, W. B. 1998 Natural disasters: Explosive volcanic eruptions and gigantic landslides. *Theor. Comput. Fluid Dyn.* **10**, 201–212.

HUPPERT, H. E. & SPARKS, R. S. J. 1980 Restrictions on the composition of mid-ocean ridge basalts: a fluid dynamical investigation. *Nature* **286**, 46–48.

HUPPERT, H. E. & SPARKS, R. S. J. 1984 Double-diffusive convection due to crystallization in magmas. *Ann. Rev. Earth Planet. Sci.* **12**, 11–37.

HUPPERT, H. E. & SPARKS, R. S. J. 1988a Melting the roof of a magma chamber containing a hot, turbulently convecting fluid. *J. Fluid Mech.* **188**, 107–131.

HUPPERT, H. E. & SPARKS, R. S. J. 1988*b* The generation of granitic magmas by intrusion of basalt into continental crust. *J. Petrol.* **29**, 599–624.

HUPPERT, H.E. & SPARKS, R. S. J. 1988*c* The fluid dynamics of crustal melting by injection of basaltic sills. *Trans. R. Soc. Edin.: Earth Sci.* **79**, 237–243.

HUPPERT, H. E., SPARKS, R. S. J. & TURNER, J. S. 1982 The effects of volatiles on mixing in calcalkaline magma systems. *Nature* **297**, 554–557.

HUPPERT, H. E., SPARKS, R. S. J., WHITEHEAD, J. A. & HALLWORTH, M. A. 1986*a* The replenishment of magma chambers by light inputs. *J. Geophys. Res.* **91**, 6113–6122.

HUPPERT, H. E., SPARKS, R. S. J., WILSON, J. R. & HALLWORTH, M.A. 1986*b* Cooling and crystallization at an inclined plane. *Earth & Planet. Sci. Lett.* **79**, 319–328.

HUPPERT, H. E. & TURNER, J. S. 1981*a* Double-diffusive convection. *J. Fluid Mech.* **106**, 299–329.

HUPPERT, H. E. & TURNER, J. S. 1981*b* A laboratory model of a replenished magma chamber. *Earth & Planet. Sci. Lett.* **54**, 144–152.

HUPPERT, H. E. & WORSTER, M. G. 1985 Dynamic solidification of a binary melt. *Nature* **314**, 703–707.

JACKSON, I. 1998 *The Earth's Mantle: Composition, Structure and Evolution.* Cambridge University Press.

JAEGER, J. C. 1968 Cooling and solidification of igneous rocks. In *Basalts. The Podervaart Treatise on Rocks of Basaltic Composition Volume 2* (ed. H. H. Hess & A. Poldervaart). Interscience.

JAUPART, C. & BRANDEIS, G. 1986 The stagnant bottom layer of convecting magma chambers *Earth Planet. Sci. Lett.* **80**, 183–199.

KERR, R. C., WOODS, A. W., WORSTER, M. G. & HUPPERT, H. E. 1989 Disequilibrium and macrosegregation during solidification of a binary melt. *Nature* **340**, 357–362.

KERR, R. C. 1994 Melting driven by vigorous compositional convection. *J. Fluid Mech.* **280**, 255–285.

LIEPMANN, H. & ROSHKO, A. 1957 *Elements of Gas Dynamics.* Wiley.

LISTER, J. R. 1990*a* Buoyancy-driven fluid fracture: the effects of material toughness and low-viscosity precursors. *J. Fluid Mech.* **210**, 263–280.

LISTER, J. R. 1990*b* Buoyancy-driven fluid fracture: similarity solutions for the horizontal and vertical propagation of fluid-filled cracks. *J. Fluid Mech.* **217**, 213–239.

LISTER, J. R. & BUFFETT, B. A. 1995 The strength and efficiency of thermal and compositional convection in the geodynamo. *Phys. Earth Planet. Inter.* **91**, 17–30.

LISTER, J. R. & DELLAR, P. J. 1996 Solidification of pressure-driven flow in a finite rigid channel with application to volcanic eruptions. *J. Fluid Mech.* **323**, 267–283.

LISTER, J. R. & KERR, R. C. 1991 Fluid-mechanical models of crack propagation and their application to magma transport in dykes. *J. Geophys. Res.* **96**, 10049–10077.

MCKENZIE, D. P. 1984 The generation and compaction of partially molten rock. *J. Petrol.* **25**, 713–765.

MARTIN, D. & NOKES, R. 1988 Crystal settling in a vigorously convecting magma chamber. *Nature* **332**, 534–536.

MILES, J. W. 1971 *Integral Transforms in Applied Mathematics.* Cambridge University Press.

MORTON, B. R., TAYLOR, G. I. & TURNER, J. S. 1956 Turbulent gravitational convection from maintained and instantaneous sources. *Proc. R. Soc. Lond.* **A** 234, 1–23.

PANTIN, H. M. 1979 Interaction between velocity and effective density in turbidity flow: phase-plane analysis, with criteria for autosuspension. *Mar. Geol.* **31**, 55–99.

PARKER, G. 1982 Conditions for the ignition of catastrophically erosive turbidity currents. *Mar. Geol.* **46**, 307–327.

PELTIER, W. R. 1989 *Mantle Convection and Plate Tectonics*. Cambridge University Press.

PHILLIPS, O. M. 1990 Flow-controlled reactions in rock fabrics. *J. Fluid Mech.* **212**, 263–278.

PHILLIPS, O. M. 1991 *Flow and Reactions in Permeable Rocks*. Cambridge University Press.

PRESS, F. & SIEVER, R. 1986 *Earth*. W. H. Freeman & Co.

SCOTT, D. R., STEVENSON, D. J. & WHITEHEAD, J. A. 1986 Observations of solitary waves in a viscously deformable pipe. *Nature* **319**, 759–761.

SIMPSON, J. E. 1997 *Gravity Currents in the Environment and the Laboratory*. Cambridge University Press.

SPARKS, R. S. J., BONNECAZE, R. T., HUPPERT, H. E., LISTER, J. R., HALLWORTH, M. A., PHILLIPS, J. & MADER, H. 1993 Sediment-laden gravity currents with reversing buoyancy. *Earth and Planet. Sci. Lett.* **114**, 243–257.

SPARKS, R. S. J., BURSIK, M. I., CAREY, S. N., GILBERT, J. S., GLAZE, L. S., SIGURDSON, H. & WOODS, A. W. 1997 *Volcanic Plumes*. Wiley and Sons.

SPARKS, R. S. J., HUPPERT, H. E. & TURNER, J. S. 1984 The fluid dynamics of evolving magma chambers. *Phil. Trans. R. Soc. Lond.* **A** 310, 511–534.

SPENCE D. A., SHARP, P. W. & TURCOTTE, D. L. 1987 Buoyancy-driven crack propagation: a mechanism for magma migration. *J. Fluid Mech.* **174**, 135–153.

SPIEGELMAN, M. 1993 Physics of melt extraction: theory, implications and applications. *Phil. Trans. R. Soc. Lond.* **342**, 23–41.

TURNER, J. S. 1979 *Buoyancy Effects in Fluids*. Cambridge University Press.

TURNER, J. S. 1986 Turbulent entrainment; the development of the entrainment assumption and its application to geophysical flows. *J. Fluid Mech.* **173**, 431–471.

TURNER, J. S. & CAMPBELL, I. H. 1986 Convection and mixing in magma chambers. *Earth-Science Rev.* **23**, 255–352.

TURNER, J. S., HUPPERT, H. E. & SPARKS, R. S. J. 1983 Experimental investigations of volatile exsolution in evolving magma chambers. *J. Volcanol. Geotherm. Res.* **16**, 263–277.

WHITHAM, G. B. 1974 *Linear and Nonlinear Waves*. Wiley.

WOODS, A. W. 1995 The dynamics of explosive volcanic eruptions. *Rev. Geophys.* **33**, 495–530.

WOODS, A. W. 1999 Liquid and vapour flow in superheated rocks. *Ann. Rev. Fluid Mech.* **31**, 171–199.

Institute for Theoretical Geophysics, Department of Applied Mathematics and Theoretical Physics, Silver Street, Cambridge CB3 9EW, UK

10

The Dynamic Ocean
CHRIS GARRETT

1 Introduction

We depend on the oceans for the role they play in determining the climate, for the food and other resources we obtain from them, for waste disposal, for transportation and for defence. They also provide for a wide range of recreational opportunities; even just looking at the sea is a source of fascination. As a fluid medium the ocean presents unique challenges beyond those with which we are qualitatively familiar in everyday life. One key factor is the Earth's rotation which has a major influence on the way in which currents such as the Gulf Stream (figure 1) arise as a response to wind stress over a much larger area. Many motions are driven by buoyancy forces, subtly complicated by the competing influences of heat and salt. This competition manifests itself in powerful ways on the large scale of the global ocean circulation and may be associated with climate variability. On the small scale it gives rise to surprising 'double-diffusive' effects associated with the different diffusion rates of heat and salt (figure 2). Physics and fluid dynamics are also involved in the interaction of the ocean with sediment or with living organisms and at high latitudes in the formation of sea ice.

Topography plays a major role in the response of the oceans, clearly through the effect of coastlines and in controlling the exchange between ocean basins (figure 3), but also in more indirect ways: the turbulence and mixing in the ocean, which control its gross properties and circulation, may in large part be generated at seamounts and over rough topographic features (figure 4).

The range of scales is enormous, with important processes spanning a range of about 10^9, from millimetres to megametres; the large-scale motions set the stage for the small-scale processes which have a significant effect back on the large-scale behaviour of the ocean. Observationally this presents a major challenge, with undersampling in space and time gradually giving

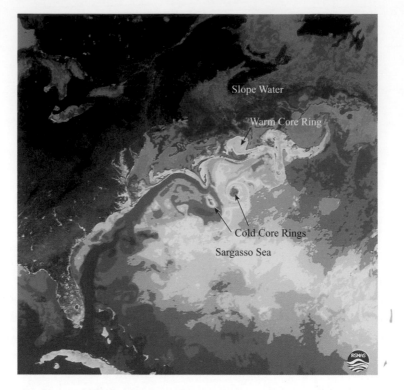

Figure 1. The Gulf Stream, shown here in a false colour image of sea surface temperature, carries warm water up the east coast of North America and across the Atlantic. (Courtesy of Otis Brown, University of Miami.)

way to the power of new technologies, such as remote sensing from Space (figure 1) or *in situ* (figure 5). Theoretically, it remains a major goal to understand the interactions of the motions on different scales; ever more powerful computers permit the development of numerical models that resolve motions to finer scales, but on the ocean-basin scale are still limited to a grid size of several tens of kilometres in the horizontal and many tens of metres in the vertical. Different resolution requirements in the lateral and vertical directions are associated with a tendency for water to move more easily laterally, unconstrained by the buoyancy forces associated with density stratification, but in all directions the effects of 'subgrid-scale' processes at smaller scales must be 'parameterized' in terms of those variables and scales that are resolved.

While laboratory investigation of some processes is useful, oceanography is essentially a field subject, with observations taking the place of controlled experiments. Nonetheless, traditions that have served well in other branches of fluid dynamics, or physics, are also powerful in the study of the oceans.

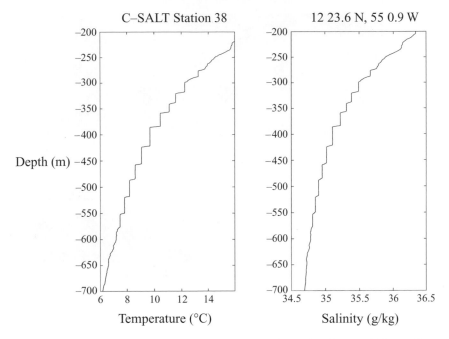

Figure 2. Vertical profiles of temperature and salinity in the Caribbean Sea near Barbados show that warm salty water over cooler fresher water has broken up into a series of layers at depths between 250 and 550 m. (From Schmitt, 1994.)

Two important ones are the use of over-idealized models to help build up understanding and intuition, and the identification of dimensionless numbers to determine a parameter space that needs to be explored.

The purpose of this chapter is to outline first the driving forces and large-scale response of the oceans, and then to focus on the small-scale processes needing parameterization. The number of these is large, with an almost biological complexity. Special consideration is necessary for boundary layers, perhaps tens of metres thick, occurring at the upper surface of the ocean and at the sea floor, with the influence of the latter depending on whether it is flat or has a significant slope. It will not be possible here to pay more than cursory attention to these important regions of the ocean, nor is there space to discuss exciting interdisciplinary areas of oceanic fluid dynamics concerned with living resources or with scientific and engineering problems of the nearshore region.

2 Ocean circulation

The ocean and atmosphere form a coupled fluid dynamical system, with each profoundly influencing the other through the exchange of momentum, heat

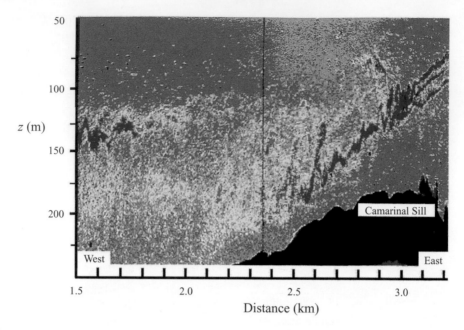

z (m)

Figure 3. Dense water flows out of the Mediterranean Sea (from right to left in this image) over a sill in the Strait of Gibraltar and beneath lighter inflowing Atlantic water near the surface. Strong acoustic backscatter (shown as red in this image) off turbulence and zooplankton at the interface shows that this undergoes an internal hydraulic jump nearly 100 m high. (After Wesson & Gregg 1994. Courtesy of Mike Gregg.)

and water. Here we focus on how the ocean is driven by surface wind stress and by the buoyancy forcing associated with heating or cooling, evaporation, precipitation and river discharge.

Accurate direct measurements of surface fluxes have been very limited in space and time, so that much of our knowledge has come from indirect observations from 'voluntary observing ships', merchant ships that have made millions of routine observations of cloud cover, air and sea temperature, atmospheric humidity, and wind direction and speed (with anemometers gradually replacing visual observations of sea state). These measurements may then be used in conjunction with standard formulae to give the surface fluxes. The wind stress, for example, is assumed to be equal to the square of the wind speed at some standard height times a drag coefficient (which may depend weakly on the wind speed as well as on factors such as the air–sea temperature difference). The data sets suffer from lack of complete spatial coverage (especially in the Southern Ocean) and from changes in observing practice which make real trends in many variables hard to detect, but they span many decades and warrant continued attention. Estimates of surface

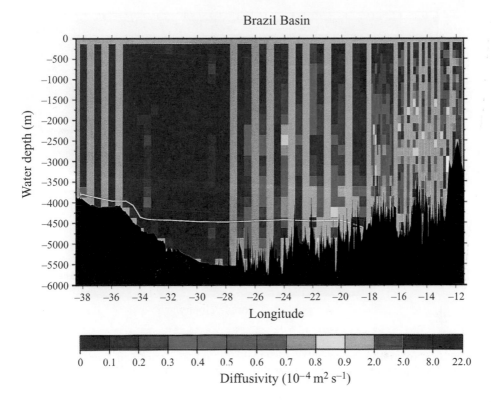

Figure 4. Measurements of the turbulent dissipation rate (expressed in terms of the equivalent vertical diffusivity) in the Brazil Basin in the South Atlantic show enhanced activity over the rough, sloping, topography on the eastern side. (From Toole *et al.* 1997. See also Ledwell *et al.* 2000.)

fluxes of momentum, heat and water vapour are also being obtained from the output of numerical models of atmospheric circulation, though with some uncertainty and sensitivity to weak assumptions. Observations from Space are now providing near-global coverage of many essential variables, but continued refinement is necessary to convert the actual measurements into estimates of the variables of interest. The basis of many techniques is described by Fu, Liu & Abbott (1990), with notable recent successes coming from altimeter measurements (Wunsch & Stammer 1998). In the future, accurate global coverage will be obtained from a combination of dedicated surface observation systems (typically from moored or drifting buoys), from voluntary observing ships and from satellites, with all the data being assimilated into numerical models of the atmosphere and ocean. Chapters in Chassignet & Verron (1998) address these topics as well as other parameterization issues.

Chris Garrett

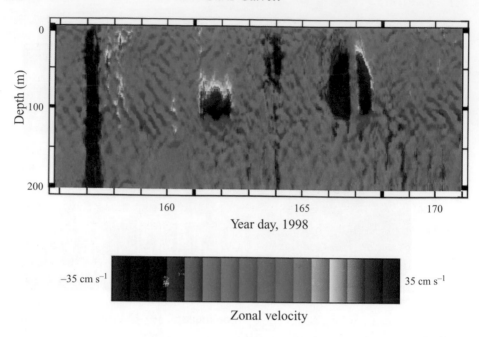

Figure 5. The eastward component of the current in the upper ocean over the Chukchi Cap in the western Arctic Ocean. Data are obtained from the Doppler shifts of a four-beam 161 kHz sonar fixed to sea ice. The vertical band on day 157 is a consequence of ice movement in response to wind. The sloping bands early in the record are near-inertial waves which have a downward phase propagation and hence upward energy propagation (see § 6.2). The feature at days 166 to 168, centred at a depth of 50 m, is an internal eddy in the ocean, with a current that changes direction as the eddy passes. The feature near day 162 at about 100 m is probably the edge of an eddy. (Courtesy of Rob Pinkel, Scripps Institution of Oceanography.)

2.1 Wind-driven circulation

The response of the ocean to surface forcing is neither purely local nor simple. One obvious problem is the existence of the Gulf Stream (figure 1) which transports many millions of cubic metres of warm water each second up the eastern seaboard of North America, without any obvious relationship to local winds. Winds over the North Atlantic do form a clockwise gyre, but of a rather symmetric nature so that the 'western intensification' of the response, as well as its strength, presents a puzzle. The main dynamical balance, as for most large-scale currents, is 'geostrophic'. This is a balance between the Coriolis force and the pressure gradient (associated with a sea surface slope), both at right-angles to the current. As is common in fluid dynamics, however, this pressure gradient is a consequence, not a cause, of the flow. Historical theories of ocean circulation rely heavily, like much of fluid dynamics, on ideas of vorticity conservation (Pedlosky 1996). Ocean dynamics is enriched by the underlying presence of 'planetary' vorticity associated with the Earth's rotation, or, rather, with the vertical component of this given the constraint towards large-scale horizontal motion imposed by gravity and geometry. The spin of this planetary vorticity is anticlockwise in the northern hemisphere. The generation of a spin of opposite sign is expected from the clockwise torque of winds in the North Atlantic; instead, a steady response involves a southward 'Sverdrup' transport to a latitude where the planetary vorticity is less. This water must return north somewhere, and the frictional force at the western boundary imposes an anticlockwise torque that opposes the inviscid tendency of a water column to acquire a clockwise relative spin as it attempts to conserve its total, relative plus planetary, vorticity.

While this theory has remained a useful building block, the actual details of the wind-driven circulation are complicated by nonlinearity (vorticity advection), bottom topography and the density stratification of the water. There is also a danger in carrying over to the large scale of the ocean the ideas of turbulent friction that are familiar in everyday life and in other areas of fluid dynamics that are uninfluenced by rotation. One might, for example, look at the eddying character of the Gulf Stream, on scales of the order of tens of kilometres, and assume that these eddies serve to carry momentum away and reduce the intensity of the stream. In fact, theory and current meter data show that so-called Rossby waves generated by the meanders on the stream tend to carry away momentum of the sign opposite to that expected, giving the mean flow a boost as they leave and producing a counterflow farther away. The parameterization of mesoscale (tens of kilometres) waves and eddies in the ocean is not just a simple matter of adding a local viscous term with a constant eddy viscosity.

2.2 Buoyancy-driven circulation

While the responses of the ocean to forcing by wind stress and buoyancy fluxes are inextricably linked, it is useful to consider the latter separately at first. Dense water tends to be created at high latitudes if the appropriate combination of heating and precipitation minus evaporation is negative. A passing curiosity is that at typical oceanic salinities the impact on density of evaporation is significantly less than that of the associated evaporative cooling (see Gill 1982). The rejection of brine during the formation of sea ice also increases the density of the liquid water. 'Deep convection' can occur in the open ocean (as in the Norwegian and Labrador Seas and in the northwest Mediterranean Sea), with convective plumes that have a horizontal scale of a kilometre or so and are as deep as two or three kilometres. The region of this deep convection, called a 'chimney', can be of the order of a hundred kilometres or more across; the scale is determined either by a region of large buoyancy loss rate or by 'preconditioning' of the underlying water column if local upwelling of a deeper dense layer has reduced the integrated buoyancy of the upper ocean. A rim current around the chimney can then develop as water is driven towards the centre by the radial density gradient and is turned by the Coriolis force. This rim current in turn can become unstable, generating eddies with scales of tens of kilometres which can carry lighter water into the centre of the region of deep convection, covering it over and allowing the dense water to sink (figure 6). The rich fluid dynamics of all this and the fascinating interaction of motions on different scales have been, and continue to be, topics of active research in the laboratory, in the field, and with theoretical and numerical models (see Marshall & Schott 1999).

Dense water may also be formed in the shallow regions of continental shelves where the convection reaches the sea floor, with the dense water then running down the slope, entraining ambient fluid as it does so. Even if the water is dense enough in the formation region to reach the bottom of the deep ocean, it may mix enough *en route* that it actually settles out at mid-depth (as, in fact, happens to the outflow from the Mediterranean Sea, see figure 3; this outflow is initially denser than the water at the bottom of the Atlantic, but it entrains so much that it settles out at a depth of about 1 km, with much of it then spreading across the North Atlantic at that depth as a rather salty, but warm, tongue).

Figure 7 suggests, however, that cold salty water formed around Antarctica does reach the bottom of the deep ocean before spreading north. The equatorward flow of dense water actually occurs in dense western boundary currents along the sides of continental slopes or ocean ridges (e.g. Warren 1981). As with the wind-driven circulation, friction can then take care of vorticity tendencies. The classical assumption, which is largely unverified and undoubtedly too simple, is that the flows peel off into the interior to upwell and slowly move poleward again; the vertical stretching of a water column permits it to adjust its vorticity to match the changing planetary vorticity. Mathematically the governing vorticity equation on this scale is taken as

$$\beta v = f \partial w / \partial z, \tag{2.1}$$

where β is the (positive) northward gradient of the Coriolis frequency f, which is itself negative in the southern hemisphere. Positive $\partial w / \partial z$ is then associated with negative v (southward flow).

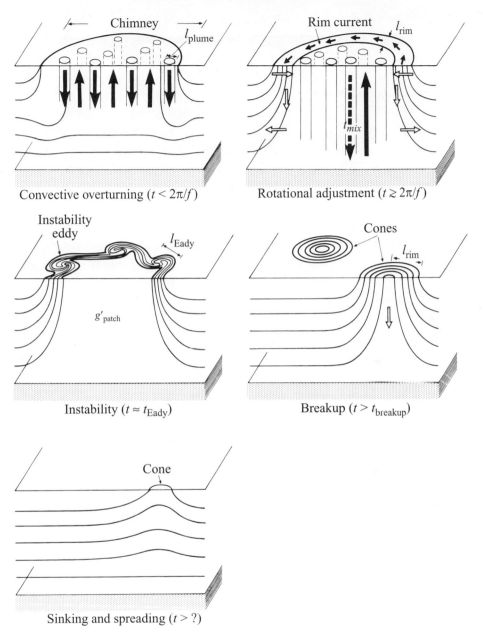

Figure 6. Schematic of various stages of deep convection in a region of the ocean, showing the influence of the Earth's rotation at each stage. (From Send & Marshall, 1995.)

Potential temperature

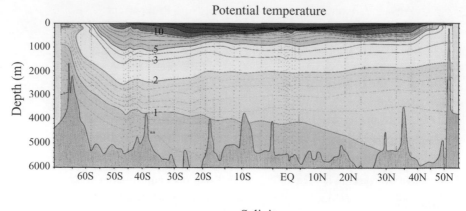

Salinity

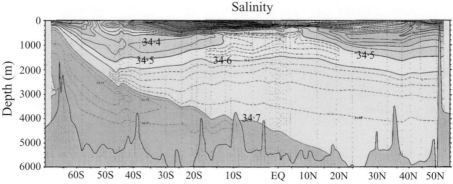

Figure 7. South to north sections of potential temperature and salinity in the Pacific Ocean between 170°W and 180°W. (From Craig, Broecker & Spencer, 1981.)

This circulation leads to a tendency for the ocean to fill up with cold dense water all the way up to the sun-lit upper few tens of metres. This tendency is offset by wind forcing which can 'spin-up' ocean circulation and push warm water down in some areas, but some of the downward penetration of heat seen in figure 7 is associated with mixing. A simple expression of this balance is

$$ w\frac{\partial \Theta}{\partial z} = \frac{\partial}{\partial z}\left(K_v \frac{\partial \Theta}{\partial z}\right), \tag{2.2} $$

where Θ is the potential temperature (the temperature corrected for compression) and K_v the vertical mixing coefficient, or 'eddy diffusivity', giving a vertical turbulent flux $K_v \partial \Theta/\partial z$.

This mixing does seem to be required in order to achieve a buoyancy-driven, or *thermohaline*, circulation. Without it the tropical ocean would become 'warmer' at a higher level than that at which the polar oceans become cooler, and there would be no tendency for warm water to rise and drive a meridional circulation. In order to generate the required buoyancy torque, the heat must be mixed down into the ocean. Scale estimates of the strength of the circulation, partly based on equations (2.1) and (2.2), suggest that, for an imposed density difference between pole and equator, the volume flux in the overturning meridional flow should be proportional to $K_v^{2/3}$, with the thickness of the thermocline (the region of significant stratification between the light upper ocean and the dense abyssal waters) scaling as $K_v^{1/3}$. This is borne out by some idealized numerical models of ocean circulation.

Different power laws result from assuming a surface buoyancy flux rather than a prescribed meridional density difference, and in reality, of course, neither is prescribed but depends on complex interactions between atmosphere and ocean. More importantly, it is possible that for small values of K_v the thickness of the thermocline is actually determined by wind forcing, with the role of K_v being only to determine the thickness of an internal boundary layer in the ocean between a light upper layer and denser abyssal ocean. Actual values of the vertical mixing rate, to be discussed later, suggest that the present-day ocean is closer to this situation than to one dominated by the buoyancy-driven forcing, but it does seem that there is some sensitivity to the value of K_v.

2.3 Multiple states

As in many fluid dynamical situations, there may be multiple equilibria of the thermohaline ocean circulation, stemming from the competing influences of temperature and salinity on density. At present, dense water formation and sinking occur in both polar regions of the Atlantic, with upwelling at low latitudes. Some simple models and numerical experiments, however, have suggested that other circulation states are possible, such as one with sinking at one pole and upwelling near the other.

The simplest model that produces two equilibrium states was originally described by Stommel (1961). He envisaged two well-mixed ocean boxes connected by horizontal pipes, one shallow and one deep, and assumed that each box has surface fluxes of heat and freshwater which work to restore the temperature and salinity to prescribed values, with the restoring time scales different for temperature and salinity. One box tends to be warm and salty (the tropics?), while the other tends to be colder and fresher (high latitudes?). A deep flow is driven from the box with the higher density to the other box, at a rate proportional to the density difference; there is an equal return flow through the shallow pipe. The main result of the fairly straightforward mathematical treatment is that there are two stable equilibrium states. In one the temperature dominates the density difference, with the deep flow being driven from the cold to the warm box, while in the other the salinity dominates and the deep flow is reversed, being driven from the warm, salty box to the colder, fresher one. The actual values of the temperature and salinity are, of course, very different in the two states.

Models with three boxes, to allow for both polar regions as well as a sub-tropical region, show four equilibrium states. Two are similar to Stommel's, showing either poleward or equatorward surface flow, but two are combinations in which the main circulation is pole-to-pole, with either direction possible. This possibility also occurs in more elaborate ocean general circulation models and in models which allow for an interacting atmosphere. The magnitude of the external stimulus required for a change of state in the real world is a major topic of research, as is the possibility that the ocean–atmosphere system is inherently unstable and can switch between states, or at least have a variable strength of the circulation in one state.

2.4 El Niño

A prime example of spontaneous variability in ocean circulation occurs in the tropical Pacific Ocean. In a normal year, westward winds drive warm water to the west, creating conditions there for intense atmospheric convection and heavy regional rainfall (figure 8). Every few years, however, the winds relax and the warm water sloshes back across the Pacific, taking the region of atmospheric convection with it and having a global impact on weather patterns. The timing in relation to the annual cycle is such that the warm water generally reaches the coast of South America near Christmas time; this has given rise to the name El Niño (the Christ-child) for the phenomenon. Good descriptions are given by Philander (1990) and up-to-date information is available on the World Wide Web at a number of sites.

Models with varying degrees of complexity give useful predictions with lead times of up to a year, but much uncertainty remains. Refined coupled ocean–atmosphere models are needed, as are simple conceptual models which describe the essential physics, build understanding and help define the processes to which detailed models might be sensitive. One such model is the 'delayed oscillator model'. Figure 9 shows a cartoon of a developing El Niño, with strong coupling between ocean and atmosphere in the central region of the equatorial Pacific. A positive temperature anomaly there would lead to enhanced atmospheric convection, sucking in more air from the west, and thus enhancing the eastward wind or at least reducing the westward

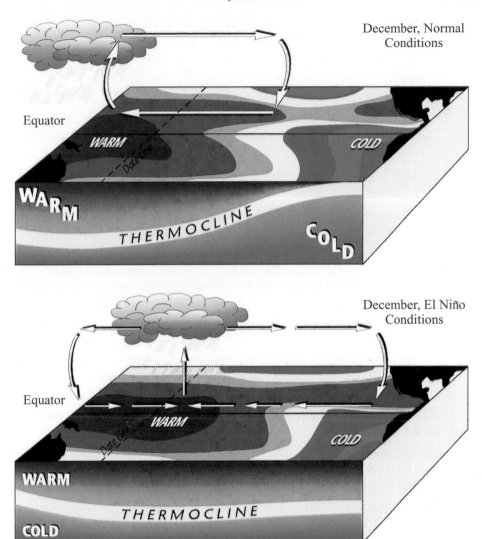

Figure 8. Schematic of the state of the equatorial Pacific during normal and El Niño conditions. (From Philander 1992. ©Jayne Doucette, Woods Hole Oceanographic Institution.)

wind. This would lead to local deepening of the thermocline and hence less mixing of colder water from below, providing a positive feedback for the local temperature anomaly. The model equation

$$\frac{\mathrm{d}T}{\mathrm{d}t} = T(t) - T(t)^3 - \alpha T(t - \delta) \tag{2.3}$$

has been proposed for suitably scaled temperature T as a function of scaled

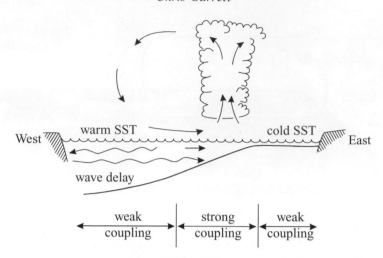

Figure 9. Schematic of a developing El Niño. The model equation (2.3) is for the sea surface temperature in the central equatorial Pacific where there is strong coupling between ocean and atmosphere. (Redrawn from Suarez & Schopf, 1988.)

time t. The first term on the right-hand side represents the positive feedback described above, the second term allows for finite-amplitude limitation (such as by increased evaporative cooling and long-wave radiation as T increases). The final term represents a delayed negative feedback associated with the westward propagation of Rossby waves generated by the dynamically unbalanced state of the local perturbation and their reflection from the western boundary as internal Kelvin waves. (Rossby and Kelvin waves are two varieties of long-period ocean waves affected by the Earth's rotation.)

If the feedback parameter $\alpha < 1$, (2.3) has two stationary states, but instability exists in a region of the (δ, α) parameter space for sufficiently large values of either. The physics of this is actually a matter of common experience, familiar, for example, from many hotel showers! The shock of initially cold water leads one to turn up the hot water tap (a negative feedback response to the temperature anomaly) and to continue to do so if the hotel plumbing causes a delay in the response. This can lead to an overshoot beyond the optimum setting, so that eventually excessively hot water arrives and the process is repeated in the reverse direction. Finite-amplitude oscillations occur and continue if either the negative feedback is too strong or the delay too long; stability requires a short delay, or, surprisingly, a weak negative feedback. It is tempting to conclude that the absence in the Atlantic of a strong, locally generated, equivalent to El Niño is associated with the smaller size of the basin and so a less delayed feedback.

The regular periodicity of the model response is obviously inconsistent with the observed irregularity, but Neelin *et al.* (1998) review simple, physically-based, modifications of the model that lead to chaotic, rather than periodic, behaviour. The basic model is, of course, vastly oversimplified in its neglect of spatial variations other than in consideration of the delayed feedback. It is, however, useful in drawing attention to the requirement for more complicated models to have accurate parameterization of processes such as the local positive feedback (and hence air–sea interaction and vertical exchange processes in the upper ocean) and the wave propagation speed and reflection properties (and hence the vertical structure of the ocean and boundary conditions at the western boundary of the ocean).

The model is also useful in its suggestion that El Niño is essentially an example (like the hotel shower) of a delayed negative feedback oscillation, in this case involving both ocean and atmosphere. Similar analogies for other long-term variability in the ocean and atmosphere would be equally useful. Coupled numerical models of the ocean and atmosphere, as well as more comprehensive data sets, will obviously play a major role in the development of understanding and predictability, with the treatment of resolvable processes limited only by the accuracy of numerical techniques. There will continue to be a need, however, to explore those processes that are not explicitly resolved in the models but which must be parameterized accurately because the model output is sensitive to them. Such processes, including mixing in the ocean and the behaviour of boundary layers, are the main topics of this review.

3 The parameterization of small-scale processes

Fluctuations in the velocity and properties of the ocean can occur at scales as small as a few millimetres, so that a numerical model used to perform a direct numerical simulation, using only the Navier–Stokes and thermodynamic equations, would need a resolution of about 1 mm and contain about 10^{26} elements. The largest numerical models currently in use have a resolution of about 10 km in the horizontal, 100 m in the vertical, perhaps 10^7 elements in all. The development of such models is a formidable achievement, but it seems very unlikely that the gap of 10^{19} will ever be entirely bridged. It is thus imperative that we learn how to parameterize the unresolved processes and use the models to investigate the sensitivity of model output to the parameterization. In the simplest situation one would proceed with a 'Reynolds decomposition' (e.g. Tennekes & Lumley 1972) of the velocity and scalar properties of the ocean, assuming a resolvable slowly varying mean and an unresolved fluctuation, and then seek parameterizations of the 'eddy fluxes' such as $\overline{u'C'}$ for velocity fluctuation u' and scalar fluctuation C', where the overbar signifies the ensemble average.

Replacing the ensemble average with a time or space average requires that there be a 'spectral gap', i.e. a band of time or space scales, separating the mean and fluctuations, with lower energy than the mean or fluctuations. If this is the case, then an average over the time or space scale at the centre of the gap will be long enough to give eddy fluxes from the fluctuations, but short enough that the low-frequency, or low-wavenumber, behaviour may be treated as slowly varying. If this is not the case it is difficult to write down equations for the effect of unresolved processes on those that are resolved. An example is three-dimensional turbulence with a continuous inertial subrange; group renormalization techniques have been applied in an effort to determine the effect of high-wavenumber fluctuations on lower-wavenumber motions, but only in specific problems, and without universal understanding or acceptance (Frisch 1995).

Wunsch (1996) has argued that there are no useful spectral gaps in ocean circulation dynamics, though it does seem that the persistence of a general feature such as the Gulf Stream implies some kind of a separation between mean and fluctuations for horizontal motions. It is possible that a gap may not exist in both space and time. Mid-latitude internal waves in the ocean, for example, can have spatial scales comparable to those of fronts and eddies, but still have shorter time scales. In equatorial regions, however, phenomena with different dynamical balances may overlap in both frequency and wavenumber, complicating the separation problem. It is possible in such cases that describing the effect on resolved motions of some physical process that one wishes just to parameterize will be achieved by a more sophisticated decomposition than just averaging. The spectral gap problem is one that must always be borne in mind, but we proceed on the assumption that it is meaningful to decompose a field into a mean and fluctuations, and first discuss some simple kinematics.

3.1 A framework

We consider the concentration of a scalar field to be made up of a slowly varying mean field $\overline{C}(x, t)$ and a fluctuation C' which may be written $C' = -X \cdot \nabla \overline{C}$, where X is the displacement of a particle from a position where its value of C matches the mean value. This assumes that the displacement is small compared with the scale over which the gradient of $\overline{C}$ varies significantly and also assumes that a fluid particle retains its value of

C, i.e. that the effect of molecular diffusivity is negligible. We will return to these issues, but for now note the consequence that the eddy flux is

$$\overline{u_i' C'} = -\overline{u_i' X_j} \, \frac{\partial \overline{C}}{\partial x_j}. \tag{3.1}$$

The tensor multiplying the gradient of $\overline{C}$ may be decomposed into its symmetric and antisymmetric parts as

$$\overline{u_i' X_j} = K_{ij} + S_{ij}, \tag{3.2}$$

where the symmetric part $K_{ij} = \frac{1}{2}(\overline{u_i' X_j} + \overline{u_j' X_i})$. In the coordinate frame in which it becomes diagonal it represents fluxes proportional to the mean gradients in the three principal directions. In the ocean it is generally assumed that one direction will be normal to the mean density surfaces, known as isopycnals, with an associated small 'diapycnal' diffusivity across them, and that the diffusivities in the other two directions, on the isopycnals, are much larger (they need not be equal).

The antisymmetric term $S_{ij} = \frac{1}{2}(\overline{u_i' X_j} - \overline{u_j' X_i})$ has an associated flux $F_{Si} = -S_{ij}\partial \overline{C}/\partial x_j$ which may be written

$$F_S - -D \wedge \nabla \overline{C}, \quad \text{where} \quad D = \tfrac{1}{2}\overline{X \wedge u'}, \tag{3.3}$$

and so is orthogonal to the gradient of the mean concentration field (hence the name 'skew flux' for F_S). This surprising result can be understood if one rewrites the flux as

$$F_S = -(\nabla \wedge D)\overline{C} + \nabla \wedge (D\overline{C}). \tag{3.4}$$

The second term on the right-hand side of (3.4) is non-divergent and so does not contribute to the equation for the evolution of the mean field $\overline{C}$. The first term is an advection of the mean concentration by a velocity $\nabla \wedge D$. As D is related to a correlation between particle velocity and displacement, it can arise if there is a systematic tendency for particles to turn in one direction as they move, as discussed by Moffatt (1983). For a statistically stationary field of waves, the advective velocity may be written $\overline{X \cdot \nabla u'}$. This is just the Stokes drift of the particles, i.e. the extra drift of Lagrangian particles over and above the mean Eulerian velocity at a point (Middleton & Loder 1989), as obtained from a Taylor series expansion of the velocity of a particle displaced by X. A simple example occurs for surface waves in which the orbital velocity is elliptical and decreases from the surface: a water particle will drift in the direction of wave motion, advecting properties with it, and hence causing a flux normal to the mean gradient if this is vertical.

This purely kinematic discussion shows that the eddy flux, in a situation in which particle displacements are small compared with the distance over which the gradient of the mean concentration changes greatly, may be represented by a diagonalizable symmetric diffusivity tensor and a mean velocity which must be added to the mean Eulerian velocity. Values of the diffusivity and advective velocity depend on the physics of the small-scale motions.

As the symmetric tensor K_{ij} from (3.2) is diagonalizable, it may be examined in just one direction, corresponding to one of the principal axes. Dropping subscripts, and recognizing that the particle velocity $u = dX/dt$, with X the displacement as before, we write

$$K = \overline{X \frac{dX}{dt}} = \int_0^t \overline{u(t')u(t)}\, dt' \qquad (3.5)$$

with t being the time elapsed since the particle was at its reference position, and after the interchange of the integration and averaging. This may be written

$$K = \overline{u^2} \int_0^t R(t-t')\, dt' = \overline{u^2} \int_0^t R(\tau)\, d\tau, \qquad (3.6)$$

where $R(\tau)$ is the autocorrelation function of the Lagrangian velocity u, with $R(0) = 1$ and $R(\tau)$ tending to zero for large τ. The integral $\int_0^\infty R(\tau)\, d\tau = T_L$, the 'Lagrangian integral time scale'. As a consequence, for a particle released from a fixed spot, the mean-square displacement averaged over an ensemble of releases is $\overline{X^2} \simeq \overline{u^2}t^2$ for $t \ll T_L$ (as the velocity does not change over a short time and so the particle will reach a distance ut from the origin) and $\overline{X^2} \sim 2\overline{u^2}T_L t$ for $t \gg T_L$. The equivalent diffusivity is $\overline{u^2}t$ for $t \ll T_L$, thus increasing with time but reaching a constant $\overline{u^2}T_L$ for $t \gg T_L$. This may be written alternatively as $K = (\overline{u^2})^{1/2}L$, where $L = (\overline{u^2})^{1/2}T_L$ is a 'mixing length'.

This theory is the extension to continuous movements of the (one-dimensional) drunkard's walk for which the expected displacement after t steps is $st^{1/2}$ for a step length s. If T_s is the time interval between steps, the equivalent diffusivity is $\frac{1}{2}s^2/T_s$, or $\frac{1}{2}u^2 T_s$ if u is the average speed during each step. This agrees with the general formula $K = \overline{u^2}T_L$ as the velocity autocorrelation decreases linearly from 1 at zero lag to 0 at lag T_s, giving $T_L = \frac{1}{2}T_s$.

As implied earlier, in more than one dimension the theory applies to each principal axis of the symmetric diffusivity tensor. It is important to recognize that, while the mean-square velocity may be derived from either Lagrangian or Eulerian data, the time scale involved is the integral decorrelation time of the Lagrangian velocity. This is, in general, shorter than the Eulerian equivalent T_E evaluated in a framework in which the mean flow is zero.

These results may be used to derive the expected average concentration field of a scalar dispersing from a continuous point source. The smaller diffusivity for the 'young' particles near the source leads to a stronger singularity than for a constant diffusivity: in three dimensions, with isotropic dispersion, the concentration falls off like $1/(\text{distance})^2$ rather than the well-known $1/\text{distance}$. As remarked earlier, however, motions in the ocean are very anisotropic. Lateral mixing along mean density surfaces is uninhibited by stratification and is associated with eddies on the scale of tens of kilometres or more, whereas the very much weaker 'diapycnal' mixing across density surfaces is inhibited by buoyancy and is generally associated with mixing events on the scale of a metre or so. The above discussion of near-source behaviour is thus relevant to horizontal stirring, but the vertical spreading over a scale of more than a few metres can be represented by a constant diffusivity. This combination still leads to a larger near-source concentration than would occur if the lateral dispersion rate attained its long-time value everywhere; the result is relevant to marine pollution problems.

For naturally occurring scalars in the ocean it is generally assumed that the typical particle has dispersed far enough from the position where its properties match the mean field that we may use the long-time limit of the theory, though application of the basic theory to the ocean is hampered by the effects of a sheared mean flow and inhomogeneity in the eddy field. The effects of the latter are intriguing and may be illustrated by considering the evolution in one dimension of the mean concentration field $\overline{C}(x,t)$ subject to a diffusivity $K(x)$. The governing equation

$$\frac{\partial \overline{C}}{\partial t} + u \frac{\partial \overline{C}}{\partial x} = \frac{\partial}{\partial x} \left(K \frac{\partial \overline{C}}{\partial x} \right) \tag{3.7}$$

may be written

$$\frac{\partial \overline{C}}{\partial t} + \left(u - \frac{\mathrm{d}K}{\mathrm{d}x} \right) \frac{\partial \overline{C}}{\partial x} = K \frac{\partial^2 \overline{C}}{\partial x^2}, \tag{3.8}$$

suggesting that the mean concentration field has a tendency to move away from regions of large $K(x)$. If, however, we define the position of the centre of mass of the scalar by $r = \int_{-\infty}^{\infty} x\overline{C}\,\mathrm{d}x$, assuming that $\overline{C} \to 0$ as $x \to \pm\infty$ and that $\int_{-\infty}^{\infty} \overline{C}\,\mathrm{d}x = 1$, then it is easy to show that

$$\frac{\mathrm{d}r}{\mathrm{d}t} = \int_{-\infty}^{\infty} \left(u + \frac{\mathrm{d}K}{\mathrm{d}x} \right) \overline{C}\,\mathrm{d}x, \tag{3.9}$$

demonstrating a tendency for movement towards regions of high diffusivity! The 'pseudo-advection' on the left-hand side of (3.8) is outweighed by the effect of the spatially varying diffusivity on the right-hand side. Further details, and complications, of the inhomogeneity of the ocean are discussed by Davis (1991).

This discussion of the spread of a scalar by an irregular velocity field really describes the dispersion of discrete particles. For a continuous scalar field the stirring leads to the development of streaks on a finer and finer scale until eventually there is a balance between lateral compression of a streak and the effects of smaller-scale, even molecular, diffusion. This happens rather quickly but does cause a small reduction of the lateral dispersion rate for the scalar, compared with particles, as the scalar diffuses out of spreading filaments of high concentration moving down-gradient and into regions of low concentration which are likely to be moving up-gradient. The issue is complex, however, and cannot be regarded as completely settled, particularly in its application to the ocean. What, for example, are the small-scale processes that eventually take over from the lateral stirring of temperature and salinity on isopycnals?

This discussion has focused on the flux of a conservative scalar. An appropriate framework for dealing with the Reynolds stresses in the momentum equation is still a topic of research. Some approaches treat potential vorticity as a conservative scalar, subject to turbulent diffusion like any other scalar, but care is needed to avoid creating unphysical sources of momentum, rather than the redistribution implied by the Reynolds-averaged equations.

The appropriate parameterization of unresolved processes clearly depends on what scales of motion are being resolved, but the need for parameterization also needs to be determined. As discussed above, the large-scale thermohaline circulation of the ocean seems to be sensitive to the strength of vertical mixing, at least above a certain value. On the other hand, in a model that resolves mesoscale eddies it is quite likely that model behaviour is insensitive to the value assigned to a 'horizontal eddy viscosity' used to represent the final viscous damping that acts on the eddies, in much the same way that the overall behaviour of turbulence in a homogeneous fluid is insensitive to the precise value of the Reynolds number provided that it is large enough. The questions remain, of course, as to whether a horizontal viscosity is a correct representation of the physical processes that prevent the indefinite sharpening of horizontal gradients of horizontal velocity components and whether these processes also involve mixing of scalars such as heat and salt.

3.2 Approaches

Historically there have been three approaches to determining the correct representation of the effects of unresolved processes in the ocean:

Inference: Models of the ocean that contain parameterizations of unresolved processes may have the associated coefficients adjusted until the model output, for the resolved features of the ocean, matches observations. This assumes, of course, that the form of the parameterization (such as local eddy vicosity and diffusivity coefficients) is appropriate.

Measurement: It is clearly desirable to measure eddy fluxes directly. This is sometimes statistically or technically impossible, but other slightly less direct measurements can provide useful answers.

Process studies: Observations and theory may lead to an understanding of the unresolved processes, and hence formulae for the eddy fluxes.

These approaches will be discussed in a little more detail shortly, with an emphasis on the particular problem of vertical mixing in the ocean, but some general points can be made. First, one expects the three approaches to give the same answer, bearing in mind that an inferred value may represent an average geographically, temporally, and over different processes. Secondly, while inferred and measured values should be representative of the ocean as it is now, models with predictive capability for different conditions (as might occur in climate change) require formulae, not just values, for the parameterizations in terms of resolved processes; these formulae can only come from an understanding of the physical processes and might give different rates under different conditions. Neglecting this point might lead to the omission of an important feedback process. Suppose, for example, that a warmer world is predicted as a consequence of anthropogenic changes to atmospheric composition or the land surface. If this were to lead to stronger winds over the ocean this could cause more vertical mixing and reduce the increase in sea surface temperature, possibly providing a negative feedback on the warming that would be excluded if one were to keep oceanic mixing rates constant.

4 Inference

One rather convincing indirect demonstration of the presence of mixing has come from the study of a number of deep oceanic basins into which water flows as a cold current through a channel, and perhaps over a sill, but leaves at a higher temperature due to the downward mixing of heat. As sketched in figure 10, an isotherm, with potential temperature Θ_1, say, then intersects the sea floor all the way around the basin except at its entrance. In a steady state

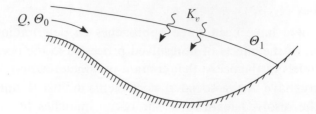

Figure 10. Schematic diagram of the heat budget of an abyssal basin.

the volume flux Q enters the basin with an average potential temperature Θ_0 and leaves across the isotherm with potential temperature Θ_1. This causes a heat loss, since $\Theta_1 > \Theta_0$, and must be compensated by downward mixing of heat. The average mixing rate K_v is then obtained from

$$K_v A \frac{\partial \Theta}{\partial z} = Q(\Theta_1 - \Theta_0), \tag{4.1}$$

where A is the area of the isotherm within the basin and $\partial \Theta / \partial z$ the vertical potential temperature gradient on it. (More properly, the coordinate and diffusivity are normal to the potential temperature surfaces rather than vertical.)

Using this formula, a vertical diffusivity of a few times $10^{-4}\,\mathrm{m^2\,s^{-1}}$ has been found for a number of deep ocean basins. Uncertainty can be associated with the difficulty in measuring the inflow rate Q and there are other complications. One is that the isotherms may not coincide with isopycnals (constant-density surfaces), so that mixing across isotherms may be accomplished by mixing along the differently inclined isopycnals as well as by mixing across the isopycnals. Even a small difference in angle can have a large influence because mixing along isopycnals, unconstrained by buoyancy forces, is generally several orders of magnitude greater than across them. The problem can be avoided if a buoyancy, rather than heat, budget is considered.

It is also most important to remark that the diapycnal (cross-isopycnal) diffusivity derived from a basin budget is an average, possibly including strong mixing in the immediate vicinity of the sloping and irregular sea floor as well as mixing in the interior of the basin. In the Brazil Basin, in fact, it seems that mixing near the rough topography of the eastern side of the basin (figure 4) may be the dominant process. In some basins the mixing across some deeper isotherms may even be dominated by vigorous mixing in internal hydraulic jumps (as illustrated in figure 3) in the inflow over a sill.

In the upper ocean a similar approach may be used for bowls of warm (or buoyant) water, near the sea surface, which are entirely enclosed by a mean isotherm (or isopycnal) that intersects only solid boundaries or the sea surface. Assuming that the volume is unchanged from year to year, then the average velocity across the surface of the bowl is zero and the heat or buoyancy flux across the sea surface must be balanced by a diffusive flux across the surface of the bowl (assuming that there is no interannual change in heat or buoyancy content). A complication is that the result may reflect processes occurring in the surface mixed layer rather than in the stratified interior, in much the same way that abyssal budgets may reflect boundary processes; the mixing across density surfaces in the ocean interior may be rather weak in both cases.

Away from these enclosed bottom or surface regions, one may still be able to derive a diapycnal diffusivity if the flow field is known. (The simple balance (2.2) should really be interpreted using a coordinate normal to the average temperature, or density, surfaces.) Matching the solution to observed profiles leads to a value of K_v/w, and hence to K_v if the upwelling speed w is known. Additional information can be obtained from profiles of radioactively decaying tracers. For example, ^{14}C is high in surface waters in contact with the atmosphere and in bottom waters that have been recently replenished. The interior of the water column, however, has a lower concentration because of the time taken for the tracer to reach it by upwelling or vertical mixing. Comparing observed profiles with solutions of (2.2) with an additional decay term leads to an estimate of K_v/w^2, and hence, in combination with K_v/w, to estimates of both $K_v \simeq 10^{-4}\,\mathrm{m^2\,s^{-1}}$ and $w \simeq 5\,\mathrm{m\,yr^{-1}}$ for the Pacific Ocean. The eight thousand years decay time of ^{14}C is rather long compared with the thousand years turnover time of the ocean, however, so the effect on the profile is small and the model does not give precise results. Nonetheless, other decaying tracers, or anthropogenic ones with a time-dependent source (such as tritium from nuclear weapons tests or chlorofluorocarbons from the chemical industry) have been powerful aids in learning about ocean circulation and mixing (e.g. Jenkins 1998).

The discussion so far has concentrated on the mixing across average isopycnals, which is much weaker than the stirring and mixing along isopycnals. The latter affects the distribution on isopycnals of tracers other than density, such as temperature and salinity in regions where they have significant compensating gradients along isopycnals (i.e. the water becomes colder and fresher, or warmer and saltier, but with the same density). In this case the temperature and salinity fields themselves are affected by the lateral stirring on isopycnals. One prominent example of this occurs in the North Atlantic in the so-called 'Mediterranean Salt Tongue'. The outflow of dense water from the Mediterranean (figure 3) entrains large amounts of ambient ocean water as it sinks down the continental slope and eventually settles out at a depth of about 1 km where its density matches the surrounding water. Much of it then spreads across the North Atlantic. The pattern of the anomaly permits an interesting determination of the vertical and horizontal mixing rates, on the assumption that these act in combination with flow westward across the Atlantic. Assuming also that the mixing rates are uniform, but different laterally and diapycnally, the governing equation for a salinity anomaly S' with respect to the far field is

$$U\frac{\partial S'}{\partial x} = K_h \nabla_h^2 S' + K_v \frac{\partial^2 S'}{\partial z^2}, \qquad (4.2)$$

where K_h and K_v in fact represent the isopycnal and diapycnal mixing coefficients, respectively, and z should be the diapycnal, not vertical, coordinate. The larger either K_h/U or K_v/U, the smaller the anomaly at a given distance downstream, whereas changing the ratio of the diffusivities changes the ratio of the scales of the anomaly pattern in the vertical and horizontally across the tongue. Needler & Heath (1975) compared observed changes in anomaly size and scale ratio between pairs of sections across the tongue, thus determining that, to within a factor of 3 or so, $K_h \simeq 2 \times 10^3$ m^2 s^{-1} and $K_v \simeq 5 \times 10^{-5}$ m^2 s^{-1} for a mean westward current U of 3 mm s^{-1}. This is a nice result in that it clearly gives internal mixing rates, not average rates that might be dominated by boundary processes. It may not be typical, however, in that the very presence of the salt tongue might lead to dynamical processes and mixing that do not occur as strongly elsewhere. The assumption of a constant westward current is also suspect and uncertain; it would be more appropriate to allow for a spatially variable current and for lateral mean flows, horizontally and vertically, that are consistent with the dynamical equations of motion. The need for these generalizations has led to the extensive use of more precise 'inverse' methods, in which observed ocean properties are fitted using a velocity field that is dynamically consistent with the observed density field.

Establishing this velocity field is one of the classical problems of oceanography. In the atmosphere the wind speed can be determined to a good approximation by the geostrophic formula which balances the Coriolis force on the wind against the measured pressure gradient. In the ocean we lack the routine pressure measurements that this requires, but can still determine the vertical changes in geostrophic flow by evaluating the horizontal gradients of the changes in hydrostatic pressure caused by horizontal gradients in density. Progressing beyond a knowledge of the vertical gradients of horizontal velocity requires the use also of vorticity constraints such as (2.1), as well as the conservation equations for heat and salt, and is discussed in detail by Wunsch (1996). Determining the local magnitude of mixing terms added to the momentum and scalar equations is prone to great uncertainty if the mixing terms, for momentum as well as scalars, are small in largely adiabatic inviscid mean flows. The results of such inversions do tend to give $K_v = O(10^{-4})$ m^2 s^{-1}, but not generally bounded away from zero except perhaps in parts of the Southern Ocean and North Atlantic. When the mixing is weak a number of other factors need careful attention, with one being that even very gradual changes with time in the ocean may give a time derivative in the scalar equations that is as large as the mixing terms.

It thus seems that uncertainty is the price of increasing objectivity, at least in determining the mixing coefficients that may be used to represent the eddy fluxes by unresolved processes in the interior of the ocean. All we can really say is that K_v is typically small, probably less than $O(10^{-4})$ m^2 s^{-1}. The lateral mixing rate for scalars seems to be $O(10^3)$ m^2 s^{-1}, but with no reason to expect a geographically constant value. On the other hand, while K_v in the ocean interior is rather uncertain from inference studies, the abyssal basin budgets in particular demonstrate significant basin-average vertical transports, though with an uncertain location. This must be clarified by direct measurement and by process studies. Also, as discussed earlier, lateral stirring is not the only consequence of unresolved mesocale eddies; there may also be a mean advection that is not included in the Eulerian mean flow.

5 Measurements

5.1 Direct techniques

In a vigorously turbulent flow field in other branches of fluid mechanics the correlation coefficient between a typical velocity component and a scalar is large enough to be determined with a record of reasonable length (e.g. Tennekes & Lumley 1972) and this is still true in vigorously mixing parts of the ocean, such as strong tidal flows in coastal regions or in surface or bottom boundary layers in the ocean.

In the ocean interior, the correlation coefficients for motion horizontally, or laterally on isopycnals, may also be large enough to be determined from records of reasonable length. In addition, valuable results may be obtained by measuring the eddy energy and Lagrangian velocity decorrelation time of drifting floats that move passively with the water, as described in §3.1. A typical lateral dispersion coefficient is $1000 \text{ m}^2 \text{ s}^{-1}$, though there are great regional variations and questions remain about spatial inhomogeneity and the reality of a spectral gap (Davis 1991).

Diapycnal, rather than isopycnal, eddy fluxes in the stratified ocean, on the other hand, are difficult to measure directly either by eddy correlation or by using the Lagrangian statistics of particles. Most of the fluctuations measured at a point can be associated with largely adiabatic displacements by internal waves. This dilutes the turbulent fluctuations and gives a very small correlation coefficient which cannot be established reliably with a record of reasonable length. (Various schemes to avoid this sampling problem have been proposed but they rely on further assumptions.) Tracking floats is also inappropriate as they will be subject to buoyancy forces after a turbulent mixing event, unlike the water itself which will adjust through mixing at the molecular scale.

Diapycnal mixing can be determined unequivocally, however, by observing the spread of an introduced 'dye'. In one especially valuable project an artificial tracer (sulphur hexafluoride) was introduced within a metre or so of a particular density surface at a depth of about 300 m in the North Atlantic and tracked as it spread vertically across density surfaces over the subsequent months (Ledwell, Watson & Law 1998). The vertical profile of the lateral average of the tracer became, and remained, fairly Gaussian, suggesting that the diapycnal eddy diffusivity did not have a strong vertical gradient, with the rate of spread giving $K_v \simeq 1.1 \times 10^{-5} \text{ m}^2 \text{ s}^{-1}$ over the first seven months from April to November 1992. This subsequently increased slightly over the winter giving an average of about $1.7 \times 10^{-5} \text{ m}^2 \text{ s}^{-1}$ for the following two years.

The lateral spread of the tracer in this experiment is also interesting. It could not be used to evaluate the Lagrangian correlation function, as no individual tracer particle could be tracked. In the long term its spread can provide a measure of the lateral diffusivity required for models, but at short times, when it has not spread over many eddy scales, it cannot provide the 'absolute' or 'one-particle' dispersion appropriate for an ensemble average of the results of many releases. Instead it merely spreads with respect to its centre of mass, in a process termed 'relative dispersion'. This is understood well for three-dimensional isotropic turbulence in which a cloud of particles will grow as if with a diffusivity that is proportional to the linear scale of the cloud to the 4/3 power.

It is important to distinguish between absolute and relative dispersion. They become equivalent when the tracer patch from a single release is much bigger than the dominant eddy scale, but before then the region occupied by the tracer is much smaller for a single release. Moreover, the actual concentration field within the domain defined by the relative dispersion may be streaky at first until the streaks wrap around each other and merge. The phenomenon is familiar in the context of cream stirred into coffee. In the Atlantic tracer experiment described above, an initial cloud of the tracer was drawn out into streaks a few hundred kilometres long and a few kilometres wide after six months, but after a further six months or so the tracer was more smoothly distributed. The width of the streaks actually presented a puzzle. They are presumed to get narrower and narrower, as a consequence of stirring by the eddies, until small-scale lateral mixing processes can balance the convergence, but the streak width in this experiment seemed to imply a larger value for the small-scale lateral mixing than had been expected, associated with processes that have yet to be clearly determined.

Returning to the diapycnal mixing rate, the value observed, albeit in just one region of the ocean, provided unequivocal evidence for diapycnal mixing in the ocean interior (rather than just near boundaries) and provided a rate with which less direct determinations could be compared. A prior release in Santa Cruz Basin and a subsequent release in the abyssal Brazil Basin (Ledwell *et al.* 2000) showed greatly enhanced vertical mixing once the tracer spread laterally and encountered more turbulent regions close to topographic features, though the interpretation of these experiments and their implications for vertical heat and salt transport are more complicated because the temperature and salinity may already be rather well-mixed near the boundary.

While observations of the spread of a tracer provide direct estimates of mixing rates, this approach is logistically complicated and expensive. Much of what is known about diapycnal mixing in the ocean has thus come from less-direct measurements involving the details of vertical profiles of temperature, salinity, and velocity.

5.2 Techniques based on fluctuations in profiles

We start with the equation for temperature, taken as

$$\partial T/\partial t + \boldsymbol{u} \cdot \nabla T = \kappa \nabla^2 T, \qquad (5.1)$$

and it is assumed that $T = \overline{T}(z) + T'(\boldsymbol{x}, t)$. The equation for the variance of the temperature fluctuation T' is

$$\frac{\partial \overline{T'^2}}{\partial t} + \nabla \cdot \left(\overline{\boldsymbol{u}}\, \overline{T'^2} + \overline{\boldsymbol{u}'\, T'^2} - \kappa \nabla \overline{T'^2} \right) + 2\overline{\boldsymbol{u}'T'} \cdot \nabla \overline{T} = -\chi, \qquad (5.2)$$

where $\chi = 2\kappa \overline{\nabla T' \cdot \nabla T'}$ is the rate of dissipation, by thermal diffusion, of temperature variance. The use of this equation depends on ignoring the terms for the time rate of change and divergence on the left-hand side of (5.2) on the grounds of stationarity and spatial homogeneity. The equation then describes a balance between production of temperature variance, by turbulent stirring of horizontal and vertical gradients, and dissipation. In particular, if the vertical production is assumed to be dominant, the vertical eddy diffusivity (for temperature), denoted K_θ, is given by

$$K_\theta = \frac{\chi}{2\Theta_z^2}, \tag{5.3}$$

where, to allow for adiabatic changes, we have now replaced $\mathrm{d}\overline{T}/\mathrm{d}z$ by the mean vertical gradient Θ_z of the potential temperature.

Since its proposal and first use by Osborn & Cox (1972) this result has provided a wealth of information about vertical mixing in the ocean. The temperature dissipation rate χ depends on the smallest scales of temperature 'microstructure', down to a scale of a few millimetres. Measuring these is technically demanding for both the resolution and response time of the sensors used. Usually χ is estimated from vertical gradients alone, with the assumption that at small scales the turbulence is isotropic so that the gradients in other directions are equal. The ratio of the vertical eddy diffusivity to the thermal diffusivity is then given by three times the ratio of the mean-square vertical temperature gradient to the mean gradient squared, with this ratio sometimes being called the Cox number. Gregg (1989) and others have found Cox numbers not much more than 10 or so to be typical of the main thermocline, corresponding to values of a few times 10^{-5} m^2 s^{-1} for K_θ. Agreement with the value from a tracer release is not expected if the tracer diffuses at the same rate as salinity and this behaves differently from temperature because of double-diffusive effects. We will discuss this later.

Another important factor is that the horizontal production of temperature variance may dominate, particularly at high latitudes in regions with a small vertical temperature gradient. This is an interesting point, as it raises the issue that much of the horizontal production is initially associated with stirring by mesoscale eddies, though this must give way ultimately to smaller-scale processes, the nature and physics of which are not established.

A further factor to consider is that the 'mean' vertical temperature gradient on a single cast may be affected by mesoscale eddies, whereas the correct vertical diffusivity for use in a model of the long-term average ocean circulation should be obtained from the ratio of the time averages of numerator and denominator of (5.3), and not from an average of the ratio. The vertical flux may be carried by both eddies and turbulence; a triple decomposition of the temperature field into a mean field, eddies and higher-frequency fluctuations is instructive (Davis 1994).

The above formalism can also be applied to salinity, but unfortunately the very small molecular diffusivity for salt (only about a per cent of the thermal diffusivity) means that turbulent fluctuations of salinity (or rather of the associated electrical conductivity, which is the quantity measured in practice) occur at scales of a millimetre or less, much too small to be resolved with current technology. It is thus often assumed that the value of K_θ applies also to salt and hence to density, though this need not be true, particularly in double-diffusive situations which will be discussed later.

Even measuring temperature microstructure is difficult, particularly if K_θ is large, making the temperature fluctuation scale very small, so much of our knowledge of vertical mixing actually comes from measurements of the velocity field, or rather of the vertical shear of horizontal currents down to scales of a few millimetres. When combined with the assumption of isotropy, this gives the turbulent energy dissipation rate

$$\varepsilon = 7.5\nu \overline{(\partial u_1/\partial z)^2} \tag{5.4}$$

in terms of the vertical gradient of one of the horizontal shears (Batchelor 1953). The turbulent kinetic energy equation contains similar time-dependent and divergence terms to those in (5.2), but if these are neglected it reduces to

$$-\overline{u_i' u_j'}(\partial \overline{u}_i/\partial x_j) = K_v N^2 + \varepsilon, \tag{5.5}$$

showing turbulent kinetic energy production by Reynolds stresses working against the mean shear, and the loss of this to increased potential energy by the action of the vertical buoyancy flux (which may be written as $K_v N^2$, where $N^2 = -(g/\rho)(\mathrm{d}\rho/\mathrm{d}z)$ is the buoyancy gradient based on the average density profile $\rho(z)$) and to viscous dissipation at a rate ε.

In the ocean interior the mean shear is usually very small, with the energy input actually coming from breaking internal waves, but it is assumed that the division between the sinks of buoyancy flux and dissipation is given by a constant $\Gamma = K_v N^2/\varepsilon$ so that

$$K_v = \Gamma \varepsilon/N^2. \tag{5.6}$$

If temperature microstructure is also measured, giving K_θ from (5.3), and if the diffusivities for temperature, salinity and density are assumed to be the same, so that $K_v = K_\theta$ in particular, then Γ is also given by

$$\Gamma = \frac{\chi N^2}{2\varepsilon\Theta_{\bar{z}}^2},\tag{5.7}$$

and is generally found to be about 0.2 or less.

Most of our knowledge of vertical mixing in the ocean has actually come from (5.6). The formula does not suffer from the uncertainties of horizontal versus vertical production that weaken the Osborn–Cox technique and it is hard to imagine that turbulent energy dissipation is occurring without some vertical mixing. On the other hand, the value of Γ is not necessarily universally constant: as will be discussed later, different values of Γ as defined by (5.7) are found in some double-diffusive regions where the diffusivities of heat and salt differ.

It is also unfortunate that measuring the velocity microstructure is technically exacting and expensive. A simpler method of assessing the vertical mixing rate can come from observations of regions of static instability in vertical temperature profiles, a technique first used in Loch Ness by Thorpe (1977). He calculated the r.m.s. vertical displacement in a re-sorting of the temperature profile into a stable state, a distance now known as the Thorpe scale L_T. It can be argued on dimensional grounds that this must be proportional to $(\varepsilon/N^3)^{1/2}$, known as the Ozmidov scale L_O. Observationally it has been found that $L_T \simeq 1.2L_O$, so that (5.6) gives $K_v \simeq 0.1NL_T^2$, a form to be expected dimensionally, and also dynamically given that the collapse time of the overturns is proportional to N^{-1}.

In the ocean it is necessary to evaluate density, rather than temperature, overturns unless the relationship between temperature and salinity is tight enough for one to represent both. If not, apparent inversions in temperature can actually be rendered hydrostatically stable by the salinity field. In any event, the required resolution in temperature (or density) and in the vertical cannot be measured with standard ocean profilers (CTDs) unless the stratification and mixing are both strong. This has ruled out the deep ocean interior except for regions of intense mixing, but improvements in CTD technology are likely to make the measurement of Thorpe scales a standard approach to quantifying mixing in many estuarine and coastal situations.

An even less direct estimate of diapycnal mixing can be obtained from

$$K_v \simeq 5 \times 10^{-6} \overline{S_{10}^4/S_{GM}^4} \text{ m}^2 \text{ s}^{-1},\tag{5.8}$$

where S_{10} is the vertical shear of the horizontal current measured over a vertical distance of 10 m, S_{GM} the value it takes in a typical situation, and the value 0.2 has been used for Γ in (5.6). The formula in (5.8) is based on theories, to be discussed in §6.2, which suggest that energy in internal waves is transferred by nonlinear processes to shorter vertical wavelengths, eventually leading to shear instability, overturning and mixing. For an internal wave spectrum with a fixed distribution of energy between different frequencies and wavenumbers, but varying overall energy level E, the flux to small scales is proportional to E^2, and S_{10}^4 is a convenient proxy for this.

Gregg (1989) has found that (5.8) comes within a factor of 2 of giving the same value for K_v as (5.6) in a number of open ocean situations with K_v varying over a range of nearly 100. Polzin, Toole & Schmitt (1995) have proposed a more general formula for K_v which involves estimates of the vertical strain (the vertical derivative of the vertical displacement, measured from the departure of vertical density profiles from smooth) as well as the vertical shear of horizontal currents, and show that it improves agreement with estimates from (5.6) using direct measurements of turbulent dissipation.

The success of these indirect formulae has led to the suggestion that a high priority be given to global surveys of shear and strain as a means of mapping global values of the diapycnal diffusivity. This would require vertical resolution of only metres, and so would be much easier than measuring the temperature or velocity microstructure on a scale one hundred times less, but places a heavy burden on the assumptions.

To summarize this short review of direct observations, a picture is emerging in which measurements of the diapycnal diffusivity are compatible with those inferred from large-scale balances: K_v is typically rather small, $O(10^{-5})\,\text{m}^2\,\text{s}^{-1}$ in the main thermocline except perhaps in special regions such as the Southern Ocean, but typically much larger, $O(10^{-4})\,\text{m}^2\,\text{s}^{-1}$ or more, in abyssal regions near rough topography. The thermocline values are probably small enough for vertical diffusion not to be an important part of the dynamics there, but the abyssal values do suggest a major role for vertical mixing in determining the strength of the ocean's thermohaline circulation. It is thus important, as stressed earlier, to understand the processes responsible for the mixing so that numerical models of ocean circulation that aspire to predictive capability can be run with appropriate formulae for the mixing rate, rather than just the current value.

6 Processes

The resolution of a model determines which processes need to be parameterized. In the atmosphere the dominant eddy (weather system) scale is of the order of 1000 km, and it is now taken for granted that a plausible and accurate model must resolve this scale. The ocean, unfortunately, has a dominant eddy scale about an order of magnitude smaller, so that an 'eddy-resolving' numerical model of ocean circulation requires a horizontal grid scale of the order of 10 km. This, especially when combined with the need for integration for hundreds of years in climate problems, demands enormous computing resources. These may become available in due course, but at present considerable attention is being paid to schemes which parameterize the effects of mesoscale eddies, so these will be discussed briefly first before turning to small-scale processes which need to be parameterized in all ocean circulation models.

6.1 Mesoscale eddies

Eddies with a horizontal scale of the order of a hundred kilometres horizontally, hundreds of metres vertically, and a time scale of the order of a month, are clearly associated with instabilities of vigorous currents such as the Gulf Stream (figure 1). They can also be smaller and/or be generated by the instability of weaker flows (figure 11), by flow over topographic features and by the wind. They occur throughout the world's oceans, though with an intensity that varies very widely (figure 12). They are clearly responsible for the lateral diffusivity of $O(10^3)$ m^2 s^{-1} cited earlier, though this value also varies widely. The main point to be made here, however, is that mesoscale eddies may also have an associated 'skew diffusion', as described earlier, which can be parameterized as an additional velocity. This velocity, like the symmetric diffusivity, needs to be prescribed in terms of the large-scale resolved properties of an ocean model.

Much eddy activity probably arises from 'baroclinic instability' in which the potential energy available in the tilted isopycnal surfaces of a vertically sheared ocean current is released adiabatically into eddies; this is the same process that gives rise to mid-latitude weather systems in the atmosphere. This conversion of mean potential energy into the kinetic and potential energy of the eddies is accompanied by an average flattening of the isopycnals. A popular scheme that seeks to represent this in the ocean adds a velocity $\boldsymbol{u}^* = (\kappa \nabla_h b / b_z)_z$, where κ is a diffusion rate for the lateral spread of the 'thickness', or isopycnal separation, b is the large-scale buoyancy field and $\nabla_h b$ is its horizontal gradient. This is related to the vertical shear of the mean, geostrophic, horizontal current, so that the total horizontal transport $\boldsymbol{U} = \boldsymbol{u}^* + \boldsymbol{u}$, where $\boldsymbol{u}$ is the mean horizontal Eulerian velocity, can be shown to behave very much as if acted upon by a vertical viscosity $\kappa f^2 / N^2$. This result confirms that the scheme does tend to flatten mean isopycnals and is also legitimate in that it just redistributes momentum without introducing a

Figure 11. Spiral patterns on the surface of the Black Sea, as photographed from the Space Shuttle in September 1992. The image is about 100 km across. The spirals are made visible by the wave damping effect of surface films, wound up by eddy motions. See Munk *et al.* (2000) for more pictures and a discussion of the predominance of anticlockwise spirals.

$Ke \sin^2 \phi$

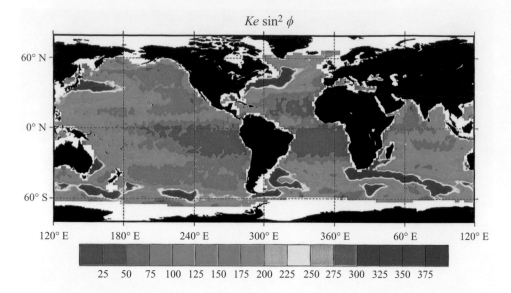

Figure 12. Map of the surface kinetic energy of ocean eddies on scales greater than 30 km, multiplied by $\sin^2 \phi$, where ϕ is the latitude. The surface currents were obtained from measurements of sea surface slope derived from satellite altimeter data, assuming a balance between the pressure gradient and Coriolis force. (From Wunsch & Stammer 1998.)

spurious source. Subsequent work has sought to determine an appropriate value of κ (often assumed to be of the order of 10^3 m^2 s^{-1}) and its vertical profile, and to generalize the formula to forms that are more consistent with ocean dynamics and the output of eddy-resolving models. Major questions remain, however, about the appropriateness of a purely local parameterization of the eddies and about the ultimate fate, and effects, of the potential energy released. A research topic which needs more attention is that of the connection between mesoscale eddies and the ultimate dissipative processes that must be in series; the problem is important for eddy-resolving as well as non-eddy-resolving models.

6.2 Internal waves

Returning now to diapycnal mixing in the ocean, the main cause seems to be the breaking of internal 'inertio-gravity' waves. At a density interface such waves are similar to surface waves, though with a much longer period due to the much reduced buoyancy restoring force, so that they are also influenced by the Earth's rotation. In a continuously stratified incompressible fluid, waves must be transverse, as $\nabla \cdot \boldsymbol{u} = 0$, so that a purely horizontal wavenumber corresponds to water columns oscillating vertically and feeling the full effect of restoring buoyancy forces associated with the basic density gradient $\rho(z)$. The wave frequency is then simply equal to the buoyancy frequency N where $N^2 = -(g/\rho)(d\rho/dz)$, as is readily seen by considering the vertical acceleration of a vertically displaced fluid column that experiences a restoring buoyancy force.

Other directions of wave phase propagation are also possible, with associated transverse oscillations of tilted fluid columns. The restoring force for a given transverse displacement is now reduced, as is the restoring force along the line of motion, by a factor $\cos\theta$, where θ is the angle that the particle motion makes to the vertical. The frequency ω of the oscillations is then given by $\omega^2 = N^2 \cos^2\theta$ if the Earth's rotation is ignored. On the other hand, if the particle motion is purely horizontal, the Coriolis force will lead to circular motion with a frequency f, with this contribution similarly reduced if the particle motion is inclined to the horizontal.

It is thus plausible, and may be readily derived from linearized perturbation equations, that the dispersion relation for internal waves in a stratified but incompressible fluid (e.g. Phillips 1977) is

$$\omega^2 = N^2\cos^2\theta + f^2\sin^2\theta = (N^2(k^2 + l^2) + f^2m^2)/(k^2 + l^2 + m^2). \qquad (6.1)$$

There are a number of remarkable properties of this dispersion relation in addition to the fact that the frequency does not depend only on the magnitude of the wavenumber $\boldsymbol{k} = (k, l, m)$. In particular, the group velocity $\nabla_k \omega$ is at right angles to the wavenumber $\boldsymbol{k}$, and hence to the phase velocity!

Internal wave energy in the ocean is distributed over the full range of frequencies from f to N, and for each frequency is distributed over a range of vertical wavenumbers, with a typical tendency for more energy near f and in low vertical modes. A cartoon by Thorpe (1975) (figure 13) may be used as a basis for discussion of wave generation, propagation, interactions with other waves, the effects of topography, and mixing caused by gravitational or shear instability. These will only be discussed briefly here in view of extensive reviews and compendia of articles in the literature (e.g. Müller & Henderson 1999).

The main generating agent of internal waves is thought to be the surface wind stress, though they may also be generated as the stratified ocean is moved over topographic features by lower-frequency currents and by the tides. Once generated, internal waves propagate through a medium with variable buoyancy frequency N. This typically decreases downwards, so that downward-propagating waves of a given frequency may encounter a level where $\omega = N$ and at which the waves undergo internal reflection. A curious

Chris Garrett

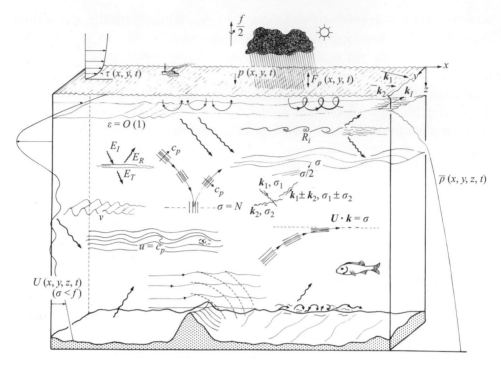

Figure 13. A cartoon of the various processes affecting internal waves in the ocean. (From Thorpe 1975.)

feature of this, shown in figure 13, is that the ray path becomes vertical at the reflection level, rather than horizontal as would occur for sound waves at a level of internal reflection.

Perhaps more interestingly, much of the energy in internal waves is near the local inertial (Coriolis) frequency f which increases poleward. 'Near-inertial' waves cannot propagate poleward and still remain in the frequency band from f to N. An inertial peak in observed spectra seems to be partly associated with this latitudinal turning point; it is inappropriate to treat near-inertial waves as if they were on an infinite plane with constant f.

As indicated in figure 13, waves may be scattered by small-scale irregularities in stratification, but important effects are also associated with interactions between waves. A single plane wave is actually an exact solution of the nonlinear equations of motion because the result $\boldsymbol{k} \cdot \boldsymbol{u} = 0$ implies that the nonlinear terms involving $\boldsymbol{u} \cdot \nabla$ vanish, but this is not true if more than one wave is present. Two waves give rise, through the nonlinear terms, to contributions at the sum and difference wavenumbers and frequencies. If the sum (or difference) frequency corresponds to that given by the dispersion relation for the sum (or difference) wavenumber, then this wave will be resonantly excited. The many such interactions that are possible have been simplified into three classes described as 'elastic scattering', 'induced diffusion' and 'parametric subharmonic instability'. The first tends to balance downward and upward travelling energy, whereas the latter two tend to transfer energy to lower frequency and higher vertical wavenumber, resulting in increased shear and likelihood of breaking. The transfer rate for a typical spectrum is one of the reasons for the proposal of (5.8) as a formula for the diapycnal mixing rate.

Unfortunately, however, the assumption of weak interactions in the above theory breaks down at a much larger vertical scale than that on which breaking occurs. An alternative approach considers the propagation of small-scale internal waves through the background shear of other waves, assumed to have a much larger scale, and also finds a flux of energy to small scales. Interestingly, a formula for this flux, and hence K_v, is similar to that from weak interaction theory. The successful formula (5.8) is thus based on two different approaches, each having limiting assumptions but giving the same form of result. Merging the two approaches into a unified theory remains to be done, but it is encouraging (though quite possibly misleading!) that internal wave theories evaluated for typically observed spectra in the thermocline, microstructure measurements and observations of the vertical spread of a tracer all give similar values, $O(10^{-5})$ m^2 s^{-1}, for K_v. Higher mixing rates are associated with elevated energy levels of internal waves.

Interactions between different waves are not the only processes that shape the internal wave spectrum: for much of the most energetic part of the spectrum interactions between waves may be less important than interactions with bottom topography. Bottom reflection is unlike that for, say, sound waves because the dispersion relation (6.1) for internal waves shows that for them to conserve their frequency on reflection from a *sloping* sea floor they must also conserve their angle to the vertical, not their angle to the normal. As a consequence, waves reflected upslope will tend to have a shorter wavelength, narrower ray tube and smaller group speed (figure 14). The wave amplitude is thus increased, particularly near the 'critical frequency' for which the wave rays are parallel to the slope. For different angles of incidence the amplification is less, or reduction may occur, but for an isotropic incident spectrum overall amplification is expected and has been well-documented by Eriksen (1998) for the steeply sloping sides of Fieberling Guyot, a seamount in the Pacific Ocean. It seems quite likely that the associated increase in shear, either in this reflection off a plane slope or by scattering off rough

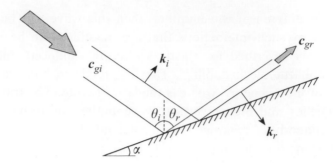

Figure 14. Schematic of a beam of internal waves being reflected off a plane slope at an equal angle to the vertical. Subscripts i and r denote incident and reflected waves.

topography, increases the rate of mixing and may account for part of the abyssal mixing that has been documented.

The internal tide (internal waves of tidal frequency, presumably generated at the sea floor as tidal currents flow over bottom topography) actually appears to be implicated in the mixing shown in figure 4, raising the possibility of a lunar influence on climate. The current global rate of dissipation of tidal energy is known from a variety of geophysical and astronomical observations (e.g. Munk 1997), including amazingly accurate laser ranging of the changing distance from the Earth to the Moon. Much of the energy losses occur through turbulence in shallow coastal seas, but it increasingly seems that a significant fraction of the dissipation occurs in the abyssal ocean and contributes to mixing there. The internal tide may, of course, be acting as a catalyst for mixing by surface-generated internal waves, by increasing the shear, as well as acting on its own.

6.3 Boundary mixing

As implied above, there is increasing recognition that much of the mixing in the ocean may take place near topographic features. The topic is one of active observational and theoretical research, complicated, of course, by the complexity of topography. One issue has been the degree to which the buoyancy flux is reduced if the local stratification is reduced; mixing water that is already mixed is ineffective. The subtlety of the problem may be demonstrated even for the simplest possible case, illustrated here in figure 15. We assume a plane sloping boundary with constant stratification away from the boundary but reduced stratification near the boundary as a consequence of mixing there with an eddy diffusivity K_v (using the earlier

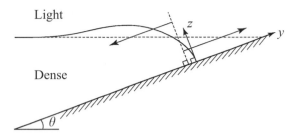

Figure 15. Schematic diagram of an isopycnal near a sloping boundary close to which there is enhanced mixing. The long arrows indicate the buoyancy-driven flow.

notation, though it in fact represents the diapycnal diffusivity; the mixing rate along isopycnals may be larger but does not affect the density field). We allow K_v to be a function of the coordinate z normal to the boundary but not of the upslope coordinate y. The buoyancy $b = -g(\rho - \rho_0)/\rho_0$, with ρ_0 a constant reference density, is a measure of the stratification. Far from the boundary the isopycnals are flat with a vertical buoyancy gradient N^2 which has components $\nabla b = (N^2 \cos\theta, N^2 \sin\theta)$ in our coordinates. Closer to the boundary $\partial b/\partial y$ is still equal to $N^2 \sin\theta$, but $\partial b/\partial z$ is different from $N^2 \cos\theta$, tending to zero at the boundary itself. The steady-state equation for the buoyancy field is

$$v N^2 \sin\theta = \frac{\partial}{\partial z}\left(K_v \frac{\partial b}{\partial z} \right), \tag{6.2}$$

balancing upslope advection against the divergence of the diffusive flux normal to the boundary. The total upslope volume flux $\int_0^\infty v \, dz$ is zero if $K_v \to 0$ as $z \to \infty$. Under these conditions it can be shown that the total vertical buoyancy flux is given by

$$F_b = \int_0^\infty K_v \left[N^2 \sin^2\theta + \left(\frac{\partial b/\partial z}{N^2 \cos\theta} \right)^2 N^2 \cos^2\theta \right] dz / \sin\theta, \tag{6.3}$$

where $dz/\sin\theta$ is just the horizontal line element (e.g. Garrett, MacCready & Rhines 1993). This is an interesting equation as the factor $(\partial b/\partial z)/(N^2 \cos\theta)$ occurs quadratically in the second term; if it entered linearly then (6.3) would just be the vertical diffusive flux of buoyancy. In practice, $(\partial b/\partial z)/(N^2 \cos\theta)$ will be less than 1 in most of the mixing region, so F_b is reduced below its diffusive value. The reason for this is that the flow v, given by (6.2), carries light water upslope near the boundary and denser water downslope farther out (see figure 15). This buoyancy-driven restratification thus causes

This point of view is less convincing if the vertical transport by other processes, such as breaking internal waves, is also weak. To establish a framework for interpreting turbulence data, it is worth considering a situation in which salt fingering is actually the dominant vertical transport process. The vertical diffusivity K_θ for heat could still be obtained from (5.3) if temperature microstructure data were available, but the effective diffusivities K_s, K_ρ for salt and density would be different! The reason is that, in the absence of the input of mechanical energy from breaking internal waves, the energy dissipation rate is derived from the release of potential energy from the basic state; both salt and heat are transported down-gradient, but the density is transported up-gradient. The salt flux releases potential energy from the basic state; and this loss is only partly balanced by the gain in potential energy associated with the heat flux if the buoyancy flux ratio $r = (K_\theta g\alpha\Theta_z)/(K_s g\beta S_z)$ is less than 1. With $\varepsilon = -K_\rho N^2 = -K_\theta g\alpha\Theta_z + K_s g\beta S_z$ and $N^2 = g\alpha\Theta_z - g\beta S_z$, and using (5.3) and (6.4), the diffusivities are related by

$$K_\theta = rR_\rho^{-1}K_s, \quad K_\rho = -(1-r)(R_\rho - 1)^{-1}K_s, \qquad (6.5)$$

showing that the effective diffusivity for heat is less than that for salt and the diffusivity for density is not only negative but may be much larger in magnitude than that for salt if R_ρ is not much greater than 1.

The scaled ratio (5.7) of temperature variance dissipation to kinetic energy dissipation now takes the value $\Gamma^{(f)}$ given by

$$\Gamma^{(f)} = -K_\theta/K_\rho = \left(\frac{R_\rho - 1}{R_\rho}\right)\left(\frac{r}{1-r}\right), \qquad (6.6)$$

which can be very different from the value of about 0.2 that seems to apply in doubly stable regions where double-diffusive processes do not occur.

In a remarkable study based on temperature and velocity microstructure data from the site of the North Atlantic Tracer Release Experiment (see § 5.1), St. Laurent & Schmitt (1999) have, in fact, found evidence (figure 16) for large values of Γ in places where the Richardson number Ri, given by the ratio of N^2 to the square of the vertical shear of the horizontal current, is greater than 1 and R_ρ less than about 2 (with all the mean gradients based on linear fits to 5 m sections of vertical profiles). At smaller values of the Richardson number, or larger R_ρ, Γ seemed to approach a value of about 0.2 or less, as also found in doubly stable regions. It thus appears that the mixing included double diffusive effects, even though shear-generated turbulence occurred with sufficient frequency and intensity to prevent the formation of layers. At the depth of the tracer release St. Laurent & Schmitt (1999) found

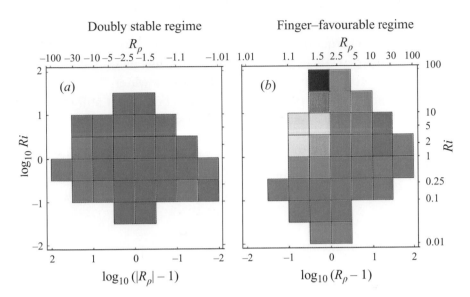

Figure 16. The ratio Γ, defined by (5.7), as a function of Richardson number Ri and density ratio R_ρ for a region with stable profiles of both temperature and salinity (a) and for a region which could permit salt fingers (b). (After St. Laurent & Schmitt (1999). Courtesy of Louis St. Laurent.)

a net vertical diffusivity for salt in close agreement with that obtained from the tracer (which should behave like salinity), but a smaller value of K_θ, and a much smaller, even slightly negative, diffusivity for buoyancy as if the down-gradient flux from mechanical mixing were almost exactly offset by the up-gradient flux by salt fingers. These results have significant implications for the mean diapycnal velocity. It appears that double-diffusive processes can be important even if they do not make their presence obvious through the appearance of layers.

Double-diffusive processes are also implicated in the formation of intrusive features that can develop when there are lateral gradients of temperature and salinity, even if the vertical stratification is stable for both. An incipient intrusion from the warm salty side of a front into fresher colder water of equal density can develop salt fingers on its underside and lose more density through them than it gains through diffusive instability at its top. It will thus tend to mix and be propelled laterally as a density current, rising slowly at the same time. Similarly, a cold fresh intrusion will tend to spread and sink.

Such intrusions have been studied in the laboratory and have been clearly implicated in the decay of 'Meddies', warm salty lenses, hundreds of metres thick and tens of kilometres in radius, that originate with water that spills out of the Mediterranean Sea, mixes with Atlantic water and then partly spreads out across the North Atlantic (as discussed earlier in §4). They persist as coherent structures for more than a year, with gradual erosion of their core by the growth of double-diffusive intrusions at their boundaries where there is a frontal transition to cooler fresher water.

Away from special features such as Meddies, one might still expect 'thermohaline' fronts, and associated double-diffusive intrusions, to develop as mesoscale eddies stir lateral temperature and salinity gradients on isopycnals, but we lack adequate observations and understanding. There is more to be understood about the subtle and intriguing physics that arises merely from the influence of both salt and temperature on the density of seawater.

6.5 Nonlinearities of the equation of state

In a discussion of the rich and idiosyncratic behaviour of seawater, it would be incomplete not to mention fascinating phenomena arising from the nonlinear dependence of its density on temperature, salinity and pressure. An important one, known as 'cabbeling', is where two water masses of the same density, but different temperature and salinity, will produce denser water on mixing. Thus mixing along isopycnals defined in terms of average properties can lead to a diapycnal velocity. The compressibility of seawater is also a function of temperature and salinity, an effect termed 'thermobaricity', again leading to diapycnal motion if eddies stir lateral gradients of temperature and salinity along sloping isopycnals. Thermobaricity is also implicated in the tendency for cold salty water formed in the Southern Ocean to descend into the abyss; it is more compressible than the ambient water and so acquires an extra buoyancy difference as it sinks. Some of these subtleties are automatically taken into account in models that treat temperature and salinity separately and use a full nonlinear equation of state, but some affect the parameterization of unresolved processes, or make for interesting phenomena that are worth studying in their own right.

6.6 Conclusions

Much has been learnt about the fascinating and important fluid dynamics that controls mixing in the ocean, but many of the theories and supporting observations are incomplete and tenuous. Provisional formulae can be provided to represent the effects of small-scale processes in the ocean, but the output of models for larger-scale behaviour seems to be sensitive to these formulae so that major efforts will be required to refine them.

7 Other problems

This survey has stressed the importance of small-scale processes which cannot be treated explicitly in regional or basin-scale models. An emphasis on mixing processes that occur in the generally stratified interior excludes any but the briefest mention of other fascinating and important problems.

7.1 Boundary layers

Boundary layers exist at the surface of the ocean and at the sea floor and connect the ocean interior with the atmosphere and the solid earth. The boundary layers are generally characterized by much higher levels of mixing than the ocean interior, becoming somewhat homogeneous in properties, and so warrant separate attention.

The surface boundary layer is subject to mixing and deepening by wind, by surface buoyancy loss associated with cooling and evaporation, and by turbulence generated by shear instability at the base of the layer. Restratification is principally associated with surface heating, though rainfall can play a role, as can flows driven by lateral gradients of density in the surface layer. Models for the layer range from simple 'slab' models, through models which assume particular profiles of the eddy diffusivity and eddy viscosity in the layer and just below it, to higher-order closure models which parameterize eddy fluxes of buoyancy and momentum in terms of mean properties. All these models have adjustable parameters which can be tuned so that the model output for the seasonal cycle of sea surface temperature, for example, matches observations. This procedure may be adequate for many operational purposes, but is scientifically unsatisfactory, particularly as the models generally ignore dominant physical processes such as Langmuir circulation, helical circulation cells in the upper ocean driven by a combination of wind and waves. Progress will come from better observations of the detailed motion in the mixing layer, enabling the direct verification of parameterizations for the mixing, and from 'large-eddy simulations' which treat the turbulence explicitly and may be used to search for appropriate parameterizations (e.g. McWilliams, Sullivan & Moeng 1997).

The boundary layer at the sea floor is similar in some ways to the surface mixed layer in that it can be mixed by turbulence associated with a stress, but it is also obviously affected by rough topography in some areas, and, most importantly, can only restratify through lateral processes. This can take place if the boundary layer is on a slope, with buoyancy-driven restratification constantly acting to offset the homogenizing influence of mixing (as discussed in §6.3). The exchange of mixed fluid on a bottom slope with stratified water away from the slope is thus not necessary for restratification, though it may occur. For all boundary layers, in fact, the mechanisms and rate of exchange with the interior constitute an important and comparatively unexplored topic.

A bottom slope also influences the way in which a no-slip boundary condition affects currents away from the bottom. If this current is geostrophic, with a balance between the Coriolis force and a pressure gradient, frictionally reduced flow near a boundary implies a reduced Coriolis force so that the pressure gradient can drive a flow in the orthogonal direction, to the left of the current in the northern hemisphere. On a flat bottom convergences and divergences of this so-called 'Ekman' flux produce secondary circulations which retard the interior flow, but on a slope the Ekman flux may be stopped by buoyancy forces, halting the back effect on the flow away from the slope. This flow then essentially feels no further frictional influence of the boundary. There is also an interesting asymmetry between Ekman layers moving upslope and downslope: the former are hydrostatically stable whereas the latter must thicken to preserve stability. The combined effects of boundary layers and complex topography are important and sometimes surprising.

One topic that should be mentioned in any discussion of boundary layers in the ocean is that in shallow seas with strong tidal currents the bottom mixed layer may reach all the way to the sea surface, preventing the establishment of stratification even in the summer heating season. A pioneering paper by Simpson & Hunter (1974) discusses how the transition from well-mixed to stratified water seems to occur at a critical value of the rate of gain of surface buoyancy times water depth divided by the dissipation rate of the tidal currents. For spatially uniform heating this reduces to a criterion for H/U^3, where H is the depth and U the r.m.s. tidal current; at mid-latitudes a current of 1 m s^{-1} will keep up to 70 m of water well-mixed. The criterion, derived from energetic considerations but also dimensionally inevitable, is an interesting illustration of the value of simple arguments, even though the actual details of mixing in stratified tidal waters are complicated.

7.2 Interdisciplinary topics

Physical oceanography is the foundation of many studies of biological, chemical and geological processes in the ocean. Fluid dynamics is also directly relevant in low Reynolds number flow for the behaviour of small organisms or particles; bubble dynamics for air–sea gas exchange; suspended matter dynamics and flow in porous media for the consideration of the shore processes. Water can also exist in solid form as sea ice, with another whole class of challenging scientific problems associated with brine rejection from growing ice, the rheology of pack ice, and ice–water interactions. All of these areas deserve chapters to themselves.

8 Summary and outlook

It is clearly interesting and scientifically valuable to understand the resolvable processes in ocean models, but one could argue that they are just a consequence of the governing equations, which are known except for the parameterization of the subgrid-scale processes, and that it is therefore the investigation of these small-scale processes that should be a priority. This is especially true for those small-scale processes, such as diapycnal mixing, to which the output of models for the large-scale behaviour of the oceans seems to be sensitive. Remarkable observational and theoretical progress has been made in the last few decades, but we still lack robust simple parameterizations in terms of variables that can be resolved. It is worth remarking that numerical models of processes in idealized situations, as well as large-scale models, are proving useful in unravelling processes and determining appropriate parameterizations.

The important role of topography has become clearer in recent years, with recognition of the significance, for mixing, of effects such as internal tide generation or internal wave scattering. The details and correct parameterization need further investigation. Appropriate sea-floor boundary conditions for the velocity field are also uncertain, and observational studies are plagued by the undersampling that almost inevitably results from the heterogeneity of the topography.

The sea surface and its boundary layer might not seem to have as much heterogeneity as expected at the bottom of the ocean over rough topography, but nonetheless have a broad spectrum of variations at all scales. The sea surface is, however, observable from Space, so the sampling problem is less formidable. For both surface and bottom boundary layers, the nature and effects of the exchange with the interior, perhaps advective from large-scale convergence and divergence, and diffusive through eddy processes, is likely to remain a major topic of investigation.

Much of fluid dynamical research in the ocean will continue to be motivated by purely physical questions, partly in response to questions of climate variability, with uncertainties in present understanding and prediction requiring major improvements in our knowledge of the physics of the ocean. On some other important practical questions, such as those associated with marine productivity and pollution, physical oceanography often provides the essential foundation.

The study of ocean dynamics offers a wonderful combination of intellectual challenge and societal relevance, but the investigator should also pause occasionally to savour the sheer beauty of many of the phenomena, whether breaking waves on the sea shore, spiral eddies on the sea surface or patterns of internal motion revealed by modern sensing techniques.

Acknowledgements

I thank Carl Wunsch, Lou St. Laurent, Keith Moffatt and Grae Worster for comments on a draft. I also thank the colleagues who have provided the illustrations used here. Space and the style of this volume have prohibited full references to the original research reviewed, but it is hoped that the papers that are cited, and the volumes in which they appear, will lead the reader to other key papers.

References

BATCHELOR, G. K. 1953 *The Theory of Homogeneous Turbulence.* Cambridge University Press.

CHASSIGNET, E. P. & VERRON, J. 1998 *Ocean Modelling and Parameterization.* Kluwer.

CRAIG, H., BROECKER, W. S. & SPENCER, D. 1981 GEOSECS Pacific Expedition, Volume 4. *International Decade of Ocean Exploration.* National Science Foundation.

DAVIS, R. E. 1991 Lagrangian ocean studies. *Ann. Rev. Fluid Mech.* **23**, 43–64.

DAVIS, R. E. 1994 Diapycnal mixing in the ocean: The Osborn–Cox model. *J. Phys. Oceanogr.* **24**, 2560–2576.

ERIKSEN, C. C. 1998 Internal wave reflection and mixing at Fieberling Guyot. *J. Geophys. Res.* **103**, 2977–2994.

FRISCH, U. 1995 *Turbulence: the Legacy of A.N. Kolmogorov.* Cambridge University Press.

FU, L.-L., LIU, W. T. & ABBOTT, M. R. 1990 Satellite remote sensing of the ocean. In *The Sea*, vol. 9B, pp. 1193–1236. Wiley.

GARRETT, C., MACCREADY, P. & RHINES, P. 1993 Boundary mixing and arrested Ekman layers: rotating stratified flow near a sloping boundary. *Ann. Rev. Fluid Mech.* **25**, 291–323.

GILL, A. E. 1982 *Atmosphere-Ocean Dynamics.* Academic.

GREGG, M. C. 1989 Scaling turbulent dissipation in the thermocline. *J. Geophys. Res.* **94**, 9686–9698.

JENKINS, W. J. 1998 Studying subtropical thermocline ventilation and circulation using tritium and ^{3}He. *J. Geophys. Res.* **103**, 15817–15831.

LEDWELL, J. R., MONTGOMERY, E. T., POLZIN, K. L., ST. LAURENT, L. C., SCHMITT, R. W. & TOOLE, J. M. 2000 Evidence for enhanced mixing over rough topography in the abyssal ocean. *Nature* **403**, 179–182.

McWILLIAMS, J. C., SULLIVAN, P. P. & MOENG, C.-H. 1997 Langmuir turbulence in the ocean. *J. Fluid Mech.* **334**, 1–30.

MARSHALL, J. & SCHOTT, F. 1999 Open-ocean convection: observations, theory and models. *Rev. Geophys.* **37**, 1–64.

MIDDLETON, J. F. & LODER, J. W. 1989 Skew fluxes in polarized wave fields. *J. Phys. Oceanogr.* **19**, 68–76.

MOFFATT, H. K. 1983 Transport effects associated with turbulence with particular attention to the influence of helicity. *Rep. Prog. Phys.* **46**, 621–664.

MÜLLER, P. & HENDERSON, D. (Eds.) 1999 *Dynamics of Oceanic Internal Gravity Waves, II. Proc. 'Aha Huliko'a Hawaiian Winter Workshop.* School of Ocean and Earth Science and Technology.

MUNK, W. H. 1997 Once again: once again – tidal friction. *Progr. Oceanogr.* **40**, 7–35.

MUNK, W. H., ARMI, L., FISCHER, K. & ZACHARIASEN, F. 2000 Spirals on the sea. *Proc. R. Soc. Lond.* A **456**, 1217–1280.

NEEDLER, G. T. & HEATH, R. A. 1975 Diffusion coefficients calculated from the Mediterranean salinity anomaly in the North Atlantic Ocean. *J. Phys. Oceanogr.* **5**, 173–182.

NEELIN, J. D., BATTISTI, D. S., HIRST, A. C., JIN, F.-F., WAKATA, Y., YAMAGATA, T. & ZEBIAK, S. E. 1998 ENSO theory. *J. Geophys. Res.* **103**, 14261–14290.

OSBORN, T. R. & COX, C. S. 1972 Oceanic fine structure. *Geophys. Fluid Dyn.* **3**, 321–345.

PEDLOSKY, J. 1996 *Ocean Circulation Theory.* Springer.

PHILANDER, S. G. 1990 *El Niño, La Niña, and the Southern Oscillation.* Academic.

PHILANDER, S. G. 1992 El Niño. *Oceanus* **35**(2), 56–61.

PHILLIPS, O. M. 1977 *The Dynamics of the Upper Ocean*, 2nd Edn. Cambridge University Press.

POLZIN, K. L., TOOLE, J. M. & SCHMITT, R. W. 1995 Finescale parameterizations of turbulent dissipation. *J. Phys. Oceanogr.* **25**, 306–328.

SCHMITT, R. W. 1994 Double diffusion in oceanography. *Ann. Rev. Fluid Mech.* **26**, 255–285.

SEND, U. & MARSHALL, J. 1995 Integral effects of deep convection. *J. Phys. Oceanogr.* **25**, 855–872.

SIMPSON, J. H. & HUNTER, J. R. 1974 Fronts in the Irish Sea. *Nature* **250**, 404–406.

ST. LAURENT, L. & SCHMITT, R. W. 1999 The contribution of salt fingers to vertical mixing in the North Atlantic Tracer Release Experiment. *J. Phys. Oceanogr.* **29**, 1404–1424.

STOMMEL, H. M. 1961 Thermohaline convection with two stable regimes of flow. *Tellus* **13**, 224–230.

SUAREZ, M. J. & SCHOPF, P. S. 1988 A delayed action oscillator for ENSO. *J. Atmos. Sci.* **45**, 3283–3287.

TENNEKES, H. & LUMLEY, J. L. 1972 *A First Course in Turbulence.* MIT Press.

THORPE, S. A. 1975 The excitation, dissipation and interaction of internal waves in the deep ocean. *J. Geophys. Res.* **80**, 328–338.

THORPE, S. A. 1977 Turbulence and mixing in a Scottish loch. *Phil. Trans. R. Soc. Lond.* A **286**, 125–181.

TOOLE, J. M., LEDWELL, J. R., POLZIN, K. L., SCHMITT, R. W., MONTGOMERY, E. T., ST.

LAURENT, L., & OWENS, W. B. 1997 The Brazil Basin tracer release experiment. *Intl WOCE Newsletter* **28**, 25–28.

TURNER, J. S. 1973 *Buoyancy Effects in Fluids.* Cambridge University Press.

WARREN, B. A. 1981 Deep circulation of the world ocean. In *Evolution of Physical Oceanography* (ed. B. A. Warren & C. Wunsch), pp. 6–41, The MIT Press.

WESSON, J. C. & GREGG, M. C. 1994 Mixing at Camarinal Sill in the Strait of Gibraltar. *J. Geophys. Res.* **99**, 9847–9878.

WUNSCH, C. 1996 *The Ocean Circulation Inverse Problem.* Cambridge University Press.

WUNSCH, C. & STAMMER, D. 1998 Satellite altimetry, the marine geoid and the oceanic general circulation. *Ann. Rev. Earth Planet. Sci.* **26**, 219–253.

Department of Physics and Astronomy,
University of Victoria, Victoria, B.C., V8W 3P6, Canada

On Global-Scale Atmospheric Circulations

MICHAEL E. McINTYRE

1 Introduction

Atmospheric and oceanic fluid flows are a crucial part of the planetary life-support system on which we and the rest of the biosphere depend. It is to fluid motions that we, and other forms of life, owe a tolerable local environment despite the continual production of waste substances. Global-scale atmospheric circulations strongly influence the amount and distribution of stratospheric ozone and of many other greenhouse gases, including man-made chlorofluorocarbons (CFCs). For instance such circulations largely control the time scale, of the order of a century, on which the CFCs are removed from the atmosphere by exposure to energetic solar ultraviolet radiation. They are part of what determines natural ozone replenishment rates, which – as a warning to would-be geoengineers – are measured in megatonnes per day. And the atmosphere and oceans reduce climatic extremes and increase biological productivity by transporting vast amounts of heat energy from the tropics to higher latitudes, at a total rate of the order of 5×10^{15} watts. This is five thousand million megawatts, over a thousand times greater than the world's total electric power generation.

I shall concentrate mainly on global-scale circulations and what controls them, emphasizing what is now recognized as the simplest and best-understood case, the global-scale mean circulation of the middle atmosphere. In round numbers, the middle atmosphere is the layer between about 10 km and 100 km altitude, whose lower half is the stratosphere, containing the bulk of the atmosphere's ozone. The fluid-dynamical principles that emerge are relevant, also, to understanding the circulation in the more massive troposphere below. And they highlight how profoundly different, and profoundly more difficult, is the corresponding ocean circulation problem, relevant not only to heat transport but also to heat uptake and future sea-level rise. All the fluid motions in question owe their existence to thermal forcing from the Sun. Yet in a strong sense to be explained these circulations are under mechanical, not thermal, control.

As we shall see, the way in which today's understanding was reached has historical interest, including a cogent illustration of what might be called the 'Michelson–Morley principle' – the importance of negative results in science. It also involved the solution of three great enigmas of the 1960s. The first was a mysterious 'negative eddy viscosity' exhibited by large-scale atmospheric fluctuations, as seen at altitudes of about 10 km by the meteorological observing network. The second was the reversal, every 13–14 months or so, of the zonal or east–west winds in the equatorial stratosphere in the altitude range 16–30 km. In the 1960s this had been observed for just long enough to be seen not to be phase-locked to the annual cycle. Ever since then it has been called the quasi-biennial oscillation or QBO. Figure 1 shows the observed phenomenon. Its cause was completely mysterious until 1968.

The third enigma was, and is, even more striking. Because of the 23.5° tilt of the Earth's axis, the sunniest place on Earth is the summer pole. At solstice the insolation at the pole, i.e. the incoming solar energy striking unit horizontal area per unit time, is nearly 40% greater than, for instance, the diurnally averaged insolation over the tropical latitude band ±20°. The time during which insolation is maximal at the summer pole is about three-fifths of the 3-month summer season. Yet, at altitudes around 80–90 km, the sunniest place on Earth is also the coldest place on Earth – colder even than the high Antarctic plateau in midwinter, which seldom if ever gets below 170 K. The cold region shows up even in heavily averaged temperature cross-sections, such as figure 2, compiled from infrared remote sensors in orbit. Note the temperature, 150 K, at the top left of the figure, where the altitude is about 85 km. There are strong fluctuations about this average, with instantaneous temperatures as low as 105 K (e.g. Lübken 1999, figure 7). The extreme cold leads to the formation, at altitudes near 83 to 85 km, of the world's highest ice clouds, known as noctilucent clouds or polar mesospheric clouds. The term mesosphere denotes the layer of upward-decreasing temperatures in the upper half of the middle atmosphere, stopping at the mesopause, above which temperatures increase again into the thermosphere, not shown in figure 2.

As I will show, the three enigmas have now been solved. We have a good qualitative understanding of all the phenomena just mentioned and of their relation to many other observed phenomena. For instance, we have achieved scientific near-certainty as to the main cause of the Antarctic ozone hole. Contrary to what is still printed in some newspapers, the ozone hole is

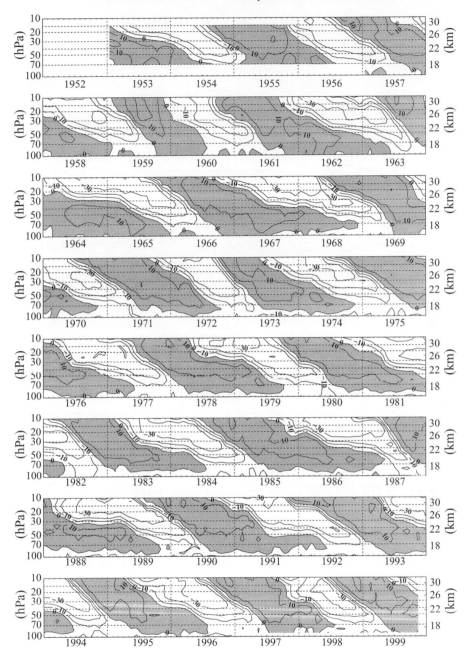

Figure 1. Time–height cross-section showing the QBO (quasi-biennial oscillation) of the east–west wind in the equatorial stratosphere between altitudes of about 16 km and 30 km. From radiosonde data at Canton Island, Gan/Maldives, and Singapore (since 1976), by kind courtesy of Drs B. Naujokat and C. Marquardt of the Free University of Berlin. Contour values are in m s^{-1}, with shaded regions representing positive, prograde or eastward flow and unshaded regions negative, retrograde or westward flow.

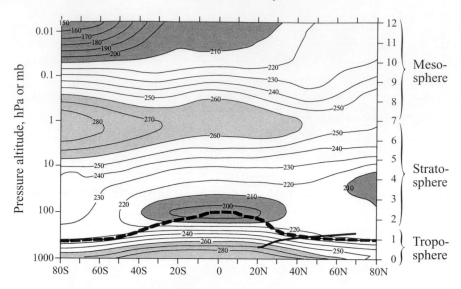

Figure 2. Mean observed temperature structure of the atmosphere for January conditions, from a standard compilation of temperatures retrieved from satellite-borne radiometers (the COSPAR International Reference Atmosphere). The light shading shows the warmest regions and the dark shading the coldest. The mean involves time averaging over the Januarys of two or three seasons, combined with averaging around latitude circles at constant pressure-altitude. The scale on the right is in nominal e-folding pressure scale heights of about 7 km (see equation (2.5)), so that for instance the temperature maximum defining the stratopause is at about 50 km. Temperatures are in degrees Kelvin.

due mainly to the buildup of CFCs. The evidence for this is overwhelming. We can unequivocally answer the lay person's question, 'Why should CFCs emitted in the north cause an ozone hole in the south?' – part of the answer being that fluid motions take CFC molecules on epic journeys. The CFC molecules are carried around the globe many times before entering the stratosphere and initiating the complex photochemical events leading to ozone destruction. All this contributes, moreover, to the patterns in long-term temperature trends now thought to show a human influence on climate.

Still remaining, as one of the greatest of all challenges for the millennium, is the task of making this picture more precise and quantitative and using it skilfully in the effort to understand, predict, and accurately monitor climate change. My aim in this chapter is a relatively modest one: to describe, as clearly as I can, according to today's best understanding, the most crucial and basic aspects of the fluid dynamics involved.

2 Some fundamentals, including anti-friction

Fluid dynamics, in the usual sense based on the standard hypotheses of continuum mechanics, is almost always an accurate model for the motion of air and water in our global environment. The most important exception to this rule is the gravitational settling of certain aerosols, such as the sulphuric acid droplets in what is called the Junge aerosol layer of the stratosphere, important both to climate and to ozone-layer chemistry since it produces a significant radiative cooling effect as well as providing reactive surfaces. The droplets have diameters $< 10^{-6}$ m, which is not large in comparison with the local mean free paths of air molecules in the stratosphere. In round numbers, mean free paths ℓ_{path} vary inversely with density from $\sim 10^{-7}$ m at sea level to $\sim 10^{-1}$ m at 100 km altitude, exceeding 10^{-6} m above about 16 km.

Aerosol settling aside, practically all other naturally occurring scales of motion greatly exceed ℓ_{path}, all the way down to the smallest (Kolmogorov) scales ℓ_{K} of turbulent motion. The same holds for the oceans, with ℓ_{path} replaced by intermolecular scales around 10^{-9} m, which again are far smaller than the smallest naturally occurring values of ℓ_{K}. As is well known, the scale ℓ_{K} is a rough estimate of the order of magnitude of the inner viscous scales of ordinary three-dimensional turbulence: from dimensional analysis, $\ell_{\mathrm{K}} \sim (v^3/\varepsilon)^{1/4}$, where v is the kinematic viscosity, i.e. the molecular diffusivity of momentum, and ε is the rate per unit mass at which the turbulence dissipates kinetic energy (see Chapter 5). The associated velocity scale $u_{\mathrm{K}} \sim (v\,\varepsilon)^{1/4}$. This is such that the Kolmogorov-scale Reynolds number

$$u_{\mathrm{K}}\ell_{\mathrm{K}}/v \sim 1. \tag{2.1}$$

For the atmosphere we have from gas kinetics, again as a rough estimate,

$$c_s \ell_{\mathrm{path}}/v \sim 1, \tag{2.2}$$

where c_s is the sound speed, of the same order as a typical molecular velocity. Thus $\ell_{\mathrm{path}}/\ell_{\mathrm{K}} \sim u_{\mathrm{K}}/c_s$, of the order of the Kolmogorov-scale Mach number, a negligibly small quantity. The validity of continuum mechanics, as expressed by the Euler and Navier–Stokes equations of fluid-dynamical textbooks, will from here on be taken for granted with no further comment.

For the atmosphere we may also make the ideal-gas assumption; thus, in the middle atmosphere, which is very dry (with water vapour mixing ratios $\lesssim 6$ p.p.m.v., i.e. 6×10^{-6} by volume), we may assume that the density ρ, pressure p and absolute temperature T are related to within a small fraction of a percent by

$$p = R\rho T \quad \text{where} \quad R = 287 \, \mathrm{m^2\,s^{-2}\,K^{-1}}, \tag{2.3}$$

the specific gas constant for dry air of present-day composition. The atmosphere – a thin shell of air held in place by an inward body force $g = 9.8\,\mathrm{m\,s^{-2}}$ per unit mass, the Earth's gravity – is normally (barring asteroid impacts or other cataclysms) in a state of small fluctuations about hydrostatic balance,

$$\frac{\partial p}{\partial z} = -g\rho = -\left(\frac{g}{RT}\right) p,\qquad (2.4)$$

implying that p and ρ decrease monotonically, and roughly exponentially, with geometric altitude z. As z ranges from 0 to 100 km, p and ρ range over six orders of magnitude while T ranges over little more than a factor 3. The monotonic decrease of p and ρ is a robust property, long relied on by air traffic control systems based on pressure altimetry. Equation (2.4) shows that the decrease would be exactly exponential if g/RT were exactly constant. Otherwise, we may still define an e-folding pressure scale height H_p as

$$H_p = -\frac{p}{\partial p/\partial z} = \frac{RT}{g}.\qquad (2.5)$$

When $g = 9.8\,\mathrm{m\,s^{-2}}$ and $T = 239\,\mathrm{K}$, well within the range of figure 2, then $H_p = 7.00\,\mathrm{km}$.

Even at the continuum-mechanics level of description we shall see that the atmosphere, viewed on a global scale, clearly illustrates one of the grand themes in physics: the organization of chaotic fluctuations in complex dynamical systems, with non-trivial, persistent mean effects. This is a theme recognized as important not just for classical textbook cases like molecular gas kinetics but also for an increasingly wide range of dynamical systems with large phase spaces, not in simple statistical-mechanical equilibrium or near-equilibrium.[1]

The phase space of the atmosphere and/or oceans considered as a dynamical system is, of course, unimaginably vast; and it is not surprising that attempts to find 'weather attractors' of extremely low dimension $\lesssim 6$ fail to produce convincing results (Ruelle 1990), even though ideas about weather 'regimes' or preferred regions of phase space are still likely to be important, e.g. for patterns in long-term temperature trends (Corti *et al.* 1999). Such preferred regions of phase space may or may not contain mathematically well-defined attractors. Little is known about such questions. What seems more surprising, indeed astonishing, is that today's numerical weather prediction systems, with sparse data inputs, and atmospheric models whose

---

[1] Biological molecular motors, now beginning to be understood, are a striking example. 'Persistent mean effects' include your own ability to hold this book up against gravity, as a result of complex but highly organized molecular-scale fluctuations whose time scales encompass those of thermal agitation, picoseconds or less.

phase-space dimensions are more like 10^8 – still minuscule in comparison with reality, $\sim 10^{30}$ – succeed as well as they do. In a fair proportion of cases, such systems now have useful predictive power over time scales of a few days, though they are far from infallible. The predictive power is good enough to be routinely exploited by the aviation and other weather-sensitive industries.

A key aspect of the atmospheric global-circulation problem is a tendency for fluctuations about a large-scale mean state to behave very differently from classical expectations. 'Classical' here refers to ideas about fluid turbulence going back to Prandtl's mixing-length theory and to statistical theories that assume small departures from spatially homogeneous turbulence. 'Mean' refers to an axisymmetric mean state defined by zonal averaging, i.e. averaging around latitude circles as in figure 2.

For a combination of reasons, the fluctuations about such a mean state tend very often to drive it away from solid rotation. This contradicts the classical idea, still enshrined in the term 'eddy viscosity', that turbulence theory should be like gas kinetics, with turbulent fluctuations acting like molecular-scale fluctuations about a nearly-homogeneous mean state. That is, the fluctuations should act, according to this idea, like a viscosity – with ℓ_{path} replaced by some 'mixing length', 'Austausch length', or other length scale representative of the irreversible local displacements of fluid elements. The classical idea therefore says that the fluctuations by themselves should drive the mean state toward, not away from, solid rotation.

But observations have long shown that the most typical real behaviour is the reverse. This began to be recognized through the work of H. Jeffreys, V. P. Starr and others on the mean effects of large-scale atmospheric fluctuations. We may call the observed behaviour anti-classical, or anti-frictional. In the 1950s and 1960s it was seen as remarkable and mysterious, and described as 'countergradient momentum transport' or 'negative eddy viscosity' even though the latter, if taken literally, would have to imply a mathematically pathological time-reversed diffusion. In 1967, in the closing pages of a landmark review by E. N. Lorenz (1967), the anti-frictional behaviour was flagged as a major enigma for atmospheric science – the first of the three enigmas mentioned earlier.

In 1967 there seems to have been no suspicion of anything in common with the second enigma, the QBO shown in figure 1. But we now know, as a matter of practical certainty, that the QBO is fluctuation-driven and that it is another example of anti-frictional behaviour. Besides this, the QBO is a remarkable instance of order emerging from chaos. What is seen in figure 1 is the atmosphere's most predictable phenomenon beyond the timespan of the seasonal cycle. A closely similar phenomenon has been produced in a laboratory experiment, described in the celebrated paper of Plumb & McEwan (1978), in which it is unequivocal that the departures from solid rotation are driven solely by fluctuations. The experiment will be discussed in §7 below.

And the solution to the third enigma is essentially similar. The cold summer-polar anomaly near 80–90 km is due to the persistent mean effects of fluctuations. Throughout a vast volume of the atmosphere, certain fluctuations drive a mechanical pumping action that powers a gigantic natural refrigerator. Air is pulled upward, causing it to expand and cool, in a region surrounding the summer pole. Besides depending on the mean effects of fluctuations, essentially as in the QBO, the pumping action also depends on Coriolis effects. For that reason it is sometimes called 'gyroscopic pumping'. It is more than the solution to the third enigma: it is the driving mechanism for the entire global-scale mean circulation of the middle atmosphere, and a major factor in shaping temperature anomalies as well as chemical transport.

3 Wave propagation and gyroscopic pumping

The anti-frictional behaviour, and the solution to all three enigmas, is related to the existence of certain wave propagation mechanisms. These enter the fluctuation dynamics along with turbulence, and give rise to systematic fluxes or transports of momentum quite different from those found in gas kinetics or in classical turbulence theories.

One important difference is that wave-induced momentum transport is a long-range process. It can be effective over distances far exceeding the fluctuating displacements of fluid elements by the fluid motion itself, regardless of whether that motion is locally wavelike or turbulent, or both at once, as on ocean beaches. By its nature, any wave propagation mechanism tends to promote systematic correlations between the different fluctuating fields. This is a case of the dynamical organization of chaotic fluctuations mentioned earlier. The correlations are almost always such that mean stresses are set up. In other words, the correlations result in momentum being systematically transported through the fluid, over distances limited only by the distances over which the waves can propagate. Momentum transports of this kind are sometimes called *radiation stresses* (e.g. Brillouin 1925). They tend to be highly anisotropic, contrary to what might be suggested by the older term 'radiation pressure' still found in the literature.

Such momentum transports can occur for the simplest non-turbulent, purely wavelike motion involving reversible oscillations of fluid elements.

For example there are simple wavemotions in which oscillating fluid elements move slantwise – e.g. upward while moving eastward then downward while moving westward. The resulting covariance or Reynolds stress implies a tendency for eastward momentum to be transported systematically upward, if the tendency is not cancelled or dominated by other effects. One example is that of internal gravity waves. These are wavemotions that owe their existence to stable stratification of the density field.

Internal gravity waves are of interest because they cause the refrigeration effect underlying the third enigma. What happens is this. The waves propagate upward and break near the top of the middle atmosphere, most conspicuously above about 80 km. The exponential falloff of density ρ with altitude z – recall (2.4) and (2.5) – implies that upward-propagating waves must encounter conditions analogous to those encountered by surface gravity waves near an ocean beach. Near a beach, the shallowing water causes incoming waves to steepen and break. The top of the summertime middle atmosphere is like a beach on which the waves are incident obliquely, for reasons connected with the so-called filtering of waves (§ 5 below), in such a way that the air at altitudes near 90 km feels a persistent prograde or eastward force. The recoil, a retrograde or westward force, is felt at lower altitudes.

But Coriolis effects are strong in the dynamics of the mean circulation, whose time scales greatly exceed the time scales of the Earth's rotation. Anything that is pushed eastward tries to turn equatorward. This is the gyroscopic pumping action. In response to it, the air near 90 km does move persistently equatorward, on average, and air is pulled up to replace it in a large area surrounding the summer pole, roughly 60° latitude poleward.

The air being pulled up not only expands and cools but also brings with it a supply of water vapour, at the low yet significant mixing ratios $\lesssim 6$ p.p.m.v. already mentioned above equation (2.3). It is this combination of circumstances that gives rise to the noctilucent or polar mesospheric clouds mentioned in § 1. In the summer polar mesosphere, solar ultraviolet radiation is so copious and energetic that water vapour is destroyed photochemically on time scales of days to weeks. The ice crystals could not form near 85 km without a supply of water substance from below. Nor could the ice crystals form without the extreme refrigeration. Radiative–photochemical calculations suffer from considerable uncertainties at these altitudes, but it is thought that the refrigeration effect holds temperatures below radiative-equilibrium temperatures T_{rad} by something of the order of 100 K. We shall use the symbol T_{rad} to denote the temperature distribution toward which actual temperatures T, in the absence of any fluid motion, would tend to relax. Typical radiative relaxation times are of the order of days to weeks.

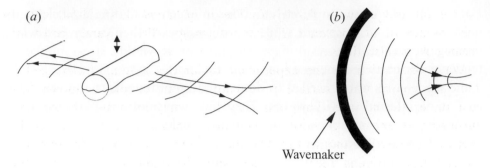

Wavemaker

Figure 3. Simple experiments with surface waves on water, illustrating wave-induced momentum transport, generating a strong mean flow that can be made visible by sprinkling a little powder such as chalk dust on the surface of the water. Experiment (*a*), my standard lecture demonstration, uses a cylinder about 10 cm long and 4 cm in diameter. In experiment (*b*), the curved wavemaker is about 60 cm in arclength and radius. Good results are obtained with capillary–gravity waves at frequencies around 5Hz.

4 Wave-induced momentum transport: experiment and theory

Wave-induced momentum transport has long been recognized as a generic phenomenon, characteristic of many types of waves in a vacuum or in fluid media. Its physical reality and robustness in fluid media can be illustrated in a simple experiment, easily performed in the bath or in the kitchen sink. I have often performed it, when lecturing on this topic, in a transparent dish on a overhead projector. First I get the audience's attention with an assertion like 'This demonstration is robust. It *always* works.' And it does. A cylindrical wavemaker is oscillated vertically at $\sim$ 5Hz in the free surface of a layer of water, radiating capillary–gravity waves as suggested in figure 3(*a*). Chalk dust or other floating powder sprinkled on the surface reveals that the water flows away from the wavemaker, in a persistent and conspicuous mean motion that sweeps the surface dust along with it. There is an irreversible transport of momentum from the wavemaker to the water on either side of it.

This irreversible momentum transport depends on the dissipation of the waves. The need for dissipation can be seen at once from Kelvin's circulation theorem. Recall that when Coriolis effects are negligible, as is the case in this experiment, Kelvin's circulation $C(\Gamma)$ is the line integral of fluid velocity $\boldsymbol{u}$,

$$C(\Gamma) = \oint_{\Gamma} \boldsymbol{u} \cdot \mathrm{d}\boldsymbol{s}, \tag{4.1}$$

over a material circuit Γ, i.e. a closed material contour, whose line element is $\mathrm{d}\boldsymbol{s}$. The relevance of $C(\Gamma)$ to problems of this general kind was first pointed out, as far as I know, in Rayleigh's *Theory of Sound* (1945, §§ 240, 352). In the absence of dissipation $C(\Gamma)$ cannot change. This severely constrains mean-flow accelerations.

Wave dissipation, in the experiment under discussion, is enhanced by the dirty, chalky water surface, producing a strong viscous boundary layer just under the surface. This boundary layer may even be intermittently turbulent, an example of *wave breaking* even when air is not entrained. The fact that irreversible momentum transport is involved can be demonstrated by carefully stopping the wavemaker and observing that the mean flow persists much longer than does the outgoing wave field. In other words, there is a contribution to the mean flow that is not a mere 'Stokes drift' or other reversible mean-flow contribution that leaves $C(\Gamma)$ unchanged. Such reversible contributions would immediately return to zero if the waves were to propagate out of the region of interest.

It is clear moreover that the observed mean flow is indeed due to wave-induced momentum transport rather than to the oscillatory boundary layer on the wavemaker itself. Because the wavemaker oscillates vertically, the horizontal branch of the Rayleigh–Schlichting boundary-layer streaming generated at the solid surface of the wavemaker is directed *toward* the wavemaker. This is opposite to what is observed. So it is the wave effects that dominate. Moreover, with a larger tank, one can easily do versions of the experiment in which the waves are generated by a larger, concave wavemaker that focuses them at a more distant spot, as suggested in figure 3(*b*). Sure enough, momentum is transported over the greater distance: the strongest mean flow is seen just where the waves are focused. Once again, the mean flow persists after the wavemaker is stopped.

A spectacular case of wave-induced momentum transport over still greater distances was demonstrated in 1966 by W. H. Munk and colleagues in a famous observational study of surface gravity waves propagating from the Southern Ocean across the entire Pacific, all the way to beaches in Alaska, where the waves, incident obliquely, generate longshore mean currents in fundamentally the same way. In other words, forces exerted on the Southern Ocean, by storms in the latitudes of the roaring forties and fifties, were shown to drive mean currents in surf zones on beaches in the far north. Wave-induced momentum transport can indeed be a long-range process.

Our theoretical understanding of such phenomena is fairly well developed today, though far from perfect. The theory needed to describe the phenomena in their full generality – applying for instance to three-dimensional mean states in a rotating, stratified atmosphere – is technically and conceptually tricky, with some of it still not in a fully satisfactory state. Amongst the conceptual difficulties are those of sorting out, in a generally valid way, long-range from short-range effects, reversible from irreversible, rotational from irrotational, and general from particular. Compounding these difficulties is a largely unspoken 'wave-momentum myth' that assumes problems of waves in material media to be universally isomorphic to problems of waves in a vacuum, as if, for momentum-transport purposes, wave packets always behaved like isolated bullets or photons possessing momentum of their own. There is a valuable *partial analogy* with photon behaviour in a certain range of circumstances (Bühler 2000), but not a universal isomorphism – a fact underlined by the existence of counterexamples (e.g. in internal wave scattering: McIntyre 1981). And there is a minefield of difficulties, or pitfalls at least, of a purely technical kind, whose debris is scattered over a vast research literature. Incomplete theoretical arguments, such as arguments that ignore boundary conditions or confine attention to a subset of the governing equations, can easily miss significant contributions.

A typical mistake is to suppose that in order to describe wave-induced momentum transport one need only consider the momentum equation. If correct, this would imply that radiation stresses are always the same as wave-induced Reynolds stresses. But there is a well known set of counterexamples. These involve what might be called the 'Eliassen–Palm effect'. This is the partial cancellation, or even dominance, of Reynolds-stress contributions by terms arising from Coriolis forces. The simplest self-consistent theoretical framework, confined to cases of translationally or rotationally invariant mean states, such as those defined by zonal averaging, and involving the so-called 'transformed Eulerian mean equations' and an associated 'effective momentum flux' or 'Eliassen–Palm flux', grew from seminal contributions by Eliassen & Palm (1961) and Charney & Drazin (1961). It is reviewed in chapter 3 of the book by Andrews *et al.* (1987).

The inadequacy of considering Reynolds stresses alone can be demonstrated in other, more direct ways, for instance from a consideration of Kelvin's circulation $C(\Gamma)$, or of the mean stress across a material surface undulated by the waves. Especially important in stably stratified fluid systems are material surfaces that would be horizontal in the absence of the waves. Fluctuations in surface slope are usually correlated with fluctuations in pressure, giving rise to a mean tangential stress across the surface. The terms *radiation stress*, and *wave resistance per unit area*, are both used to denote this mean stress. To a first approximation it can be shown to be equal to the vertical component of the Eliassen–Palm flux. It is a more reliable measure of wave-induced momentum transport than the wave-induced Reynolds stress. The two can differ because of Coriolis forces on a varicose layer sandwiched between the undulating material surface and a neighbouring fixed, non-material surface. There can be a mean Coriolis force associated with correlations between velocity fluctuations and fluctuations in the thickness of the varicose layer, related to what is called bolus transport in the oceanographic literature. To my knowledge this point, implicit in the original results of Eliassen & Palm (1961) and Charney & Drazin (1961), was first explicitly made by Bretherton (1969). In the oceanographic literature the radiation stress itself is sometimes called 'form drag'. Being an essentially frictionless phenomenon, it has to be sharply distinguished from what aerodynamicists call form drag – the drag contribution due to the turbulent or viscous wake behind a solid body moving through fluid.

Any theory of wave-induced momentum transport needs to be more accurate, of course, than the standard linearized wave theories of physics textbooks. The linearized theories are correct only to $O(a)$ in wave amplitude a and, within their regimes of validity, can describe only the temporary or reversible oscillations of fluid elements characteristic of purely wavelike motion in a material medium. The momentum-transporting correlations – more precisely, covariances – are $O(a^2)$ as $a \to 0$, as are the Coriolis–thickness covariances just mentioned. Therefore, $O(a^2)$ is the minimum accuracy for a self-consistent asymptotic description of the consequent mean effects.

At such accuracies there can be substantial differences, moreover, between the results of applying Eulerian, Lagrangian or other averaging operators to define the mean flow. Lagrangian-mean velocities $\bar{u}^{\mathrm{L}}$ are relevant to mean particle motion, such as the chalk dust in the experiments of figure 3, whereas Eulerian means $\bar{u}$ are taken at a fixed point. Lagrangian-mean velocities $\bar{u}^{\mathrm{L}}$ are theoretically important because of their relation to Kelvin's circulation $C(\Gamma)$ and to the radiation stress or wave resistance across an undulating material surface. When $\bar{u}^{\mathrm{L}}$ is well defined, $C(\Gamma)$ can be rewritten as

$$C(\Gamma) = \oint_{\overline{\Gamma}} (\bar{u}^{\mathrm{L}} - \hat{p}) \cdot \mathrm{d}s \qquad (4.2)$$

(Bühler 2000 and references therein), where $\overline{\Gamma}$ is a suitably defined mean material circuit and where $\hat{p}$ is an intrinsic, $O(a^2)$ wave property called the quasimomentum, or pseudomomentum, per unit mass (whose conflation with momentum in the literature contributes to the wave-momentum myth). Once $\overline{\Gamma}$ and $\bar{u}^{\mathrm{L}}$ are defined, (4.2) can be regarded as the exact definition of $\hat{p}$. In simple cases its magnitude $|\hat{p}|$ is approximately equal to the intrinsic wave energy $\hat{E}$ per unit mass divided by an intrinsic phase speed $\hat{c}$. Formulae for $\hat{E}$ and $\hat{p}$ in such cases will be given in (11.17a,b) below. Intrinsic means, by convention, intrinsic to the wave dynamics as described to leading order,

$O(a)$, by linearized wave theory. Thus the intrinsic phase speed $\hat{c}$ means the phase speed relative to the fluid, i.e. relative to any $O(1)$ mean motion. Intrinsic wave properties are Galilean invariant by definition. The above-mentioned radiation stress or vertical Eliassen–Palm flux equals the vertical group velocity times the horizontal projection $\hat{p}_{\text{horiz}}$ of $\hat{p}$ – *as if* the waves had momentum $\hat{p}$.

The Stokes drift mentioned earlier is defined to be just the $O(a^2)$ difference

$$\overline{u}^{\text{Stokes}} = \overline{u}^{\text{L}} - \overline{u} \tag{4.3}$$

between the Lagrangian and Eulerian mean velocities. It is another intrinsic wave property. Note incidentally that $\overline{u}^{\text{Stokes}} \simeq \hat{p}$ in two special cases: surface capillary–gravity waves and sound waves, but not in other cases. Capillary–gravity and sound waves are exceptional in this regard because the wave-motion is irrotational.

Lagrangian-mean velocities and Stokes drifts, even when well defined, can have counterintuitive properties. An example is the so-called divergence effect, $\nabla \cdot \overline{u}^{\text{L}} \neq 0$ when $\nabla \cdot u = 0$, which arises from wave amplitude growth or turbulent particle dispersion. The divergence effect is implicit in equation (3.9) of Chapter 10, and is further discussed in McIntyre (1980, 1988).

5 Wave breaking, wave filtering, and critical layers

The theoretical difficulties are compounded, of course, when dimensionless wave amplitudes a reach values large enough for the waves to begin to break. Wave breaking of one kind or another is commonplace in the real atmosphere. The ocean-beach analogy is almost always relevant. Lagrangian means become complicated, ill-defined, or both, or at best crudely definable as a 'surf-zone average' if particle dispersion is sufficiently confined to some 'surf zone' or wave-breaking region (Thuburn & McIntyre 1997). All the relevant kinds of waves will usually have started to break by the time the amplitude a reaches values of order unity with a defined in a suitable, dimensionless way, for instance as follows, or as in (11.15) below.

The simplest way is to define a as the ratio of the amplitude u'_{max} of a fluctuating horizontal fluid velocity component u' to an intrinsic horizontal phase speed $\hat{c}$, representing propagation of phase surfaces relative to the fluid. More precisely, $\hat{c}$ is defined here to be a positive quantity and to be the relative speed of travel of the intersection of such phase surfaces with a fixed horizontal surface, while u'_{max}, also positive, is the amplitude of the fluctuating component of u in the direction of travel. (Then $\hat{E}/\hat{c}$ can be shown to approximate the magnitude of $\hat{p}_{\text{horiz}}$, which also points in the

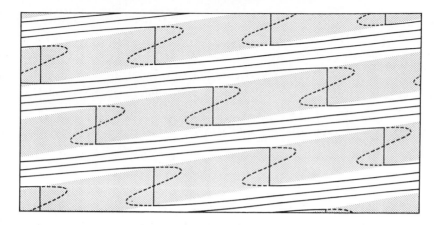

Figure 4. Material surfaces, which are also passive-tracer isopleths (curves, partly dashed and partly solid), in a monochromatic, large-amplitude, convectively unstable internal gravity wave viewed from the side. The dimensionless amplitude a defined by equation (5.1) is 1.66, and Coriolis effects are assumed negligible in the dynamics of the wave motion. The solid curves, with the vertical portions included, show the passive-tracer isopleths after wave breaking, in an idealized thought experiment in which the waves arrive and then break by convective overturning. More precisely, the thought experiment assumes (*a*) that the passive tracer isopleths are initially horizontal, (*b*) that they are undulated when the waves begin to arrive, taking the shapes of the solid–dashed curves, and then (*c*) instantaneously and perfectly mixed in the shaded regions. Diagram taken from McIntyre (1989).

direction of travel.) The definition

$$a = u'_{max}/\hat{c} \tag{5.1}$$

is apt for the simplest forms of wave breaking, which involve fluid elements overtaking wave crests. From the kinematics of the situation, one can see that such overtaking begins when a just exceeds 1.

Figure 4 shows some of the material surfaces (dashed curves and non-vertical solid curves) in a finite-amplitude solution, viewed from the side, representing simple internal gravity waves with $a > 1$. In fact $a = 1.66$ in this case: relative to the undisturbed state, fluid elements are travelling up to 66% faster than the waves themselves. The shaded regions contain heavy fluid over light and can therefore overturn convectively, leading to turbulence and mixing. This is the simplest conceptual model of internal-gravity-wave 'surf'. Evidence that such mixing actually occurs has been seen in high-resolution vertical profiles of certain chemical tracers, such as atomic oxygen, measured from rockets near 90 km altitude (C. R. Philbrick, P. J. Espy, personal communication). The observed atomic-oxygen profiles were close to what they would be in an idealized thought experiment assum-

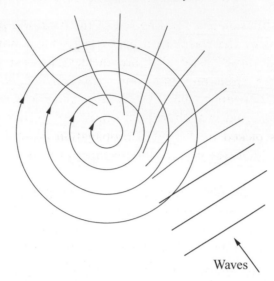

Waves

Figure 5. Sketch of surface capillary–gravity waves being selectively refracted into a bathplug vortex, an everyday example of wave filtering.

ing perfect instantaneous vertical mixing, depicted by the solid curves in figure 4, which show the isopleths of atomic-oxygen mixing ratio for such a thought experiment. The atomic oxygen is assumed to behave as an inert, passive tracer over the short time scales involved, and its isopleths are assumed horizontal before the waves arrive. Further observational and laboratory evidence for the breaking of internal gravity waves by such convective overturning and mixing, at mesospheric altitudes, can be found scattered throughout a vast literature.

Even without such observational evidence there would be strong expectations, on theoretical grounds, that values of a as defined in (5.1) should often reach values of order unity, leading to wave breaking. Sometimes this may happen via chance constructive interference, or even via subtle nonlinearities such as resonant-triad interactions (e.g. McEwan 1971). But there are two robust and systematic reasons for a to increase, which tend to dominate the pattern of persistent, long-range momentum transport. First, there is the exponential falloff of density ρ with height. Second, there is the selective refraction and Doppler shifting of waves by background mean shear (more detail in §11), which systematically changes values of $\hat{c}$.

This last is the filtering effect already mentioned. It is so called because it can act very differently on different components of the wave field depending on whether $\hat{c}$ is Doppler shifted upward or downward. A similar phenomenon can be seen in an ordinary bathplug vortex surrounded by random surface waves, the usual situation after getting out of the bath and pulling the plug. The vortex is seen by the waves as a background shear flow. It selectively refracts into itself those waves that are incident against its rotation, as sketched in figure 5. (This may test your powers of observation. If at first glance the vortex appears to be rotating, say, anticlockwise – as popular mythology says it should in the northern hemisphere – then the chances are that your eye has picked up the wave propagation around the vortex and that the background vortex flow is actually clockwise, as revealed by specks of dirt on the surface.)

Extreme cases of the filtering effect for internal gravity waves in background shear are discussed in the literature under the headings 'critical line', 'critical level', and 'critical layer', following a simple, linear-theoretic example, with no Coriolis effects and with vertical shear only, first elucidated in full detail by Booker & Bretherton (1967). The critical line or level – more generally *surface*, in the real three-dimensional world – is defined with respect to a hypothetical monochromatic wave train like that illustrated in figure 4, except that the waves are considered to have infinitesimal amplitude. The critical line, level, or surface is most usefully and generally defined as a locus at which the wavelength would be Doppler shifted to zero if the waves were to reach it. In fact the waves never reach it, according to the standard linearized ray theory (eikonal, wave-packet, or group-velocity theory); more precisely, a ray point never reaches it in a finite time if background shear and other background gradients are finite. A ray point or wave-packet centroid is a point moving with the group velocity. The ray equations predict that the wavenumber $|k|$ cannot increase faster than linearly with time, following a ray point. Regardless of all other details, therefore, the wavelength, $\propto |k|^{-1}$, cannot reach zero in a finite time, and the ray point or wave packet can never reach its critical surface.

What is more important, though, especially as ray theory is not valid in all cases, is that besides increasing $|k|$ the Doppler shifting reduces $\hat{c}$. In practice, therefore, as (5.1) suggests, real waves approaching a critical surface will often start breaking or otherwise dissipating some finite distance away from the surface. For, whenever non-turbulent viscous or thermal dissipation is negligible, it can be shown that although u'_{max} may decrease it does so much less steeply than $\hat{c}$, as $t \to \infty$. Hence (5.1) says that if $\hat{c} \to 0$, as happens in Booker & Bretherton's example, then $a \to \infty$. Sooner or later, therefore, the wave field must locally resemble that in figure 4, and wave breaking will begin. Thus the linearized theory predicts its own breakdown, as pointed out in Booker & Bretherton (§6, equations (6.4)ff), even though, according to that theory, $\hat{E}$ may evanesce like u'^2_{max} as $t \to \infty$.

The linear-theoretic prediction that $a \to \infty$ if $\hat{c} \to 0$ can be verified, in a fairly general set of cases, from a conservation equation satisfied by the horizontal projection $\hat{\boldsymbol{p}}_{\text{horiz}}$ of $\hat{\boldsymbol{p}}$ for a horizontally homogeneous background. This is related to what is called wave-action conservation, (11.18) below. The group velocity as well as the phase velocity goes to zero as $t \to \infty$, and the conservation equation predicts that, asymptotically, $|\hat{\boldsymbol{p}}_{\text{horiz}}| \simeq |\hat{E}/\hat{c}| \propto |u_{\text{max}}'^2/\hat{c}| \gtrsim$ constant. The precise behaviour depends of course on details of how strongly the rays converge, as discussed briefly following (11.18), as well as on whether $\hat{c} \to 0$. As also shown in §11, Coriolis effects may cause $\hat{c}$ to approach a small but finite limit.[1]

The term *critical layer* has no universally defined meaning. But for real, finite-amplitude waves it can simply and usefully be identified with the region where the waves are breaking or otherwise dissipating as a result of the Doppler shifting. In terms of the ocean-beach analogy, therefore, the critical line, level or surface is analogous to the edge of the beach where your toes barely get wet, whereas the critical layer is analogous to the surf zone extending seaward. For real, finite-amplitude ocean-beach waves, most of the surf and most of the wave dissipation is far from the edge unless the beach is a very steep one, more like a sea wall. For internal gravity waves, a background shear $\partial \bar{u}/\partial z$ corresponding to such steepness would have to be so strong, or other conditions so non-uniform, as to invalidate ray theory. A standard dimensionless measure of $|\partial \bar{u}/\partial z|$, or rather of its inverse square, is the gradient Richardson number

$$Ri = N^2 / |\partial \bar{u}/\partial z|^2, \tag{5.2}$$

where N is the buoyancy or Brunt–Väisälä frequency of the stable stratification defined as $\sup |\hat{\omega}|$, the supremum or lowest upper bound on intrinsic frequencies $\hat{\omega}$ of internal gravity waves that can exist in that stratification (§§ 10–11 below). For typical mean wind fields in the middle atmosphere, values of Ri are in the tens to hundreds or more. Much smaller values are sometimes attained, most notably, above mountain ranges in the so-called downslope windstorm conditions that produce chinook or bora winds. There, wave amplitudes a increase steeply with altitude and cause a sudden onset of convective or shear-unstable wave breaking, producing a critical layer or surf zone in which the vertical scale is so small that the distinction be-

---

[1] There are other cases exhibiting so-called valve effects, in which a surface exists that is critical for waves approaching from one side but not the other, in the sense that a ray point can cross the surface keeping $|\boldsymbol{k}|$ finite – i.e. not seeing the surface as critical – but can then turn back and approach from the other side with $|\boldsymbol{k}| \to \infty$, i.e. then seeing the surface as critical. For instance this can happen with internal gravity waves when horizontal as well as vertical shear is important (Acheson 1973). In practice these linear-theoretic effects are usually pre-empted by wave breaking.

tween waves and background is lost, and within which strong mixing brings N^2 close to zero with Richardson numbers effectively small (e.g. Farmer & Armi 1999, and references therein). Like sea walls, critical layers can then be significant wave reflectors.

Besides its significance for wave-induced momentum transport, convective wave breaking also has importance, or potential importance, as a mechanism for mixing heat and chemical constituents vertically. As such, it has the curious property of failing to conform to the usual mixing-efficiency rule, equation (5.6) of Chapter 10, namely that in some average sense the mixing can be described by a vertical eddy diffusivity of around $0.2\varepsilon/N^2$ where ε is an average turbulent energy dissipation rate. McIntyre (1989) discusses an idealized case in which the energy changes and average diffusivity can be computed unambiguously, essentially the thought experiment in which the shaded regions in figure 4 are perfectly mixed. The results show that the coefficient of ε/N^2 is not even roughly constant, but is highly sensitive to the value of a just before mixing. By implication, therefore, the coefficient is sensitive to the gentleness or violence of the wave breaking. For gently breaking waves in 'shallow beach' conditions, with $(a-1) \ll 1$, the coefficient is asymptotically proportional to $(a-1)$ and can be far smaller than 0.2. It increases monotonically with a, exceeding 0.2 when a exceeds 2.4, which would require the waves to be breaking very suddenly and violently, as in 'steep beach' or 'sea wall' conditions. In the case of figure 4, which has $a = 1.66$, the coefficient 0.2 is replaced by 0.11.

6 The general definition of wave breaking

From the viewpoint of the general theory discussed in §4, 'wave breaking' is most appropriately defined in the manner suggested by (4.1) and (4.2). Breaking is then, by definition, any wave-induced process leading to the rapid and irreversible deformation of otherwise wavy material contours Γ (McIntyre & Palmer 1985). Such rapid and irreversible deformation must take place for instance in regions of turbulent mixing, like the shaded regions in figure 4, where the material contour shapes (not shown) become exceedingly complicated. 'Otherwise wavy' is a reminder that the linearized wave theory within its regime of validity would describe the same material contour as undulating reversibly, as might happen in physical reality if $a < 1$.

Wave breaking, then, in the sense thus defined, is a process that can efficiently remove the constraint on mean-flow change imposed by (4.1) and (4.2) and allow irreversible momentum transport to take place. For historical reasons this is sometimes called 'violating the non-acceleration constraint'

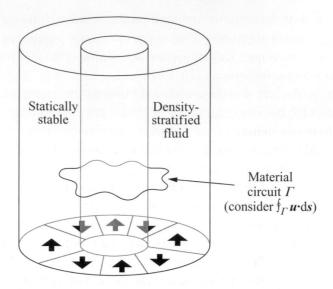

Figure 6. The experiment of Plumb & McEwan (1978), illustrating how fluctuations can behave *anti-frictionally* in the sense of driving a fluid system away from (zero) solid rotation. A reversing mean flow, first clockwise then anticlockwise around the annulus, is generated from rest by nothing more than standing oscillations of a flexible membrane at the bottom. The annulus is filled with a stably stratified salt solution. For a useful set of notes on how the experiment was successfully repeated at Kyoto University, after some failures, see the 'Inside Stories' and 'Tech Tips' at http://www.gfd-dennou.org/library/gfd_exp/ under the heading 'QBO', where an .avi movie of the experiment running can also be seen.

(McIntyre 1980). The general definition of wave breaking in terms of suitable material contours Γ has another advantage besides direct relevance to (4.1) and (4.2): it circumvents any need to decide which of an infinite variety of shapes or forms of persistent or intermittent wave breaking does, or does not, involve dynamical instability, or partly or fully developed turbulence in one sense or another, or whitecapping or even the spectacular air entrainment by ocean-beach surf. It is the irreversible deformation of otherwise wavy material contours Γ that is common to the whole panoply of cases, and directly relevant to (4.1) and (4.2).

As I have tried to indicate, one can reach a useful level of understanding without solving all the outstanding problems. One saving grace is the frequent usefulness of linearized wave theory, including ray theory, and its predictions of long-range momentum transport *outside* surf zones, as in the trans-Pacific case of Munk and colleagues. The main uncertainties tend to be about the precise distribution of momentum-flux convergence within surf zones and the extent to which such zones reflect or re-radiate waves.

7 The Plumb–McEwan experiment and the refrigeration effect

There are beautiful experimental cross-checks on our understanding, going beyond the crudely qualitative experiments of figure 3. One of the most beautiful is the Plumb–McEwan (1978) experiment. Figure 6 is a schematic of the set-up. A flexible membrane is oscillated in a standing wave, at the bottom of an annular container filled with a stably stratified salt solution, saltier and therefore denser at the bottom. Because the wave is a standing wave, the experimental conditions are close to mirror-symmetric. There is no externally imposed difference between the clockwise and anticlockwise directions around the annulus. Coriolis forces are negligible. When the bottom boundary is oscillated at a suitable frequency, internal gravity waves are generated. In this case they can be regarded as a superposition of two progressive wave components travelling clockwise and anticlockwise. They propagate up into the fluid and dissipate, in this case, not by breaking but almost exclusively by viscous friction. Salt diffusivity is negligible for this purpose. It would of course be interesting, though expensive, to repeat the experiment in an apparatus big enough for wave breaking to take over as the predominant wave dissipation mechanism.

With the wavemaker oscillating at sufficient amplitude, the mirror symmetry is broken spontaneously and a conspicuous mean flow appears, clockwise or anticlockwise around the annulus, depending on time t and altitude z in the manner of figure 1. For instance at given t there could be clockwise flow above, and anticlockwise below, some altitude z_0, with z_0 descending as t increases. In this case the mean flow is much stronger than the Stokes drift: $\bar{u}^L$ and $\bar{u}$ are practically the same. The system quickly evolves into a more or less periodic sequence of asymmetric states, in which the mean flow keeps reversing. The mirror symmetry survives only in the long-time average. The space–time signature of the mean-flow evolution is much more regular than that in figure 1 if the wavemaker amplitude is kept accurately constant. Sufficient amplitude, in the laboratory case, means sufficient to overcome viscous drag on the mean flow.

The mechanism underlying the observed behaviour is well understood. It depends on a wave–mean interaction in a nonlinear feedback loop. Irreversible wave-induced transport of angular momentum cumulatively changes the mean flow by changing Kelvin's circulation $C(\Gamma)$, for material circuits like the circuit Γ sketched in figure 6. This again reminds us that viscosity is one way of 'violating the nonacceleration constraint'. The mean flow, in turn, filters the waves. Selective refraction and Doppler shifting leads to reduced $\hat{c}$ values, enhanced dissipation, reduced vertical group velocity, and reduced vertical penetration of the clockwise and anticlockwise wave components alternately. This produces the characteristic space–time signature in which features such as zero crossings in the mean velocity and mean shear profiles move inexorably downward.

Of course the term filtering refers to a view of things from some chosen altitude well above the wavemaker. Depending on the mean velocity profile below that altitude, one of the wave components, the one with reduced vertical penetration at that instant, has been more or less 'filtered out' relative to the other. The wave that has been filtered out is the wave whose intrinsic phase speed $\hat{c}$ has been Doppler shifted toward low rather than high values by the intervening mean velocity profile, as if there were a real or virtual critical level or surface at some higher altitude – though in fact no real critical surfaces occur in the experiment. The wave components experience metaphorical 'shallows' rather than 'beaches'.

Such filtering is involved, too, in the refrigeration effect already described. But, in that case, the pattern of Doppler shifting is dominated by a more or less steady westward or retrograde mean flow that is present at lower altitudes for other reasons, connected with strong Coriolis forces and with radiative influences on temperatures in the summertime extratropical stratosphere. There, the filtering effect stays more or less fixed and does not evolve interactively as in the Plumb–McEwan experiment. That is why the wave-induced force near 90 km is persistently eastward, the gyroscopic pumping persistently equatorward, and the polar summer mesopause persistently cold.

Note well that the entire motion in the Plumb–McEwan experiment is driven by nothing but the oscillations of the wavemaker. This indeed is a fluid system driven away from solid rotation – in this case, away from zero rotation – entirely by a field of fluctuations behaving anti-frictionally. Note also that, once the symmetry is broken, the wavemaker can exert net mean torques on the fluid system through correlations between pressure fluctuations and boundary slopes, just as there can be radiation stresses hence mean torques across undulating material surfaces at any altitude. Mean torques can similarly be exerted on the real middle atmosphere by disturbances originating in the troposphere beneath, including waves generated by flow over mountains or by tropical thunderstorms.

8 Historical note: the Michelson–Morley principle

As already stated, we can be practically certain that the real QBO shown in figure 1 is also fluctuation-driven. The evidence for this is overwhelming, even though there are still, even today, large uncertainties as to exactly which wave types are contributing most significantly on average (Baldwin *et al.*

2000). It is likely that we need to think in terms of a more or less continuous spectrum of eastward and westward phase speeds rather than the simple, discrete pair of phase speeds in the experiment. Real wave sources such as tropical thunderstorms are complicated, chaotic, and ill-understood.

But if wave sources are so ill-understood, then how can we be so confident that the real QBO is fluctuation-driven? Couldn't it be an entirely different mechanism, not yet thought of? Might not 'anti-friction' be a myth, or a mere theoreticians' plaything? The answer illustrates what I am calling the Michelson–Morley principle, the importance of negative results in science – a principle neglected today but neglected at our peril.

Today's understanding can be summarized as follows. First, irreversible momentum transport is generically a robust and almost inevitable accompaniment to real wave propagation in real fluid media, as already illustrated. The general theories referred to in §4, despite their difficulties, give us great confidence on this point, especially when set alongside laboratory experiments. Second, many suitable types of wavelike fluctuation are available. The atmosphere, practically anywhere one looks, is observed to be full of a variety of propagating waves, internal gravity and other types, many of which have amplitudes big enough to make substantial contributions to the required rates of momentum transport. Third, and very importantly, we have on record, from the late 1960s, a clear failure of serious and insightful attempts to model the QBO *without* invoking wave-induced momentum transport.

That modelling failure is a clear example of an important negative result in science, like the Michelson–Morley experiment to detect an electromagnetic aether wind that proved to be non-existent. I am thinking especially of the work of Wallace & Holton (1968). This was the most thorough attempt, before wave effects were thought of in this context, to construct a realistic numerical model of the QBO using a complete set of fluid-dynamical equations. That is, the numerical model was built in a dynamically self-consistent manner, and was made consistent also with the known orders of magnitude of relevant physical processes such as radiative heat transfer, and consistent with available observational information about large-scale fluctuations. The attempt failed spectacularly: angular momentum changes near the equator following the space–time pattern of figure 1 simply could not be explained. At least the pattern could not be explained without assuming, *ad hoc*, the existence of mysterious eastward and westward downward-moving forces. These had to be quite unlike any forces attributable to a classical eddy viscosity, such as was customarily assumed at the time to be associated with unresolved fluctuations. Indeed, the forces plainly had to be what I am calling anti-frictional, systematically tending to reinforce rather than attenuate departures from solid rotation. To have hypothesized that such a pattern of forces was due to an eddy viscosity would have implied a pattern of eddy-viscosity values that not only moved downward in time but also exhibited nonsensical negative and infinite values.

Despite being unable to guess what physical processes might give rise to such mysterious east–west forces, Wallace and Holton rather courageously published their negative result, arguing that the combination of observation and modelling implied that the forces must exist despite their strange behaviour, and despite the lack of any direct observation, at the time, of fluctuation patterns that could somehow produce the implied momentum transport. It was only then recognized, with progressively increasing clarity, through the work of R. S. Lindzen and others, that there is just one kind of physical process in the atmosphere, wave-induced momentum transport, that can and should produce just such a force pattern through the filtering effect.

Had the history been different, with the possibility of wave effects recognized from the outset, there might not have been so thorough an investigation into other possible explanations. We should then be less confident that wave effects are essential. It can be added that a growing number of new observations, theoretical arguments, and numerical modelling studies suggest that the real QBO must be driven by rather complicated fields of wavelike fluctuations involving a number of different wave types (e.g. Andrews & McIntyre 1976; Dunkerton 1983; Baldwin *et al.* 2000). Modelling studies have shown that wave fields with a broadband spectrum of eastward and westward phase speeds tend to produce anti-frictional behaviour in fundamentally the same way as in the experiment. The wave dissipation mechanisms include infrared radiative damping and several varieties of wave breaking.

9 Material invariants and stratification surfaces

Fundamental to understanding the various wave types and the associated restoring mechanisms are two Lagrangian or material invariants of the frictionless, adiabatic (ideal, Euler) fluid-dynamical equations. These invariants will be denoted by Q and θ. Both are scalar fields. Material invariance means that Q and θ are constant on fluid elements, i.e. that they satisfy

$$DQ/Dt = 0 \qquad \text{and} \qquad D\theta/Dt = 0, \qquad (9.1a,b)$$

where $D/Dt = \partial/\partial t + \boldsymbol{u} \cdot \nabla$, the material derivative for a fluid with velocity field $\boldsymbol{u}$. The invariant Q is related to Kelvin's circulation and is called the (Rossby–Ertel) potential vorticity; θ is a thermodynamic variable measuring stratification and buoyancy effects and is usually, in the atmospheric case, taken to be what is called the potential temperature. The following applies equally to the atmosphere and ocean and is fundamental to both.

The properties of Q and θ are fundamental, in particular, to understanding the two wave types most important for the global atmospheric circulation problem. These are internal gravity waves and another type called Rossby waves, significant at larger scales. Rossby waves are the key to the first enigma. They share with internal gravity waves a tendency to drive the system away from solid rotation, though for different reasons as we shall see. Their restoring mechanism is fundamental, moreover, to nearly every aspect of atmosphere–ocean dynamics, including the stability or robustness of large-scale eddies in the form of cyclonic or anticyclonic vortices. This robustness is sometimes called 'vortex coherence' and is important for weather regimes such as 'blocking', which at certain times of year can produce successive sunny days or even weeks in Western Europe. The Rossby-wave restoring mechanism is fundamental, also, to the so-called baroclinic and barotropic shear instabilities that can produce vortices from simpler flows.

Rossby-wave dynamics, as described by the standard theories of frictionless, adiabatic wave motion, is bound up with the behaviour of certain special material contours $\Gamma_{Q\theta}$. These contours are distinguished by constancy of both Q and θ; in other words they mark the intersections of the two sets of special material surfaces defined as the surfaces of constant Q and θ. Rossby-wave *propagation* involves the sideways, approximately horizontal undulations of the contours $\Gamma_{Q\theta}$. Rossby-wave *breaking* involves the sideways irreversible deformation of the same contours, which are the 'otherwise wavy' material contours appropriate to Rossby waves. Figure 7 illustrates such deformation. The contours bounding the different colours are a subset of the contours $\Gamma_{Q\theta}$ lying on a single constant-θ surface, located near the cruising altitudes of passenger jetliners. Parts, at least, of these contours are deforming irreversibly, though in a geometrically different way from the three-dimensional contour deformations implicit in figure 4. The contour shapes in figure 7 were estimated from observational meteorological data on the assumption that material invariance, (9.1a, b), is a sufficient approximation to reality over a four-day timespan.

For the atmosphere, θ can be taken to be any function of the specific entropy S. The surfaces of constant θ are also, therefore, surfaces of constant S, and are often called isentropic surfaces. Usually we take

$$\theta = T(p_0/p)^\kappa \propto \exp(S/C_p),\qquad (9.2)$$

where p_0 is a constant reference pressure and $\kappa = R/C_p = (\gamma-1)/\gamma = 2/7 =$

14/5/92   12GMT

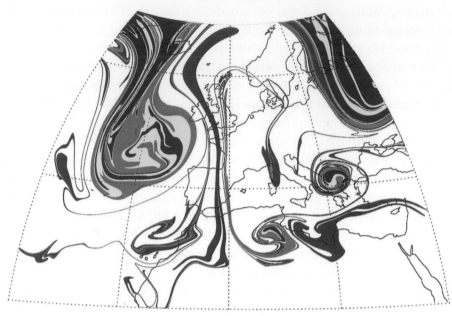

Figure 7. A subset of the special contours $\Gamma_{Q\theta}$, defined as the boundaries of the differently coloured regions, on 14 May 1992 at 1200 Z (Greenwich mean time) on the 320 K isentropic surface, estimated from observations by assuming material invariance (9.1*a, b*) over 4 days, from Appenzeller *et al.* (1996). The contours correspond to values 1, 2, 3, 4, and 5, in the standard potential-vorticity units of $10^{-6}\mathrm{m}^2\mathrm{s}^{-1}\,\mathrm{K}\,\mathrm{kg}^{-1}$; see equation (9.3). Values from 1 unit upwards are coloured rainbow-wise from dark blue to red and represent stratospheric or transitional (tropopause) air. The physical reality of small-scale features has been directly verified, in similar cases (though not this one), by special airborne measurements of their chemical signatures (e.g. Waugh *et al.* 1994).

0.286, C_p being the specific heat at constant pressure and $\gamma = 7/5$ the ratio of specific heats for a diatomic ideal gas. From standard thermodynamics and the ideal-gas equation (2.3) it is a straightforward exercise to verify that θ, thus defined, is the *potential temperature* referred to pressure p_0. This means the temperature that a fluid element would have if compressed dry-adiabatically to pressure p_0. It follows at once that θ is indeed a material invariant, for dry-adiabatic motion. Dry-adiabatic, hereafter 'adiabatic' for brevity, means that there are neither heat exchanges nor phase changes, such as condensation of water vapour when moist air expands, and also that frictional heating at turbulent energy dissipation rates ε is negligible. As a rule, such frictional heating can safely be neglected at altitudes $\lesssim 50\,\mathrm{km}$, though not always at higher altitudes.

For convenience we usually take $p_0 = 1000\,\mathrm{hPa}$, a nominal sea-level pressure. With this convention, the θ value for figure 7 is 320 K, corresponding to an isentropic surface whose typical altitudes are roughly indicated by the heavy curve in figure 2. The 320 K surface slopes gently upward toward the pole, and intersects the extratropical tropopause near the 230 K isotherm at lower right in figure 2. The tropopause, the surface or thin layer marking the transition between the troposphere and the stratosphere (Holton *et al.* 1995), slopes the opposite way for the most part, upward toward the equator as suggested by the heavy dashed curve in figure 2. It has a highly convoluted geometry, and its intersections with isentropic surfaces in the extratropics are often marked by a crowding together of the special contours $\Gamma_{Q\theta}$, corresponding to the edges of the large coloured regions in figure 7. The coloured regions correspond to air that is either at the tropopause or in the stratosphere (large coloured regions) or in the troposphere but of recent stratospheric or near-tropopause origin (smaller coloured filaments and vortices), where 'recent' means of the order of days.

By a strange accident, the special contours $\Gamma_{Q\theta}$ that mark the extratropical tropopause, poleward of about $20°$ latitude, all tend to have Q values in a similar numerical range on different isentropic surfaces, around 1–3 standard atmospheric 'potential vorticity units' of $10^{-6}\mathrm{m}^2\mathrm{s}^{-1}\,\mathrm{K\,kg}^{-1}$; see (9.3) below. Thus, for instance, it is often practical to take the surface $Q = 2$ in these units as marking the extratropical tropopause.

The oceanic counterparts of isentropic surfaces are variously called isopycnic, isopycnal, or neutral surfaces. There are non-trivial complications associated with the equation of state that replaces (2.3). Because of the simultaneous importance of pressure, temperature, and salinity, the neutral surfaces are not uniquely definable to perfect accuracy, though it happens that they are uniquely definable to a very good approximation, for today's oceans, as surfaces of constant 'neutral density' ρ_{neut} (Jackett & McDougall 1997). For ocean dynamics, therefore, θ in the following discussion is best identified with $\rho_0 - \rho_{\mathrm{neut}}$, where ρ_0 is a constant reference value arbitrarily chosen for convenience. The price paid is that ρ_{neut} depends on latitude and longitude as well as on pressure, temperature, and salinity and is therefore a thermodynamic variable in a geographically local sense only. Laboratory cases like the Plumb–McEwan experiment have no such problems: θ can be identified with $\rho_0 - \rho$ to more than sufficient accuracy, where ρ is the ordinary density. With the foregoing definitions, $\partial\theta/\partial z > 0$ always signifies stable stratification, also called static stability.

The isopleths of θ, whether defined by (9.2) or in terms of ρ_{neut} or ρ, as appropriate, will be called the 'stratification surfaces' of the system and denoted by $\mathscr{S}$. Their most important general property is that the density ρ and pressure p on any given $\mathscr{S}$ are functionally constrained such that $\oint dp/\rho = 0$ for any circuit lying on $\mathscr{S}$. This means that for frictionless, adiabatic motion Kelvin's circulation theorem holds, in the stratified system, for all material circuits Γ that lie on a single stratification surface $\mathscr{S}$ and remain on that $\mathscr{S}$ as time goes on. Such circuits can remain on $\mathscr{S}$ as long as $D\theta/Dt = 0$. From Kelvin's circulation theorem and mass conservation there follows the material invariance of Q, as noted next.

The standard definition of Q, first given in the famous paper by Ertel (1942) – generalizing earlier definitions by Rossby (1936) that depended on assuming hydrostatic balance – can be written in two equivalent ways,

$$Q = \rho^{-1}\boldsymbol{\zeta} \cdot \nabla\theta = \sigma^{-1}\boldsymbol{\zeta} \cdot \boldsymbol{n}, \tag{9.3}$$

where $\boldsymbol{\zeta}$ is the absolute vorticity vector, or curl of the fluid velocity field in an inertial (sidereal) reference frame. On the right, $\boldsymbol{n} = \nabla\theta/|\nabla\theta|$, the upward-directed unit normal to the stratification surface $\mathscr{S}$ on which Q is being evaluated, and $\sigma^{-1} = \rho^{-1}|\nabla\theta|$. The scalar field σ^{-1} is a strictly positive quantity for a stably stratified system, as is its reciprocal σ, which can be regarded as a stratification-related mass density. More precisely, $\sigma\,d\theta$ is the mass per unit area between neighbouring surfaces $\mathscr{S}$, so that if dA is the surface area element then $\sigma\,dA\,d\theta$ is the mass element. From mass conservation we therefore have

$$\iint_{\mathscr{S}(\Gamma)} \sigma\,dA = \text{constant} \tag{9.4}$$

whenever $D\theta/Dt = 0$, where $\mathscr{S}(\Gamma)$ denotes the part of $\mathscr{S}$ enclosed by any material circuit Γ in $\mathscr{S}$ such that $\mathscr{S}(\Gamma)$ is simply connected. But from Kelvin's circulation theorem and Stokes' theorem we also have

$$C(\Gamma) = \iint_{\mathscr{S}(\Gamma)} \boldsymbol{\zeta} \cdot \boldsymbol{n}\,dA = \iint_{\mathscr{S}(\Gamma)} Q\,\sigma\,dA = \text{constant}. \tag{9.5}$$

The material invariance of Q, equation (9.1a), now follows from (9.4), (9.5), and the arbitrariness of Γ. Note that $C(\Gamma)$ is still given by (4.1) provided that $\boldsymbol{u}$ is interpreted as the absolute velocity, i.e. velocity in the inertial frame. Equation (9.1a) is often called Ertel's theorem.

Equation (9.1a) can also be regarded as expressing, and making precise, the tendency of the normal component $\boldsymbol{\zeta} \cdot \boldsymbol{n}$ of absolute vorticity to be preserved through vortex tilting when the surfaces $\mathscr{S}$ tilt away from the horizontal, and to increase through vortex stretching when the surfaces $\mathscr{S}$ move apart – a generalization of angular momentum conservation, the 'ballerina effect', or 'ice-skater's spin'. Such vortex stretching often contributes to the spin-up of cyclonic vortices, such as the small cyclonic vortex over the Balkans in figure 7, whose core, shown coloured in the figure, consists of high-Q air that has undergone stretching while moving equatorward out of the stratosphere. Cyclonic means with the Earth's rotation, i.e. anticlockwise in the northern hemisphere when viewed from above, as shown by the spiralling of the surrounding contours $\Gamma_{Q\theta}$.

Whenever material invariance (9.1) holds, we have not only (9.5) but also

$$\iiint \varphi(Q, \theta)\, \sigma\, \mathrm{d}A\, \mathrm{d}\theta = \text{constant} \qquad (9.6)$$

taken over the whole fluid domain, with $\varphi(Q, \theta)$ an arbitrary function. These domain integrals are called 'Casimir invariants'.

In the atmosphere, many of the stratification surfaces $\mathscr{S}$ span the globe: they are topologically spherical and have no boundary. For instance this holds for the $\mathscr{S}$ of figure 7 and for all those above it. For each such topologically spherical $\mathscr{S}$, Stokes' theorem implies that the mass-weighted global integral

$$\iint_{\mathscr{S}} Q\, \sigma\, \mathrm{d}A = 0, \qquad (9.7)$$

regardless of whether material invariance holds or not. The relation (9.7) imposes a severe constraint on the possible evolution of the global-scale Q distribution on $\mathscr{S}$. We shall see that (9.7) is almost enough, by itself, to guarantee that fluctuations involving Rossby-wave dynamics will behave anti-frictionally. The point will become clear when we discuss what is sometimes called the *invertibility principle* for Q, which is also the simplest way of understanding the dynamics of the waves themselves.

Notice that θ may be replaced by any monotonic function of θ alone, for instance specific entropy S in the case of the atmosphere, in (9.3) and elsewhere, without significantly affecting the foregoing theoretical developments. The stratification surfaces $\mathscr{S}$ remain the same, the special contours $\Gamma_{Q\theta}$ remain the same, and (9.1a, b) together with (9.4)–(9.7) still hold for the Q and θ thus redefined. Only the practical utility of Q as a tropopause marker is affected, underlining the point that the near-constancy of Q at the extratropical tropopause is, indeed, nothing but an accident, peculiar to the standard definition of Q with θ taken as the potential temperature (9.2).

The integral constraint (9.7) is related to certain general conservation properties of 'potential-vorticity substance', the notional 'substance' or 'charge' whose amount per unit mass is Q. For thorough discussions see Haynes & McIntyre (1990) and McIntyre (1992, §11). Regardless of whether (9.1a, b)

hold, the surfaces $\mathcal{S}$ are impermeable to the notional 'charged particles' of the substance, which moreover has no source or sinks within the fluid.

10 Stable stratification and balanced flow

Stable stratification or static stability, $\partial\theta/\partial z > 0$, means stability to adiabatic vertical displacements that are also quasi-static, i.e. non-acoustic. More precisely, consider a thought experiment in which a small fluid element is displaced adiabatically upward through a stably stratified fluid otherwise at rest, in such a way that the pressure p within the fluid element decreases to match the pressure in its undisturbed surroundings. If the fluid is compressible, the element must expand and do work on its surroundings and therefore cool, as indicated in the atmospheric case by the material invariance of θ and the definition (9.2). It is this adiabatic cooling that produces the refrigeration effect at the summer polar mesopause. Such a fluid element will find itself colder and denser than its surroundings if $\partial\theta/\partial z > 0$ in the surroundings. It will then feel a negative Archimedean buoyancy force. Gyroscopic pumping has to do work against this force in order to hold local temperatures below radiative equilibrium (radiative-equilibrium values of $\partial\theta/\partial z$ being positive, in fact, throughout the middle atmosphere).

The same buoyancy force provides the restoring mechanism giving rise to internal gravity waves, where it acts in a more temporary way as the springlike component or 'stiffness' of a freely oscillating system, fluctuating about zero as fluid elements are displaced alternately up and down. Thus the waves can be visualized as free oscillations or undulations of the stratification surfaces $\mathcal{S}$. In the case of figure 4, the surfaces $\mathcal{S}$ coincide with the tracer isopleths shown.

We could reasonably use the positive quantity $\partial\theta/\partial z$ as a measure of the strength of the stable stratification, with the appropriate choice of θ for the atmosphere, oceans, or laboratory. But it is useful to normalize $\partial\theta/\partial z$ by a factor, $g/\tilde{\theta}$ say, proportional to gravity g, defined in such a way that the result is equal to minus the buoyancy force per unit mass per unit vertical displacement, for small displacements in the above-mentioned thought experiment. It is standard practice to carry out such a normalization and to denote the result by the symbol N^2. So we write

$$N^2 = \frac{g}{\tilde{\theta}}\frac{\partial\theta}{\partial z}. \tag{10.1}$$

Thus defined, N^2 has dimensions of frequency squared. It is easy to verify that $\tilde{\theta} = \theta$ in the atmospheric case, with θ defined by (9.2), and that $\tilde{\theta} = \rho$ in the laboratory case, for which $\partial\theta/\partial z = -\partial\rho/\partial z$. In the oceanic case, with $\partial\theta/\partial z = -\partial\rho_{\text{neut}}/\partial z$, $\tilde{\theta}$ differs from ρ_{neut} and is a complicated function of latitude and longitude as well as of pressure, temperature, and salinity, for the reasons already referred to. By virtue of the foregoing definition in terms of the buoyancy force, N is equal to the radian frequency with which the small fluid element in the thought experiment *would* oscillate vertically when released, *if* its own inertia were the only inertia involved. But because the surrounding fluid has additional inertia, and because the displacements in a real disturbance are never all exactly vertical, real internal gravity waves have intrinsic radian frequencies of magnitude $|\hat{\omega}| < N$, as will be verified shortly; hence the alternative definition $N = \sup|\hat{\omega}|$ used in §5. The frequency N can be thought of as the frequency of an actual vertical oscillation in the limiting case of an infinitesimally thin, yet frictionless, vertically oriented needle or slab of fluid. This can be imagined to oscillate freely without disturbing its surroundings. Typical values of N are, for instance, $2\pi/(1 \text{ hour})$ in the main thermoclines of the oceans, $2\pi/(10 \text{ minutes})$ in the troposphere, and $2\pi/(5 \text{ minutes})$ in the stratosphere.

The time scales N^{-1} are very short in comparison with the time scales of figure 7 and related atmospheric weather systems (days), of ocean eddies and Gulf stream meanders (weeks to months), and of global-scale mean circulations (months to centuries). For all such flows, stable stratification is an immensely powerful constraint on vertical motion. A famous consequence, basic to the technology of numerical weather prediction and of numerical modelling in general, is that the partial differential equations describing the motion are stiff in the numerical-analytic sense – like the ordinary differential equations for a 'springy pendulum' consisting of a mass suspended from a pivot by a stiff elastic spring.

Such a pendulum has slow, rotating or swinging, modes of motion in which the relatively fast, compressional, modes of the mass and spring are hardly excited. A first approximation in describing such slow motions is simply to set the length of the spring to be constant. More accurate approximations would allow the spring to change its length in a quasi-static way. The fast modes are then, so to speak, 'slaved' to the slow modes. These approximations and their ultimate limitations can be studied mathematically via techniques ranging all the way from formal two-timing (multiple scales) to bounded-derivative theory and KAM (Kolmogorov–Arnol'd–Moser) theory and other dynamical-systems techniques; there is an enormous literature. We may describe such slow springy-pendulum motion as balanced with respect to fast compressional oscillations. Similarly, we may describe stratification-constrained flows like that illustrated in figure 7 as balanced with respect to internal gravity wave oscillations.

Such flows can also be described as layerwise-two-dimensional. If internal gravity waves are not excited, then the flow on each $\mathscr{S}$ has the qualitative character of a horizontal two-dimensional incompressible flow. One consequence is that vertical accelerations are extremely small, so that hydrostatic balance (2.4) holds rather accurately. Such flows are possible if Richardson numbers are not too small.

More generally, we often use the idea of internal gravity wave oscillations *about* a stratification-constrained, layerwise-two-dimensional mean flow. The wave-induced mean effects are seldom strong enough to stop the mean flow, as such, from being balanced to good approximation.[1] The Plumb–McEwan experiment provides a simple illustration. The total flow includes an oscillatory vertical velocity, but the mean flow is constrained by the stratification to be purely horizontal.

To check numerical values of N^2 from cross-sections like figure 2, it is useful to note that, under hydrostatic balance, in the atmospheric case,

$$N^2 = g \, \frac{\partial (\ln \theta)}{\partial z} = \frac{g}{T} \left(\frac{\partial T}{\partial z} + \frac{g}{C_p} \right). \qquad (10.2)$$

This can be seen from (2.4), (9.2), (10.1), and the identity $R = \kappa C_p$. Therefore N^2 is positive, i.e. $\partial \theta / \partial z > 0$ and the atmosphere is stably stratified, whenever $\partial T / \partial z > -g/C_p$. Numerically, we have $g/C_p = g\kappa/R = \kappa T/H_p = 9.8 \, \mathrm{K/km}$, the dry-adiabatic 'lapse rate' characterizing a fictitious unstratified terrestrial atmosphere, one in which the refrigeration effect would vanish and internal gravity waves could not exist. The condition $\partial T / \partial z > -g/C_p$, hence $\partial \theta / \partial z > 0$, can now be seen to hold throughout figure 2. Here, as already hinted, the stable stratification is mostly a consequence of radiative heating and cooling, dependent upon, and interactive with, the distributions of carbon dioxide, ozone, water vapour, CFCs and many other gases whose molecules are triatomic or bigger, i.e. are greenhouse gases. The main exception is the stable stratification of the tropical troposphere, at bottom centre of figure 2, which is tightly controlled by deep compositional (cumulonimbus) convection. The tropical troposphere has positive N^2 but is close to having zero static stability in the moist-convective or compositional-convective sense.

In a coarse-grained, global-scale view like that in figure 2, the largest values of N^2 found anywhere in the atmosphere are in the stratosphere, justifying

---

[1] There are a few exceptions to this statement, though not in the parameter regimes of greatest interest here (Bühler & McIntyre 1998).

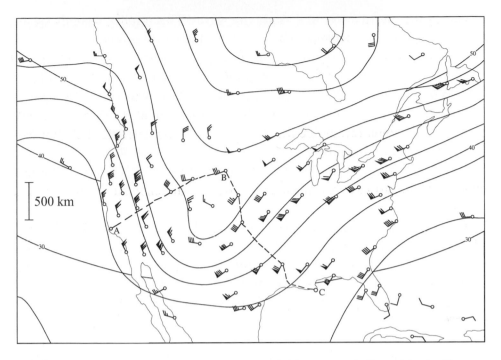

Figure 8. Observed winds and proxy pressure field at around 6 km altitude, illustrating approximate geostrophic balance as typically observed in large-scale atmospheric flows. The winds, from routine meteorological balloon (radiosonde) measurements taken at the approximate positions marked by small circles, are coded symbolically with triangular barbs representing $25 \, \mathrm{m \, s^{-1}}$, and thin full and half-length barbs representing 5 and $2.5 \, \mathrm{m \, s^{-1}}$ respectively. (Thus the maximum observed wind in this case, near the left of the picture, shown by the symbol with two triangular and three thin barbs, is $65 \, \mathrm{m \, s^{-1}}$ south-south-eastward.) The contour interval is such that a contour spacing of 500 km corresponds to a horizontal pressure-gradient force per unit mass that will balance the horizontal Coriolis force per unit mass of a $25 \, \mathrm{m \, s^{-1}}$ wind at 45° latitude. From Phillips (1963).

its name. Ozone, famously, absorbs solar ultraviolet radiation most strongly at wavelengths around 310 nm or less, and the consequent radiative heating accounts for the temperature increase up to the stratopause at about 50 km altitude, and for the stratopause temperature maximum at the summer pole. This temperature maximum (within the light shading) is one of the most persistent features of the observed temperature field.

The equations describing flows like that of figure 7 have an additional stiffness from Coriolis effects, associated with so-called inertia waves, also called Coriolis waves. The time scales of inertia waves are longer than buoyancy time scales – hours rather than minutes, but still short in comparison with days, outside the tropics. We can usually regard such flows as balanced with respect to inertia waves as well as gravity waves. This implies, roughly speaking, a simple balance between the Coriolis and pressure-gradient terms in the horizontal momentum equations, called geostrophic balance, or geostrophy (figure 8). But geostrophy is to be regarded as merely a first approximation to a more subtle and more accurate balance, just as, in the springy pendulum problem, constant spring length is merely a first approximation. The idea of 'balanced motion' thus generalized is of great importance in atmosphere–ocean dynamics, and can apply, in some circumstances at least, with astonishing accuracy. Its ultimate limitations are analysed and discussed by Ford *et al.* (2000). These limitations have yet to be fully understood; they are related to fluid-dynamical concepts such as Lighthill radiation (aerodynamic sound generation) and to dynamical-systems concepts such as Poincaré homoclinic tangles and the breakup of KAM tori.

With a suitable balance assumption, we can obtain a coarse-grained, global-scale view of the mean zonal winds $\bar{u}$ corresponding to the mean temperatures $\bar{T}$ shown in figure 2, extending the wind field upward from the troposphere. The troposphere and lower stratosphere are relatively well observed by the meteorological network, which provides direct wind observations such as those used for figures 1, 7 and 8. In the global-scale picture corresponding to figure 2, cross-sections of $\bar{u}$ (reckoned relative to the Earth's rotating reference frame) show broad jets peaking at magnitudes $|\bar{u}| \simeq 60\,\mathrm{m\,s^{-1}}$ to $70\,\mathrm{m\,s^{-1}}$ in the mid-latitude mesosphere. The sense is eastward or prograde ('westerly') in the northern, winter hemisphere ($\bar{u} > 0$ by convention) and westward or retrograde ('easterly') in the summer hemisphere ($\bar{u} < 0$). Balance implies that $\partial \bar{u}/\partial z > 0$ wherever $\bar{T}$ decreases poleward at constant p, i.e. wherever the stratification surfaces $\mathscr{S}$ tilt upward-poleward. This is called the 'thermal wind relation' and is made more precise in (11.6) below.

Associated values of the Richardson number $N^2/|\partial \bar{u}/\partial z|^2$ are typically in the tens to hundreds, as mentioned earlier, again reflecting the strength of the stratification constraint. The summertime westward jet is rather steady, and provides the persistent filtering that underlies the refrigeration effect, both literally and metaphorically.

The gyroscopically pumped mean circulation is itself an example of balanced motion. One can think of the response to an eastward or westward force as a continual adjustment to keep the thermal-wind relation satisfied.

11 Oscillations about balance: inertia–gravity wave dynamics

To justify the foregoing and to complete our description of the frictionless, adiabatic fluid dynamics, balanced and unbalanced, we need to supplement

(9.1*a*, *b*) and the equation of state by the mass-conservation equation,

$$\partial \rho / \partial t + \nabla \cdot (\rho \boldsymbol{u}) = 0, \tag{11.1}$$

and two momentum or vorticity equation components. A suitable form of the three-dimensional vorticity equation, valid whenever rotational forces such as viscosity are negligible, is

$$\frac{\mathrm{D}}{\mathrm{D}t}\left(\frac{\zeta}{\rho}\right) = \left(\frac{\zeta}{\rho}\right) \cdot \nabla \boldsymbol{u} + \frac{\nabla p \wedge \nabla \theta}{\rho^2 \tilde{\theta}}. \tag{11.2}$$

This is valid in a rotating reference frame if ζ is again interpreted as the absolute vorticity vector, so that in a rotating frame we have

$$\zeta = 2\boldsymbol{\Omega} + \nabla \wedge \boldsymbol{u}, \tag{11.3}$$

where $\boldsymbol{\Omega}$ is the Earth's angular velocity, taken as constant. Then $\boldsymbol{u}$ is the velocity relative to the rotating frame, both on the right-hand side of (11.2) and within the material derivative $\mathrm{D}/\mathrm{D}t = \partial/\partial t + \boldsymbol{u} \cdot \nabla$ on the left. The second term on the right of (11.2) is often written with $-\nabla \rho / \rho$ in place of $\nabla \theta / \tilde{\theta}$. The two forms are equivalent even for compressible motion because, in all cases, $-\nabla \rho / \rho = \nabla \theta / \tilde{\theta} +$ (terms $\propto \nabla p$). For instance in the ideal-gas case this last relation follows at once, with $\tilde{\theta} = \theta$, from (9.2) and (2.3).

The component of (11.2) normal to the stratification surfaces $\mathscr{S}$ merely duplicates the information in (9.1*a*). Picking out that component by taking the scalar product of (11.2) with $\nabla \theta$, we annihilate the second term on the right and quickly recover (9.1*a*), with the help of $\nabla(\mathrm{D}\theta/\mathrm{D}t) = 0$. Any other two components of (11.2) can be used to complete the set of equations, allowing us to describe balanced flow, as well as departures from balance in the form of inertia waves and internal gravity waves.

The two wave types are associated, respectively, with the first and second terms on the right of (11.2). These terms are what make the equations stiff. They tend to cancel each other in balanced motion. A first approximation to balance would set the right-hand side of (11.2) to zero taking ∇p as hydrostatic, $\nabla p = -g\rho \widehat{z}$, where $\widehat{z}$ is a unit vertical vector.

The first term on the right of (11.2) represents the Coriolis stiffness, which arises when $2\boldsymbol{\Omega} \cdot \nabla$ is the dominant contribution to $\zeta \cdot \nabla$. Such dominance is typical when the Rossby number

$$Ro = U/fL \ll 1, \tag{11.4}$$

where U and L are horizontal velocity and length scales, with $L \gg H$ for height scale H and where

$$f = 2\boldsymbol{\Omega} \cdot \widehat{z} = |2\boldsymbol{\Omega}| \sin \phi, \tag{11.5}$$

where ϕ is latitude; f is called the Coriolis parameter. The Coriolis stiffness is essentially the stiffness you feel when you twist a spinning gyroscope. When $Ro \ll 1$ and $L \gg H$ we have $2\boldsymbol{\Omega} \cdot \nabla \boldsymbol{u} \approx f \partial \boldsymbol{u}/\partial z$, so that the approximation to balance in which the right-hand side of (11.2) is set to zero takes the simple form

$$f \partial \boldsymbol{u}/\partial z \approx g\widehat{\boldsymbol{z}} \wedge \nabla\theta/\tilde{\theta}. \tag{11.6}$$

This is the 'thermal wind relation'.

The second term on the right of (11.2), representing the stratification stiffness, responds directly to undulations of the stratification surfaces $\mathscr{S}$. It is called the baroclinic term – 'baro-' to signify pressure and '-clinic' to signify slope or inclination – because $\nabla p \wedge \nabla \theta$ becomes non-zero whenever the surfaces $\mathscr{S}$ tilt away from the horizontal, or, more precisely, whenever they tilt away from the isobaric or constant-p surfaces. The isobaric surfaces may themselves tilt, though usually much less. Indeed, isobaric and horizontal are precisely equivalent in the simplest form of the wave theory, based on linearizing the equations for small, $O(a)$, disturbances about a state of rest and assuming quasi-static, non-acoustic disturbances. For we can then neglect altogether the disturbance pressure field in the baroclinic term and, as in (11.6), replace the vector $(\rho^2\tilde{\theta})^{-1}\nabla p$ by its undisturbed, $O(1)$, hydrostatic value $-(\rho\tilde{\theta})^{-1}g\widehat{\boldsymbol{z}}$, which points exactly downward. We may also treat ρ and $\tilde{\theta}$ as constants. It is well known that these approximations – often lumped under the heading 'Boussinesq approximation' – are self-consistent if $|\hat{\omega}| \leqslant N$ and if disturbance height scales are small in comparison with the density scale height, of the order of $H_p \sim 7\,\mathrm{km}$ for the atmosphere and $200\,\mathrm{km}$, practically infinite, for the oceans.

If we imagine ourselves viewing the surfaces $\mathscr{S}$ from the side, as in figure 4, and if for instance they tilt gently upward to the right, i.e. are rotated anticlockwise, then the isobaric component of $\nabla\theta$ is directed to the left. Then $\nabla p \wedge \nabla\theta$ is directed forward, away from the viewer, corresponding to a clockwise rotational acceleration – more precisely, to a positive rate of change of the forward horizontal component of the vorticity ζ – tending to untilt the surfaces $\mathscr{S}$, i.e. to rotate them back toward the horizontal.

This way of viewing the gravity-wave restoring mechanism is consistent with the previous definition of N^2, as minus the buoyancy force per unit mass per unit displacement for quasi-static, non-acoustic disturbances. Displacement, for this purpose, means the small vertical displacement of a surface $\mathscr{S}$ due to its $O(a)$ tilting relative to an isobaric surface. In the example considered, the displacement increases rightward at a rate equal to the slope of $\mathscr{S}$. Correct to $O(a)$ this slope or displacement gradient is $|\nabla p \wedge \nabla\theta| / \left(|\nabla p| \, |\nabla\theta|\right)$.

Multiplying it by N^2 must convert it into the corresponding leftward rate of increase of the (upward) buoyancy force per unit mass. The latter rate, being the force-curl that changes the horizontal vorticity, must be equal to ρ times the baroclinic term in (11.2). This may be verified from hydrostatic balance (2.4) and the definition (10.1) of N^2. For, in the circumstances assumed, we have, correct to $O(1)$ and with error $O(a)$, $|\nabla p| = g\rho$ and $|\nabla\theta| = N^2\tilde{\theta}/g$.

It is useful to define an upward 'buoyancy acceleration' $b' = g\theta'/\tilde{\theta}$, associated with the local departure θ' of θ from its undisturbed hydrostatic value in a state of rest with the surfaces $\mathscr{S}$ horizontal. Then, under the Boussinesq approximation, the baroclinic term in (11.2) becomes simply $-\rho^{-1}\hat{z}\wedge\nabla b'$ with ρ constant; and to comparable accuracy we may also take ρ to be constant in the mass-conservation equation (11.1), simplifying the equation to

$$\nabla \cdot \boldsymbol{u} = 0. \tag{11.7}$$

It is now a straightforward exercise to linearize the equations about rest and find $\hat{\omega}$ for a plane inertia–gravity wave or Fourier component of a small disturbance, with wave-vector $\boldsymbol{k}$, and fluctuating fields $\boldsymbol{u}'$, θ', etc., $\propto \exp\{\mathrm{i}(\boldsymbol{k}\cdot\boldsymbol{x} - \hat{\omega}t)\}$. The small-height-scale condition now reads $|m|^{-1} \ll H_p$, where m is the vertical component of $\boldsymbol{k}$. The linearized form of (9.1b) is

$$\partial b'/\partial t + N^2 w' = 0, \tag{11.8}$$

where $w' = \hat{z}\cdot\boldsymbol{u}'$. Mass conservation (11.7) is already linear, giving $\nabla\cdot\boldsymbol{u}' = 0$, hence $\boldsymbol{k}\cdot\boldsymbol{u}' = 0$. Finally

$$\frac{\partial \zeta'}{\partial t} = 2\boldsymbol{\Omega}\cdot\nabla\boldsymbol{u}' - \hat{z}\wedge\nabla b', \tag{11.9}$$

the linearized form of (11.2).

For the plane wave to be a consistent approximate solution we must also assume that not only $\boldsymbol{\Omega}$ but also $\hat{z}$ is constant (flat-Earth approximation), and that $|\boldsymbol{k}|^{-1} \ll H_\theta$, where H_θ is defined to be the smallest height scale for background inhomogeneities in N^2. Now replacing $\partial/\partial t$ by $-\mathrm{i}\hat{\omega}$ and ∇ by $\mathrm{i}\boldsymbol{k}$, and noting that $\zeta' = \mathrm{i}\boldsymbol{k}\wedge\boldsymbol{u}'$, we can easily verify that the plane-wave solutions satisfy the linearized equations if and only if $\hat{\omega}$ satisfies the 'dispersion relation'

$$\hat{\omega}^2 = \frac{|N\hat{z}\wedge\boldsymbol{k}|^2}{|\boldsymbol{k}|^2} + \frac{|2\boldsymbol{\Omega}\cdot\boldsymbol{k}|^2}{|\boldsymbol{k}|^2} = \hat{\omega}_N^2 + \hat{\omega}_\Omega^2, \qquad \text{say,} \tag{11.10}$$

the two terms on the right being the squared frequencies for, respectively, pure internal gravity waves in a non-rotating system and pure inertia waves in an unstratified system. The squared frequencies add because the fluctuating forces add. More precisely, the condition $\boldsymbol{k} \cdot \boldsymbol{u}' = 0$ implies that the dynamics of a single plane wave is essentially the dynamics of an inclined-plane motion with slabs of fluid moving parallel to themselves, supported not by an inclined solid surface but instead by an $O(a)$ disturbance-pressure-fluctuation field p' whose gradient $\nabla p'$ is parallel to $\boldsymbol{k}$. (This is still negligible in the baroclinic term, as already implied.) The force per unit displacement due to buoyancy effects adds scalarly to that for Coriolis effects, when components in the inclined planes of the phase surfaces are taken, i.e. components perpendicular to $\boldsymbol{k}$. Note that $\min_{|\boldsymbol{k}|}|\hat{\omega}| > 0$ except at the equator, where $\boldsymbol{\Omega} \cdot \hat{\boldsymbol{z}} = 0$ so that taking $\boldsymbol{k}$ parallel to $\hat{\boldsymbol{z}}$ gives $\hat{\omega} = 0$.

The single plane-wave solution is a finite-amplitude solution of the equations after the Boussinesq approximations are made. The condition $\boldsymbol{k} \cdot \boldsymbol{u}' = 0$, from $\nabla \cdot \boldsymbol{u}' = 0$, implies that the operator $\boldsymbol{u}' \cdot \nabla$ from $\mathrm{D}/\mathrm{D}t$ annihilates any disturbance field belonging to the same plane wave. The solution displayed in figure 4 is just such a finite amplitude solution, for a simple internal gravity wave, i.e. for a case $\hat{\omega}_N^2 \gg \hat{\omega}_\Omega^2$ with negligible Coriolis effects.

When $N^2 \gg |2\boldsymbol{\Omega}|^2$, the most important case in practice, we may simplify (11.10) to its standard form

$$\hat{\omega}^2 = \frac{N^2(k^2 + l^2) + f^2 m^2}{k^2 + l^2 + m^2} = N^2 \cos^2 \vartheta + f^2 \sin^2 \vartheta, \qquad (11.11)$$

where ϑ is the angle of elevation of the wave-vector $\boldsymbol{k} = (k, l, m)$ from the horizontal. The approximation (11.11) says that stratification is so strong that the dynamics notices only the local vertical component, f, of $2\boldsymbol{\Omega}$ – a principle that applies much more generally as well, for instance to balanced, layerwise-two-dimensional flows like that of figure 7, as already illustrated by (11.6). Notice that there are upper and lower bounds $\sup_{|\boldsymbol{k}|}|\hat{\omega}| = N$ and $\inf_{|\boldsymbol{k}|}|\hat{\omega}| = |f|$ and that $(N^2 - \hat{\omega}^2)/(\hat{\omega}^2 - f^2) = m^2/(k^2 + l^2) = \tan^2 \vartheta$. The limiting case mentioned below (10.1) corresponds to letting $\boldsymbol{k}_{\mathrm{horiz}} = (k, l) \to \infty$ holding m finite (and keeping $|m^{-1}| \ll H_p$).

For fixed ϑ, (11.11) emphasizes that $\hat{\omega}$ is independent of $|\boldsymbol{k}|$ – a consequence of the inclined-plane dynamics – implying in turn that the intrinsic group velocity $\hat{\boldsymbol{c}}_g = \nabla_{\boldsymbol{k}}\hat{\omega} = (\partial\hat{\omega}/\partial k, \; \partial\hat{\omega}/\partial l, \; \partial\hat{\omega}/\partial m)$ is perpendicular to $\boldsymbol{k}$, i.e. that $\hat{\boldsymbol{c}}_g \cdot \boldsymbol{k} = 0$. This strange though well-known result reminds us that we are dealing with an anisotropic propagating medium. From (11.10) with no approximation we see that

$$\hat{\boldsymbol{c}}_g = \frac{\hat{\omega}_N}{\hat{\omega}} \hat{\boldsymbol{c}}_{gN} + \frac{\hat{\omega}_\Omega}{\hat{\omega}} \hat{\boldsymbol{c}}_{g\Omega}, \qquad (11.12)$$

where $\hat{\boldsymbol{c}}_{gN} = \nabla_{\boldsymbol{k}}\hat{\omega}_N$ and $\hat{\boldsymbol{c}}_{g\Omega} = \nabla_{\boldsymbol{k}}\hat{\omega}_\Omega$, each perpendicular to $\boldsymbol{k}$, with directions lying respectively in the $(\hat{\boldsymbol{z}}, \boldsymbol{k})$- and $(\boldsymbol{\Omega}, \boldsymbol{k})$-planes, and with respective

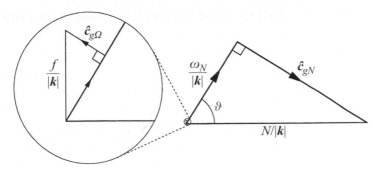

Figure 9. The large triangle on the right shows the relation between phase and group velocities for a simple internal gravity wave, in which only the first term of (11.11) is significant. The hypotenuse (horizontal) gives the horizontal intrinsic phase speed $\hat{c}_N$. The left-hand sloping side of the triangle is taken parallel to the wavevector k, and its magnitude gives the phase velocity in that direction. The other sloping side gives the magnitude and direction of the group velocity. The smaller triangle on the left, magnified as suggested by the zoom circle, shows the corresponding relation for a pure inertia (Coriolis) wave, in which only the second term of (11.11) is significant.

magnitudes $|\hat{c}_{gN}| = N|\hat{z} \cdot k|/|k|^2$ and $|\hat{c}_{g\Omega}| = |2\Omega \wedge k|/|k|^2$. The large right-angled triangle in figure 9 illustrates the geometric relation between k and $\hat{c}_{gN}$ in the $(\hat{z}, k)$-plane. The hypotenuse is drawn horizontal with magnitude $\hat{c}_N = N/|k|$. One of the two shorter sides is drawn parallel to k, the left-hand side in this example. It represents the full phase velocity $\hat{\omega}_N k/|k|^2$. The other of the two shorter sides then represents the group velocity. The corresponding triangle for $\hat{c}_{g\Omega}$ lies in the (Ω, k)-plane and has a hypotenuse of magnitude $|2\Omega|/|k|$ drawn parallel to Ω. The triangle on the left of figure 9 illustrates this for cases in which either Ω is parallel to $\hat{z}$, or the approximation (11.11) is valid. In these cases $\hat{c}_{gN}$ and $\hat{c}_{g\Omega}$ have opposite directions, as suggested by the arrows: the two terms in (11.12) oppose each other. In the practically important case $N^2 \gg |2\Omega|^2$, suggested by the zoom circle at the left of figure 9, we have $\hat{c}_g = \hat{c}_{gN}[1 + O(|2\Omega|^2/N^2)]$. The error term is often negligible.

As suggested by the arrows in figure 9, the sense of $\hat{c}_{gN}$ is given by the rule that upward group propagation corresponds to downward phase propagation, and vice versa. It is worth noting that when $N^2 \gg |2\Omega|^2$ the same rule, upward group with downward phase, holds for the intrinsic properties of every other possible wave type in stratified, rotating fluid systems with the sole exception of acoustic waves. During the 1950s and 1960s, ignorance of this point caused some confusion among observers studying certain 'travelling ionospheric disturbances', detectable by radio methods at altitudes $\gtrsim 100$ km. These disturbances showed systematic descent and were thought at first to have extraterrestrial origin. Today, however, it is clear that they are a manifestation of internal gravity waves arriving from below. The conclusion was not widely accepted until around 1967 when a laboratory experiment by D. Fultz and C. O. Hines demonstrated, to many researchers' astonishment, phase surfaces propagating vertically *toward* a wavemaker radiating internal gravity waves into a stratified fluid.

For waves on a non-zero background flow $\bar{\boldsymbol{u}}$, as in the Plumb–McEwan experiment and typically in the real atmosphere, it can be shown that the dispersion relations (11.10) and (11.11) still apply as consistent first approximations, when $Ri = N^2/|\partial\bar{\boldsymbol{u}}/\partial z|^2 \gg 1$, provided that we view the wave dynamics in a frame of reference moving with the local background flow $\bar{\boldsymbol{u}}$. If the frequency of a wave relative to the ground is ω, then its intrinsic frequency, i.e. the frequency $\hat{\omega}$ seen in the moving frame and satisfying (11.10) or (11.11), is related to ω by

$$\hat{\omega} = \omega - \boldsymbol{k} \cdot \bar{\boldsymbol{u}}, \qquad (11.13)$$

the right-hand side arising from the linearized material derivative $\partial/\partial t + \bar{\boldsymbol{u}} \cdot \nabla$ for disturbances $\propto \exp\{i(\boldsymbol{k} \cdot \boldsymbol{x} - \omega t)\}$. This describes the Doppler shifting and filtering mentioned earlier. The simplest case is that of horizontally homogeneous $N^2 = N^2(z)$ and steady, horizontal, horizontally homogeneous $\bar{\boldsymbol{u}} = \bar{\boldsymbol{u}}(z)$. Then k, l, and ω are all constant following a ray point, and (11.13) is enough to determine the Doppler shifting of $\hat{\omega}$. Along with (11.11) this forms the basis for the standard ray-theoretic models of filtering and critical-layer effects.

If $|\hat{\omega}|$ is Doppler shifted up to N, then the vertical group velocity $\hat{\boldsymbol{c}}_g \cdot \hat{\boldsymbol{z}}$ is reversed in a finite time, giving total internal reflection. If $|\hat{\omega}|$ is Doppler shifted downward, then the behaviour is ocean-beach-like, as discussed in §5. If f is neglected (Booker & Bretherton's case), then (11.11) allows $\hat{\omega}$ to approach zero with $\boldsymbol{k}_{\mathrm{horiz}} = (k, l, 0)$ constant and $|m| \to \infty$, hence $\vartheta \to \frac{1}{2}\pi$ and $|\boldsymbol{k}| \to \infty$, as asserted in §5. Otherwise, $\hat{\omega}^2$ is bounded below by f^2, as $m \to \infty$ and $|\boldsymbol{k}| \to \infty$. The standard ray equation for the rate of change of m following a ray point is

$$\dot{m} = -\boldsymbol{k}_{\mathrm{horiz}} \cdot \partial\bar{\boldsymbol{u}}/\partial z - \hat{\omega}^{-1} \cos^2\vartheta \, N \, \partial N/\partial z, \qquad (11.14)$$

whether or not f is negligible, the right-hand side being minus the partial derivative, with respect to z, of ω defined by (11.11) and (11.13) and regarded as a Hamiltonian function $\omega(\boldsymbol{k}; \boldsymbol{x}) = \omega(k, l, m; x, y, z)$. This shows that m cannot change faster than linearly with time, both confirming the possibility of total internal reflection in a finite time, with m passing through zero and

$|\hat{\omega}|$ through N, and also confirming that the ray time to a critical surface, with $m \to \infty$ and $|k| \to \infty$, is necessarily infinite, as asserted in § 5.

To take account of wave-breaking behaviour when f is not negligible, the definition (5.1) of wave amplitude can usefully be generalized to

$$a = a_N = u'_{max}/\hat{c}_N \qquad (11.15)$$

where again

$$\hat{c}_N = N/|k| = |\hat{\omega}_N|/|\hat{z} \wedge k|, \qquad (11.16)$$

the intrinsic horizontal phase speed that the wave would have if it were a simple internal gravity wave. Note well that $\hat{c}_N \to 0$ when m and $|k| \to \infty$, even when $f \neq 0$ and $\hat{c} \not\to 0$. The reason why $\hat{c}_N$ is then relevant rather than $\hat{c}$ is that waves reaching frequencies close to f before breaking can no longer break by convective overturning as in figure 4. As soon as a_N reaches values $\gtrsim 2$, however, the waves can break in an entirely different way, by Kelvin–Helmholtz shear instability (figure 1 of Chapter 4 and figure 8 of Chapter 5). In such cases (11.15) and (11.16) show that a_N^{-2} takes on the significance of a Richardson number for the maximum wave-induced shear $|\partial u'/\partial z|_{max} = |m|u'_{max} \approx |k|u'_{max}$. As is well known, values $a_N^{-2} < \frac{1}{4}$ or $a_N > 2$ usually signify shear instability. Such shear-associated breaking of low-frequency inertia–gravity waves is one of the mechanisms giving rise to the intermittent clear-air turbulence familiar to air travellers.

The intermittency, associated with shallow, sparsely distributed, pancake-like layers or 'blini' of three-dimensional turbulence – observed to be tens to hundreds of metres thick in the stratosphere for instance – is typical of situations with large background Richardson number Ri. It marks an important difference between the real atmosphere and oceans and most numerical models of them. For reasons of numerical stability, most models explicitly or implicitly assume fictitious, relatively uniform, all-pervasive eddy viscosities and diffusivities, with values no smaller than some fixed dimensionless number times (grid size)^2/(time step).

The mean intrinsic energy $\hat{E}$ and pseudomomentum $\hat{p}$ of the plane inertia–gravity wave, per unit mass, can be shown to be given by the formulae

$$\hat{E} = \tfrac{1}{2}\overline{|u'|^2} + \tfrac{1}{2}\overline{b'^2}/N^2 = O(a^2) \qquad \text{and} \qquad \hat{p} = \hat{E}k/\hat{\omega} = O(a^2), \quad (11.17)$$

where the overbar signifies the average over a period or wavelength. The contribution $\tfrac{1}{2}\overline{b'^2}/N^2$ to $\hat{E}$ is the 'available potential energy', defined as the excess potential energy due to undulating the stratification surfaces $\mathscr{S}$. It can be shown to equal the kinetic contribution $\tfrac{1}{2}\overline{|u'|^2}$ when Coriolis effects are negligible, as expected from classical small-oscillations theory. Then the wave-induced fluctuations u' are alternately upslope and downslope. These are the slantwise velocity fluctuations leading to the momentum-transporting correlation, and covariance, the Reynolds stress first mentioned in § 3.

Coriolis effects complicate the picture by introducing an out-of-phase sideways or horizontal component of inclined-plane motion, outside the scope of the classical small-oscillations theory and leading to excess kinetic energy, $\frac{1}{2}\overline{|u'|^2} > \frac{1}{2}\overline{b'^2}/N^2$. This sideways velocity fluctuation produces the 'Eliassen–Palm effect' through its correlation with fluctuations in the thickness of the varicose layer described in § 4, causing the wave-induced momentum transport or radiation stress to differ from the wave-induced Reynolds stress. Each slab moves parallel to itself in an ellipse of upslope-to-sideways aspect ratio $(f/\hat{\omega})\operatorname{cosec}\vartheta$, lying in the inclined phase surface. When the approximate dispersion relation (11.11) applies, $(\frac{1}{2}\overline{b'^2}/N^2)/\frac{1}{2}\overline{|u'|^2} \to 0$ in the limit $\vartheta \to \frac{1}{2}\pi$. In that limit the sloping ellipse becomes a horizontal circle. The buoyancy acceleration b' vanishes, the motion is in horizontal slabs under the Coriolis acceleration $-f\hat{z} \wedge u'$ alone, so that $\hat{\omega} = f$, and $\hat{c}_g = 0$. All the intrinsic wave energy is kinetic. This is somewhat misleadingly called 'pure inertial motion'.[1] The vanishing of $\hat{c}_g$ means that such motion can be very persistent, following a fluid element, at amplitudes below the wave-breaking threshold $a_N \sim 2$. Motion approximating pure inertial motion has often been observed in the atmosphere and oceans.

With background shear, even if steady, $\hat{E}$ is not conserved, as illustrated by its evanescence in Booker & Bretherton's problem – energy can be exchanged between waves and background – but a horizontal component of $\hat{p}$ is conserved if the background is homogeneous in the corresponding direction. The 'wave action' $\hat{A} = \hat{E}/\hat{\omega}$ per unit mass is conserved for any slowly varying background, even if unsteady and horizontally inhomogeneous:

$$\partial(\rho\hat{A})/\partial t + \nabla \cdot \{\rho(\overline{u} + \hat{c}_g)\hat{A}\} = 0. \tag{11.18}$$

This follows from the standard energy–momentum tensor formalism of classical field theory applied to the linearized wave problem, with space or time translation replaced by wave phase shift; the simplest derivation is that of Andrews & McIntyre (1978, their case $0 \leqslant \alpha < 2\pi$). Equation (11.18) is the simplest way to predict wave amplitudes when using ray theory.

For instance (11.18) can be used, as hinted in § 5, to show in a fairly general

---

[1] Because of the factor 2 in (11.5), the period of pure inertial motion at the North Pole, for instance, is 12 hours and not 24 hours as the term 'inertial' might seem to suggest. Seeing why this does not correspond to a particle displaced off the pole and at rest relative to the cosmos – why 'inertial' is a misnomer – is an instructive little puzzle. The key is to remember what 'horizontal' means, as with a billiard table or ice rink at the North Pole. 'Horizontal' means along the level surfaces of the total, gravitational plus centrifugal, geopotential, and not the level surfaces of gravitation alone.

way that the frictionless, adiabatic, linear critical-layer theory predicts its own breakdown. In the standard critical-layer models assuming horizontal homogeneity and based on (11.11), with finite background shear $\partial \bar{u}/\partial z$ of a horizontal mean flow $\bar{u}$, the asymptotic behaviour $|m| \to \infty$ implies that each ray asymptotes to its critical level, with $\widehat{c}_g \cdot \widehat{z} \to 0$. This in turn implies that $|\hat{A}| \gtrsim$ constant. For strictly monochromatic waves (line spectrum in $\hat{c}$) the rays asymptote to a single critical level implying – if ray theory remains valid, as it does in Booker & Bretherton's case when $Ri \gg 1$ – that the ray points pile up and that $|\hat{A}| \to \infty$. For a continuous spectrum in $\hat{c}$ there is a spread of critical levels, so that the ray points remain spread out in the vertical (and ray theory remains valid in a wider range of cases), implying that $|\hat{A}| \sim$ constant. So the statement $|\hat{A}| \gtrsim$ constant covers practically every case, as does $|\hat{p}_{\text{horiz}}| = |\hat{A}k_{\text{horiz}}| \gtrsim$ constant. In the continuous-spectrum case, the asymptotic behaviour exhibits a Kelvin sheared-disturbance structure with u', $\theta' \propto \exp\{ik_{\text{horiz}} \cdot (x - \bar{u}t)\}$ times a complex function of t alone. This sheared disturbance is an exact, finite-amplitude solution of the Boussinesq equations if $k_{\text{horiz}} \cdot \partial \bar{u}/\partial z = $ constant, as can be straightforwardly verified by substitution into the equations, noting that $\partial/\partial t + \bar{u} \cdot \nabla$ exactly annihilates the factor $\exp\{ik_{\text{horiz}} \cdot (x - \bar{u}t)\}$.

A less drastic, though sometimes important, tendency toward wave breaking occurs when waves propagate from low to high background N^2 values, even without Doppler shifting. Seeing why this is so is a not-quite-trivial exercise in using the theory sketched above. The beach effect from the evanescence of ρ is, by contrast, a straightforward consequence of the presence of the factor ρ in (11.18) and its absence in (5.1) and (11.15).

For inertia–gravity waves of large vertical scale $|m^{-1}|$ violating the condition $|m|^{-1} \ll H_p$, a first correction to (11.11) can easily be obtained in the case $|k_{\text{horiz}}| \ll |k| \sim |m|$, which observations suggest to be the most important case in practice. The correction is simply to replace m^2 in (11.11) by $m^2 + \frac{1}{4}H_p^{-2}$ (e.g. Andrews *et al.* 1987). This correction always reduces $|\hat{\omega}|$, and rotates $\widehat{c}_g$ toward the horizontal, essentially because the kinematics of the wave motion is such as to involve more inertia than before.

12 Balanced oscillations: PV inversion and Rossby wave dynamics

The focus now shifts back to balanced motion, as distinct from oscillations about balance. Rossby waves, the key to the first enigma, are an example of balanced motion. As already suggested by figure 7, it is here that the properties of Q, the Rossby–Ertel potential vorticity, become especially important, along with the behaviour of the special contours $\Gamma_{Q\theta}$. From here on I shall use the acronym 'PV' as a synonym for the phrase 'Rossby–Ertel potential vorticity', as is conventional in the literature. Recall that the PV was defined in (9.3) and (11.3), and that it may be redefined with θ replaced by any monotonic function of θ.

The simplest way to understand the dynamics of Rossby waves, and of many other balanced motions, is to recognize the property known as PV invertibility, also called the 'invertibility principle' for PV. Its far-reaching importance was first fully appreciated, independently it seems, by J. G. Charney and E. Kleinschmidt. Roughly speaking, it says that if the motion is balanced then the distributions of PV on the stratification surfaces $\mathscr{S}$ contain all the dynamical information. A more careful and comprehensive statement is given in Hoskins *et al.* (1985). At each instant, the PV can be 'inverted' to obtain the velocity field $\boldsymbol{u}$, together with other dynamical quantities such as p and T. The main points are illustrated by a simple limiting case, very like textbook two-dimensional vortex dynamics, that arises as follows.

Imagine a layerwise-two-dimensional Boussinesq motion in the limit $N^2 \to \infty$. The stratification constraint is so strong that it makes the surfaces $\mathscr{S}$ effectively rigid and strictly horizontal. Then the flow on each $\mathscr{S}$ is not merely *like* a horizontal two-dimensional incompressible flow, in the qualitative sense of figure 7. It is, exactly, such a flow. It is strictly horizontal and strictly incompressible: $\boldsymbol{u} = \hat{\boldsymbol{z}} \wedge \nabla \psi$ for some stream function ψ, and, from (9.3),

$$Q = \sigma^{-1}(f + \nabla^2_{\text{horiz}}\psi) \tag{12.1}$$

with σ strictly constant, where ∇^2_{horiz} is the horizontal two-dimensional Laplacian on the surface of the sphere.

For simplicity we may think of the motion in this limiting case as independent of z and as taking place above an idealized model Earth with no topography. But that is not essential, because z now enters the problem only as a parameter. There is no derivative $\partial/\partial z$ anywhere in the problem, either in the horizontal Laplacian or in the material derivative $D/Dt = \partial/\partial t + \boldsymbol{u} \cdot \nabla$. There is, of course, an implicit assumption that Richardson numbers $Ri = N^2/|\partial\boldsymbol{u}/\partial z|^2 \to \infty$, implying some restriction on the magnitude of $\partial/\partial z$ as the limit $N^2 \to \infty$ is taken.

We may regard (12.1) as a Poisson equation to be solved for ψ when Q is given. This is a well-defined, and well-behaved, operation on the sphere provided that the given Q field satisfies the global constraint (9.7) on each $\mathscr{S}$. Thus we may rewrite the above symbolically as

$$\boldsymbol{u} = \hat{\boldsymbol{z}} \wedge \nabla \psi \qquad \text{where} \qquad \psi = \nabla^{-2}_{\text{horiz}}(\sigma Q - f). \tag{12.2a}$$

This expresses PV invertibility in the present limiting case. Q does indeed contain all the dynamical information. If we were to watch a moving picture of the Q field, whether or not Q is materially conserved, then we would be following everything about the dynamics because, at each instant, by (12.2a), knowledge of Q implies knowledge of $\boldsymbol{u}$.

The dynamical system is completely specified, therefore, by (12.2a) together with a single equation for the rate of change of Q. That equation can be (9.1a), or its generalization to motion that is not frictionless and adiabatic, symbolically

$$D Q / D t = \text{frictional and diabatic terms.} \qquad (12.2b)$$

This way of writing the dynamical equations points directly to the generalizations needed to describe real flows like those of figures 7 and 8, and to describe various kinds of Rossby waves at various levels of idealization or realism. In all cases the dynamical system takes the same generic form, with just two components: first, an inversion operator, which acts on the PV distribution at a single instant to yield all other dynamical quantities, including $\boldsymbol{u}$, and second, a single prognostic equation of the form (12.2b). There is therefore just one time derivative in the problem. This has the immediate implication that Rossby waves, quite unlike acoustic and inertia–gravity waves, exhibit what might be called one-way propagation. That is, Rossby-wave dispersion relations do not permit intrinsic frequencies $\hat{\omega}$ and phase speeds $\hat{c}$ to take either sign. Associated with this one-signedness, moreover, is what might be called one-way, or ratchet-like, wave-induced momentum transport, with important consequences for the global-scale circulation.

All that distinguishes the various cases from one another, therefore, is their different inversion operators. Generally, of course, these operators are less simple than (12.2a). When N^2 and Ri are no longer infinite, $\partial/\partial z$ reappears in the problem, and the two-dimensional inverse Laplacian is replaced by an inverse elliptic operator that is more like a three-dimensional inverse Laplacian in a stretched coordinate Nz/f. There is a tradeoff between accuracy and mathematical simplicity. In the most accurate cases, the inversion operator is weakly nonlinear, and complicated to write down, let alone realize computationally. The study of accurate inversion operators is still at or beyond the research frontier, both computationally and conceptually; some idea of what is involved can be gained from the 'shallow-water' idealizations studied in detail by McIntyre & Norton (2000). With finite stratification N^2, there is vertical coupling between the layerwise-two-dimensional flows on different stratification surfaces $\mathscr{S}$. When the motion can be characterized by a horizontal length scale L and when the Boussinesq approximations apply, the coupling is usually significant over vertical scales of order

$$H \sim fL/N, \tag{12.3}$$

consistent with the stretching of the z-coordinate. This vertical scale is called the Rossby height for the given L. The aspect ratio $H/L \sim f/N$ is called Prandtl's ratio of scales ($\lesssim 10^{-2}$ in the middle atmosphere). If H is given, then the implied $L \sim NH/f$ is called the Rossby length or, for historical reasons, the Rossby radius of deformation. The simplest though least accurate version of the above is the standard 'quasi-geostrophic theory' (e.g. Phillips 1963; Hoskins *et al.* 1985), which depends on assuming that Rossby numbers are small and Richardson numbers large. The formula (12.3) for the Rossby height is valid only when $H \lesssim H_p$; generalizations are given in the review by Hoskins *et al.* (1985).

Three aspects common to (12.2a) and all its generalizations deserve special emphasis, or re-emphasis:

(1) Local knowledge of Q does not imply local knowledge of $\boldsymbol{u}$; inversion is a non-local computational process, because it involves the solution of elliptic partial differential equations.

(2) Invertibility depends on balance with respect to acoustic and inertia–gravity waves. In the case of (12.2a) such waves have been completely excluded from the dynamics, by the assumptions implicit in the limiting process.

(3) There is a scale effect, whereby small-scale features in the Q field have a relatively weak effect on the ψ and $\boldsymbol{u}$ fields, while large-scale features have a relatively strong effect. In particular, ψ and $\boldsymbol{u}$ are to varying degrees insensitive to fine-grained structure in the Q field, such as the thin filaments in figure 7. Inverse elliptic operators like the inverse Laplacian in (12.2a) and its three-dimensional generalizations are smoothing operators; and some of the smoothing survives even when followed by the single differentiations in the formula $\boldsymbol{u} = \widehat{\boldsymbol{z}} \wedge \nabla \psi$ and its generalizations.

Equations (12.2a) and (12.2b) epitomize, with remarkable succinctness, and simplicity, the peculiar way in which fluid elements push each other around in a balanced flow. The non-localness of the inversion operator and the implied action at a distance, aspect (1), are related of course to the balance condition, aspect (2), in an essential way. The invertibility principle holds exactly in the limiting case of (12.2a), because in that limit the waves representing departures from balance propagate infinitely fast and have infinitely strong restoring mechanisms, as with the springy pendulum in the limit of an infinitely stiff spring. This permits exact action at a distance. Otherwise, with finite stiffness characterizing departures from balance, and correspondingly finite wave propagation speeds, the whole picture of balanced flow – in which PV inversion still formally involves action at a distance – has to be inherently approximate and subject to ultimate limitations, as hinted earlier (e.g. Ford *et al.* 2000) even though it can be astonishingly accurate (McIntyre & Norton 2000) – far more accurate than, for instance, the standard quasi-geostrophic theory – essentially because PV inversion operators tend to acquire a short-range character as stiffness diminishes and wave speeds become slower. In other words the distances over which the inversion operators 'act' tend to shrink.

Let us look a little more closely at the limiting case (12.2*a*), and use it to illustrate first the concept of 'vortex' and then the Rossby-wave restoring mechanism. Notice first that diagnosing the ψ field from the Q field is almost the same thing mathematically, according to (12.2*a*), as calculating the electrostatic potential induced by a given charge distribution, or calculating the static displacement of a stretched membrane induced by a given pressure distribution on it. (In the spherical geometry, we can visualize the undisturbed membrane as a spherical soap-film bubble.) Thus strong local anomalies in Q make, so to speak, distinct dimples in the membrane, implying that they tend to 'induce' strong circulations around themselves. Such a Q anomaly, together with its induced velocity field (the electric field vectors rotated through right angles), is nothing other than the coherent structure that fluid dynamicists call a 'vortex'. Examples of the corresponding layer-wise-two-dimensional coherent structures in a stratified, rotating atmosphere, long familiar to synoptic meteorologists as 'cyclones' and 'anticyclones', may be found in the review by Hoskins *et al.* (1985). The term 'vortex' is used interchangeably to mean either the entire coherent structure, or simply the PV anomaly that induces it.

The power of the viewpoint exemplified by (12.2*a*) has long been recognized, of course, and made extensive use of, by aerodynamicists as well as by theoretical meteorologists and oceanographers. Indeed the idea of invertibility is built into classical low-Mach-number aerodynamical language, as with the idea that a strong vorticity anomaly can roll 'itself' up into a nearly circular vortex. With 'vorticity' replaced by 'PV', the small cyclonic vortex over the Balkans in figure 7 nicely illustrates this last point.

Now Rossby-wave propagation, as mentioned earlier, involves sideways undulation of the special contours $\Gamma_{Q\theta}$. Just as the restoring mechanism giving rise to internal gravity waves depends on a background vertical gradient of θ, so does the restoring mechanism giving rise to Rossby waves depend on background isentropic gradients of Q, or, in the oceans, isopycnic

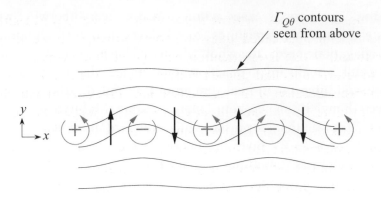

Figure 10. Special material contours $\Gamma_{Q\theta}$ on a single stratification surface $\mathscr{S}$, viewed from above, showing the restoring mechanism to which Rossby waves owe their existence (see text).

gradients of Q – that is, in both cases, on quasi-horizontal gradients of PV on the stratification surfaces $\mathscr{S}$. Such gradients are marked by sets of special contours $\Gamma_{Q\theta}$, as sketched in figure 10. The figure is a view from above showing several of the contours $\Gamma_{Q\theta}$ on a single $\mathscr{S}$, with Q increasing in the direction labelled y. Such gradients arise, for instance, in the simplest example from our limiting case (12.2a), in which we linearize about a state of relative rest, $\psi = 0$, $\sigma Q = f$. Then the background PV gradient is entirely due to the latitudinal gradient of f, conventionally denoted by the symbol β. From (11.5) we have

$$\beta = df/dy = r_0^{-1}|2\boldsymbol{\Omega} \wedge \widehat{\boldsymbol{z}}| = r_0^{-1}|2\boldsymbol{\Omega}| \cos\phi, \qquad (12.4)$$

where r_0 is the Earth's radius, and y is identified with $r_0\phi$, the latitudinal distance from the equator. In the undisturbed state, the contours $\Gamma_{Q\theta}$ lie along latitude circles.

Figure 10 shows how the Rossby-wave restoring mechanism works. The contours $\Gamma_{Q\theta}$ undulate sideways, i.e. northward and southward. Because of the background PV gradient, the undulations create a pattern of PV anomalies, the plus and minus signs, from which inversion, (12.2a), yields a velocity field qualitatively as shown by the solid arrows. As the electrostatic and membrane analogies show, this velocity field is a quarter wavelength out of phase with the displacement field, implying phase propagation. If one makes a mental movie of the resulting motion, one sees that the undulations will propagate westward in this case. Eastward propagation is impossible. This is how the one-way character associated with the single time derivative manifests itself. Figure 10 shows that Rossby waves always propagate with the higher Q values on their right, when Q is defined in right-handed axes: this is sometimes called 'quasi-westward propagation', i.e. propagation as if westward. If one fixes attention on a given fluid element, one sees it tending to return to its undisturbed latitude. This is the Rossby-wave restoring mechanism. Because of the scale effect, item (3) above, the mechanism is strongest at large and weakest at small scales. It is further discussed in the review by Hoskins *et al.* (1985), including the implications for group-velocity behaviour, omitted here for brevity except to say that the component of group velocity in the y-direction is generally non-zero.

The qualitative properties just sketched, including the one-way property, are easy to confirm from C.-G. Rossby's original version of the wave theory, in which equations (12.2a, b) are further simplified by means of a local Cartesian approximation with x taken eastward and y again northward. The Earth's curvature is now ignored except for its most important consequence, the background PV gradient, which is now approximated as constant. This is Rossby's famous 'nearly flat Earth' or 'beta-plane' approximation. With these simplifications it is easy to verify that disturbances of the form $\psi \propto \exp(ikx + ily - i\hat{\omega}t)$ are solutions provided that

$$\hat{\omega} = -\beta k/(k^2 + l^2), \qquad (12.5)$$

where β is now constant and where the denominator $(k^2 + l^2)$ comes from the inverse Laplacian, $\nabla^{-2} = -(k^2 + l^2)^{-1}$. The minus sign is essential; the intrinsic phase speed in the x-direction, $\hat{c} = -\beta/(k^2 + l^2)$, is always negative if β is positive, confirming the one-way behaviour deduced from figure 10. The scale effect manifests itself through the increasing magnitudes of $\hat{\omega}$ and $\hat{c}$ as the lengths k^{-1} and l^{-1} are scaled up proportionately to each other.

The same qualitative properties are found when we generalize to the three-dimensional case with finite stratification N^2 which, however, introduces one important new feature: there can now be a non-zero component of group velocity in the z-direction (Charney & Drazin 1961). The associated vertical propagation is found to be associated with significant rates of vertical transport of horizontal momentum, crucial to the stratospheric global circulation. Almost all of this comes from the Eliassen–Palm effect, with a relatively small, usually negligible, contribution from the wave-induced Reynolds stress, essentially because of the very small vertical velocities characteristic of balanced motion (e.g. Andrews *et al.* 1987).

The key to understanding many aspects of the global circulation is once again to extend the above theory not only to finite N^2 but also to accommodate background shear. Because of the scale effect, the most important waves have scales too large for ray theory to be plausibly valid in the most interesting cases. But ray theory still has a certain range of validity, and it is useful and suggestive, as a first step, to make the extension to background shear in the manner of §11. Evidently, $\hat{c} = -\beta/(k^2 + l^2)$ can be Doppler shifted to zero with k constant and $l \to \infty$, suggesting a critical-layer behaviour to some extent analogous to that for internal gravity waves. As before, this suggests what more accurate wave theories and numerical models confirm, namely that linearized theory predicts its own breakdown, both in the monochromatic and continuum cases, i.e. the cases respectively involving line and continuous $\hat{c}$ spectra. The prediction from frictionless, adiabatic dynamics is that the waves will always break, in the general sense of §6. That is, the special contours $\Gamma_{Q\theta}$ will always deform irreversibly, in one way or another. This in turn is part of what makes the wave-induced momentum transport ratchet-like, as will be seen in the next section.

In the continuum case, the wave field again takes on Kelvin sheared-disturbance structure, leading this time to small-scale shear instability (Haynes 1987). In the monochromatic case, ray theory always breaks down, as well as linearization, and the resulting nonlinear critical layer or surf zone has a behaviour intermediate between the 'beach' and 'sea wall' extremes noted earlier. Both wave absorption and reflection can be significant. By contrast with the internal gravity and inertia–gravity wave cases, we can say more about this case theoretically, for two reasons. One is the existence of a beautiful analytic solution to (12.2a, b) on the beta-plane, in the monochromatic case, due to K. Stewartson, H. Warn and T. Warn, hereafter 'SWW' (for a review giving full details, see Killworth & McIntyre 1985). This has given us a very precise characterization of fully nonlinear Rossby-wave critical-layer behaviour in certain idealized cases. The second is the existence of certain finite-amplitude 'momentum–Casimir' wave-activity conservation equations, based on exploiting the arbitrary functions appearing in Casimir invariants of the type mentioned in §9, and related to the stability theorems of V. I. Arnol'd and others. They are also, more distantly, related to the pseudomomentum conservation equation obtained by multiplying (11.18) by $\boldsymbol{k}_{\text{horiz}}$. In conjunction with PV invertibility they provide important constraints on possible behaviour even when the fluid motion is more complicated than in the SWW case, as in fact it usually is (Haynes 1989).

13 Rossby waves and anti-friction

Figure 11 shows the nonlinear Rossby-wave critical layer or surf zone described by the SWW solution. In the SWW problem, monochromatic, small-amplitude Rossby waves are excited, in a background flow having constant

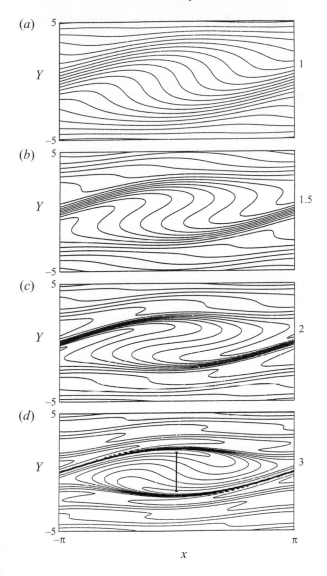

Figure 11. Contours $\Gamma_{Q\theta}$ for the inner region (nonlinear critical layer or surf zone) of the SWW solution, a special but clear-cut example of Rossby-wave breaking. The y-scale has been expanded using the re-scaled coordinate $Y = y/\delta$ with $\delta = a^{1/2}(\partial\bar{u}/\partial y)_0/\beta$, where $(\partial\bar{u}/\partial y)_0$ is the initial, constant mean shear, and a is a measure of the (small) wave amplitude outside the nonlinear critical layer or surf zone, not given by (5.1) but dependent on certain Bessel functions that describe the outer wave field. Four successive stages in the evolution are shown, at times 1, 1.5, 2, 3 in units of $2^{1/2}(k\,b\,(\partial\bar{u}/\partial y)_0)^{-1}$, where k is the x-wavenumber. The vertical bar in panel (d) gives the effective mixing scale $\delta_{\mathrm{mix}} \simeq \delta$ such that the perfectly mixed zone in figure 12 gives exactly the same net momentum deficit. From Killworth & McIntyre (1985).

shear $\partial \bar{u}/\partial y$, by a sinusoidally undulating boundary well outside the domain of the figure. For these waves there is a critical line, $Y = 0$ in the figure, that is, a line where linear ray theory predicts $l \rightarrow \infty$. A narrow critical layer or surf zone forms, in which the waves break strongly and in which both linearization and ray theory fail completely.

The prognostic equation is (12.2*b*) with right-hand side zero; thus the contours $\Gamma_{Q\theta}$ are also material contours. Figure 11 shows them deforming irreversibly in a recirculating flow, a so-called Kelvin cat's-eye pattern, straddling the critical line. (Y is a stretched coordinate on the scale of the surf zone.) As the contours wrap around, the Q distribution changes and, in part through the action-at-a-distance implied by inversion, induces a change in the flow outside the surf zone. This affects phase gradients with respect to y in the outer flow in such a way that the surf zone appears as an absorber to the Rossby waves outside it during the early stages of wave breaking, but later becomes a reflector owing to the finite scope for PV rearrangement within a surf zone of finite width. This is how the surf zone evolves from 'beach' to 'sea wall' conditions.

In fact, in this case it overshoots by temporarily becoming an 'over-reflector' or wave emitter, then going through a decaying cycle of absorption and emission. In this respect the SWW case is not very typical. What is generic is that the finite scope for PV rearrangement permits only a finite amount of wave activity to be permanently absorbed by a surf zone of bounded width, in any model in which the right-hand side of (12.2*b*) is zero. In more accurate models with finite N^2, the phrase 'PV rearrangement' is to be interpreted as shorthand for 'rearrangement of PV substance' (recall the end of §9) on each stratification surface $\mathscr{S}$.

The way in which the surf zone interacts with its surroundings is precisely expressed, in the SWW solution, by the use of matched asymptotic expansions. Full details of the solution may be found in Killworth & McIntyre (1985), in which there is also established a general theorem confirming that the absorption–reflection behaviour just described is generic, and not dependent on the particularly simple circumstances that permit the problem to be solved analytically. Wave absorption is quantified most conveniently, in this problem, via a momentum–Casimir wave-activity equation.

Also generic is the ratchet-like, one-way character of the momentum transport induced by Rossby waves. The essence of the matter is captured in a very simple way if we ignore the weaker wave breaking due to sheared disturbances in the outer regions of figure 11, and assume (as was proven rigorously in the paper just cited) that the dominant process is indeed the PV rearrangement in the central region of strong wave breaking. Figure 12(*a*)

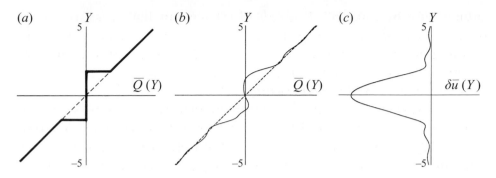

Figure 12. The equivalence between mean force and the rearrangement of potential-vorticity substance in the simplest relevant model system; see text, and equations (13.1) and (13.2). Courtesy P. H. Haynes.

further idealizes the situation by assuming that the rearrangement takes the form of perfect mixing. The graph on the left shows the background PV profile, $Q(y) = \text{constant} + \sigma^{-1}\beta y$ (shown dashed), together with the result of mixing it perfectly over a finite y-interval (shown heavy). This mimics, in idealized form, what happens in the surf zone as seen in a view based on Eulerian averaging with respect to x.

The middle graph, figure 12(b), is the x-averaged $Q(y)$ profile from an actual solution, in another case of a nonlinear Rossby-wave critical layer or surf zone that is less simple than the SWW case, and in some ways more typical (Haynes 1989 and personal communication). The detailed flow is more chaotic, because of the onset of secondary instabilities. Figure 12(b) is qualitatively similar to figure 12(a), though the effects of peripheral sheared disturbances are also discernible. To get the corresponding zonally averaged momentum changes, we need the inversion implied by the beta-plane counterpart of (12.2a). In the x-averaged view, this inversion is trivial: one merely has to integrate the PV change once with respect to y:

$$\delta \bar{u}(y) = \int_{y}^{\infty} \delta \overline{Q}(\tilde{y})\,\sigma\,\mathrm{d}\tilde{y}. \tag{13.1}$$

The result is shown in the right-hand graph, figure 12(c). The momentum change corresponding to the more idealized $Q(y)$ profile in figure 12(a) is a simple parabolic shape (not shown), qualitatively similar to figure 12(c), as is evident from a moment's consideration of (13.1). It is plain from (13.1) that any PV rearrangement that creates a surf-zone-like feature with weakened mean gradient will give rise to a net momentum deficit.

It is easy to show from the beta-plane counterparts of (12.1)–(12.2) and *x*-homogeneity, with right-hand side of (12.2*b*) zero, that

$$\frac{\partial \overline{u}}{\partial t} = -\frac{\partial}{\partial y}\left(\overline{u'v'}\right) = \sigma\,\overline{v'Q'}\,, \tag{13.2}$$

where the overbars denote the Eulerian mean with respect to x, and primes fluctuations about that mean. This makes explicit the relation between mean-flow change, wave-induced Reynolds stress, and wave-induced transport of PV substance. The second equality, easily derivable from (12.1), is often called the 'Taylor identity' after G. I. Taylor's famous 1915 paper on 'Eddy motion in the atmosphere'; this, too, generalizes to finite stratification N^2, with $-\partial/\partial y\,(\overline{u'v'})$ replaced by the divergence of the Eliassen–Palm flux (Andrews *et al.* 1987). The Taylor identity holds for general (12.2*b*).

The essential point, then, is that whenever PV is thus rearranged, a momentum deficit appears, signalling the presence of an effective force in the $-x$-direction. This makes evident the ratchet-like or one-way character of the momentum-transporting process associated with Rossby waves. More generally, the wave-induced force due to formation of a Rossby-wave surf zone is always quasi-westward, in the same sense as the term is used in connection with Rossby-wave phase propagation, that is, the force is in the direction you face when you have high PV values on the right. If the PV gradient is dominated by the y-gradient of f, then this force is always westward. The gyroscopic pumping action of such a force is therefore *always poleward*. If there is some forcing effect that persistently tends restore the background Q gradient – and this requires nonzero terms on the right of equation (12.2*b*) – then the absorption of Rossby waves, and the gyroscopic pumping, can also be persistent. In the stratosphere, the non-zero terms are provided mostly by radiative thermal relaxation.

A feature that is robust, and carries over to more realistic cases with larger wave amplitudes, is that the flow is highly inhomogeneous, being more wavelike in some places (the outer flow in the SWW case) and more 'turbulent' in others, meaning that the material contours deform irreversibly (the surf zone), with each region strongly affecting the other. More realistic cases typically show a similar inhomogeneity, the main difference being that the flow in the surf zone or zones is more, intuitively speaking, 'turbulent' in the sense of being more chaotic-looking. Figure 13 illustrates a large amplitude case on the sphere, showing contours of a passive tracer released as a small blob into the surf zone of a fairly realistic model of the winter stratosphere at altitudes around 25 km (Norton 1994). The inhomogeneity is extreme. The rapid mixing evident in the surf zone, which is marked by passive-tracer contours and, as is typical, occupies a range of middle

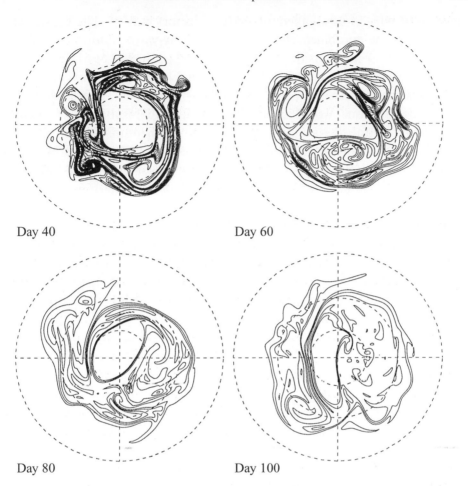

Day 40   Day 60

Day 80   Day 100

Figure 13. Model flow (shallow-water with mean depth 4 km) closely resembling flow in the real wintertime stratosphere at altitudes around 25 or 30 km. Polar stereographic projection with 0°, 30°, and 60° latitude circles shown dashed. Details of the model and the model run are given in Norton (1994). The flow is visualized by a passive tracer released in the midlatitude stratospheric surf zone. The isolation of the (core of the) polar vortex recalls classic smoke rings and is of great importance to stratospheric polar chemistry. Courtesy of Dr W. A. Norton, from whom an animated video of the model run is available (Dept of Atmospheric, Oceanic, and Planetary Physics, Clarendon Laboratory, Parks Road, Oxford, OX1 3PU, UK).

latitudes, and the almost complete isolation of the central, or polar-vortex, region from its surroundings, where no passive-tracer contours penetrate, are both important for ozone-layer chemistry. The inner border of the surf zone, marking the edge of the polar vortex – whose resilence makes it appear to behave, in animated versions of figure 13, almost like an elastic band – is an example of what is now called an 'eddy-transport barrier'. The barrier effect results from the combined effects of the Rossby-wave restoring mechanism and strong shear just outside the polar vortex (e.g. Juckes & McIntyre 1987). This has recently been called 'shear sheltering' by Hunt & Durbin (1999).

From what has already been said about PV inversion operators and the Taylor identity, one would expect the qualitative results just outlined to carry over to stratified atmospheres. On the momentum-transport aspects, for instance, Robinson (1988) has performed numerical experiments for stratified atmospheres that show the behaviour in some examples of interest. The dynamical system used is a realistic model of the wintertime stratosphere on a spherical Earth. In these experiments, PV substance is rearranged along stratification surfaces $\mathscr{S}$ in surf zones of varying widths confined to the middle stratosphere, in the manner of figures 12 and 13. As expected, the effect is just the same as that of exerting, at latitudes and altitudes corresponding to the location of the surf zone, a negative mean force, hence retrograde torque about the Earth's axis. The momentum–Casimir theorem to quantify this is given by Haynes (1988).

It is worth noting that the foregoing picture – some components of which date back to the work of G. I. Taylor already mentioned, and his later attempts to construct a 'vorticity transfer theory' of turbulence – has occasionally been regarded with suspicion, or even stated authoritatively to be plain wrong on the grounds that the scenario of figure 12 is impossible because it would violate momentum conservation. What is overlooked is the possibility, and physical reality, of wave-induced momentum transport from outside the surf zone, for instance originating at the undulating boundary in the SWW case. Both the 'wavelike' and the 'turbulent' aspects of the problem, concerning the exterior and interior of the surf zone and the way they interact dynamically, and the implications for the amount of momentum transported, are essential features, as the matched asymptotic analysis clearly emphasizes in the SWW case. These fluid systems – the SWW model, and by implication the real stratosphere as well – confront us with what McIntyre & Palmer (1984) called a highly inhomogeneous 'wave–turbulence jigsaw puzzle'. Once again we see how profoundly misleading it would be to categorize the flow as just turbulent and to think in classical turbulence-theoretic terms. This bears on the discussion near the end of § 3.1 of Chapter 10.

We may note finally that there is a tendency for the wave–turbulence inhomogeneity to be self-reinforcing. Where the contours $\Gamma_{Q\theta}$ are crowded together on a stratification surface $\mathscr{S}$, typically at tropopauses or the edges of surf zones – idealized as the 'corners' in the Q profile in figure 12(*a*) – one has a strengthening of the Rossby restoring mechanism, and even more importantly a weakening within surf zones. Thus mixing becomes easier within surf zones once they begin to be established, and harder outside. Other things being equal, then, the inhomogeneity tends to perpetuate itself. This seems to be analogous to the 'Phillips effect' in which vertically uniform stirring can create mixed layers in a stratified fluid (e.g. Ruddick *et al.* 1989).

The tendency toward such inhomogeneity is further favoured by the integral constraint (9.7). That constraint tells us that the only way to mix Q to homogeneity on the surface $\mathscr{S}$ is to have zero Q everywhere. Invertibility then says that the the entire surface $\mathscr{S}$ would have to be at rest relative to the stars, apart from acoustic and inertia–gravity oscillations. That would be a fantastically improbable state on a planet as rapidly rotating as the Earth. One can imagine a thought experiment in which the surface $\mathscr{S}$ is rotating solidly with the Earth, and then has its angular velocity uniformly reduced, corresponding to a uniformly reduced pole-to-pole latitudinal profile of PV, by some incident Rossby-wave field. But the tailoring of a Rossby-wave field to do this would be a more delicate affair than standing a pencil on its tip; the natural occurrence of such a wave field would be another fantastically improbable thing. To summarize, then, these fluid systems naturally exhibit a strong wave–turbulence inhomogeneity, and with it a marked tendency for the associated wave-induced momentum transport to behave anti-frictionally, i.e. a tendency to drive the system away from, not toward, solid rotation. And, as with the QBO, the sense of wave-induced momentum and angular momentum transport is related to the locations of wave sources and sinks, and not, in anything like the classical way, to local mean shears. In short, the first enigma is no longer an enigma. This is also highly relevant, as noted in § 16, to the so-called 'parametrization of baroclinic instability'.

14 The global-mean circulation of the middle atmosphere

We want to understand, among other things, how greenhouse gases are moved around and chemically transformed. This in turn suggests thinking in terms of Lagrangian means. But Lagrangian means are not always well defined, as noted earlier. In fact, the dispersion and mixing of fluid elements in surf zones like that illustrated in figure 13 precludes any simple use of Lagrangian means in analysing real data, except perhaps in some very broadbrush, surf-zone-averaged sense (Thuburn & McIntyre 1997). Compromise solutions to this problem include the so-called TEM or 'transformed Eulerian mean' circulation (Andrews & McIntyre 1976) and 'transport' circulation (Plumb & Mahlman 1987). Although these are not Lagrangian means they

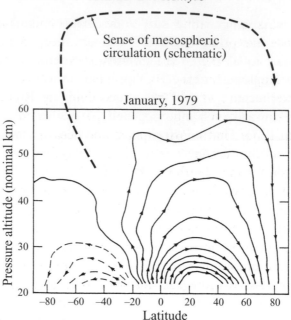

Figure 14. Mass transport streamlines of the global-scale TEM (transformed Eulerian-mean) circulation for January 1979 (light curves), from Solomon *et al.* (1986). This was estimated from satellite radiometer data in conjunction with detailed and elaborate calculations of radiative heating and cooling rates. The heavy dashed streamline (schematic only) indicates the qualitative sense of the mesospheric branch or 'Murgatroyd–Singleton circulation', deduced from other observational and theoretical evidence. The left-hand scale is pressure altitude $z = H_p \ln(p_0/p)$ in nominal kilometres, assuming a pressure e-folding scale height $H_p = 7$ km. Thus the vertical domain shown corresponds roughly to the middle third of figure 2 (multiply the right-hand scale in figure 2 by 7 km). The northward mean velocities at top right (not counting the heavy dashed streamline) are of the order of 2 or 3 m s^{-1}. The time for a marked fluid element to rise from the tropical tropopause to, say, 40 km is generally of the order of two years. The lower pair of circulation cells is often collectively referred to as the 'Brewer–Dobson circulation'.

have been found to provide, to some extent, useful practical substitutes. One reason is that to good approximation they represent the refrigeration effect, and other dynamical effects on the global-scale temperature field, in a manner similar to that of a Lagrangian mean.

Figure 14 shows an estimate, from Solomon *et al.* (1986), of the TEM circulation for January 1979 but thought to be typical of other years in its general qualitative features. It is based largely on a uniform dataset, from the LIMS infrared radiometer on board the Nimbus 7 research satellite. There is considerable interannual variability in the strength and shape of the circulation, especially in the winter stratosphere. For example, the winter stratospheric circulation in January 1979, the case of figure 14, was somewhat stronger than average probably by a factor of order 1.5 to 2. Part of this variability appears to be due to the modulation of the winter stratospheric 'wave–turbulence jigsaw' by the QBO (Baldwin *et al.* 2000 and references therein) – where because of surf-zone reflectivity there is scope for weak Rossby-wave resonant-cavity effects – and part to sources of interannual variability centred more in the troposphere and oceans, such as the irregular but strong variability on time scales of several years associated with the so-called ENSO (El Niño, Southern Oscillation) cycle.

The heavy dashed line extending outside the frame of figure 14 is schematic only, and indicates the qualitative sense of the mesospheric mean circulation, deduced from much sparser observational evidence. The gross sense of the circulation seems to be in no real doubt when one takes all the evidence into consideration. It is consistent with a pioneering study by R. J. Murgatroyd and F. Singleton and with the existence of noctilucent clouds as already emphasized in § 3.

The lower part of the circulation, in the lower half of the frame of figure 14, is now known to be driven mostly by Rossby waves in essentially the manner discussed in the last section. Many detailed observational, modelling, and data-assimilation studies support this conclusion. Because of the scale effect noted in § 12, it is the so-called 'planetary-scale' Rossby waves that can penetrate furthest upward into the stratosphere. Planetary-scale Rossby waves correspond by convention to the largest-scale displacements of the polar-vortex edge illustrated in figure 13, those with zonal wavenumbers 1 to 3. The planetary-scale Rossby waves most strongly forced from below are quasi-stationary, being tied in various ways to geography (for instance through being generated nonlinearly by the fluctuations in oceanic 'storm tracks' of intense cyclogenesis and anticyclogenesis). With phase speeds close to zero relative to the ground, quasi-stationary waves have negative $\hat{c}$ values, hence the ability to propagate as free Rossby waves, only in the wintertime eastward background flow (recall § 12). Such waves cannot propagate freely in the summertime westward background flow. In summer, the upward penetration is therefore more diffractive in nature, and relatively shallow, on height scales of the order of a Rossby height. This nicely accounts for the summer–winter asymmetry seen in figure 14. The points about upward penetration were first made, on the basis of linearized, quasi-geostrophic Rossby-wave theory, in the famous pioneering paper of Charney & Drazin (1961).

The cold summer mesopause is not the only place where temperatures T differ significantly from radiative equilibrium temperatures T_{rad}. Temperatures over the winter polar cap, for instance, are pushed above radiative equilibrium by the systematic descending motion seen on the right of figure 14. In fact it is the pattern of departures of T from T_{rad} that enables infrared radiometry to be used to deduce the circulation. The summer stratopause, the temperature maximum near centre left of figure 2, at altitudes of about 50 km or 7 scale heights, has a much weaker circulation and is relatively close to radiative equilibrium, probably within 10 K or so (Andrews *et al.* 1987). The high temperatures at the summer stratopause are well explained by the absorption, by ozone, of solar ultraviolet at wavelengths around 200–300 nm. Such wavelengths can penetrate down to 50 km or lower (Andrews *et al.* 1987).

A circulation like that in figure 14 and its seasonal variants is also consistent with the observed behaviour of many chemical tracers. For instance, man-made CFCs are chemically inert under everyday conditions. Together with their low water solubility and relatively small rate of uptake by the oceans, this means that, as is well verified by observations, the CFCs are mixed fairly uniformly throughout the troposphere. As far as the CFCs are concerned, therefore, the troposphere can be thought of as a massive tank or reservoir whose contents are recirculated through the middle atmosphere at rates of the order suggested by figure 14, or a fraction less. This accounts rather well for the observed destruction rates of CFCs, which correspond to e-folding atmospheric lifetimes of order a century, somewhat shorter for CFC11 ($CFCl_3$), by a factor ~ 2, and somewhat longer for CFC12 (CF_2Cl_2), by a factor ~ 1 to 1.5. It is mainly at altitudes $\gtrsim 25$ km that the CFCs are destroyed, by ultraviolet photolysis; and so rates of CFC destruction are closely tied to rates of circulation. In terms of mass flow that reaches altitudes $\gtrsim 25$ km, these rates amount to a few percent of the tropospheric mass per year, because of the falloff of mass density with altitude by a factor e^{-1} every 7 km or so; recall (2.5). Mass densities at 30 km, for instance, are about 1% of those at sea level. This is consistent both with the circulation times, which are of the order of several years, and with the CFC e-folding lifetimes of the order of a century.

The orders of magnitude in this picture turn out to require upward velocities at bottom centre of figure 14 close to 0.2 mm s^{-1} or 6 km per year, or a little more in an active January (Yulaeva *et al.* 1994). This sort of velocity is far too small to observe directly – or so it was thought until the mid-1990s.

Figure 15, from Mote *et al.* (1998), is very nearly a direct observation. It gives a powerful independent check on the whole picture. It was derived from satellite data obtained during the 1990s, long after results like figure 14 had

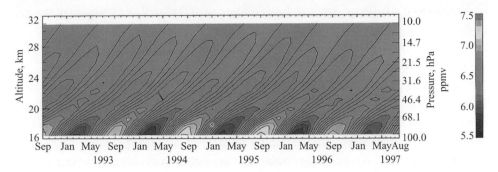

Figure 15. The tropical stratosphere as a 'tape recorder', from Mote *et al.* (1998); *q.v.* for a careful discussion of the data analysis and interpretation. The time–altitude plot is from HALOE, a solar occultation limb sounder on the Upper Atmosphere Research Satellite (UARS) launched in September 1991. Soundings in the deep tropics were used, between 14°N and 14°S. The quantity plotted is water vapour plus twice methane in p.p.m.v. (parts per million by volume), a notional mixing ratio proportional to the ratio of numbers of hydrogen atoms to numbers of air molecules. Because of methane oxidation chemistry this is to excellent approximation a material invariant, a passive tracer, in the altitude range shown. The raw data have been fitted to a suitable set of empirical orthogonal functions to extract the annual cycle in an objective way; but the tape-recorder signal is still easily visible through the noise without any such processing – see Plate 1 of Mote *et al.* (1998) – and in data from other instruments, notably water-vapour data from the microwave limb sounder on UARS.

been obtained with the help of an earlier generation of satellites. Figure 15 shows the water vapour signature of the annual cycle being carried up in the rising flow in the tropical lower stratosphere, just like a signal recorded on a moving magnetic tape. The annually varying temperature minimum at the tropical tropopause near 16 km acts as the recording head, modulating water-vapour mixing ratios by controlling the freeze-drying of air passing upwards into the stratosphere, a process originally suggested by A. W. Brewer in 1949 to account for the observed dryness of the stratosphere. At altitudes around 20 km, the velocity apparent in figure 15 is the same 0.2 mm s^{-1} deduced earlier from infrared radiometry and from Rossby-wave dynamics.

The observed tape-recorder signal is also one of our strongest checks, today, that Lagrangian-mean surrogates such as the TEM circulation used in, or implicit in, estimates like that of figure 14 are indeed relevant to the transport of greenhouse gases, despite their tenuous and necessarily imperfect relation with Lagrangian means. As pointed out by Holton *et al.* (1995), this is, in part, simply because of the typical configuration of stratospheric surf zones suggested in figure 13 – the implied existence of subpolar and subtropical eddy-transport barriers – leaving a relatively isolated, and relatively less disturbed, tropical stratosphere, which is plainly, also, a prerequisite for seeing any tape-recorder signal at all.

15 The response to gyroscopic pumping

The response to gyroscopic pumping is an interesting and non-trivial dynamical problem. In a thought experiment in which one switches on a given, zonally symmetric westward force in some region occupying a finite altitude and latitude range away from the equator, and asks how the model responds, one sees an initial transient after which the motion settles down to a steady poleward flow whose mass streamlines close downward (Haynes *et al.* 1996). The steady-state response looks like the lower part of figure 14, if the westward force is applied wherever the streamlines indicate poleward flow. The Coriolis force of the poleward flow comes into balance with the applied westward force as the steady state is approached. The streamline pattern burrows downward until it reaches the Earth's surface, where it closes off in a frictional boundary layer. This is sometimes discussed under the heading 'downward control', the idea being to emphasize the control of long-time-mean circulations by given zonal forces, in the thought experiment just described. It is the essential reason why, for instance, the refrigeration effect and noctilucent cloud formation depend mainly on the breaking of internal gravity waves *above*, and equatorward of, the cloud level.

The approach to the steady state and the downward closure of the streamlines is related to the finiteness of the pressure and density scale heights, i.e. to the exponential falloff of pressure and density with altitude. In essence, we have a situation in which the finite mass of atmosphere above the forcing region is incapable of absorbing infinite amounts of angular momentum. The system solves this problem, so to speak, by sending all the angular momentum downward, and ultimately into the solid Earth, whose moment of inertia is regarded as infinite. That the steady state is, indeed, robustly approached, if the force is applied far enough from the equator – weaker forces can be applied closer – has been well demonstrated both in analytical-mathematical and in numerical versions of the thought experiment. For realistic values of the force, 'far enough from the equator' means further than $15°$ latitude or so. In such cases the response problem is well described as balanced flow satisfying the thermal-wind relation (11.6), and is linear to good approximation. For a more thorough analysis and discussion, the reader may consult Haynes *et al.* (1996), Holton *et al.* (1995), and Haynes (1998). The timescales of the initial transient depend on thermal radiative relaxation times and on spatial scales, as well as on f and N. The adjustment is quickest and most robust at the largest, or near-global scales, those of figure 14, on which the finiteness of the scale height H_p is most strongly felt.

If one gets too close to the equator, then the response problem for a steadily applied force (as distinct from the oscillating force involved in the QBO) becomes nonlinear, and robustness may be lost. This is at the research frontier and is not yet well understood, though progress is currently being made. But mass continuity (11.1) still says, of course, that the circulation must close somehow, as illustrated by the rising branch of the circulation in figure 14. Solar heating has a role, here, that can be described in terms of the tape-recorder metaphor as that of 'guidance wheels' for the tape: that is, the distribution of $T_{\rm rad}$ across the tropics can influence the upward path take by the central rising branch in figure 14, though not the total mass flux being pumped into the extratropics.

It is now possible to see what I meant by fluid motions taking CFC molecules on epic journeys. Imagine yourself as a CFC molecule that has just leaked into the atmosphere from an old air conditioner in, say, New York. You will probably begin your journey by being carried off across the Atlantic toward Norway, and then spend twenty years or so wandering all over the globe in the troposphere. You might get caught in a measuring instrument at the South Pole or at Cape Grim, Tasmania. You might dissolve in the ocean. But it is much more likely that sooner or later you will find yourself gently rising in the tropical stratosphere at an average speed of 0.2 mm s^{-1}, being battered by increasingly energetic solar ultraviolet photons. Your chances of survival and return to the troposphere now depend, crucially, on whether you get into the higher or lower branches of the circulation shown in figure 14. At this point, there is about one chance in five that you will be carried above 25 km and then blasted into chemically active fragments by an ultra-violet photon. Your remains, including chlorine atoms, will end up wandering generally poleward and downward under the ratchet-like gyroscopic pumping action of Rossby waves – travelling on tortuous pathways through surf-zone chaos, participating in countless chemical changes – some of those remains arriving, a few years later, in the polar lower stratosphere either in the north or in the south. There, if conditions are suitable, as they tend to be especially in the south during spring, the chlorine atoms participate in catalytic ozone destruction, with one such atom destroying thousands of ozone molecules.

16 Postlude: the oceans, the troposphere, and climate feedback

The oceans and the troposphere are fundamentally more complicated than the middle atmosphere, partly because of (*a*) the more complicated buoyancy effects, dependent respectively on salinity and on moist latent heat (compositional convection), (*b*) the non-global spans of the stratification surfaces $\mathscr{S}$, vitiating (9.7), (*c*) the fact that a zonally averaged picture is less useful, if applicable at all, and (*d*) even more fundamentally, in the case of the oceans, because there is no counterpart to radiative thermal relaxation. The ocean circulation, heat transport, and heat uptake are therefore, in the long-time mean, under a very different kind of mechanical control: that of wind and tide induced stirring and mixing across mean stratification surfaces (Munk & Wunsch 1998). This makes the ocean circulation problem profoundly different from, and more difficult than, any atmospheric circulation problem. Gyroscopic pumping must occur, wave-driven as well as Ekman, but cannot by itself push water across mean stratification surfaces.

The tropical troposphere, the climate system's 'thermal powerhouse', has stable stratification under moist-convective control – temperature profiles clamped to the moist adiabat – and a persistent deep-convective upward mass flux in the intertropical convergence zone, or trade-wind confluence zone, consisting of two contributions having somewhat comparable magnitudes. One of these, sometimes called the 'Hadley' contribution, has no significant counterpart in the middle atmosphere. It is controlled by radiative cooling of the whole tropics outside the relatively small area of the intertropical convergence zone, but within the regions of effectively small Coriolis parameter f. The clamping of the stratification to the moist adiabat, together with the distribution of water vapour, determines the radiative cooling rate and hence the mass flux downward across the stratification surfaces $\mathscr{S}$ in these regions. The intertropical convergence zone can freely accept the upward return flow of this radiatively determined mass flux.

The other contribution to the upward mass flux fed into the intertropical convergence zone is that controlled by gyroscopic pumping from the extratropics. This now includes the extratropical troposphere, where the fluctuations doing the pumping include so-called 'baroclinic instabilities'. These are large-scale potential-energy-converting balanced motions that depend on large-scale, near-surface gradients of θ and that can be regarded as pairs of phase-locked, counterpropagating Rossby waves that induce each other to grow in amplitude and go strongly nonlinear, leading to cyclogenesis and anticyclogenesis, the production of large-scale, layerwise-two-dimensional coherent vortices, and to upward and equatorward Rossby-wave radiation producing an irreversible momentum transport, near 10 km, very like that higher up in the winter stratosphere (Hoskins *et al.* 1985, §6*d*). This explains not only the first enigma but also why parts of a system ultimately driven by solar heating should be under mechanical rather than thermal control, in the sense analysed above. Bates (1999) has shown that the momentum transport near 10 km provides a crucial feedback in the stabilization of climate, hence crucial to include in the 'parameterizations' of baroclinic instability used in simplified climate models. Its impact on surface wind distributions – another illustration of fluctuations driving the system away from solid rotation – strongly affects evaporation and cooling at the tropical sea surface. As far as I know this feedback had not been previously recognized.

It is remarkable and exciting that we can understand anything at all about so complex and chaotic a system as the Earth's atmosphere and oceans. But humility is called for: our understanding is very incomplete and very imperfect. Will sea levels be zero, one, or two metres higher at the end of this century? The scientific journey to understand and predict our fluid environment has, in crucial respects, barely begun.

Acknowledgements

Many of the ideas that are central to this survey grew out of material originally developed for an essay that shared the 1981 Adams Prize in the University of Cambridge and was subsequently developed further in various invited lectures and surveys. The development of the ideas involved has been influenced, in countless ways, by a number of scientific collaborators with whom it has been a privilege to work closely, and by other friends and colleagues too numerous to mention. However, I want to mention G. K. Batchelor, without whose strong early influence, and whose encouragement of my urge to understand, I would not have got started at all; and I want to record that my early interest in the systematic effects of waves on mean flows – which has proved to be central to today's understanding of the atmosphere – was greatly stimulated by the pioneering work of my doctoral thesis supervisor F. P. Bretherton, and in different ways by D. O. Gough and E. A. Spiegel. My research in recent years has been generously sponsored through a Senior Research Fellowship from the UK Engineering and Physical Sciences Research Council and through various grants from the Natural Environment Research Council, the UK Meteorological Office, the US Office of Naval Research, the Nuffield Foundation, the UK Department of the Environment, and the European Commission. For direct help in the preparation of this chapter I am especially grateful to D. Broutman, O. Bühler, L. J. Drath, P. J. Espy, C. J. R. Garrett, M. Hallworth, B. Haßler, P. H. Haynes, J. R. Holton, P. Kushner, A. Lee, T. J. McDougall, H. K. Moffatt, B. Naujokat, W. A. Norton, T. J. Pedley, C. R. Philbrick, J. A. Pyle, N. J. Shackleton, T. G. Shepherd, A. J. Simmons, W. E. Ward, D. W. Waugh, M. G. Worster, C. Wunsch, M. Yamada and S. Yoden.

References

ACHESON, D. J. 1973 Valve effect of inhomogeneities on anisotropic wave propogation. *J. Fluid Mech.* **58**, 27–37.

ANDREWS, D. G., HOLTON, J. R. & LEOVY, C. B. 1987 *Middle Atmosphere Dynamics.* Academic.

ANDREWS, D. G. & McINTYRE, M. E. 1976 Planetary waves in horizontal and vertical shear: the generalized Eliassen–Palm relation and the mean zonal acceleration. *J. Atmos. Sci.* **33**, 2031–2048.

ANDREWS, D. G. & McINTYRE, M. E. 1978 On wave-action and its relatives. *J. Fluid Mech.* **89**, 647–664 (and Corrigendum **95**, 796).

APPENZELLER, C., DAVIES, H. C. & NORTON, W. A. 1996 Fragmentation of stratospheric intrusions. *J. Geophys. Res.* **101**, 1435–1456.

BALDWIN, M. P., *et al.*, 2000 The quasi-biennial oscillation. *Revs. Geophys.* in press.

BATES, J. R. 1999 A dynamical stabilizer in the climate system: a mechanism suggested by a simple model. *Tellus* **51 A**, 349–372.

BOOKER, J. R. & BRETHERTON, F. P. 1967 The critical layer for internal gravity waves in a shear flow. *J. Fluid Mech.* **27**, 513–539.

BRETHERTON, F. P. 1969 Momentum transport by gravity waves. *Q. J. R. Met. Soc.* **95**, 213–243.

BRILLOUIN, L. 1925 On radiation stresses. *Annales de Physiques* **4**, 528–586.

BÜHLER, O. 2000 On the vorticity transport due to dissipating or breaking waves in shallow-water flow. *J. Fluid Mech.* **407**, 235–263.

BÜHLER, O. & McINTYRE, M. E. 1998 On non-dissipative wave–mean interactions in the atmosphere or oceans. *J. Fluid Mech.* **354**, 301–343.

CHARNEY, J. G. & DRAZIN, P. G. 1961 Propagation of planetary-scale disturbances from the lower into the upper atmosphere. *J. Geophys. Res.* **66**, 83–109.

CORTI, S., MOLTENI, F. & PALMER, T. N. 1999 Signature of recent climate change in frequencies of natural atmospheric circulation regimes. *Nature* **398**, 799–802. Also *News and Views* commentary by Klaus Hasselmann, 755–756.

DUNKERTON, T. J. 1983 Laterally-propagating Rossby waves in the westward acceleration phase of the quasi-biennial oscillation. *Atmos.–Ocean* **21**, 55–68.

ELIASSEN, A. & PALM, E. 1961 On the transfer of energy in stationary mountain waves. *Geofysiske Publ.* **22(3)**, 1–23.

ERTEL, H. 1942 Ein Neuer hydrodynamischer Wirbelsatz. *Met. Z.,* **59**, 277–281.

FARMER, D. & ARMI, L. 1999 Stratified flow over topography: the role of small-scale entrainment and mixing in flow establishment. *Proc. R. Soc. Lond.* A **455**, 3221–3258.

FORD, R., McINTYRE, M. E. & NORTON, W. A. 2000 Balance and the slow quasi-manifold: some explicit results. *J. Atmos. Sci.* **57**, 1236–1253.

HAYNES, P. H. 1987 On the instability of sheared disturbances. *J. Fluid Mech.* **175**, 463–478.

HAYNES, P. H. 1988 Forced, dissipative generalizations of finite-amplitude wave-activity conservation relations for zonal and non-zonal basic flows. *J. Atmos. Sci.* **45**, 2352–2362.

HAYNES, P. H. 1989 The effect of barotropic instability on the nonlinear evolution of a Rossby wave critical layer. *J. Fluid Mech.* **207**, 231–266.

HAYNES, P. H. 1998 The latitudinal structure of the quasi-biennial oscillation. *Q. J. R. Met. Soc.* **124**, 2645–2670.

HAYNES, P. H. & McINTYRE, M. E. 1990 On the concervation and impermeability theorems for potential vorticity. *J. Atmos. Sci.* **47**, 2021–2031.

HAYNES, P. H., McINTYRE, M. E. & SHEPHERD, T. G. 1996 Reply to comments J. Egger on 'On the "downward control" of extratropical diabatic circulations by eddy-induced mean zonal forces'. *J. Atmos. Sci* **53**, 2105–2107.

HOLTON, HAYNES, P. H., McINTYRE, M. E., DOUGLASS, A. R., ROOD, R. B. & PFISTER, L. 1995 Stratosphere–troposphere exchange. *Rev. Geophys.* **33**, 403–439.

HOSKINS, B. J., McINTYRE, M. E. & ROBERTSON, A. W. 1985 On the use and significance of isentropic potential-vorticity maps. *Q. J. R. Met. Soc.* **111**, 877–946. Also *Corrigendum*, etc., **113**, 402–404 (1987).

HUNT, J. C. R. & DURBIN, P. A. 1999 Perturbed vortical layers and shear sheltering. *Fluid Dyn. Res.* **24**, 375–404.

JACKETT, D. R. & McDOUGALL, T. J. 1997 A neutral density variable for the world's oceans. *J. Phys. Oceanog.* **27**, 237–263.

JUCKES, M. N. & McINTYRE, M. E. 1987 A high resolution, one-layer model of breaking planetary waves in the stratosphere. *Nature*, **328**, 590–596.

KILLWORTH, P. D. & McINTYRE, M. E. 1985 Do Rossby-wave critical layers absorb, reflect or over-reflect? *J. Fluid Mech.* **161**, 449–492.

LORENZ, E. N. 1967 *The Nature and Theory of the General Circulation of the Atmosphere.* Geneva, World Met. Org.

LÜBKEN, F.-J. 1999 Thermal structure of the Arctic summer mesosphere. *J. Geophys. Res.* **104**, 9135–9149.

McEWAN, A. D. 1971 Degeneration of resonantly-excited internal gravity waves. *J. Fluid Mech.* **50**, 431–448.

McINTYRE, M. E. 1980 Towards a Lagrangian-mean description of stratospheric circulations and chemical transports. *Phil. Trans. R. Soc. Lond.* **A296**, 129–148.

McINTYRE, M. E. 1981 On the "wave momentum" myth. *J. Fluid Mech.* **106**, 331–347.

McINTYRE, M. E. 1988 A note on the divergence effect and the Lagrangian-mean surface elevation in water waves. *J. Fluid Mech.* **189**, 235–242.

McINTYRE, M. E. 1989 On dynamics and transport near the polar mesopause in summer. *J. Geophys. Res.* **94**, 14617–14628.

McINTYRE, M. E. 1992 Atmospheric dynamics: some fundamentals, with observational implications. In *Proc. Intl School Phys. 'Enrico Fermi'. CXV Course* (ed. J. C. Gille & G. Visconti), pp. 313–386. North-Holland.

McINTYRE, M. E. & NORTON, W. A. 2000 Potential-vorticity inversion on a hemisphere. *J. Atmos. Sci.* **57**, 1214–1235.

McINTYRE, M. E. & PALMER, T. N. 1984 The 'surf zone' in the stratosphere. *J. Atmos. Terr. Phys.* **46**, 825–849.

McINTYRE, M. E. & PALMER, T. N. 1985 A note on the general concept of wave breaking for Rossby and gravity waves. *Pure Appl. Geophys.* **123**, 964–975.

MOTE, P. W., DUNKERTON, T. J., McINTYRE, M. E., RAY, E. A., HAYNES, P. H., RUSSELL, J. M. III 1998 Vertical velocity, vertical diffusion, and dilution by midlatitude air in the tropical lower stratosphere. *J. Geophys. Res.* **103**, 8651–8666.

MUNK, W. H. & WUNSCH, C. 1998 Abyssal recipes II: energetics of tidal and wind mixing. *Deep Sea Res.* **45**, 1977–2010.

NORTON, W. A. 1994 Breaking Rossby waves in a model stratosphere diagnosed by a vortex-following coordinate system and a technique for advecting material contours. *J. Atmos. Sci.* **51**, 654–673.

PHILLIPS, N. A. 1963 Geostrophic motion. *Rev. Geophys.* **1**, 123–176.

PLUMB, R. A. & MAHLMAN, J. D. 1987 The zonally averaged transport characteristics of the GFDL general circulation/transport model. *J. Atmos. Sci.* **44**, 298–327.

PLUMB, R. A. & MCEWAN, A. D. 1978 The instability of a forced standing wave in a viscous stratified fluid: a laboratory analogue of the quasi-biennial oscillation. *J. Atmos. Sci.* **35**, 1827–1839.

RAYLEIGH, LORD 1945 *The Theory of Sound.* Dover.

ROBINSON, W. A. 1988 Analysis of LIMS data by potential vorticity inversion. *J. Atmos. Sci.* **45**, 2319–2342.

ROSSBY, C.-G. 1936 Dynamics of steady ocean currents in the light of experimental fluid mechanics. *Papers in Physical Oceanography and Meteorology (Mass. Inst. of Technology and Woods Hole Oceanogr. Inst.),* vol. 5 (1). See equation (75).

RUDDICK, B. R., MCDOUGALL, T. J. & TURNER, J. S. 1989 The formation of layers in a uniformly stirred density gradient. *Deep-Sea Res.* **36**, 597–609.

RUELLE, D. 1990 Deterministic chaos: the science and the fiction (Claude Bernard Lecture). *Proc. R. Soc. Lond.* **A427**, 241–248.

SOLOMON, S., KIEHL, J. T., GARCIA, R. R. & GROSE, W. 1986 Tracer transport by the diabatic circulation deduced from satellite observations. *J. Atmos. Sci.* **43**, 1603–1617.

THUBURN, J. & MCINTYRE, M. E. 1997 Numerical advection schemes, cross-isentropic random walks, and correlations between chemical species. *J. Geophys. Res.* **102**, 6775–6797.

WALLACE, J. M. & HOLTON, J. R. 1968 A diagnostic numerical model of the quasi-biennial oscillation. *J. Atmos. Sci.* **25**, 280–292.

WAUGH, D. W., PLUMB, R. A., ATKINSON, R. J., SCHOEBERL, M. R., LAIT, L. R., NEWMAN, P. A., LOEWENSTEIN, M., TOOHEY, D. W., AVALLONE, L. M., WEBSTER, C. R. & MAY, R. D. 1994 Transport of material out of the stratospheric Arctic vortex by Rossby wave breaking. *J. Geophys. Res.* **99**, 1071–1078.

YULAEVA, E., HOLTON, J. R. & WALLACE, J. M. 1994 On the cause of the annual cycle in tropical lower stratospheric temperatures. *J. Atmos. Sci.* **51**, 169–174.

Centre for Atmospheric Science at the Department of Applied Mathematics and Theoretical Physics, Silver Street, Cambridge CB3 9EW, UK; http://www.atm.damtp.cam.ac.uk/people/mem/

Index